Principles of the Theory of Lattice Dynamics

HARALD BÖTTGER

Principles of the Theory of Lattice Dynamics

With 106 Figures

AKADEMIE-VERLAG · BERLIN 1983

Author:

Professor Dr. sc. Harald Böttger

Technische Hochschule „Otto von Guericke“
Magdeburg, Sektion Mathematik und Physik

Erschienen im Akademie-Verlag, DDR - 1086 Berlin, Leipziger Straße 3—4

Lizenznummer: 202 · 100/432/82
Gesamtherstellung: VEB Druckhaus „Maxim Gorki“, 7400 Altenburg
Bestellnummer: 762968 2 (6628) · LSV 1184
Printed in GDR
DDR 60,— M

Dedicated to Professor O. Stasiw

PREFACE

Much progress has been achieved during the last two decades in the theoretical understanding of dynamical properties of solids, especially of disordered and anharmonic solids. This progress is based on the application of new effective analytical and numerical methods, such as the Green function method and the molecular dynamics technique. The Green function method, for example, renders it possible to study solids with arbitrary defect concentrations and even, with fair success, to study solids with glass-like disorder. The Green function method is also a powerful tool for investigating anharmonic solids, including strongly anharmonic solids which cannot be treated by perturbation theory.

Among the problems of lattice dynamics having attracted particular interest in recent years are dynamical properties of disordered solids (localization of vibrational states, breakdown of quasi-momentum selection rules, role of short-range order) and anharmonic solids (strong anharmonicity, phonon hydrodynamics), dynamical aspects of anharmonicity and electron-phonon driven phase transitions (phonon softening, lattice solitons) and the development of a microscopic theory of lattice dynamics.

On a 'quick trip through lattice dynamics' I present here basic ideas and theoretical methods which have been developed in recent years for describing dynamical properties of solids, in particular, of disordered and anhormonic solids. I hope that this book will help the researcher in lattice dynamics to make himself acquainted with recent advances in fields which are somewhat remote from his special field of research. But I believe that the present book may also serve as introductory textbook to the theory of lattice vibrations. To a large extent, the book is based on lectures I gave for advanced students from 1972 to 1979 in Berlin at the Humboldt University.

Chapter 1 contains a concise compilation of basic elements of the theory of lattice dynamics knowledge of which is a necessary prerequisite for an understanding of the following chapters. This chapter also contains the microscopic approach to lattice dynamics which is based on a study of the response of the crystal electrons to the field produced by the vibrating ions (dielectric function approach).

Chapter 2 is devoted to compositionally disordered systems, including solids with single isolated defects. In this chapter the main emphasis is on the introduction of the Green function technique and on the use of this technique for evaluating configuration averaged quantities.

Chapter 3 is devoted to structurally disordered systems. Using the Green function method, in this chapter the problem of localized and plane-wave-like states is studied. Owing to the lack of a periodic reference lattice, the investigation of structurally disordered solids turns out to be in general much more complicated than that of compositionally disordered solids.

Chapter 4 is devoted to anharmonic systems and systems exhibiting structural phase transitions. Perturbation theory is developed for weakly anharmonic solids, and a self-consistent harmonic approximation for strongly anharmonic solids. Further topics treated in this chapter are lattice solitons, the Peierls transition and incommensurate structures.

Some topics with more formal character (application of group theory to lattice dynamics, response of crystal electrons to an electromagnetic field, calculation of experimental quantities) are given in the appendix.

It should be noted that in order to bring this book up-to-date I also included some problems (such as phonon-like excitations in structurally disordered solids, certain problems of incommensurate systems, etc.) where no full consensus is evident from present literature. Furthermore, the choice of the material presented here is to some extent arbitrary. Some fields (such as vibrational properties of surfaces, interaction of phonons with other elementary excitations, etc.) I have completely ignored.

The list of references is not complete. It contains, above all, only review articles and more recent original papers.

This book was written while I was at the Zentralinstitut für Elektronenphysik of the Academy of Sciences of the GDR in Berlin. I would like to thank my former colleagues Dr. M. Giehler, Dr. E. Jahne, Dr. P. Kleinert, R. Leihkauf, Dr. W. Ulrici and Dr. D. Wruck for helpful conversations. Furthermore, I am indepted to Prof. W. Kuhn for valuable suggestions, to Dr. T. Nattermann for critical remarks concerning section 4.4., to B. Gärtner for critically reading the manuscript, and to Mrs. Trautmann, Akademie-Verlag, for efficient cooperation. Finally, I wish to thank my wife who gave me enormous help with the preparation of the manuscript.

H. Böttger

CONTENTS

CHAPTER 1
BASIC ELEMENTS OF THE THEORIE OF LATTICE DYNAMICS

This chapter is an introductory one. We summarize here some elements of the theory of lattice dynamics which will be needed in the following chapters. We describe the phenomenological approach to lattice dynamics[1]. This approach is based on the idea that the motion of the atoms in a solid is governed by force constants which may be regarded as parameters resulting from experimental data. Furthermore, we outline the microscopic approach to lattice dynamics. This approach is based on a microscopic examination of the forces between the ions. It provides a useful tool for studying the relationship between the different force constant models used in the phenomenological theory of lattice dynamics.

1.1. The adiabatic approximation

According to Born and Oppenheimer in a solid the dynamical aspects of nuclear and electron motions may be decoupled (Born and Oppenheimer, 1927, see also Born and Huang, 1954). This decoupling is equivalent to the assumption that the electrons see the nuclei as being fixed in position at any given moment, since they move much more rapidly than the nuclei due to their smaller masses.

It is easy to visualize the electronic charge distribution of the ion core moving with the nucleus as it vibrates. However, such behaviour is less evident for the valence electrons. Nevertheless, it turns out that something like that also occurs for the valence electrons.

According to the Born-Oppenheimer or adiabatic approximation in a solid there exists a many-body potential Φ governing the motions of the atoms. A calculation of Φ requires that we freeze the ions in each instantaneous configuration, and calculate the energy of the electrons. Φ is then, in essence, this energy plus the electrostatic energy of the ion-ion interaction. The phenomenological theory of lattice dynamics is based on the existence of Φ. But also the microscopic theory of lattice dynamics often uses the adiabatic approximation. Moreover, the adiabatic approximation is an important tool on describing the dynamical response of a lattice to external radiation (see appendix 3). It seems therefore reasonable to outline this approximation here.

---

[1] Here we closely follow the standard monographs on lattice dynamics by Born and Huang (1954) and Maradudin et al. (1971) (see also, e. g., Brüesch, 1982).

To do this we start from the following total Hamiltonian for a crystal containing ions and (valence) electrons:

$$\mathscr{H} = T_{\text{el}} + V_{\text{el}} + T_{\text{ion}} + V_{\text{ion}} + H_{\text{int}}. \tag{1.1.1}$$

Here T_{el} is the kinetic energy of the electrons, V_{el} is the Coulomb energy of the electrons, T_{ion} is the kinetic energy of the ions, V_{ion} is the interaction energy of the ions and H_{int} describes the interaction between the electrons and the ions.

From the Hamiltonian (1.1.1) we abstract an electronic part

$$H_{\text{el}} = T_{\text{el}} + V_{\text{el}} + V_{\text{ion}} + H_{\text{int}} \tag{1.1.2}$$

depending parametrically through V_{ion} and H_{int} on the ionic coordinates $\boldsymbol{R}$. Now we expand the solution Ψ of the full Schrödinger equation

$$\mathscr{H}\Psi = \mathscr{E}\Psi \tag{1.1.3}$$

in terms of the solutions ψ of

$$H_{\text{el}}\psi_n = E_n\psi_n. \tag{1.1.4}$$

That is, we seek a solution of (1.1.3) in the form

$$\Psi(\boldsymbol{r}, \boldsymbol{R}) = \sum_n \chi_n(\boldsymbol{R})\, \psi_n(\boldsymbol{r}, \boldsymbol{R}), \tag{1.1.5}$$

where $\boldsymbol{r}$ denotes the electronic coordinates. $\chi_n(\boldsymbol{R})$ is a function only of $\boldsymbol{R}$ and is to be determined.

Substituting (1.1.5) into (1.1.3), and using (1.1.4) and the orthonormality of the ψ's we obtain

$$(T_{\text{ion}} + E_n)\, \chi_n + \sum_m C_{nm}\chi_m = \mathscr{E}\chi_n, \tag{1.1.6}$$

where

$$C_{nm} = A_{nm} + B_{nm} \tag{1.1.7}$$

with

$$\left.\begin{aligned} A_{nm} &= -\sum_l \frac{\hbar^2}{M_l} \int \mathrm{d}\boldsymbol{r} \left(\psi_n^*(\boldsymbol{r}, \boldsymbol{R}) \frac{\partial}{\partial \boldsymbol{R}_l} \psi_m(\boldsymbol{r}, \boldsymbol{R})\right) \frac{\partial}{\partial \boldsymbol{R}_l}, \\ B_{nm} &= -\sum_l \frac{\hbar^2}{2M_l} \int \mathrm{d}\boldsymbol{r} \left(\psi_n^*(\boldsymbol{r}, \boldsymbol{R}) \frac{\partial^2}{\partial \boldsymbol{R}_l^2} \psi_m(\boldsymbol{r}, \boldsymbol{R})\right), \end{aligned}\right\} \tag{1.1.8}$$

where M_l and $\boldsymbol{R}_l$ denote the mass and the position of the lth ion, respectively.

In absence of a magnetic field the ψ's can be chosen to be real, i.e. we have $A_{nn} = 0$, since the functions $\psi_n(\boldsymbol{r}, \boldsymbol{R})$ have been assumed to be normalized to unity for all $\boldsymbol{R}$.

When the terms involving nondiagonal elements of A and B (i.e. A_{nm} and B_{nm} for $n \neq m$) are dropped, the vibrational wave function is determined by the properties of the nth electronic state only, and no electronic transitions accompany the ionic motion. In this case (1.1.6) becomes

$$(T_{\text{ion}} + E_n + C_n)\,\chi_{nv} = \mathscr{E}_{nv}\chi_{nv}, \tag{1.1.9}$$

where $C_n \equiv C_{nn} = B_{nn}$. In this equation v can be regarded as a vibrational quantum number. Adopting (1.1.9) the wave function of the combined system of electrons and ions reads for a state with energy $\mathscr{E}_{nv}$

$$\Psi_{nv}(\boldsymbol{r}, \boldsymbol{R}) = \chi_{nv}(\boldsymbol{R})\,\psi_n(\boldsymbol{r}, \boldsymbol{R}). \tag{1.1.10}$$

The first factor of the wave function (1.1.10) describes the ionic motion, and the second factor shows that during the ionic motion the electrons move as if the ions were fixed in their instantaneous positions. The electrons are said to follow adiabatically the motion of the ions. The approximation (1.1.10) for the total wave function is called adiabatic approximation.

To see what the adiabatic approximation implies we use the Born-Oppenheimer expansion in the following. We begin by writing

$$\boldsymbol{R}_l = \boldsymbol{R}_l^0 + \varkappa \boldsymbol{u}_l, \tag{1.1.11}$$

where $\boldsymbol{R}_l^0$ denotes the equilibrium position of the lth ion, $\boldsymbol{u}_l$ the displacement of the lth ion, and $\varkappa$ the expansion parameter. In the limiting case $M \to \infty$ (M denotes a characteristic mass of an ion) we must have $\varkappa \to 0$. Since $\varkappa$ is a dimensionless parameter, this suggests $\varkappa = (m/M)^\alpha$ with $\alpha > 0$, where m denotes the electronic mass.

To determine the exponent α, the Hamiltonian (1.1.2) is expanded formally in powers of the ionic displacements

$$H_{\text{el}}(\boldsymbol{r}, \boldsymbol{R}) = H_{\text{el}}(\boldsymbol{r}, \boldsymbol{R}^0 + \varkappa \boldsymbol{u}) = \sum_s \varkappa^s H_{\text{el}}^{(s)}(\boldsymbol{r}, \boldsymbol{R}^0), \tag{1.1.12}$$

where $H_{\text{el}}^{(s)}(\boldsymbol{r}, \boldsymbol{R}^0)$ is a homogeneous function of degree s in $\boldsymbol{u}$. We can fix α by noting that if $\alpha = 1/4$, $\varkappa^2 H_{\text{el}}^{(s)}$ is of the same order in $\varkappa$ as

$$T_{\text{ion}} = -\varkappa^{(1/\alpha)-2} \sum_l (M/M_l)\,(\hbar^2/2m)\,\nabla^2_{\boldsymbol{u}_l} \equiv \varkappa^{(1/\alpha)-2} H_1(\boldsymbol{u}). \tag{1.1.13}$$

As will be shown below, to this order we have a harmonic Hamiltonian.

Expanding the wave function $\psi_n(\boldsymbol{r}, \boldsymbol{R}) = \psi_n(\boldsymbol{r}, \boldsymbol{R}^0 + \varkappa \boldsymbol{u})$ as a series in $\varkappa \boldsymbol{u}$, it can be readily seen from (1.1.8) that A_{nm} begins with a term $O(\varkappa^3)$, while B_{nm} begins with a term $O(\varkappa^4)$. Thus the operator C_n in (1.1.9) begins with a constant term $O(\varkappa^4)$.

In the adiabatic approximation the wave function (1.1.10) turns out to be exact to $O(\varkappa^2)$ and the energy $\mathscr{E}_{nv}$ to $O(\varkappa^5)$. To see this we expand χ_m in terms of the functions

$$\chi_m = \sum_v c_v^{(m)} \chi_{mv}. \tag{1.1.14}$$

When this expansion is substituted into (1.1.6) and use is made of (1.1.9) we obtain the following equation for $c_v^{(m)}$:

$$(\mathscr{E}_{mv} - \mathscr{E})\,c_v^{(m)} + \sum_{n(\neq m)} \sum_{v'} \langle mv|\,C_{mn}\,|nv'\rangle\,c_{v'}^{(n)} = 0. \tag{1.1.15}$$

On iterating (1.1.15) all the $c_{v'}^{(n)}$'s $(nv' \neq mv)$ may be expressed by $c_v^{(m)}$. In this manner we obtain to the lowest order of approximation

$$\mathscr{E} = \mathscr{E}_{mv} + \sum_{n(\neq m)} \sum_{v'} \langle mv| \, C_{mn} \, |nv'\rangle \langle nv'| \, C_{mn} \, |mv\rangle / (\mathscr{E}_{mv} - \mathscr{E}_{nv'}). \quad (1.1.16)$$

The second term on the right hand side of this equation describes the change of the energy due to C_{mn} $(n \neq m)$, i.e. due to transitions between different electronic states. Since C_{mn} is at least of $O(\varkappa^3)$, this term begins with $O(\varkappa^6)$, i.e. $\mathscr{E}_{mv}$ is the exact energy to $O(\varkappa^5)$.

For a state with the energy (1.1.16) it holds $c_v^{(m)} \approx 1$ and for the elements $c_{v'}^{(n)}$ $(n \neq m)$ we obtain from (1.1.15)

$$c_{v'}^{(n)} = \langle nv'| \, C_{nm} \, |mv\rangle / (\mathscr{E}_{mv} - \mathscr{E}_{nv'}) . \quad (1.1.17)$$

Since C_{mn} is of $O(\varkappa^3)$, C_{mn} yields a contribution to χ_m of $O(\varkappa^3)$. Accordingly, χ_{mv} from (1.1.9) describes the ionic motion exact to $O(\varkappa^2)$.

Hence, in the Born-Oppenheimer expansion non-adiabatic terms appear only in a relatively high order in $\varkappa = (m/M)^{1/4}$. As $\varkappa$ never exceeds 1/10, the expansion may be expected to converge quickly.

As stated above, the expansion of C_n begins with a constant term of $O(\varkappa^4)$. It follows a term of $O(\varkappa^5)$ linear in $\boldsymbol{u}$. Thus, to $O(\varkappa^4)$ in the adiabatic approximation the ionic motion is governed solely by the electronic energy $E_n(\boldsymbol{R})$. An account of terms of $O(\varkappa^5)$ gives a contribution linear in $\boldsymbol{u}$ arising from C_n, but the motion remains adiabatic.

Now let us solve (1.1.9) by perturbation theory. To this end we expand E_n, χ_{nv}, and $\mathscr{E}_{nv}$ in a series using $\varkappa$ as small parameter. We use the same notation as in (1.1.12), i.e. an expansion coefficient of sth order is labelled by an upper index s. Note that $\mathscr{E}^{(s)}$ is a constant, while $E_n{}^{(s)}$ is a quantity of sth degree in $\boldsymbol{u}$. Comparing the expression resulting from (1.1.9) term by term we obtain

$$\mathscr{E}_{nv}^{(0)} = E_n{}^{(0)}(\boldsymbol{R}^0) , \quad (1.1.18)$$

$$\mathscr{E}_{nv}^{(1)} = E_n{}^{(1)} = \sum_l \left(\partial E_n(\boldsymbol{R})/\partial \boldsymbol{R}_l\right)_{\boldsymbol{R}^0} \boldsymbol{u}_l \equiv 0 . \quad (1.1.19)$$

The identical vanishing of $E_n{}^{(1)}$ results from the fact that $\mathscr{E}_{nv}^{(1)}$ is independent of $\boldsymbol{u}$. From (1.1.19) follows that in the nth electronic state the equilibrium configuration $\boldsymbol{R}^0$ is defined by

$$\left(\partial E_n(\boldsymbol{R})/\partial \boldsymbol{R}_l\right)_{\boldsymbol{R}^0} = 0 . \quad (1.1.20)$$

On account of (1.1.19) we obtain from the second order expression

$$[H_1 + E_n{}^{(2)} - \mathscr{E}_{nv}^{(2)}] \, \chi_{nv}^{(0)} = 0 \quad (1.1.21)$$

or

$$\left[T_{\text{ion}} + (\varkappa^2/2) \sum_{ll'} \boldsymbol{u}_l \left(\frac{\partial^2 E_n}{\partial \boldsymbol{R}_l \, \partial \boldsymbol{R}_{l'}}\right)_{\boldsymbol{R}^0} \boldsymbol{u}_{l'} - \varkappa^2 \mathscr{E}_{nv}^{(2)}\right] \chi_{nv}^{(0)} = 0 . \quad (1.1.22)$$

(1.1.22) obviously describes harmonic vibrations of the ions about their equilibrium positions. The corresponding approximation is called the harmonic approximation. To this approximation the wave function of the crystal reads

$$\Psi_{nv}(\boldsymbol{r}, \boldsymbol{u}) = \chi_{nv}^{(0)}(\boldsymbol{u})\, \psi_n^{(0)}(\boldsymbol{r}, \boldsymbol{R}^0) \tag{1.1.23}$$

and the corresponding eigenvalue is given by

$$\mathscr{E}_{nv} = E_n^{(0)}(\boldsymbol{R}_0) + \varkappa^2 \mathscr{E}_{nv}^{(2)}. \tag{1.1.24}$$

The latter two equations have been obtained by substituting (1.1.10) into (1.1.3), accounting for (1.1.4), and expanding the quantities with respect to $\varkappa$. We see that the harmonic approximation is a special case of the adiabatic approximation. When we continue the expansion procedure beyond the second order, anharmonic terms occur in (1.1.9).

Concluding this section, let us briefly discuss the validity of the adiabatic approximation. A corresponding criterion may be obtained from (1.1.17) by requiring that $|c_{v'}^{(n)}| \ll 1$ for $n \neq m$. Using (1.1.8), the virial theorem, and the fact that in a solid the mean square amplitude of an atom is proportional to ω^{-1} (ω — characteristic vibrational frequency; cf. (1.4.51)), we obtain the criterion $\hbar\omega / |\mathscr{E}_{mv} - \mathscr{E}_{nv'}| \ll 1$. This criterion is quite well satisfied in the case of insulators and semiconductors due to the large gaps in the electronic spectra. Non-adiabatic effects have been found in narrow gap semiconductors (Sherrington, 1971). The above criterion is violated in metals. Nevertheless, it turns out that in metals in general the adiabatic approximation is valid (see Brovman and Kagan, 1974). Non-adiabaticity is governed in metals by the parameter $\hbar\omega/\varepsilon_F$ (ε_F — Fermi energy). Note that there is some indication that non-adiabaticity may occur in materials with strong electron correlation (cf. Kondo and Yamadji, 1977).

1.2. Force constants and their properties

In the preceding section we found that the motion of the atoms in a solid may be described in general by a many-body potential depending on the instantaneous atomic positions.

On dissociating the instantaneous position $\boldsymbol{R}(l)$ of the lth atom into a rest position $\boldsymbol{x}(l)$ and a displacement $\boldsymbol{u}(l)$ according to

$$\boldsymbol{R}(l) = \boldsymbol{x}(l) + \boldsymbol{u}(l) \tag{1.2.1}$$

we can formally expand Φ in powers of the displacements:

$$\Phi = \Phi_0 + \sum_{n=1}^{\infty} \frac{1}{n!} \sum_{\substack{\alpha_1 \ldots \alpha_n \\ l_1 \ldots l_n}} \Phi_{\alpha_1 \ldots \alpha_n}(l_1 \ldots l_n)\, u_{\alpha_1}(l_1) \ldots u_{\alpha_n}(l_n). \tag{1.2.2}$$

Here $u_\alpha(l)$ is the αth cartesian component of $\boldsymbol{u}(l)$, Φ_0 the potential energy of the static lattice and

$$\Phi_{\alpha_1 \ldots \alpha_n}(l_1 \ldots l_n) = \left. \frac{\partial^n \Phi}{\partial u_{\alpha_1}(l_1) \ldots \partial u_{\alpha_n}(l_n)} \right|_0, \tag{1.2.3}$$

where the subscript 0 means that the derivatives are evaluated with all the atoms in their rest positions.

From (1.2.3) it follows that the coefficients $\Phi_{\alpha_1\ldots\alpha_n}(l_1 \ldots l_n)$ are completely symmetric in the indices $l_1\alpha_1, l_2\alpha_2, \ldots$.

On considering the force acting on the lth atom in α-direction

$$\begin{aligned} F_\alpha(l) &= -\partial\Phi/\partial u_\alpha(l) \\ &= -\Phi_\alpha(l) - \sum_{n=1}^{\infty} \sum_{\substack{\alpha_1\ldots\alpha_n \\ l_1\ldots l_n}} \frac{1}{n!} \Phi_{\alpha\alpha_1\ldots\alpha_n}(ll_1 \ldots l_n)\, u_{\alpha_1}(l_1) \ldots u_{\alpha_n}(l_n) \end{aligned} \quad (1.2.4)$$

the physical meaning of the coefficients $\Phi_{\alpha_1\ldots\alpha_n}(l_1 \ldots l_n)$ becomes evident: $-\Phi_\alpha(l)$ is the force in α-direction acting on the lth atom when it and the other atoms in the crystal are in their rest positions, $-\Phi_{\alpha\alpha_1}(ll_1)$ is, to the first order of accuracy, the α-component of the force on the lth atom due to a unit displacement of the l_1th atom in α_1-direction, etc. Consequently, the coefficients $\Phi_\alpha(l)$, $\Phi_{\alpha\alpha_1}(ll_1)$, $\Phi_{\alpha\alpha_1\alpha_2}(ll_1l_2)$, ... are called atomic force constants of the first, second, third, ... order, respectively.

If the rest positions of the atoms are also equilibrium positions, the first order force constant $\Phi_\alpha(l)$ vanishes identically. While in a finite crystal the absence of a net force acting on any atom, i.e. $\Phi_\alpha(l) = 0$, is the only condition for a coincidence of the rest and equilibrium positions, in an infinite crystal the equilibrium condition is two-fold (Born and Huang, 1954, section 23; see also Sarkar and Sengupta, 1977): (i) $\Phi_\alpha(l) = 0$ for any atom, (ii) the configuration corresponds to vanishing stresses. In a finite crystal the latter condition is automatically fulfilled, if the former one is satisfied (for any atom including those near the surface).

For the study of certain problems, e.g. of effects caused by external stresses or strains on dynamical properties of a crystal, it is convenient to expand the lattice potential around rest positions which differ from equilibrium positions (see also section 4.2.).

Apart from specific features characterizing a special lattice, the force constants $\Phi_{\alpha_1\ldots\alpha_n}(l_1 \ldots l_n)$ are subjected to a number of general restrictions resulting from certain invariance conditions every solid must obey. There are two types of such restrictions. The first type follows from the invariance of the lattice potential and its derivatives with respect to the atomic displacements, when the lattice is subjected to infinitesimal rigid displacements (translations, rotations). The second type is connected with the symmetry and structure of a particular crystal.

If the lattice as a whole is subjected to an infinitesimal rigid body displacement $\boldsymbol{v}$ the lattice potential and its derivatives must remain unchanged. On replacing in (1.2.2) $\boldsymbol{u}$ by $\boldsymbol{v}$ this condition yields

$$\sum_{\substack{\alpha_1\ldots\alpha_n \\ l_1\ldots l_n}} \Phi_{\alpha_1\ldots\alpha_n}(l_1 \ldots l_n)\, v_{\alpha_1} \ldots v_{\alpha_n} = 0 \quad (n = 1, 2, \ldots), \quad (1.2.5)$$

where v_α is the α-component of $\boldsymbol{v}$. Since (1.2.5) must hold for all values of $\boldsymbol{v}$, we have

$$\sum_{l_1 \ldots l_n} \Phi_{\alpha_1 \ldots \alpha_n}(l_1 \ldots l_n) = 0 \qquad (n = 1, 2, \ldots). \tag{1.2.6}$$

Quite analogously we obtain from the invariance of the derivatives of the lattice potential against an infinitesimal rigid body displacement the relation

$$\sum_{l_n} \Phi_{\alpha\alpha_1 \ldots \alpha_n}(ll_1 \ldots l_n) = 0 \qquad (n = 1, 2, \ldots). \tag{1.2.7}$$

From the invariance of the lattice potential and its derivatives against an infinitesimal rigid body rotation

$$\boldsymbol{x}(l) \to \boldsymbol{x}'(l) = \Omega \boldsymbol{x}(l) \tag{1.2.8}$$

with

$$\Omega_{\alpha\alpha'} = \delta_{\alpha\alpha'} + \omega_{\alpha\alpha'}, \tag{1.2.9}$$

where $\omega_{\alpha\alpha'}$ is an infinitesimally small antisymmetric matrix

$$\omega_{\alpha\alpha'} = -\omega_{\alpha'\alpha}, \tag{1.2.10}$$

one obtains the restriction, that the expression

$$\sum_{l_{n+1}} \Phi_{\alpha_1 \ldots \alpha_n \alpha'}(l_1 \ldots l_n l_{n+1}) \, x_\alpha(l_{n+1}) + \sum_{\lambda=1}^{n} \sum_{\alpha'_\lambda} \Phi_{\alpha_1 \ldots \alpha'_\lambda \ldots \alpha_n}(l_1 \ldots l_\lambda \ldots l_n) \, \delta_{\alpha\alpha_\lambda} \delta_{\alpha'\alpha'_\lambda} \tag{1.2.11}$$

must be symmetrical with respect to an interchange of α and α' ($x_\alpha(l)$ is the α-component of $\boldsymbol{x}(l)$), i.e. the force constants of different order are not independent of each other. Here we do not prove the relation (1.2.11), but we verify it for $n = 0, 1$. For a general proof of (1.2.11) the reader is referred, e.g., to LEIBFRIED and LUDWIG (1961).

To verify (1.2.11) for $n = 0$ we expand the potential to the first order (cf. (1.2.2)) and use in accordance with (1.2.8) and (1.2.9) the following expression for the displacements:

$$u_\alpha(l) = \sum_{\alpha_1} \omega_{\alpha\alpha_1} x_{\alpha_1}(l). \tag{1.2.12}$$

From the invariance of the potential against a transformation (1.2.8) we obtain

$$\sum_{l} \Phi_\alpha(l) \, x_{\alpha'}(l) = \sum_{l} \Phi_{\alpha'}(l) \, x_\alpha(l) \tag{1.2.13}$$

in accordance with (1.2.11).

For the case $n = 1$ we use the first order expression for the force acting on an atom (cf. (1.2.4)) and substitute into this expression the displacement (1.2.12). The resulting expression for the force reads

$$F_\alpha'(l) = -\Phi_\alpha(l) - \sum_{l_1\alpha_1} \Phi_{\alpha\alpha_1}(ll_1) \, \omega_{\alpha_1\alpha_2} x_{\alpha_2}(l_1). \tag{1.2.14}$$

On the other hand a rotation (1.2.8) transforms the force $F_\alpha(l)$ as a vector according to

$$F_\alpha'(l) = \sum_{\alpha_1} (\delta_{\alpha\alpha_1} + \omega_{\alpha\alpha_1}) F_{\alpha_1}(l). \tag{1.2.15}$$

On comparing (1.2.14) and (1.2.15) and accounting for $F_\alpha(l) = -\Phi_\alpha(l)$ we obtain the relation

$$\sum_{l_1} \Phi_{\alpha_1\alpha'}(ll_1) x_\alpha(l_1) + \delta_{\alpha_1\alpha}\Phi_{\alpha'}(l) = \sum_{l_1} \Phi_{\alpha_1\alpha}(ll_1) x_{\alpha'}(l_1) + \delta_{\alpha_1\alpha'}\Phi_\alpha(l) \tag{1.2.16}$$

in accordance with (1.2.11).

Let us add some remarks on the force constant restrictions considered before. In a system with N atoms Φ will be a function of $3N - 6$ independent internal coordinates, such as distances between the atoms and the corresponding angles, which are translationally and rotationally invariant quantities. Since the internal coordinates are all independent of each other, the expansion coefficients of different orders of Φ with respect to these coordinates are independent quantities. An arbitrary choice of these coefficients will define a potential Φ. On the other hand, if we express Φ by the $3N$ cartesian coordinates the corresponding expansion coefficients must satisfy the conditions considered above, in order to guarantee the invariance of Φ with respect to rigid body rotation and translation. In particular, owing to the latter conditions the expansion with respect to cartesian coordinates will contain anharmonic terms, even if the corresponding expansion with respect to internal coordinates is strictly harmonic.

Now we discuss the restrictions imposed on the force constants by the symmetry of a particular crystal. The crystal is assumed to be infinitely extended. The position vector $\boldsymbol{x}(l\varkappa)$ of the $\varkappa$th atom in the lth primitive unit cell is given by

$$\boldsymbol{x}(l\varkappa) = \boldsymbol{x}(l) + \boldsymbol{x}(\varkappa). \tag{1.2.17}$$

Here $\boldsymbol{x}(l)$ denotes the position vector of the lth unit cell:

$$\boldsymbol{x}(l) = l_1\boldsymbol{a}_1 + l_2\boldsymbol{a}_2 + l_3\boldsymbol{a}_3, \tag{1.2.18}$$

where l_1, l_2, l_3 are integers, labelled collectively by l, and $\boldsymbol{a}_1$, $\boldsymbol{a}_2$, $\boldsymbol{a}_3$ are primitive translation vectors of the lattice. The lattice vectors (1.2.18) define the so-called Bravais lattice: $\boldsymbol{x}(\varkappa)$ describes the position of the $\varkappa$th atom in the unit cell with respect to the origin of the cell. The boundaries of the symmetric or Wigner-Seitz primitive unit cell are constructed as planes bisecting lines to nearest, next nearest neighbour Bravais lattice points and so on.

The space group of the crystal is formed by the totality of the symmetry operations $\{\boldsymbol{S} \mid \boldsymbol{v}(\boldsymbol{S}) + \boldsymbol{x}(m)\}$ with the property

$$\{\boldsymbol{S} \mid \boldsymbol{v}(\boldsymbol{S}) + \boldsymbol{x}(m)\}\, \boldsymbol{x}(l\varkappa) = \boldsymbol{S}\boldsymbol{x}(l\varkappa) + \boldsymbol{v}(\boldsymbol{S}) + \boldsymbol{x}(m) \equiv \boldsymbol{x}(LK) \tag{1.2.19}$$

(cf. appendix 1). Here $\boldsymbol{S}$ is a real, orthogonal matrix representation of proper and improper rotations of the point group of the space group. The vector $\boldsymbol{v}(\boldsymbol{S})$ is smaller than any primitive translation vector. It describes symmetry elements called screw axes and glide planes. According to the definition of a symmetry operation the lattice site $\boldsymbol{x}(LK)$ is occupied by the same kind of atom as $\boldsymbol{x}(l\varkappa)$.

The potential energy Φ of the crystal must remain unchanged under a symmetry operation (1.2.19), i.e. it holds

$$\Phi\big(\ldots \boldsymbol{x}(LK) + \boldsymbol{S}\boldsymbol{u}(l\varkappa) \ldots\big) = \Phi\big(\ldots \boldsymbol{x}(l\varkappa) + \boldsymbol{u}(l\varkappa) \ldots\big). \tag{1.2.20}$$

Expanding both sides of (1.2.20) in powers of the displacement vector $\boldsymbol{u}$ and comparing the coefficients at the same powers we obtain the following relations describing the transformation of the force constants when a crystal is subjected to an operation of its space group:

$$\Phi_\alpha(LK) = \sum_{\alpha_1} S_{\alpha\alpha_1} \Phi_{\alpha_1}(l\varkappa), \tag{1.2.21}$$

$$\Phi_{\alpha\alpha'}(LK, L'K') = \sum_{\alpha_1\alpha_2} S_{\alpha\alpha_1} S_{\alpha'\alpha_2} \Phi_{\alpha_1\alpha_2}(l\varkappa, l'\varkappa'), \tag{1.2.22}$$

$$\Phi_{\alpha\alpha'\alpha''}(LK, L'K', L''K'') = \sum_{\alpha_1\alpha_2\alpha_3} S_{\alpha\alpha_1} S_{\alpha'\alpha_2} S_{\alpha''\alpha_3} \times \Phi_{\alpha_1\alpha_2\alpha_3}(l\varkappa, l'\varkappa', l''\varkappa''), \quad \text{etc.} \tag{1.2.23}$$

By means of these relations the independent nonzero elements of the force constants can be determined by restricting oneself to symmetry operations which leave the arguments of the force constants uneffected or which interchange the arguments.

In the special case of a symmetry operation $\{\boldsymbol{E} \mid \boldsymbol{x}(m)\}$, where $\boldsymbol{E}$ denotes the 3×3 unit matrix, a lattice site $\boldsymbol{x}(l\varkappa)$ is sent into the site $\boldsymbol{x}(LK) = \boldsymbol{x}(l + m\varkappa)$ and the equations (1.2.21) to (1.2.23) become

$$\Phi_\alpha(l + m\varkappa) = \Phi_\alpha(l\varkappa), \tag{1.2.24}$$

$$\Phi_{\alpha\alpha'}(l + m\varkappa, l' + m\varkappa') = \Phi_{\alpha\alpha'}(l\varkappa, l'\varkappa'), \tag{1.2.25}$$

$$\Phi_{\alpha\alpha'\alpha}\,(l + m\varkappa, l' + m\varkappa', l'' + m\varkappa'') = \Phi_{\alpha\alpha'\alpha\alpha''}(l\varkappa, l'\varkappa', l''\varkappa''), \quad \text{etc.} \tag{1.2.26}$$

Putting $m = -l$ in (1.2.24), we see that $\Phi_\alpha(l\varkappa)$ is independent of l. In a similar manner we see that, with respect to l and l', $\Phi_{\alpha\alpha'}(l\varkappa, l'\varkappa')$ depends only on $l - l'$ and $\Phi_{\alpha\alpha'\alpha''}(l\varkappa, l'\varkappa', l''\varkappa'')$ depends only on any two of the differences $l' - l$, $l'' - l$, $l'' - l'$, etc.

Concluding this section let us add some remarks on the atomic force constants. The force constants may be regarded as parameters of the theory which can be determined from experimental data. For instance, the harmonic force constants are usually derived from the lattice frequencies. To obtain a unique set of harmonic force constants one needs also information on the eigenvectors of the lattice (LEIGH et al., 1971). — Alternatively, one may assume an analytic expression for the interaction energy of two or more atoms (cf., for example, LUDWIG, 1967, chapter 3), which allows the force constants of all orders to be expressed by the parameters involved in this expression. In the former approach the number of parameters to be determined is greatly reduced by the restrictions for the force constants derived above. In the latter approach the atomic force constants automatically satisfy these restrictions.

1.3. Equations of motion, their solution and phonons

In the harmonic approximation the potential energy Φ is quadratic in the atomic displacements, and the vibrational Hamiltonian H takes the form

$$H = \sum_{l\alpha} p_\alpha^2(l)/2M_l + \frac{1}{2} \sum_{ll'\alpha\alpha'} \Phi_{\alpha\alpha'}(ll')\, u_\alpha(l)\, u_{\alpha'}(l'), \tag{1.3.1}$$

where $u_\alpha(l)$ and $p_\alpha(l)$ are the αth cartesian component of the displacement and the momentum of the lth atom, respectively; M_l is the mass of the lth atom, and the $\Phi_{\alpha\alpha'}(ll')$'s are the harmonic force constants.

The equation of motion for the displacements in the harmonic approximation reads

$$M_l \ddot{u}_\alpha(l) = -\sum_{l'\alpha'} \Phi_{\alpha\alpha'}(ll')\, u_{\alpha'}(l'). \tag{1.3.2}$$

This equation may be obtained either from Hamilton's classical equations of motion

$$\dot{u}_\alpha(l) = \partial H/\partial p_\alpha(l) = p_\alpha(l)/M_l, \tag{1.3.3}$$

$$\dot{p}_\alpha(l) = -\partial H/\partial u_\alpha(l) = -\sum_{l'\alpha'} \Phi_{\alpha\alpha'}(ll')\, u_{\alpha'}(l') \tag{1.3.4}$$

or from the quantum-mechanical equations of motion

$$\dot{u}_\alpha(l) = \frac{i}{\hbar} [H, u_\alpha(l)] = p_\alpha(l)/M_l, \tag{1.3.5}$$

$$\dot{p}_\alpha(l) = \frac{i}{\hbar} [H, p_\alpha(l)] = -\sum_{l'\alpha'} \Phi_{\alpha\alpha'}(ll')\, u_{\alpha'}(l'), \tag{1.3.6}$$

where we have used the Hamiltonian (1.3.1), and in (1.3.5) and (1.3.6) the commutation relations

$$[u_\alpha(l), p_{\alpha'}(l')] = i\hbar\delta_{ll'}\, \delta_{\alpha\alpha'}, \; [u_\alpha(l), u_{\alpha'}(l')] = [p_\alpha(l), p_{\alpha'}(l')] = 0. \tag{1.3.7}$$

When we assume solutions of (1.3.2) of the form

$$u_\alpha(l) = \frac{B_\alpha(l)}{(M_l)^{1/2}}\, e^{-i\omega t} \tag{1.3.8}$$

the equation satisfied by the time-independent amplitudes $B_\alpha(l)$ reads

$$\omega^2 B_\alpha(l) = \sum_{l'\alpha'} D_{\alpha\alpha'}(ll')\, B_{\alpha'}(l'), \tag{1.3.9}$$

where

$$D_{\alpha\alpha'}(ll') = \Phi_{\alpha\alpha'}(ll')/\sqrt{M_l M_{l'}}. \tag{1.3.10}$$

The matrix $D_{\alpha\alpha'}(ll')$ is called the dynamical matrix.

If there are N atoms in the solid, the dynamical matrix (1.3.10) is a real, symmetric $3N$ matrix. The eigenvalue equation (1.3.9), therefore, has $3N$ solutions labelled by an index s which runs from 1 to $3N$:

$$\omega_s^2 B_\alpha^{(s)}(l) = \sum_{l'\alpha'} D_{\alpha\alpha'}(ll')\, B_{\alpha'}^{(s)}(l'). \tag{1.3.11}$$

Because the dynamical matrix is a real, symmetric matrix its eigenvalues ω_s^2 are real. In order that the system is stable in the harmonic approximation ω_s must be real, since according to (1.3.8) an imaginary frequency corresponds to atomic motions whose amplitudes depend exponentially on the time. Further, the eigenvectors $B_\alpha^{(s)}(l)$ can be chosen to be real and mutually orthogonal, so that they satisfy the orthonormality and closure conditions

$$\sum_{l\alpha} B_\alpha^{(s)}(l)\, B_\alpha^{(s')}(l) = \delta_{ss'}, \tag{1.3.12}$$

$$\sum_{s} B_\alpha^{(s)}(l)\, B_{\alpha'}^{(s)}(l') = \delta_{ll'}\delta_{\alpha\alpha'}. \tag{1.3.13}$$

An eigenmode with frequency ω_s and wave vector $B_\alpha^{(s)}(l)$ is called the sth normal mode of the system.

When we assume the solutions of (1.3.11) to be known we can simplify the Hamiltonian (1.3.1) by replacing the operators $u_\alpha(l)$ and $p_\alpha(l)$ by new operators b_s and b_s^+ which are defined by

$$u_\alpha(l) = \left(\frac{\hbar}{2M_l}\right)^{1/2} \sum_s \frac{B_\alpha^{(s)}(l)}{\sqrt{\omega_s}}\, (b_s + b_s^+), \tag{1.3.14}$$

$$p_\alpha(l) = \frac{1}{i}\left(\frac{\hbar M_l}{2}\right)^{1/2} \sum_s \sqrt{\omega_s}\, B_\alpha^{(s)}(l)\, (b_s - b_s^+). \tag{1.3.15}$$

By making use of (1.3.12) the inverse transformation is readily found:

$$b_s = \left(\frac{1}{2\hbar}\right)^{1/2} \sum_{l\alpha} B_\alpha^{(s)}(l) \left\{\sqrt{M_l\omega_s}\, u_\alpha(l) + \frac{ip_\alpha(l)}{\sqrt{M_l\omega_s}}\right\}, \tag{1.3.16}$$

$$b_s^+ = \left(\frac{1}{2\hbar}\right)^{1/2} \sum_{l\alpha} B_\alpha^{(s)}(l) \left\{\sqrt{M_l\omega_s}\, u_\alpha(l) - \frac{ip_\alpha(l)}{\sqrt{M_l\omega_s}}\right\}. \tag{1.3.17}$$

By means of (1.3.7) the following commutation relations for the new operators may be varified:

$$[b_s, b_{s'}^+] = \delta_{ss'}, \qquad [b_s, b_{s'}] = [b_s^+, b_{s'}^+] = 0. \tag{1.3.18}$$

Replacing the operators in the Hamiltonian (1.3.1) according to (1.3.14) and (1.3.15) we obtain

$$H = \sum_{s=1}^{3N} \hbar\omega_s(b_s^+ b_s + 1/2), \tag{1.3.19}$$

which represents the sum of $3N$ independent harmonic oscillators.

The operator $N_s \equiv b_s^+ b_s$ occurring in the Hamiltonian (1.3.19) is a hermitian operator and therefore has real eigenvalues. It is called the particle number operator.

Using the representation on which the operator N_s is diagonal

$$N_s \, |n_s\rangle = n_s \, |n_s\rangle \tag{1.3.20}$$

it is seen from the relation

$$\begin{aligned} n_s &= \langle n_s| \, N_s \, |n_s\rangle = \sum_{m_s} \langle n_s| \, b_s^+ \, |m_s\rangle \, \langle m_s| \, b_s \, |n_s\rangle \\ &= \sum_{m_s} |\langle m_s| \, b_s \, |n_s\rangle|^2 \geqq 0 \end{aligned} \tag{1.3.21}$$

that the eigenvalues n_s of N_s are greater than or equal to zero.

Using the commutation relations (3.1.18) we obtain

$$[N_s, b_s^+] = b_s^+, \tag{1.3.22}$$

$$[N_s, b_s] = -b_s, \tag{1.3.23}$$

and from (1.3.22) and (1.3.23)

$$N_s(b_s \, |n_s\rangle) = (n_s - 1) \, (b_s \, |n_s\rangle), \tag{1.3.24}$$

$$N_s(b_s^+ \, |n_s\rangle) = (n_s + 1) \, (b_s^+ \, |n_s\rangle), \tag{1.3.25}$$

i.e. b_s and b_s^+ act as annihilation and creation operators, respectively.

Repeated application of b_s to any eigenstate must eventually give zero, since otherwise (1.3.24) could produce a state with a negative eigenvalue of N_s, in contradiction to (1.3.21). Consequently, N_s has the eigenvalues

$$n_s = 0, 1, 2, \ldots \tag{1.3.26}$$

According to (1.3.25) the operator b_s^+ raises the eigenvalue by one. The state $|n_s\rangle$ normalized to unity is therefore

$$|n_s\rangle = \frac{1}{\sqrt{n_s!}} \, (b_s^+)^{n_s} \, |0\rangle, \tag{1.3.27}$$

where $|0\rangle$ denotes the ground state function with the property

$$b_s |0\rangle = 0. \tag{1.3.28}$$

Thus the eigenfunctions and eigenvalues of H in particle number representation read

$$\left.\begin{aligned} |n_1, n_2, \ldots, n_{3N}\rangle &= |n_1\rangle |n_2\rangle \ldots |n_{3N}\rangle \\ &= \left(\prod_{s=1}^{3N} n_s!\right)^{-1/2} \prod_{s=1}^{3N} (b_s^+)^{n_s} |0\rangle, \\ E_{n_1 n_2 \ldots n_{3N}} &= \sum_{s=1}^{3N} \hbar\omega_s(n_s + 1/2), \end{aligned}\right\} \tag{1.3.29}$$

and the operators b_s and b_s^+ act on the states (1.3.29) as follows:

$$\begin{aligned} b_s |n_1 \ldots n_s \ldots n_{3N}\rangle &= \sqrt{n_s}\, |n_1 \ldots (n_s - 1) \ldots n_{3N}\rangle, \\ b_s^+ |n_1 \ldots n_s \ldots n_{3N}\rangle &= \sqrt{n_s + 1}\, |n_1 \ldots (n_s + 1) \ldots n_{3N}\rangle. \end{aligned} \tag{1.3.30}$$

A stationary state of the system described by the Hamiltonian (1.3.19) is characterized by the set of integers n_s which denote the number of quantums of energy $\hbar\omega_s$ connected with the sth oscillator. The quantum of energy $\hbar\omega_s$ is called a phonon.

In practice the solution of the eigenvalue equation (1.3.11) is generally a very intricate mathematical problem, as (1.3.11) is of order $3N$, where N is a large number. For crystalline materials, however, the problem becomes tractable because of simplifying features in the dynamical matrix. In this case, due to the lattice periodicity, the dynamical matrix is a simple recurrent pattern, which allows the problem to be reduced to one associated with a set of matrices of a much lower order. As will be shown below, these matrices are of order $3r$ only, where r denotes the number of atoms in a primitive unit cell.

In a crystal with perfect periodicity the atomic masses are independent of the unit cell index l, and according to (1.2.25) the atomic force constants $\Phi_{\alpha\alpha'}(l\varkappa, l'\varkappa')$ depend on the cell indices l and l' only through their difference, i.e. the Hamiltonian (1.3.1) commutes with the operator describing a rigid body translation of the crystal through a lattice translation vector. From the latter fact it follows that the atomic displacements $u_\alpha(l\varkappa)$ must simultaneously be eigenfunctions of the translation operator and the Hamiltonian. Thus, $u_\alpha(l\varkappa)$ must have the form

$$u_\alpha(l\varkappa) = \frac{e_\alpha(\varkappa)}{\sqrt{M_\varkappa}}\, \mathrm{e}^{i(\boldsymbol{k}\boldsymbol{x}(l) - \omega t)}, \tag{1.3.31}$$

where $\boldsymbol{x}(l)$ is the position vector of the lth unit cell (cf. (1.2.18)).

To determine the vector $\boldsymbol{k}$ in (1.3.31) we subdivide the infinitely extended crystal into parallelepipeds whose edges are defined by lattice vectors $L\boldsymbol{a}_1$, $L\boldsymbol{a}_2$,

La_3 ($L^3 = N$ is the number of primitive unit cells in such a parallelepiped) and impose on the $u(l\varkappa)$'s the cyclic boundary conditions

$$u(l_1l_2l_3\varkappa) = u(l_1 + m_1L, l_2 + m_2L, l_3 + m_3L\varkappa), \tag{1.3.32}$$

where m_1, m_2, m_3 are integers. From (1.3.31) and (1.3.32) we find the vector k to be given by

$$k = \frac{n_1}{L} b_1 + \frac{n_2}{L} b_2 + \frac{n_3}{L} b_3, \qquad n_1, n_2, n_3 = 1, 2, \ldots, L, \tag{1.3.33}$$

with

$$b_1 = 2\pi \frac{a_2 \times a_3}{v_a}, \quad b_2 = 2\pi \frac{a_3 \times a_1}{v_a}, \quad b_3 = 2\pi \frac{a_1 \times a_2}{v_a}, \tag{1.3.34}$$

where

$$v_a = a_1(a_2 \times a_3). \tag{1.3.35}$$

Here b_1, b_2, b_3 are the unit vectors of the reciprocal lattice, i.e. it holds that

$$a_ib_j = 2\pi\delta_{ij} \qquad (i, j = 1, 2, 3) \tag{1.3.36}$$

and v_a is the volume of a primitive unit cell of the crystal.

To obtain all of the distinct solutions of the equation of motion satisfying the condition (1.3.32) we have to restrict the vector k to lie in a unit cell of the reciprocal lattice. We use a symmetric unit cell, the first Brillouin zone, which is constructed analogously as the Wigner-Seitz cell. The usefulness of this choice lies in the property of the first Brillouin zone to be invariant under operations of the point group of the space group of the crystal, which is also the point group of the reciprocal lattice.

When we substitute (1.3.31) into (1.3.2) we obtain the eigenvalue equation determining the relation between the frequency ω and the wave vector k

$$\omega^2 e_\alpha(\varkappa) = \sum_{\varkappa'\alpha'} D_{\alpha\alpha'}(\varkappa\varkappa' \mid k)\, e_{\alpha'}(\varkappa'), \tag{1.3.37}$$

where

$$D_{\alpha\alpha'}(\varkappa\varkappa' \mid k) = (M_\varkappa M_{\varkappa'})^{-1/2} \sum_{l'} \Phi_{\alpha\alpha'}(l\varkappa, l'\varkappa')\, e^{-ik(x(l)-x(l'))}. \tag{1.3.38}$$

For a crystal with r atoms in a primitive unit cell the dynamical matrix $D(k)$ (1.3.38) is a $3r \times 3r$ hermitian matrix,

$$D_{\alpha\alpha'}(\varkappa\varkappa' \mid k) = D^*_{\alpha'\alpha}(\varkappa'\varkappa \mid k), \tag{1.3.39}$$

which according to (1.3.38) has the property

$$D_{\alpha\alpha'}(\varkappa\varkappa' \mid -k) = D^*_{\alpha\alpha'}(\varkappa\varkappa' \mid k). \tag{1.3.40}$$

The eigenvalue equation (1.3.37) has for each k $3r$ solutions labelled by an index j, called the branch index, which runs from 1 to $3r$:

$$\omega_j^2(k)\, e_\alpha(\varkappa \mid kj) = \sum_{\varkappa'\alpha'} D_{\alpha\alpha'}(\varkappa\varkappa' \mid k)\, e_{\alpha'}(\varkappa' \mid kj). \tag{1.3.41}$$

Thus, as mentioned above, for a periodic crystal the eigenvalue equation (1.3.41) is of order $3r$ rather than of order $3N$ (cf. (1.3.9)), which appreciably simplifies the solution of the eigenvalue problem.

Because of the hermicity of the dynamical matrix (1.3.38) its eigenvalues $\omega_j^2(\boldsymbol{k})$ are real and the stability of the crystal requires positive $\omega_j^2(\boldsymbol{k})$, i.e. real $\omega_j(\boldsymbol{k})$. Further, replacing $\boldsymbol{k}$ in (1.3.41) by $-\boldsymbol{k}$ and taking the complex conjugate of the resulting equation we find that $\omega_j^2(-\boldsymbol{k})$ and $e_\alpha^*(\varkappa \mid -\boldsymbol{k}j)$, respectively, are also eigenvalues and eigenvectors of $\boldsymbol{D}(\boldsymbol{k})$, i.e. with appropriate choice of the phase, we may set

$$e_\alpha^*(\varkappa \mid -\boldsymbol{k}j) = e_\alpha(\varkappa \mid \boldsymbol{k}j) \tag{1.3.42}$$

and

$$\omega_j^2(-\boldsymbol{k}) = \omega_j^2(\boldsymbol{k}). \tag{1.3.43}$$

Note that for Bravais lattices the eigenvectors $\boldsymbol{e}(\boldsymbol{k}j)$ are real, because for these lattices inversion is a symmetry operation so that we have $\Phi_{\alpha\alpha'}(l) = \Phi_{\alpha\alpha'}(-l)$ from (1.2.22), and thus $D^*_{\alpha\alpha'}(\boldsymbol{k}) = D_{\alpha\alpha'}(\boldsymbol{k})$ from (1.3.38).

The eigenvectors $e_\alpha(\varkappa \mid \boldsymbol{k}j)$ describe the polarization of the $3r$ eigenmodes of the crystal associated with a given wave vector $\boldsymbol{k}$. We may assume that they satisfy the orthonormality and closure conditions

$$\sum_{\varkappa\alpha} e_\alpha^*(\varkappa \mid \boldsymbol{k}j)\, e_\alpha(\varkappa \mid \boldsymbol{k}j') = \delta_{jj'}, \tag{1.3.44}$$

$$\sum_{j} e_\alpha^*(\varkappa \mid \boldsymbol{k}j)\, e_{\alpha'}(\varkappa' \mid \boldsymbol{k}j) = \delta_{\varkappa\varkappa'}\delta_{\alpha\alpha'}, \tag{1.3.45}$$

since $\boldsymbol{D}(\boldsymbol{k})$ is a hermitian operator and the eigenfunctions of it are determined only to within an arbitrary constant factor.

The branches of the vibrational modes are in general nondegenerate. However, they may touch on certain points or lines in the Brillouin zone. To investigate this degeneracy and the symmetry properties of $\omega_j^2(\boldsymbol{k})$ we now consider the consequence of transforming $\boldsymbol{k}$ to $\boldsymbol{S}\boldsymbol{k}$, where, as in (1.2.19), $\boldsymbol{S}$ is a real orthonormal matrix representation of one of the proper or improper rotations of the point group of the space group of the crystal. For simplicity we restrict ourselves here to a crystal with only one atom per primitive unit cell. The general case of an arbitrary number of atoms in a unit cell will be considered in appendix 1.

On account of the fact that $\Phi_{\alpha\alpha'}(l, l')$ depends on l and l' in a periodic crystal only through their difference we get from (1.3.39)

$$\left.\begin{aligned} D_{\alpha\alpha'}(\boldsymbol{S}\boldsymbol{k}) &= \frac{1}{M} \sum_l \Phi_{\alpha\alpha'}(0, l)\, \mathrm{e}^{i(\boldsymbol{S}\boldsymbol{k})\boldsymbol{x}(l)} \\ &= \frac{1}{M} \sum_l \Phi_{\alpha\alpha'}(0, l)\, \mathrm{e}^{i\boldsymbol{k}(\boldsymbol{S}^{-1}\boldsymbol{x}(l))} \\ &= \frac{1}{M} \sum_l \Phi_{\alpha\alpha'}(0, L)\, \mathrm{e}^{i\boldsymbol{k}\boldsymbol{x}(l)}, \end{aligned}\right\} \tag{1.3.46}$$

where $(S^{-1})_{\alpha\alpha'} = S_{\alpha'\alpha}$ and L labels the lattices site $\boldsymbol{Sx}(l)$. The last line in (1.3.46) holds since $\boldsymbol{Sx}(l)$ is a lattice vector.

From (1.3.46) and (1.2.22) we obtain

$$D_{\alpha\alpha'}(\boldsymbol{Sk}) = \sum_{\alpha_1\alpha_1'} S_{\alpha\alpha_1} D_{\alpha_1\alpha_1'}(\boldsymbol{k}) \, (S^{-1})_{\alpha_1'\alpha'}, \tag{1.3.47}$$

i.e. $\boldsymbol{D}(\boldsymbol{Sk})$ and $\boldsymbol{D}(\boldsymbol{k})$ are connected by an orthogonal transformation, and so have the same eigenvalues

$$\omega_j^2(\boldsymbol{Sk}) = \omega_j^2(\boldsymbol{k}). \tag{1.3.48}$$

Thus $\omega_j^2(\boldsymbol{k})$ has the symmetry of the point group of the space group of the crystal. As will be shown in appendix 1, this also holds for a crystal with more than one atom per unit cell. If the point group coincides with the holoedric group, i.e. with the group of all rotations leaving the Bravais lattice of the crystal invariant, then $\omega_j^2(\boldsymbol{k})$ has the symmetry of the Brillouin zone. This is not always the case when there is more than one atom in a primitive unit cell (see appendix 1). Furthermore, it should be noticed that according to (1.3.43) $\omega_j^2(\boldsymbol{k})$ is an even function of $\boldsymbol{k}$, even if the point group of the crystal does not contain the inversion.

Let us now consider the degeneracy of vibrational branches at points $\boldsymbol{k}$ in the Brillouin zone which are of high symmetry (fig. 1.1). For these points there are operations $\boldsymbol{S}$ other than the unit operator such that

$$\boldsymbol{Sk} = \boldsymbol{k} + \boldsymbol{K}, \tag{1.3.49}$$

where $\boldsymbol{K}$ is a vector of the reciprocal lattice, i.e.

$$\boldsymbol{K} = \lambda_1\boldsymbol{b}_1 + \lambda_2\boldsymbol{b}_2 + \lambda_3\boldsymbol{b}_3 \tag{1.3.50}$$

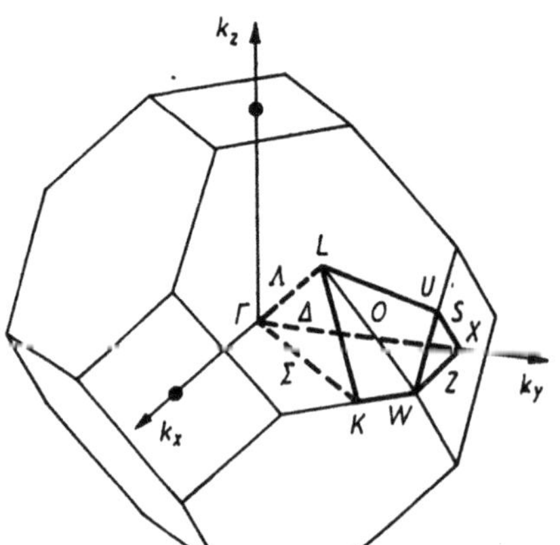

Fig. 1.1. Brillouin zone of a crystal with diamond structure. The points and lines of high symmetry, defined by (1.3.49), are labelled by Greek and Latin capitals

with $\boldsymbol{b}_1, \boldsymbol{b}_2, \boldsymbol{b}_3$ given in (1.3.34) and integers λ_1, λ_2, λ_3. For operations $\boldsymbol{S}$ with the property (1.3.49) equation (1.3.47) reads

$$[\boldsymbol{D}(\boldsymbol{k}), \boldsymbol{S}] = 0. \tag{1.3.51}$$

Thus, if we apply the operation $\boldsymbol{S}$ to an eigenfunction of $\boldsymbol{D}(\boldsymbol{k})$ we may obtain a new function linear independent of the former one, but both the functions have the same eigenvalue. A study of this so-called essential degeneracy requires that we determine the irreducible representations of the subgroup formed by operations $\boldsymbol{S}$ satisfying (1.3.51) (for more details see appendix 1).

The eigenvalue equation (1.3.41) has three solutions with zero frequency. This is readily seen by substituting (1.3.38) into (1.3.41) and using the relation

$$\sum_{l'\varkappa'} \Phi_{\alpha\alpha'}(\varkappa l, \varkappa' l') = 0, \tag{1.3.52}$$

which results from (1.2.7). It turns out that the three solutions with zero frequency are given by

$$e_\alpha(\varkappa \mid 0j) = \sqrt{\frac{M_\varkappa}{M_\mathrm{T}}}\, e_\alpha(j), \tag{1.3.53}$$

where $\boldsymbol{e}(j)$ $(j = 1, 2, 3)$ are any three mutually perpendicular unit vectors and M_T is the total mass of the atoms in a primitive unit cell of the crystal. The modes (1.3.53) with the frequency $\omega_j(\boldsymbol{0}) = 0$ $(j = 1, 2, 3)$ are not isolated modes but a part of a continuum. It holds that

$$\lim_{\boldsymbol{k}\to\boldsymbol{0}} \omega_j(\boldsymbol{k}) = c_j(\hat{\boldsymbol{k}})\, |\boldsymbol{k}| \qquad (j = 1, 2, 3), \tag{1.3.54}$$

where $\hat{\boldsymbol{k}} = \boldsymbol{k}/|\boldsymbol{k}|$. To see this we expand the function

$$f(\omega^2, \boldsymbol{k}) = \det |\boldsymbol{D}(\boldsymbol{k}) - \omega^2 \mathbf{1}| \tag{1.3.55}$$

in powers of ω^2 and $\boldsymbol{k}$. If we account for (1.3.40) we get

$$\left(\frac{\partial f(\omega^2, \boldsymbol{k})}{\partial \boldsymbol{k}}\right)_0 = \boldsymbol{0}, \tag{1.3.56}$$

where the subscript 0 means that the derivative is evaluated at $\omega = 0$, $\boldsymbol{k} = \boldsymbol{0}$. Further, it holds that $f(0, \boldsymbol{0}) = 0$, since the eigenvalue equation (1.3.37) must have a solution for $\omega = 0$, $\boldsymbol{k} = \boldsymbol{0}$. Expanding now (1.3.55) the eigenvalue equation takes the form

$$\omega^2 \left(\frac{\partial f}{\partial \omega^2}\right)_0 + \frac{1}{2!} \sum_{\alpha\alpha'} k_\alpha k_{\alpha'} \left(\frac{\partial^2 f}{\partial k_\alpha\, \partial k_{\alpha'}}\right)_0 + \text{higher order terms} = 0, \tag{1.3.57}$$

from which we obtain (1.3.54) with

$$c_j(\hat{\boldsymbol{k}}) = \sum_{\alpha\alpha'} \left[\hat{k}_\alpha \hat{k}_{\alpha'} \left(\frac{\partial^2 f_j}{\partial k_\alpha\, \partial k_{\alpha'}}\right)_0 \Big/ 2 \left(\frac{\partial f_j}{\partial(-\omega^2)}\right)_0\right]^{1/2}, \tag{1.3.58}$$

where f_j denotes that branch of the multivalued function f which yields $\omega_j(\boldsymbol{k})$.

Similarly, the eigenvectors belonging to the eigenvalues (1.3.54) become independent of $\boldsymbol{k}$ in the limit $\boldsymbol{k} \to \boldsymbol{0}$. The corresponding atomic motions closely resemble a rigid body translation of the crystal (cf. (1.3.53)).

A wave vector dependence of eigenfrequencies of the type (1.3.54) is characteristic for an elastic continuum.

Therefore, the three vibrational branches of a crystal with the long wavelength property (1.3.54) are called acoustic branches. In the limit $\boldsymbol{k} \to 0$ the eigenfrequencies of the remaining $3r-3$ branches of a crystal have a finite limit, and the atoms in different sublattices vibrate rigidly against each other. In an ionic crystal such a motion gives rise to a macroscopic fluctuating dipole moment, through which the crystal can interact with electromagnetic radiation. Because of this property the latter $3r-3$ branches are called optical branches. For a linear diatomic lattice the wave vector dependences of the acoustic and optical modes, i.e. the so-called dispersion curves, are shown in figure 1.2 and the displacements of the atoms in such a lattice in figure 1.3. For illustration, figure 1.4 shows the dispersion curves of an actual solid (silicon). Note that neutron spectroscopy is a powerful technique for determining dispersion curves (for a review see DOLLING, 1974, cf. also appendix 3).

It is noteworthy that the eigenvalue equation (1.3.11) has also three solutions with zero frequency. These solutions read

$$B_\alpha^{(s)}(l) = \sqrt{\frac{M_l}{M_T'}}\, B_\alpha^{(s)} \tag{1.3.59}$$

where $\boldsymbol{B}^{(s)}$ ($s = 1, 2, 3$) are any three mutually perpendicular unit vectors and M_T' is the total mass of the crystal. They describe a rigid body translation of the crystal.

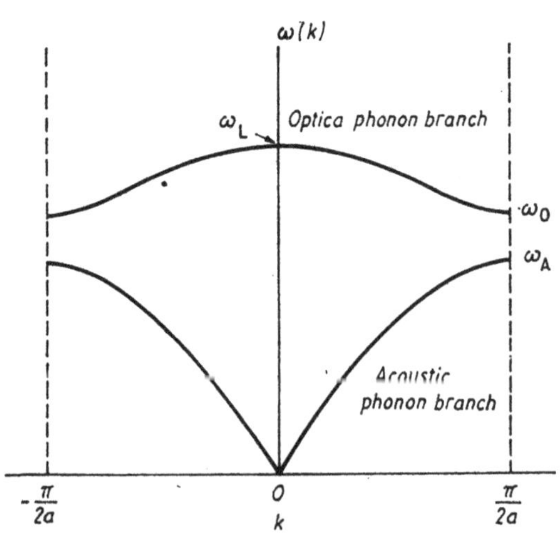

Fig. 1.2. Optical and acoustic phonon branches of a diatomic linear lattice with nearest neighbour spacing a. For a lattice with nearest neighbour force constant γ and masses M_1 and M_2, with $M_1 \geqq M_2$, the limiting frequencies are $\omega_A = (2\gamma/M_1)^{1/2}$, $\omega_O = (2\gamma/M_2)^{1/2}$ and $\omega_L = (2\gamma(1/M_1 + 1/M_2))^{1/2}$

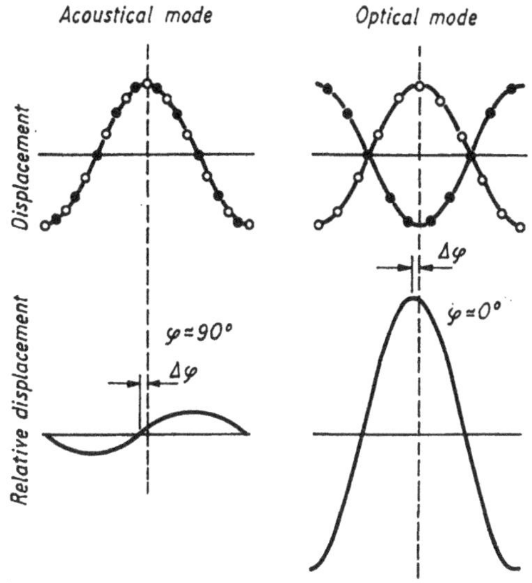

Fig. 1.3. Acoustic and optical modes of a linear diatomic lattice with atoms of equal mass, illustrated by the displacements (upper curves) and the relative displacements (lower curves) of the atoms for the two modes at the same wavelength. The relative displacement curve for the optical mode is approximately in phase with the displacement of the atoms, while the relative displacement curve for the acoustic mode is spatially displaced by approximately 90°. The departure $\Delta\varphi$ of the phases from the idealized values 0° and 90°, respectively, arises from the finite lattice spacing (after DUTHLER, 1976)

The Hamiltonian of a periodic crystal can be diagonalized quite analogously to above (cf. (1.3.14) to (1.3.17)) by introducing new operators $b_{\boldsymbol{k}j}$, $b^+_{\boldsymbol{k}j}$ according to

$$u_\alpha(l\varkappa) = \left(\frac{\hbar}{2M_\varkappa N}\right)^{1/2} \sum_{\boldsymbol{k}j} \frac{e_\alpha(\varkappa \mid \boldsymbol{k}j)}{\sqrt{\omega_j(\boldsymbol{k})}}\, e^{i\boldsymbol{k}\boldsymbol{x}(l)}(b_{\boldsymbol{k}j} + b^+_{-\boldsymbol{k}j}), \qquad (1.3.60)$$

$$p_\alpha(l\varkappa) = \frac{1}{i}\left(\frac{\hbar M_\varkappa}{2N}\right)^{1/2} \sum_{\boldsymbol{k}j} \sqrt{\omega_j(\boldsymbol{k})}\, e_\alpha(\varkappa \mid \boldsymbol{k}j)\, e^{i\boldsymbol{k}\boldsymbol{x}(l)}(b_{\boldsymbol{k}j} - b^+_{-\boldsymbol{k}j}), \qquad (1.3.61)$$

where we have used (1.3.42) and (1.3.43). The inverse relations read

$$b_{\boldsymbol{k}j} = \left(\frac{1}{2N\hbar}\right)^{1/2} \sum_{l\varkappa\alpha} e_\alpha^*(\varkappa \mid \boldsymbol{k}j)\, e^{-i\boldsymbol{k}\boldsymbol{x}(l)}\Big\{\sqrt{M_\varkappa\omega_j(\boldsymbol{k})}\, u_\alpha(l\varkappa) + ip_\alpha(l\varkappa)/\sqrt{M_\varkappa\omega_j(\boldsymbol{k})}\Big\}, \qquad (1.3.62)$$

$$b^+_{\boldsymbol{k}j} = \left(\frac{1}{2N\hbar}\right)^{1/2} \sum_{l\varkappa\alpha} e_\alpha(\varkappa \mid \boldsymbol{k}j)\, e^{i\boldsymbol{k}\boldsymbol{x}(l)}\Big\{\sqrt{M_\varkappa\omega_j(\boldsymbol{k})}\, u_\alpha(l\varkappa) - ip_\alpha(l\varkappa)/\sqrt{M_\varkappa\omega_j(\boldsymbol{k})}\Big\}. \qquad (1.3.63)$$

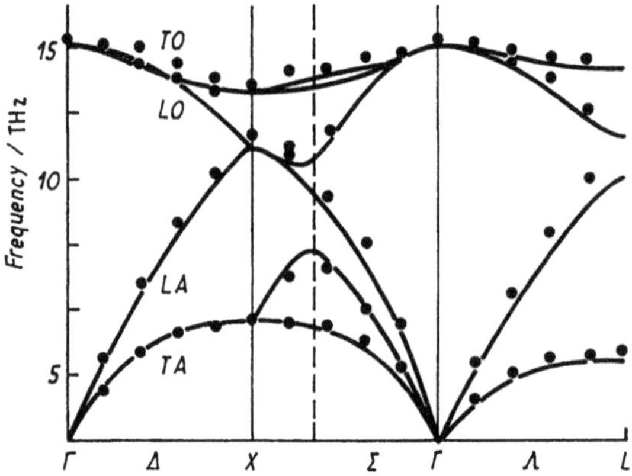

Fig. 1.4. Dispersion curves of the lattice vibrations for silicon. The points and lines of high symmetry in the Brillouin zone are labelled by Greek and Latin capitals (cf. fig. 1.1). The solid circles represent the experimental values from inelastic coherent scattering of cold neutrons (cf. appendix A3.3.). The solid lines show results obtained by the bond charge model (cf. section 1.6.1.) TO, LO (TA, LA) denote transverse and longitudinal optical (acoustic) modes (after Weber, 1977, experimental points from Dolling, 1963, and Nilsson and Nelin, 1972)

By means of (1.3.7) we find the commutation relations

$$[b_{\boldsymbol{k}j}, b^+_{\boldsymbol{k}'j'}] = \delta_{jj'}\Delta(\boldsymbol{k} - \boldsymbol{k}'),\ [b_{\boldsymbol{k}j}, b_{\boldsymbol{k}'j'}] = [b^+_{\boldsymbol{k}j}, b^+_{\boldsymbol{k}'j'}] = 0 \tag{1.3.64}$$

with

$$\Delta(\boldsymbol{k}) = \frac{1}{N} \sum_{l} e^{i\boldsymbol{k}\boldsymbol{x}(l)} = \sum_{\boldsymbol{K}} \delta_{\boldsymbol{k},\boldsymbol{K}} \tag{1.3.65}$$

where $\boldsymbol{K}$ is a vector of the reciprocal lattice (cf. (1.3.50)). In the present case $\boldsymbol{K}$ equals zero, since in (1.3.64) $\boldsymbol{k}$ and $\boldsymbol{k}'$ are restricted to lie inside the first Brillouin zone.

Thus the Hamiltonian takes the form (cf. (1.3.19))

$$H = \sum_{\boldsymbol{k}}^{N} \sum_{j=1}^{3r} \hbar\omega_j(\boldsymbol{k}) \, [b^+_{\boldsymbol{k}j} b_{\boldsymbol{k}j} + 1/2], \tag{1.3.66}$$

and a periodic crystal is described by equations analogous to (1.3.20) to (1.3.31), i.e. $b^+_{\boldsymbol{k}j}$ and $b_{\boldsymbol{k}j}$ are phonon creation and destruction operators, respectively, and the eigensolutions of the Hamiltonian (1.3.66) read

$$|n_{j_1}(\boldsymbol{k}_1) \dots n_{j_{3r}}(\boldsymbol{k}_1) \dots n_{j_1}(\boldsymbol{k}_N) \dots n_{j_{3r}}(\boldsymbol{k}_N)\rangle$$

$$= \left(\prod_{\boldsymbol{k}}^{N} \prod_{j=1}^{3r} n_j(\boldsymbol{k})! \right)^{-1/2} \prod_{\boldsymbol{k}}^{N} \prod_{j=1}^{3r} (b^+_{\boldsymbol{k}j})^{n_j(\boldsymbol{k})} \, |0\rangle, \tag{1.3.67}$$

$$E_{n_{j_1}(\boldsymbol{k}_1)\dots n_{j_{3r}}(\boldsymbol{k}_1)\dots n_{j_1}(\boldsymbol{k}_N)\dots n_{j_{3r}}(\boldsymbol{k}_N)} = \sum_{\boldsymbol{k}}^{N} \sum_{j=1}^{3r} \hbar\omega_j(\boldsymbol{k}) \left(n_j(\boldsymbol{k}) + 1/2\right), \tag{1.3.68}$$

where

$$n_j(\boldsymbol{k}) = 1, 2, \ldots \tag{1.3.69}$$

denotes the number of phonons of frequency $\omega_j(\boldsymbol{k})$ present in the state (1.3.67). Like photons, the phonons are bosons, as the commutation relations (1.3.64) and (1.3.18) show. In absence of any phonons the energy (1.3.68) is that of the so-called zero point vibrations, i.e. equals $\frac{1}{2} \sum_{\boldsymbol{k}j} \hbar\omega_j(\boldsymbol{k})$ of $\frac{1}{2} \sum_s \hbar\omega_s$ (cf. (1.3.30)), respectively.

It should be noticed that in disordered systems, which are governed by the eigenvalue equation (1.3.11), the terminus phonon is often not applied to the quanta $\hbar\omega_s$ but, as suggested by neutron spectroscopy, to quanta of plane-wave-like excitations, provided that these excitations are not too strongly damped by disorder (see section 3.2.).

Anharmonic terms of the lattice potential (1.2.2) give rise to an interaction between phonons. To see this, we replace the displacement operators in the potential (1.2.2) by phonon creation and destruction operators by means of (1.3.14) or (1.3.60), respectively. The anharmonic part of the Hamiltonian, H_A, thus obtained reads for the case of a periodic crystal

$$\begin{aligned} H_A = {} & \sum_{\boldsymbol{k}j} V(\boldsymbol{k}j)\, A_{\boldsymbol{k}j} \\ & + \sum_{n=3}^{\infty} \sum_{\boldsymbol{k}_1 j_1} \cdots \sum_{\boldsymbol{k}_n j_n} V(\boldsymbol{k}_1 j_1, \ldots, \boldsymbol{k}_n j_n)\, A_{\boldsymbol{k}_1 j_1} \cdots A_{\boldsymbol{k}_n j_n}, \end{aligned} \tag{1.3.70}$$

where

$$\begin{aligned} V(\boldsymbol{k}_1 j_1, \ldots, \boldsymbol{k}_n j_n) = {} & \frac{(1/n!)\,(\hbar/2N)^{n/2}}{\left(\omega_{j_1}(\boldsymbol{k}_1) \ldots \omega_{j_n}(\boldsymbol{k}_n)\right)^{1/2}} \\ & \times \sum_{l_1\varkappa_1\alpha_1} \cdots \sum_{l_n\varkappa_n\alpha_n} \Phi_{\alpha_1\ldots\alpha_n}(l_1\varkappa_1, \ldots\, l_n\varkappa_n)\, \frac{e_{\alpha_1}(\varkappa_1 \mid \boldsymbol{k}_1 j_1)}{\sqrt{M_{\varkappa_1}}} \cdots \frac{e_{\alpha_n}(\varkappa_n \mid \boldsymbol{k}_n j_n)}{\sqrt{M_{\varkappa_n}}} \\ & \times \exp\{i[\boldsymbol{k}_1\boldsymbol{x}(l_1) + \cdots + \boldsymbol{k}_n\boldsymbol{x}(l_n)]\}, \end{aligned} \tag{1.3.71}$$

and

$$A_{\boldsymbol{k}j} = b_{\boldsymbol{k}j} + b^{+}_{-\boldsymbol{k}j}. \tag{1.3.72}$$

The first term on the right hand side of (1.3.70) vanishes identically, if the lattice potential is expanded around rest positions of the atoms which are also equilibrium positions (cf. section 1.2.).

According to (1.3.71) the coefficients $V(\boldsymbol{k}_1 j_1, \ldots, \boldsymbol{k}_n j_n)$ are completely symmetric in the indices $\boldsymbol{k}_1 j_1, \ldots, \boldsymbol{k}_n j_n$ and have the property

$$V(\boldsymbol{k}_1 j_1, \ldots, \boldsymbol{k}_n j_n) = V^*(-\boldsymbol{k}_1 j_1, \ldots, -\boldsymbol{k}_n j_n). \tag{1.3.73}$$

Another important property of these coefficients results from the invariance of the force constants of a periodic crystal against a displacement of the lattice by any lattice vector $\boldsymbol{x}(m)$ (cf. (1.2.26)):

$$\Phi_{\alpha_1\ldots\alpha_n}(l_1\varkappa_1, \ldots, l_n\varkappa_n) = \Phi_{\alpha_1\ldots\alpha_n}(l_1 + m\varkappa_1, \ldots, l_n + m\varkappa_n), \tag{1.3.74}$$

since if we substitute (1.3.74) into (1.3.71) the latter equation takes the form

$$V(\boldsymbol{k}_1 j_1, \ldots, \boldsymbol{k}_n j_n) = \Delta(\boldsymbol{k}_1 + \cdots + \boldsymbol{k}_n) \frac{(1/n!)\,(\hbar/2N)^{n/2}}{\left(\omega_{j_1}(\boldsymbol{k}_1) \ldots \omega_{j_n}(\boldsymbol{k}_n)\right)^{1/2}}$$
$$\times \sum_{l_1\varkappa_1\alpha_1} \cdots \sum_{l_n\varkappa_n\alpha_n} \Phi_{\alpha_1\ldots\alpha_n}(l_1\varkappa_1, \ldots, l_n\varkappa_n)$$
$$\times \frac{e_{\alpha_1}(\varkappa_1 \mid \boldsymbol{k}_1 j_1)}{\sqrt{M_{\varkappa_1}}} \cdots \frac{e_{\alpha_n}(\varkappa_n \mid \boldsymbol{k}_n j_n)}{\sqrt{M_{\varkappa_n}}}$$
$$\times \exp\{i[\boldsymbol{k}_1\boldsymbol{x}(l_1) + \cdots + \boldsymbol{k}_n\boldsymbol{x}(l_n)]\}, \tag{1.3.75}$$

where $\Delta(\boldsymbol{k})$ is given by (1.3.65). Thus

$$V(\boldsymbol{k}_1 j_1, \ldots, \boldsymbol{k}_n j_n) = 0 \quad \text{unless} \quad \boldsymbol{k}_1 + \boldsymbol{k}_2 + \cdots + \boldsymbol{k}_n = \boldsymbol{K}, \tag{1.3.76}$$

where $\boldsymbol{K}$ is a reciprocal lattice vector.

Equation (1.3.76) suggests that a phonon of wave vector $\boldsymbol{k}$ be ascribed the so-called crystal momentum or quasi-momentum $\hbar\boldsymbol{k}$. This equation then corresponds to the law of conservation of crystal momentum. According to this equation in phonon interaction processes described by H_{A} (1.3.70) the crystal momentum is conserved only to within a quantity $\hbar\boldsymbol{K}$, which may be assumed to be the momentum transferred to the crystal as a whole. If $\boldsymbol{K} = 0$ we say we have a normal process; if $\boldsymbol{K} \neq 0$ we have an umklapp process. The latter process may be thought of as normal scattering plus Bragg reflexion.

Because of the possibility $\boldsymbol{K} \neq 0$ in (1.3.76) there is an essential difference between the crystal momentum and the momentum of a real particle. This difference originates from the fact that the momentum conservation in the case of real particles follows from the invariance of the Hamiltonian under infinitesimal translations, whereas the law (1.3.76) has been derived from the invariance of the Hamiltonian under translation by a lattice vector.

Generally, particle interaction and particle statistics are closely related to each other. This is due to the fact that a system which is in a nonequilibrium state relaxes to equilibrium by interaction processes between particles. In the case of a phonon system the phonon interaction described by H_{A} (1.3.70) has the property that the particle number is not a conserved quantity. This results from the fact, that H_{A} does not commute with the phonon number operator $\hat{N} = \sum_{\boldsymbol{k}j} b_{\boldsymbol{k}j}^{+} b_{\boldsymbol{k}j}$. Accordingly, phonons behave as Bose particles with vanishing chemical potential.

1.4. Thermodynamic properties and frequency spectrum

The thermodynamic functions of a crystal in the harmonic approximation can be obtained from the partition function

$$Z = \operatorname{Tr}\{e^{-\beta H}\}, \tag{1.4.1}$$

where H is Hamiltonian in the harmonic approximation; $\beta = 1/kT$, k is the Boltzmann constant and T the absolute temperature. If we evaluate the trace in (1.4.1) in the representation in which H is diagonal, then we find for a periodic crystal by (1.3.66) and (1.3.67)

$$Z = \sum_{n_{j_1}(\boldsymbol{k}_1)=0}^{\infty} \cdots \sum_{n_{j_{3r}}(\boldsymbol{k}_N)=0}^{\infty} \exp\left\{-\beta \sum_{\boldsymbol{k}j} \hbar\omega_j(\boldsymbol{k}) \left[n_j(\boldsymbol{k}) + 1/2\right]\right\}$$

$$= \prod_{\boldsymbol{k}j} \frac{\exp\left[-\frac{1}{2}\beta\hbar\omega_j(\boldsymbol{k})\right]}{1 - \exp\left[-\beta\hbar\omega_j(\boldsymbol{k})\right]}. \tag{1.4.2}$$

The Helmholtz free energy is

$$F = -kT \ln Z = kT \sum_{\boldsymbol{k}j} \ln\left[2 \sinh\left(\hbar\omega_j(\boldsymbol{k})/2kT\right)\right]. \tag{1.4.3}$$

The internal energy E, the specific heat at constant volume C_v, and the entropy S of the crystal become

$$E = F - T\left(\frac{\partial F}{\partial T}\right)_V = \sum_{\boldsymbol{k}j} \frac{1}{2} \hbar\omega_j(\boldsymbol{k}) \coth\left(\hbar\omega_j(\boldsymbol{k})/2kT\right), \tag{1.4.4}$$

$$C_v = \left(\frac{\partial E}{\partial T}\right)_V = k \sum_{\boldsymbol{k}j} \left(\frac{\hbar\omega_j(\boldsymbol{k})}{2kT}\right)^2 \Big/ \sinh^2\left(\hbar\omega_j(\boldsymbol{k})/2kT\right), \tag{1.4.5}$$

$$S = -\left(\frac{\partial F}{\partial T}\right)_V = k \sum_{\boldsymbol{k}j} \left\{\frac{\hbar\omega_j(\boldsymbol{k})}{2kT} \coth\left(\hbar\omega_j(\boldsymbol{k})/2kT\right) - \ln\left[2 \sinh\left(\hbar\omega_j(\boldsymbol{k})/2kT\right)\right]\right\}. \tag{1.4.6}$$

We see that in the harmonic approximation the thermodynamic functions are additive functions of the eigenfrequencies $\omega_j(\boldsymbol{k})$. Therefore, all these functions can be expressed as averages over the frequency spectrum $g(\omega)$, defined such that $g(\omega)\,\mathrm{d}\omega$ is the fraction of eigenfrequencies in the interval $(\omega, \omega + \mathrm{d}\omega)$. We obtain

$$F = 3rNkT \int \mathrm{d}\omega \ln\left[2 \sinh\left(\hbar\omega/2kT\right)\right] g(\omega), \tag{1.4.7}$$

$$E = 3rN \frac{\hbar}{2} \int \mathrm{d}\omega \coth\left(\hbar\omega/2kT\right) \omega g(\omega), \tag{1.4.8}$$

$$C_v = 3rNk \int \mathrm{d}\omega \left[\left(\frac{\hbar\omega}{2kT}\right)^2 \Big/ \sinh^2\left(\hbar\omega/2kT\right)\right] g(\omega), \tag{1.4.9}$$

$$S = 3rNk \int \mathrm{d}\omega \left\{\frac{\hbar\omega}{2kT} \coth\left(\hbar\omega/2kT\right) - \ln\left[2 \sinh\left(\hbar\omega/2kT\right)\right]\right\} g(\omega). \tag{1.4.10}$$

The formulae (1.4.7) to (1.4.10) can be readily verified by making use of the following expression for $g(\omega)$:

$$g(\omega) = \frac{1}{3rN} \sum_{\boldsymbol{k}j} \delta\big(\omega - \omega_j(\boldsymbol{k})\big) \tag{1.4.11}$$

with

$$\int_0^{\omega_L} \mathrm{d}\omega\, g(\omega) = 1, \tag{1.4.12}$$

where ω_L is the largest eigenfrequency of the crystal.

It is sometimes more convenient to work with the distribution function of the squares of the eigenfrequencies $G(\omega^2)$ than with $g(\omega)$, since the eigenvalue equation (1.3.41) involves the squares of the eigenfrequencies. $G(\omega^2)$ is defined such that $G(\omega^2)\,\mathrm{d}\omega^2$ is the fraction of eigenfrequencies in the interval $(\omega^2, \omega^2 + \mathrm{d}\omega^2)$. It is given by

$$G(\omega^2) = \frac{1}{3rN} \sum_{\boldsymbol{k}j} \delta\big(\omega^2 - \omega_j^2(\boldsymbol{k})\big) \tag{1.4.13}$$

with

$$\int_0^{\omega_L^2} \mathrm{d}\omega^2\, G(\omega^2) = 1. \tag{1.4.14}$$

From a comparison of (1.4.11) and (1.4.13) we obtain the following relation between $g(\omega)$ and $G(\omega^2)$:

$$g(\omega) = 2\omega G(\omega^2). \tag{1.4.15}$$

The frequency distribution function $G(\omega^2)$ may also be expressed as

$$G(\omega^2) = \frac{1}{3rN} \frac{V}{(2\pi)^3} \sum_j \int \frac{\mathrm{d}S}{|\nabla_{\boldsymbol{k}} \omega_j^2(\boldsymbol{k})|}, \tag{1.4.16}$$

where the integration is over a constant frequency surface defined by

$$\omega_j^2(\boldsymbol{k}) = \omega^2 \tag{1.4.17}$$

and V is the volume of the crystal.

To derive (1.4.16) from (1.4.13) we replace the sum over $\boldsymbol{k}$ in (1.4.13) by an integral (see, for example, Ziman, 1972, chapter 1),

$$\sum_{\boldsymbol{k}} \to \frac{V}{(2\pi)^3} \int \mathrm{d}\boldsymbol{k}, \tag{1.4.18}$$

and choose as volume element $\mathrm{d}\boldsymbol{k}$ a small cylinder with ground area $\mathrm{d}S$ standing perpendicular on the surface (1.4.17), since according to the δ-function in (1.4.13) only a small layer of $\boldsymbol{k}$ values about this surface contributes to the integral over $\boldsymbol{k}$. If we now expand the argument of the δ-function in (1.4.13) around the zeros of (1.4.17) to the first power of the wave vector, the integral over the direction perpendicular to the surface (1.4.17) may be readily calculated since a δ-function is involved in this integration. The remaining surface integral is that contained in (1.4.16).

From (1.4.16) information on singularities in $G(\omega^2)$ and $g(\omega)$, respectively, can be obtained. This equation suggests that some sort of singularity in the frequency spectrum occurs, if $\nabla_{\boldsymbol{k}}\omega_j^2(\boldsymbol{k}) = 0$ or discontinuously changes sign. The latter situation occurs at points of degeneracy. Points in the Brillouin zone are called analytical critical points, if they hold: $\nabla_{\boldsymbol{k}}\omega_j^2(\boldsymbol{k}) = 0$, $\omega_j^2(\boldsymbol{k})$ may be expanded in a Taylor series, and the functional determinant $|\partial^2\omega_j^2(\boldsymbol{k})/\partial k_\alpha\, \partial k_{\alpha'}|$ does not vanish. Nonanalytic critical points can also occur, but we will not consider them here. At critical points the frequency spectrum has a maximum, minimum, or saddle-point.

To investigate the effect of a critical point on $G(\omega^2)$ we may confine our attention to the contribution from a given branch $G_j(\omega^2)$, since according to (1.4.16) these contributions are additive. We choose the principal axes $\boldsymbol{k}_1$, $\boldsymbol{k}_2$, $\boldsymbol{k}_3$ with the origin at the critical point, so that in the neighbourhood of an analytical critical point we have for $\omega_j^2(\boldsymbol{k})$ the expansion

$$\omega_j^2(\boldsymbol{k}) = \omega_c^2 + \beta_1 k_1^2 + \beta_2 k_2^2 + \beta_3 k_3^2 + \cdots. \tag{1.4.19}$$

We classify the critical points as types 0, 1, 2, and 3. The type-number is given by the number of negative β's. We see that type 0 is a minimum, types 1 and 2 saddle-points and type 3 a maximum.

Let us consider $G_j(\omega^2)$ for a critical point of type 0. In this case the surface of constant frequency $\omega_j^2(\boldsymbol{k}) = \omega^2$, with $\omega_j^2(\boldsymbol{k})$ from (1.4.19), is an ellipsoid with volume

$$\mathscr{V} = \frac{4\pi}{3(\beta_1\beta_2\beta_3)^{1/2}}\,[\omega^2 - \omega_c^2]^{3/2}. \tag{1.4.20}$$

On the other hand this volume can be expressed by $G_j(\omega^2)$ as

$$\mathscr{V} = 3r\,\frac{N}{V}\,(2\pi)^3 \int\limits_0^{\omega^2} \mathrm{d}\omega'^2\, G_j(\omega'^2), \tag{1.4.21}$$

where we have accounted for (1.4.13) and (1.4.14) and used $V/(2\pi)^3$ as the density of $\boldsymbol{k}$-points in the Brillouin zone. If we differentiate (1.4.21) with respect to ω^2 and make use of (1.4.20) we get

$$G_j(\omega^2) = \frac{V}{12r\pi^2 N(\beta_1\beta_2\beta_3)^{1/2}}\,|\omega^2 - \omega_c^2|^{1/2} \quad \text{for } \omega^2 > \omega_c^2. \tag{1.4.22}$$

In a similar way $G_j(\omega^2)$ can be obtained for the other types of critical points and for critical points in one- ($1d$) and two- ($2d$) dimensional systems, too.

The results can be summarized as follows (see, for example, MARADUDIN et al., 1971, chapter 4):

1*d*:

$$\left.\begin{aligned} &\text{type 0 (minimum)}\\ &G_j(\omega^2) = \frac{v_a}{4\pi r(\beta_1)^{1/2}} |\omega^2 - \omega_c^2|^{-1/2} [1 + \operatorname{sgn}(\omega^2 - \omega_c^2)],\\ &\text{type 1 (maximum)}\\ &G_j(\omega^2) = \frac{v_a}{4\pi r(|\beta_1|)^{1/2}} |\omega^2 - \omega_c^2|^{-1/2} [1 - \operatorname{sgn}(\omega^2 - \omega_c^2)], \end{aligned}\right\} \quad (1.4.23)$$

2*d*:

$$\left.\begin{aligned} &\text{type 0 (minimum)}\\ &G_j(\omega^2) = \frac{v_a}{16\pi r(\beta_1\beta_2)^{1/2}} \operatorname{sgn}(\omega^2 - \omega_c^2),\\ &\text{type 1 (saddle-point)}\\ &G_j(\omega^2) = -\frac{v_a}{8\pi^2 r(|\beta_1\beta_2|)^{1/2}} \ln |\omega^2 - \omega_c^2|,\\ &\text{type 2 (maximum)}\\ &G_j(\omega^2) = -\frac{v_a}{16\pi r(\beta_1\beta_2)^{1/2}} \operatorname{sgn}(\omega^2 - \omega_c^2), \end{aligned}\right\} \quad (1.4.24)$$

3*d*:

$$\left.\begin{aligned} &\text{type 0 (minimum)}\\ &G_j(\omega^2) = \frac{v_a}{24\pi^2 r(\beta_1\beta_2\beta_3)^{1/2}} |\omega^2 - \omega_c^2|^{1/2} [1 + \operatorname{sgn}(\omega^2 - \omega_c^2)],\\ &\text{type 1 (saddle-point)}\\ &G_j(\omega^2) = \frac{v_a}{24\pi^2 r(|\beta_1\beta_2\beta_3|)^{1/2}} |\omega^2 - \omega_c^2|^{1/2} [-1 + \operatorname{sgn}(\omega^2 - \omega_c^2)],\\ &\text{type 2 (saddle-point)}\\ &G_j(\omega^2) = -\frac{v_a}{24\pi^2 r(\beta_1\beta_2\beta_3)^{1/2}} |\omega^2 - \omega_c^2|^{1/2} [1 + \operatorname{sgn}(\omega^2 - \omega_c^2)],\\ &\text{type 3 (maximum)}\\ &G_j(\omega^2) = \frac{v_a}{24\pi^2 r(|\beta_1\beta_2\beta_3|)^{1/2}} |\omega^2 - \omega_c^2|^{1/2} [1 - \operatorname{sgn}(\omega^2 - \omega_c^2)], \end{aligned}\right\} \quad (1.4.25)$$

where $v_a = V/N$ is the volume of a primitive unit cell. The singular behaviour described by equations (1.4.23) to (1.4.25) is shown in figure 1.5. To illustrate this behaviour, in figure 1.6 $G(\omega^2)$ is given for a one-dimensional alternating diatomic chain (cf. MARADUDIN et al., 1971, chapter 4). It should be noted that equations (1.4.23) to (1.4.25) give only the singular part of $G_j(\omega^2)$ for $\omega^2 \approx \omega_c{}^2$. The total frequency spectrum $G_j(\omega^2)$ for $\omega^2 \approx \omega_c{}^2$ is the sum of this part and a part which is analytic for $\omega^2 \approx \omega_c{}^2$. The latter part may be expanded in Legendre polynomials and evaluated by means of the moments of the frequency spectrum. Before we describe this in more detail we must still dwell to some extent on the problem of critical points.

The periodicity of a crystal involves the existence of critical points on surfaces of constant frequency in the k-space and hence of singularities in the

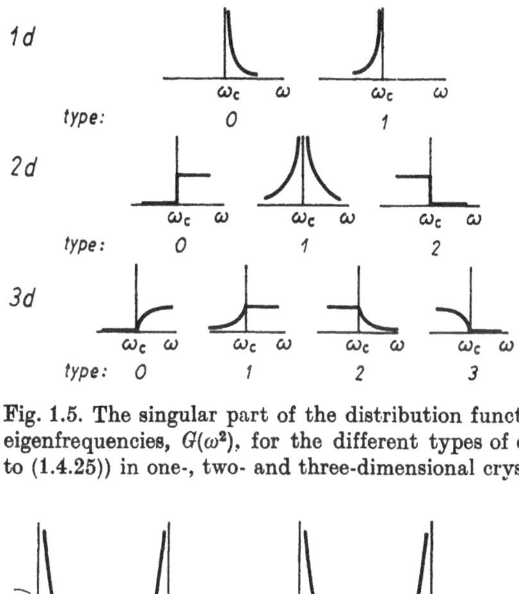

Fig. 1.5. The singular part of the distribution function of the squares of the eigenfrequencies, $G(\omega^2)$, for the different types of critical points (cf. (1.4.23) to (1.4.25)) in one-, two- and three-dimensional crystals

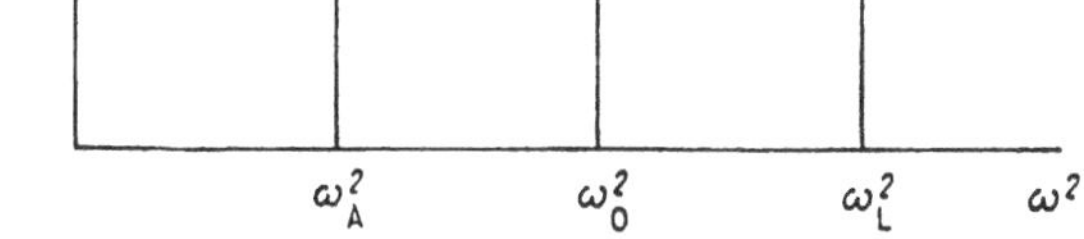

Fig. 1.6. The spectrum of the squared frequencies for a one-dimensional diatomic lattice. For a lattice with nearest neighbour force constant and masses M_1 and M_2, with $M_1 \geq M_2$, the limiting frequencies are $\omega_A = (2\gamma/M_1)^{1/2}$, $\omega_O = (2\gamma/M_2)^{1/2}$ and $\omega_L = (2\gamma(1/M_1 + 1/M_2))^{1/2}$

frequency spectrum. This was pointed out by VAN HOVE (1953) on the basis of topological considerations by making use of a general theorem of Morse. Since a formal proof of Morse's theorem lies outside the scope of this book, we restrict ourselves here to a heuristic derivation of the minimal number of critical points, in particular for a two-dimensional system. For more details on van Hove's work we refer the reader to ROSENSTOCK (1955) and PHILLIPS (1956).

In the case of a one-dimensional system the periodicity obviously implies the existence of at least one maximum and one minimum in each branch $\omega_j^2(\boldsymbol{k})$. In a two-dimensional system $\omega_j^2(\boldsymbol{k})$ must have at least one maximum and one minimum in each of its periods. In figure 1.7 we assume the maximum

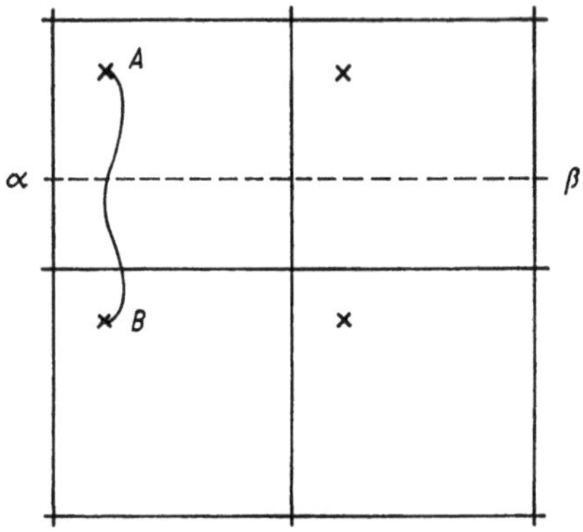

Fig. 1.7. $\boldsymbol{k}$-space of a two dimensional lattice for illustrating how the existence of one critical point implies the existence of others (see text)

of $\omega_j^2(\boldsymbol{k})$ in one cell (Brillouin zone) of the $\boldsymbol{k}$-space to be located at the point A and the equivalent maximum in an adjacent cell at the point B. Let us now consider the variation of $\omega_j^2(\boldsymbol{k})$ along any line $(\alpha\beta)$. Because of the periodicity there must be a maximum in each cell at some point on this line. The position of these maxima forms a continuous curve, represented by a solid line in figure 1.7, which connects A and B. Clearly, somewhere along this line the maximum must be smaller than any of the other maxima. At this point we have a saddle-point. From an analogous consideration for the minimum mentioned above we get a second saddle-point. A similar heuristic discussion can also be given for three-dimensional systems. It leads to the result, that there are in each branch $\omega_j^2(\boldsymbol{k})$ at least one maximum, one minimum, three saddle-points of type 1, and three saddle-points of type 2. By topological considerations certain relations between the numbers N^i of analytical critical points of type i can be derived. In the case of a three-dimensional system these so-called Morse relations read

$$N^0 \geqq 1, \; N^1 - N^0 \geqq 2, \; N^2 - N^1 + N^0 \geqq 1,$$
$$N^3 - N^2 + N^1 - N^0 = 0. \tag{1.4.26}$$

The relations (1.4.26) imply the relations mentioned above: $N^0 \geqq 1$, $N^1 \geqq 3$, $N^2 \geqq 3$, $N^3 \geqq 1$. But they can also be used to prove whether or not they are satisfied. If they do not, further critical points must be looked for. The smallest number of critical points satisfying topological (cf. (1.4.26)) and symmetry requirements is called the minimal set of critical points.

Singularities introduced into $G_j(\omega^2)$ by certain types of non-analytical critical points have been analyzed by PHILLIPS (1956), who also investigated the application of relations of type (1.4.26) in the presence of such critical points.

Let us now turn to the moments of the frequency spectrum $g(\omega)$. The nth moment is defined as

$$\mu_n = \int_0^{\omega_L} d\omega\, g(\omega)\, \omega^n, \tag{1.4.27}$$

where ω_L is the maximum allowed frequency. The even moments can be expressed equivalently as

$$\mu_{2n} = \frac{1}{3rN} \sum_{kj} \omega_j^{2n}(\boldsymbol{k}) = \frac{1}{3rN} \sum_{\boldsymbol{k}} \operatorname{Tr} \boldsymbol{D}^n(\boldsymbol{k}). \tag{1.4.28}$$

Here the first form results from (1.4.27) and (1.4.11) and the second form from the fact that $\boldsymbol{D}^n(\boldsymbol{k})$ can be diagonalized by a unitary transformation which leaves the trace invariant and which yields a diagonal matrix with elements $\omega_j^n(\boldsymbol{k})$. Thus, the even moments of $g(\omega)$ can be evaluated without knowing $g(\omega)$ itself. It is necessary only to calculate the trace of powers of the dynamical matrix.

The even moments (1.4.28) can be used to investigate the analytical part of the frequency spectrum $g(\omega)$. Taking into account that $g(\omega)$ vanishes identically for frequencies greater than ω_L and regarding $g(\omega)$ as an even function of ω, then $g(\omega)$ may be in the interval $(-\omega_L, \omega_L)$ expanded in a series of Legendre polynomials as

$$g(\omega) = \sum_{n=0}^{\infty} a_{2n} P_{2n} \left(\frac{\omega}{\omega_L}\right), \tag{1.4.29}$$

with

$$a_{2n} = \frac{1}{2} (4n + 1) \int_{-1}^{1} dx\, g(\omega_L x)\, P_{2n}(x). \tag{1.4.30}$$

Thus, in terms of the dimensionless moments

$$u_{2n} = \mu_{2n}/\omega_L^{2n}$$

we get the following symbolic expression for a_{2n}:

$$\omega_L a_{2n} = (4n + 1)\, P_{2n}(x)|_{x^{2n}=u_{2n}}. \tag{1.4.31}$$

In practice one keeps only a finite number of terms in the expansion (1.4.29), so that the behaviour of $g(\omega)$ in the vicinity of singular points cannot be described by this method. However, if the structure of the singularities is known the moment method can be extended to include them (see MARADUDIN et al., 1971, chapter 4, and references therein).

Let us now consider the thermodynamic functions (1.4.7) to (1.4.10) and, in particular, the specific heat at low temperatures. The low temperature behaviour of these functions is governed, roughly speaking, by $g(\omega)$ at low ω, and the high temperature behaviour by $g(\omega)$ at high ω. Thus on calculating these functions by making use of $g(\omega)$, low temperatures offer the least difficulty, because there are no singularities in $g(\omega)$ at small ω. For higher temperatures, however, owing to these singularities one has to seek methods of expressing the thermodynamic functions in terms of moments.

At low frequencies $g(\omega)$ depends on ω in a three-dimensional system as

$$g(\omega) = b_2\omega^2 + O(\omega^4) \tag{1.4.32}$$

with

$$b_2 = \frac{v_a}{3r(2\pi)^3} \sum_{j=1}^{3} \int \mathrm{d}\Theta \, \mathrm{d}\varphi \sin \Theta / c_j^3(\Theta, \varphi), \tag{1.4.33}$$

which results from (1.4.11) together with the low frequency dispersion relation for the three acoustical branches (cf. (1.3.54))

$$\omega_j(\boldsymbol{k}) = c_j(\Theta, \varphi) \, |\boldsymbol{k}| + O(|\boldsymbol{k}|^3), \qquad j = 1, 2, 3, \tag{1.4.34}$$

where Θ and φ are the polar and azimuthal angles of the wave vector $\boldsymbol{k}$. The quantity $c_j(\Theta, \varphi)$ is the speed of sound for an acoustic wave in the jth branch propagating in the direction given by Θ and φ. Since the eigenfrequencies of the optical modes have a finite limit for $\boldsymbol{k} \to 0$ they do not contribute to (1.4.32).

On inserting (1.4.32) into (1.4.9) and extending the upper limit in the integral to infinity, since $kT \ll \hbar\omega_{\mathrm{L}}$, we obtain

$$C_v(T) = 3rNkb_2 \frac{4\pi^4}{15} \left(\frac{kT}{\hbar}\right)^3 + O(T^5). \tag{1.4.35}$$

It is customary to write the first term of the right hand side of (1.4.35) as

$$C_v(T) = \frac{12\pi^4 rNk}{5} \left(\frac{T}{\Theta_{\mathrm{D}}}\right)^3, \tag{1.4.36}$$

where

$$\Theta_{\mathrm{D}} = \frac{\hbar}{k} \left(\frac{3}{b_2}\right)^{1/3} \tag{1.4.37}$$

is the so-called Debye temperature.

Equation (1.4.36) is exactly the low temperature expression for the specific heat obtained by the Debye model (DEBYE, 1912), which assumes continuum behaviour for a solid throughout the whole frequency range. This model yields a frequency spectrum

$$g(\omega) = \begin{cases} \dfrac{1}{3rN} \dfrac{V}{2\pi^2} \left(\dfrac{2}{c_t^3} + \dfrac{1}{c_l^3}\right) \omega^2 & \text{for} \quad \omega \leqq \omega_D, \\ 0 & \text{for} \quad \omega > \omega_D, \end{cases} \tag{1.4.38}$$

where c_t and c_l are the speeds of transverse and longitudinal waves, respectively; and ω_D is an artifical cut-off frequency (Debye frequency) defined by the normalization

$$\int_0^{\omega_D} d\omega \, g(\omega) = 1. \tag{1.4.39}$$

However, ω_D is not simply related to the true maximum frequency of the crystal. According to (1.4.38) and (1.4.39) ω_D is given by

$$\omega_D^3 = 18\pi^2 r \frac{N}{V} \left(\frac{2}{c_t^3} + \frac{1}{c_l^3}\right)^{-1}. \tag{1.4.40}$$

For the specific heat we have, using (1.4.9) and (1.4.38),

$$C_v(T) = 9rNk \left(\frac{T}{\Theta_D}\right)^3 \int_0^{\Theta_D/T} dx \frac{x^4 e^x}{(e^x - 1)^2}, \tag{1.4.41}$$

where the Debye temperature Θ_D is defined by

$$k\Theta_D = \hbar\omega_D. \tag{1.4.42}$$

For low temperatures, that is $T \ll \Theta_D$, the upper limit of the integral (1.4.41) may be taken as infinity and we thus find (1.4.36). Consequently, although the Debye spectrum (1.4.38) does not much resemble a real spectrum, it yields the correct low temperature dependence of the specific heat. This is due to the fact that at low temperatures the specific heat is dominated by the behaviour of $g(\omega)$ at low frequencies, i.e. by long wavelength modes, for which the discrete lattice behaves as a continuum. The latter fact is illustrated in figure 1.8.

From (1.4.37), (1.4.29) and (1.4.31) it is easy to get an expression for Θ_D^3 in terms of moments. If we write the Legendre polynomials as

$$P_n(x) = \sum_{\nu=0}^{n} p_n^\nu x^\nu \tag{1.4.43}$$

and pick out solely the coefficients of ω^2 on the right hand side of (1.4.29) then

$$\frac{1}{\Theta_D^3} = \frac{1}{3} \left(\frac{k}{\hbar\omega_L}\right)^3 \sum_{n=0}^{\infty} \sum_{\nu=0}^{n} u_{2\nu} p_{2n}^{2\nu} p_{2n}^2 (4n + 1). \tag{1.4.44}$$

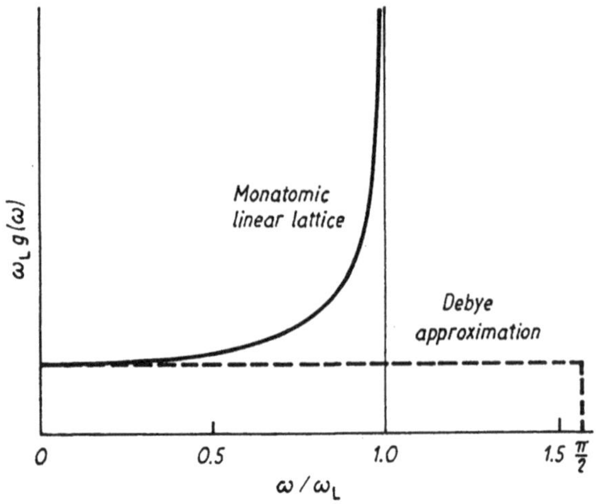

Fig. 1.8. Frequency spectrum, $g(\omega)$, for a monatomic chain of atoms with nearest neighbour interaction as compared with the frequency spectrum for such a chain in Debye approximation

However, the series (1.4.44) may not be very suitable for determining Θ_D, since it is not expected to converge rapidly.

Let us now investigate the temperature dependence of the square vibrational amplitude. The stability of a crystal depends essentially on the vibrational amplitude. Because the amplitude increases with increasing temperature, phase transitions to new energetically more favourable structures can occur. At sufficiently high temperatures the amplitude can reach such a magnitude that the crystal lattice cannot further resist the atomic vibrations, i.e. the solid begins to melt. With increasing amplitude the anharmonicity of the potential affects the vibrational motion of the atoms more and more. To illustrate this we assume for the lattice potential Φ a so-called central force potential, which describes the interatomic interactions pairwise through a potential function which is a function only of the magnitude of the separation between the atoms:

$$\Phi = \frac{1}{2} \sum_{ll'\varkappa\varkappa'} \varphi(|\boldsymbol{R}(l\varkappa) - \boldsymbol{R}(l'\varkappa')|). \tag{1.4.45}$$

The pair interaction potential φ is composed of a short-range repulsive and a long-range attractive part. In the case of the so-called Lennard-Jones potential we have

$$\varphi(r) = \frac{A}{r^{12}} - \frac{B}{r^6}, \tag{1.4.46}$$

where A and B are constants. The potential (1.4.46) is shown in figure 1.9. From this figure the increasing role of anharmonicity with increasing temperature becomes evident. This figure also suggests that in the case of a large zero point vibrational amplitude, anharmonic effects can be important down to lowest temperatures.

We restrict ourselves here to an investigation of the mean square vibrational amplitude in the harmonic approximation. It is convenient to calculate the mean square amplitude averaged over the atoms of a primitive unit cell.

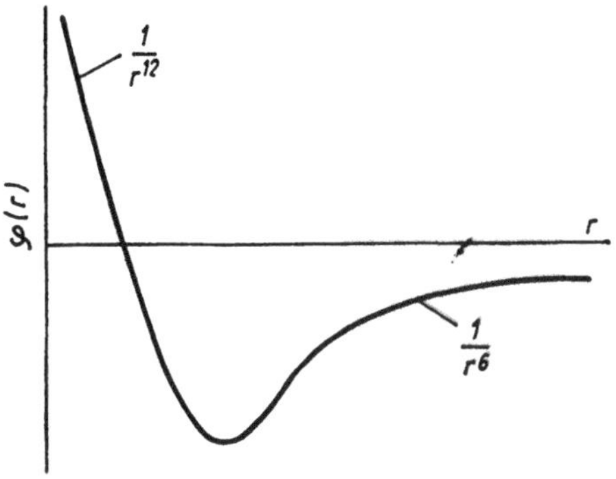

Fig. 1.9. The Lennard-Jones potential for describing the interaction between two neutral atoms

From (1.3.60) we obtain

$$\langle \boldsymbol{u}^2 \rangle \equiv \frac{1}{r} \sum_{\varkappa\alpha} \langle u_\alpha{}^2(l\varkappa) \rangle_0$$
$$= \sum_{\varkappa\alpha} \frac{\hbar}{2M_\varkappa N r} \left\langle \left[\sum_{\boldsymbol{k}j} \frac{e_\alpha(\varkappa \mid \boldsymbol{k}j)}{\sqrt{\omega_j(\boldsymbol{k})}} \, e^{i\boldsymbol{k}\boldsymbol{x}(l)} (b_{\boldsymbol{k}j} + b^+_{-\boldsymbol{k}j}) \right]^2 \right\rangle_0 \qquad (1.4.47)$$

where $\langle \cdots \rangle_0$ denotes the thermal average defined by

$$\langle \cdots \rangle_0 = Z^{-1} \operatorname{Tr} \{ e^{-\beta H} \ldots \}. \qquad (1.4.48)$$

Here H is the Hamiltonian (1.3.66) and Z the partition function (1.4.1).

If we now take into account that in the representation in which $b^+_{\boldsymbol{k}j} b_{\boldsymbol{k}j}$ is diagonal the operator $b^+_{\boldsymbol{k}j} b_{\boldsymbol{k}'j}$ $(\boldsymbol{k} \neq \boldsymbol{k}')$ has no diagonal elements, we obtain

$$\langle b^+_{\boldsymbol{k}j} b_{\boldsymbol{k}'j'} \rangle_0 = \Delta(\boldsymbol{k} - \boldsymbol{k}') \, \delta_{jj'} \bar{n}_j(\boldsymbol{k}),$$
$$\langle b_{\boldsymbol{k}j} b^+_{\boldsymbol{k}'j'} \rangle_0 = \Delta(\boldsymbol{k} - \boldsymbol{k}') \, \delta_{jj'} (\bar{n}_j(\boldsymbol{k}) + 1), \qquad (1.4.49)$$
$$\langle b_{\boldsymbol{k}j} b_{\boldsymbol{k}'j'} \rangle_0 = \langle b^+_{\boldsymbol{k}j} b^+_{\boldsymbol{k}'j'} \rangle_0 = 0,$$

where $\bar{n}_j(\boldsymbol{k})$ is the mean number of phonons in the vibrational mode $(\boldsymbol{k}j)$ at temperature T, which is given by

$$\bar{n}_j(\boldsymbol{k}) = [e^{\beta\hbar\omega_j(\boldsymbol{k})} - 1]^{-1}. \qquad (1.4.50)$$

The function (1.4.50) represents, as mentioned above, a Bose distribution with vanishing chemical potential. Thus, accounting for (1.4.49) and (1.4.50) we get from (1.4.47)

$$\begin{aligned}\langle \boldsymbol{u}^2 \rangle &= \frac{\hbar f}{2M_\mathrm{T}N} \sum_{\boldsymbol{k}j} \frac{1}{\omega_j(\boldsymbol{k})} \left(2\bar{n}_j(\boldsymbol{k}) + 1\right) \\ &= \frac{\hbar f}{2M_\mathrm{T}N} \sum_{\boldsymbol{k}j} \frac{1}{\omega_j(\boldsymbol{k})} \coth\left[\frac{1}{2}\,\beta\hbar\omega_j(\boldsymbol{k})\right],\end{aligned} \tag{1.4.51}$$

where

$$f = \frac{1}{r} \sum_{\varkappa\alpha} \frac{M_\mathrm{T}}{M_\varkappa}\, \mathrm{e}_\alpha{}^*(\varkappa \mid \boldsymbol{k}j)\, \mathrm{e}_\alpha(\varkappa \mid \boldsymbol{k}j), \tag{1.4.52}$$

and M_T is the total mass of the atoms in a unit cell. Because of the orthonormality condition (1.3.44) we have $f = 1$, if $r = 1$, or if $M_\varkappa$ does not depend on the index $\varkappa$.

The high temperature ($T > \Theta_\mathrm{D}$) limiting expression for $\langle \boldsymbol{u}^2 \rangle$ is obtained from (1.4.51) by neglecting the zero point energy contribution and by approximating

$$\mathrm{e}^{\beta\hbar\omega_j(\boldsymbol{k})} \approx 1 + \beta\hbar\omega_j(\boldsymbol{k}). \tag{1.4.53}$$

We get

$$\langle \boldsymbol{u}^2 \rangle = \frac{f}{M_\mathrm{T}N} \sum_{\boldsymbol{k}j} \frac{kT}{\omega_j{}^2(\boldsymbol{k})} = \frac{3kTrf}{M_\mathrm{T}}\, \mu_{-2}, \tag{1.4.54}$$

where μ_{-2} is the negative second moment (1.4.27). If we use the Debye spectrum (1.4.38) to evaluate μ_{-2} then the mean square amplitude (1.4.54) reads

$$\langle \boldsymbol{u}^2 \rangle = \frac{9rf\hbar^2 T}{k\Theta_\mathrm{D}{}^2 M_\mathrm{T}}. \tag{1.4.55}$$

According to Lindemann (1910), melting occurs if the mean square amplitude reaches a critical value or, in other words, it holds a relation

$$\frac{\langle \boldsymbol{u}^2 \rangle_\mathrm{melt}}{v_a{}^{2/3}} = C_\mathrm{Lind} \tag{1.4.56}$$

where C_Lind is the so-called Lindemann constant, which can be expressed by the melting temperature T_melt by means of (1.4.55) as

$$C_\mathrm{Lind} = \frac{9rf\hbar^2 T_\mathrm{melt}}{k\Theta_\mathrm{D}{}^2 M_\mathrm{T} v_a{}^{2/3}}. \tag{1.4.57}$$

Within given classes of solids, relations of type (1.4.57) are experimentally rather well established. The problem of melting is more complicated for solids with $T_\mathrm{melt} < \Theta_\mathrm{D}$. In this case the approximation (1.4.53) does not hold and anharmonic effects are important. Note that the actual melting process of a crystal is not completely understood at present (e.g. its dependence on the size of the sample and on the density of dislocations). It is still a problem of vital current interest (cf. AMINI et al., 1979, COUCHMAN, 1979, and RUFFA, 1981).

At low temperatures zero point vibrations dominate the mean square amplitude, for which we get from (1.4.51)

$$\langle \boldsymbol{u}^2 \rangle = \frac{\hbar f}{2M_{\mathrm{T}}N} \sum_{\boldsymbol{k}j} \frac{1}{\omega_j(\boldsymbol{k})} = \frac{3\hbar f r}{2M_{\mathrm{T}}} \mu_{-1}, \tag{1.4.58}$$

where μ_{-1} is the negative first moment (1.4.27). The Debye model yields $\mu_{-1} = \frac{3}{2} \omega_{\mathrm{D}}^{-1}$ and thus we obtain from (1.4.58)

$$\langle \boldsymbol{u}^2 \rangle = \frac{9}{4} \frac{\hbar f r}{\omega_{\mathrm{D}} M_{\mathrm{T}}} = \frac{9}{4} \frac{\hbar^2 f r}{k \Theta_{\mathrm{D}} M_{\mathrm{T}}}. \tag{1.4.59}$$

According to (1.4.59) the amplitude of the zero point vibrations is especially large in solids with light atoms and weak interatomic forces. Generally, the harmonic approximation (more strictly speaking: the quasi-harmonic approximation) turns out to be a reasonable approximation at temperatures $T < \Theta_{\mathrm{D}}$, if the ratio of the square root of the mean square zero point amplitude and the nearest-neighbour distance between the atoms is less than 0.15.

In the harmonic approximation the equilibrium positions of the atoms are defined by the minimum of the potential energy (cf. section 1.1.). Because these positions are independent of temperature the effect of the thermal expansion of a crystal cannot be described by this approximation. To account for the latter effect we now expand the lattice potential by starting with unknown rest positions $\boldsymbol{x}(l\varkappa)$ (cf. (1.2.2)) and regard these positions as parameters to be determined later from the minimum of the free energy, which represents the equilibrium condition of the crystal in the absence of stresses. If we now calculate the Helmholtz free energy and retain from the vibrational part the harmonic one only, we obtain

$$F = \Phi_0\big(\boldsymbol{x}(l\varkappa)\big) + kT \sum_{\boldsymbol{k}j} \ln \left[2 \sinh \big(\hbar\omega_j(\boldsymbol{k})/2kT\big)\right], \tag{1.4.60}$$

where we have also included the potential energy of the static lattice Φ_0. Equation (1.4.60) represents the harmonic free energy (cf. (1.4.3)), except that Φ_0 and the frequencies $\omega_j(\boldsymbol{k})$ now depend on the parameter $\boldsymbol{x}(l\varkappa)$. Therefore, the approximation yielding (1.4.60) is called the quasi-harmonic approximation.

To describe the temperature variation of the atomic positions we now introduce the tensor of finite thermal strain $u^{\mathrm{T}}_{\alpha\alpha'}$. The derivative of F (1.4.60) with respect to $u^{\mathrm{T}}_{\alpha\alpha'}$ has to be zero in case of vanishing stresses in the equilibrium state:

$$\frac{\partial F}{\partial u^{\mathrm{T}}_{\alpha\alpha'}} = 0 = \frac{\partial \Phi_0}{\partial u^{\mathrm{T}}_{\alpha\alpha'}} + \sum_{\boldsymbol{k}j} \hbar \frac{\partial \omega_j(\boldsymbol{k})}{\partial u^{\mathrm{T}}_{\alpha\alpha'}} [\bar{n}_j(\boldsymbol{k}) + 1/2], \tag{1.4.61}$$

where $\bar{n}_j(\boldsymbol{k})$ is the mean phonon number of the oscillator $(\boldsymbol{k}j)$ (1.4.50). Introducing the mean thermal energy of the oscillator $(\boldsymbol{k}j)$ (cf. (1.3.66), (1.4.50), (1.4.4))

$$\bar{\varepsilon}_j(\boldsymbol{k}) = \hbar\omega_j(\boldsymbol{k}) \left(\bar{n}_j(\boldsymbol{k}) + 1/2\right) \tag{1.4.62}$$

(1.4.61) can be rewritten as

$$\frac{\partial \Phi_0}{\partial u^{\mathrm{T}}_{\alpha\alpha'}} + \sum_{\boldsymbol{k}j} \frac{1}{\omega_j(\boldsymbol{k})} \frac{\partial \omega_j(\boldsymbol{k})}{\partial u^{\mathrm{T}}_{\alpha\alpha'}} \bar{\varepsilon}_j(\boldsymbol{k}) = 0. \tag{1.4.63}$$

Analogous to the case of mechanical strain, the change of Φ_0 with thermal strain, i.e. the difference of Φ_0 taken once at the minimum positions and taken once at the positions after introducing thermal strain, will be of quadratic form. Therefore, with zero stresses we have

$$\frac{\partial \Phi_0}{\partial u^{\mathrm{T}}_{\alpha\alpha'}} = N v_a \sum_{\beta\beta'} c_{\alpha\alpha',\beta\beta'} u^{\mathrm{T}}_{\beta\beta'}, \tag{1.4.64}$$

where the $c_{\alpha\alpha',\beta\beta'}$'s are the elastic constants in the harmonic approximation. On solving (1.4.64) for $u^{\mathrm{T}}_{\alpha\alpha'}$ with

$$\sum_{\beta\beta'} c_{\alpha\alpha',\beta\beta'} c^{-1}_{\beta\beta',\gamma\gamma'} = \delta_{\alpha\gamma}\delta_{\alpha'\gamma'}, \tag{1.4.65}$$

and making use of (1.4.63) we obtain

$$u^{\mathrm{T}}_{\alpha\alpha'} = \frac{1}{N v_a} \sum_{\beta\beta'} c^{-1}_{\alpha\alpha',\beta\beta'} \sum_{\boldsymbol{k}j} \gamma^j_{\beta\beta'}(\boldsymbol{k})\, \bar{\varepsilon}_j(\boldsymbol{k}), \tag{1.4.66}$$

where we have introduced by

$$\gamma^j_{\alpha\alpha'}(\boldsymbol{k}) = -\frac{\partial \ln \omega_j(\boldsymbol{k})}{\partial u^{\mathrm{T}}_{\alpha\alpha'}} \tag{1.4.67}$$

a so-called generalized Grüneisen parameter. With (1.4.66) we get the thermal expansion coefficient

$$\alpha^{\mathrm{T}}_{\alpha\alpha'} = \frac{\partial u^{\mathrm{T}}_{\alpha\alpha'}}{\partial T} = \frac{1}{N v_a} \sum_{\beta\beta'} c^{-1}_{\alpha\alpha',\beta\beta'} \sum_{\boldsymbol{k}j} \gamma^j_{\beta\beta'}(\boldsymbol{k}) \frac{\partial \bar{\varepsilon}_j(\boldsymbol{k})}{\partial T}. \tag{1.4.68}$$

In cubic crystals it holds that

$$\gamma_{\alpha\alpha'} = \gamma \delta_{\alpha\alpha'} \tag{1.4.69}$$

and

$$u^{\mathrm{T}}_{\alpha\alpha'} = u^{\mathrm{T}} \delta_{\alpha\alpha'} \tag{1.4.70}$$

which leads to the following expression for the linear thermal expansion coefficient:

$$\alpha^{\mathrm{T}} = \frac{\partial u^{\mathrm{T}}}{\partial T} = \frac{\varkappa}{3 N v_a} \sum_{\boldsymbol{k}j} \gamma^j(\boldsymbol{k}) \frac{\partial \varepsilon_j(\boldsymbol{k})}{\partial T}, \tag{1.4.71}$$

where

$$\varkappa = 3 \sum_\beta c^{-1}_{\alpha\alpha,\beta\beta} \tag{1.4.72}$$

is the compressibility. At high temperatures from (1.4.71) with (1.4.62), (1.4.50) and (1.4.53) we get a constant expansion coefficient. At low temperatures the expansion coefficient becomes zero. For intermediate temperatures the temperature dependence of α^{T} is more complicated.

On neglecting the dependence on $(\boldsymbol{k}j)$ in (1.4.67) for a cubic crystal with lattice constant a, we obtain using the relation

$$\sum_{\alpha=1}^{3} \frac{\partial}{\partial u^{\mathrm{T}}_{\alpha\alpha}} = 3V \frac{\mathrm{d}}{\mathrm{d}V} = a \frac{\mathrm{d}}{\mathrm{d}a} \tag{1.4.73}$$

the following expression for the Grüneisen parameter:

$$\gamma = -\frac{1}{3} \sum_{\alpha=1}^{3} \frac{\partial \ln \omega}{\partial u^{\mathrm{T}}_{\alpha\alpha}} = -V \frac{\mathrm{d} \ln \omega}{\mathrm{d}V} = -\frac{\mathrm{d} \ln \omega}{\mathrm{d} \ln V} = -\frac{\mathrm{d} \ln \Theta_{\mathrm{D}}}{\mathrm{d} \ln V}, \tag{1.4.74}$$

where in the last form we have replaced ω by its average, which is proportional to the Debye temperature in the Debye approximation. Eq. (1.4.74) is identical with the usual definition of the Grüneisen parameter. The minus sign in (1.4.74) reflects the usual behaviour that the frequencies decrease when the lattice expands.

If we ignore in (1.4.71) the dependence of $\gamma^j(\boldsymbol{k})$ on $(\boldsymbol{k}j)$ we obtain, on account of (1.4.62), (1.4.50) and (1.4.5),

$$\gamma = \frac{3\alpha^{\mathrm{T}} V}{C_v \varkappa}. \tag{1.4.75}$$

According to this equation the Grüneisen parameter can be expressed by macroscopic thermodynamic quantities, which can be used to prove the validity of models for calculating γ. For more details on thermal expansion see, for example, Ludwig (1967), chapter 5, and Barron et al. (1980).

1.5. Lattice vibrations in ionic crystals

If in an ionic crystal the ions are displaced from their equilibrium positions a dipole moment occurs giving rise to a Coulomb field which is composed of a macroscopic field and a Lorentz field associated with the local field corrections (cf. Bonneville, 1980). The macroscopic field on its part affects the motion of the ions by acting as an external force and, moreover, it can polarize the ions and thus enhance their dipole moments. If we ignore the polarizability of the ions, i.e. if we treat the ions as rigid or point ions, the vibrational properties of an ionic crystal can be studied by the theory developed in the preceding sections. In this case, the equations of motion of the ions can be derived from a potential function consisting of a long-range Coulomb part and a short-range repulsive part (see LACINA and PERSHAN, 1970, and MARADUDIN et al., 1971, chapter 6).

However, an account of the deformability and polarizability of the ions requires another approach to the dynamical problem, since these properties are essentially electronic in origin and cannot be incorporated into the theory of lattice vibrations by starting from an expansion of a lattice potential in powers of the ionic displacements alone. On deriving the equations of motion in a phenomenological approach to the lattice vibrations of ionic crystals, beside the components of the ionic displacements $u_\alpha(l\varkappa)$ the components of the macroscopic electric field in the crystal $E_\alpha(l\varkappa)$ are treated as independent dynamical variables. However, by using the Maxwell equations the latter variables can be expressed by the former, so that the final equations of motion, which determine the eigenmodes, again contain the displacement variables only.

The starting point for such a phenomenological theory of lattice vibrations in ionic crystals can be the following expression for the potential energy (cf. MARADUDIN et al., 1971, chapter 6):

$$\begin{aligned}\Phi = \frac{1}{2} \sum_{\substack{l\varkappa\alpha \\ l'\varkappa'\alpha'}} \Phi_{\alpha\alpha'}(l\varkappa, l'\varkappa')\, u_\alpha(l\varkappa)\, u_{\alpha'}(l'\varkappa') \\ - \sum_{\substack{l\varkappa\alpha \\ l'\varkappa'\alpha'}} M_{\alpha\alpha'}(l\varkappa, l'\varkappa')\, E_\alpha(l\varkappa)\, u_{\alpha'}(l'\varkappa') \\ - \frac{1}{2} \sum_{\substack{l\varkappa\alpha \\ l'\varkappa'\alpha'}} P_{\alpha\alpha'}(l\varkappa, l'\varkappa')\, E_\alpha(l\varkappa)\, E_{\alpha'}(l'\varkappa').\end{aligned} \tag{1.5.1}$$

Here the force constants $\Phi_{\alpha\alpha'}(l\varkappa, l'\varkappa')$ describe the short-range forces, including that connected with the Lorentz field. Effects of the macroscopic electric field are contained in the second and third terms on the right hand side of (1.5.1). $M_{\alpha\alpha'}(l\varkappa, l'\varkappa')$ is the so-called transverse effective charge tensor. It gives the αth component of the dipole moment at site $(l\varkappa)$ arising from a unit displacement of the ion at site $(l'\varkappa')$ in α'-direction. $P_{\alpha\alpha'}(l\varkappa, l'\varkappa')$ is the electric polarizability tensor. It gives the αth component of the dipole moment at site $(l\varkappa)$ arising from a unit macroscopic electric field acting on the ion at site $(l'\varkappa')$ in the α'-direction. Force constants, charge tensor, and polarizability tensor are nonlocal quantities, which reflects the fact that, due to the final spatial extent of the charge distribution about each ion and the overlap between these charge distributions, a force (displacement or macroscopic field) acting on a given ion can affect a different ion.

Analogous to the force constants in non-ionic solids (cf. section 1.2.) the coefficients $\Phi_{\alpha\alpha'}(l\varkappa, l'\varkappa')$, $M_{\alpha\alpha'}(l\varkappa, l'\varkappa')$ and $P_{\alpha\alpha'}(l\varkappa, l'\varkappa')$ in ionic crystals can be shown to have several general properties. According to (1.5.1) the coefficients $\Phi_{\alpha\alpha'}(l\varkappa, l'\varkappa')$ and $P_{\alpha\alpha'}(l\varkappa, l'\varkappa')$ are symmetric in the indices $l\varkappa\alpha$, $l'\varkappa'\alpha'$. From the invariance of the force acting on an ion and of the dipole moment under a rigid body displacement of the crystal as a whole we get

$$\sum_{l'\varkappa'} \Phi_{\alpha\alpha'}(l\varkappa, l'\varkappa') = 0, \tag{1.5.2}$$

$$\sum_{l'\varkappa'} M_{\alpha\alpha'}(l\varkappa, l'\varkappa') = 0. \tag{1.5.3}$$

Further, from the invariance of the force on an ion and the dipole moment under an infinitesimal rigid body rotation of the crystal as a whole it follows that

$$\sum_{l'\varkappa'} \Phi_{\alpha\alpha'}(l\varkappa, l'\varkappa')\, x_{\alpha''}(l'\varkappa') \tag{1.5.4}$$

and

$$\sum_{l'\varkappa'} M_{\alpha\alpha'}(l\varkappa, l'\varkappa')\, x_{\alpha''}(l'\varkappa') \tag{1.5.5}$$

are symmetric in α' and α''. Finally, from the invariance of the potential (1.5.1) under symmetry operations of the space group of the crystal (1.2.19) we obtain

$$\Phi_{\alpha\alpha'}(LK, L'K') = \sum_{\alpha_1\alpha_2} S_{\alpha\alpha_1} S_{\alpha'\alpha_2} \Phi_{\alpha_1\alpha_2}(l\varkappa, l'\varkappa'), \tag{1.5.6}$$

$$M_{\alpha\alpha'}(LK, L'K') = \sum_{\alpha_1\alpha_2} S_{\alpha\alpha_1} S_{\alpha'\alpha_2} M_{\alpha_1\alpha_2}(l\varkappa, l'\varkappa'), \tag{1.5.7}$$

$$P_{\alpha\alpha'}(LK, L'K') = \sum_{\alpha_1\alpha_2} S_{\alpha\alpha_1} S_{\alpha'\alpha_2} P_{\alpha\alpha'}(l\varkappa, l'\varkappa'), \tag{1.5.8}$$

where we have used the same symbols and notation as in section 1.2. From (1.5.6) to (1.5.8) it follows in particular that $\Phi_{\alpha\alpha'}(l\varkappa, l'\varkappa')$, $M_{\alpha\alpha'}(l\varkappa, l'\varkappa')$, and $P_{\alpha\alpha'}(l\varkappa, l'\varkappa')$ depend on l and l' only through their difference (cf. (1.2.25)).

Making use of the potential (1.5.1) we obtain the equation of motion

$$\begin{aligned} M_\varkappa \ddot{u}_\alpha(l\varkappa) &= -\partial\Phi/\partial u_\alpha(l\varkappa) \\ &= -\sum_{l'\varkappa'\alpha'} \Phi_{\alpha\alpha'}(l\varkappa, l'\varkappa')\, u_{\alpha'}(l'\varkappa') + \sum_{l'\varkappa'\alpha'} M_{\alpha'\alpha}(l'\varkappa', l\varkappa)\, E_{\alpha'}(l'\varkappa') \end{aligned} \tag{1.5.9}$$

and the dipole moment induced at site $(l\varkappa)$

$$\begin{aligned} p_\alpha(l\varkappa) &= -\partial\Phi/\partial E_\alpha(l\varkappa) \\ &= \sum_{l'\varkappa'\alpha'} M_{\alpha\alpha'}(l\varkappa, l'\varkappa')\, u_{\alpha'}(l'\varkappa') + \sum_{l'\varkappa'\alpha'} P_{\alpha\alpha'}(l\varkappa, l'\varkappa')\, E_{\alpha'}(l'\varkappa'). \end{aligned} \tag{1.5.10}$$

Because of the periodicity of the crystal we may write (cf. (1.3.31))

$$u_\alpha(l\varkappa) = \frac{v_\alpha(\varkappa)}{\sqrt{M_\varkappa}}\, \mathrm{e}^{i\boldsymbol{k}\boldsymbol{x}(l\varkappa) - i\omega t}, \tag{1.5.11}$$

$$p_\alpha(l\varkappa) = p_\alpha(\varkappa)\, \mathrm{e}^{i\boldsymbol{k}\boldsymbol{x}(l\varkappa) - i\omega t}, \tag{1.5.12}$$

$$E_\alpha(l\varkappa) = E_\alpha\, \mathrm{e}^{i\boldsymbol{k}\boldsymbol{x}(l\varkappa) - i\omega t}, \tag{1.5.13}$$

where $\boldsymbol{x}(l\varkappa)$ denotes the position vector of the $\varkappa$th atom in the lth unit cell (cf. (1.2.17)). E_α in (1.5.13) is independent of $\varkappa$, since $\boldsymbol{E}$ is a macroscopic field, i.e. it is nearly constant on the scale of a primitive unit cell. Inserting (1.5.11) to (1.5.13) into (1.5.9) and (1.5.10) we obtain

$$\omega^2 v_\alpha(\varkappa) = \sum_{\varkappa'\alpha'} C_{\alpha\alpha'}(\varkappa\varkappa' \mid \boldsymbol{k})\, v_{\alpha'}(\varkappa') - \frac{1}{\sqrt{M_\varkappa}} \sum_{\varkappa'\alpha'} M^*_{\alpha'\alpha}(\varkappa'\varkappa \mid \boldsymbol{k})\, E_{\alpha'}, \tag{1.5.14}$$

$$p_\alpha(\varkappa) = \sum_{\varkappa'\alpha'} \frac{1}{\sqrt{M_{\varkappa'}}} M_{\alpha\alpha'}(\varkappa\varkappa' \mid \boldsymbol{k})\, v_{\alpha'}(\varkappa') + \sum_{\varkappa'\alpha'} P_{\alpha\alpha'}(\varkappa\varkappa' \mid \boldsymbol{k})\, E_{\alpha'}, \tag{1.5.15}$$

$$C_{\alpha\alpha'}(\varkappa\varkappa' \mid \boldsymbol{k}) = (M_\varkappa M_{\varkappa'})^{-1/2} \sum_{l'} \Phi_{\alpha\alpha'}(l\varkappa, l'\varkappa')\, \mathrm{e}^{-i\boldsymbol{k}(\boldsymbol{x}(l\varkappa) - \boldsymbol{x}(l'\varkappa'))}, \tag{1.5.16}$$

$$M_{\alpha\alpha'}(\varkappa\varkappa' \mid \boldsymbol{k}) = \sum_{l'} M_{\alpha\alpha'}(l\varkappa, l'\varkappa')\, \mathrm{e}^{-i\boldsymbol{k}(\boldsymbol{x}(l\varkappa) - \boldsymbol{x}(l'\varkappa'))}, \tag{1.5.17}$$

$$P_{\alpha\alpha'}(\varkappa\varkappa' \mid \boldsymbol{k}) = \sum_{l'} P_{\alpha\alpha'}(l\varkappa, l'\varkappa')\, \mathrm{e}^{-i\boldsymbol{k}(\boldsymbol{x}(l\varkappa) - \boldsymbol{x}(l'\varkappa'))}. \tag{1.5.18}$$

Let us now derive the expression for the macroscopic polarization. We need this quantity to express $\boldsymbol{E}$ in (1.5.14) by $\boldsymbol{v}$ by means of the Maxwell equations and (1.5.15), so that we obtain an equation of motion containing the displacements alone. To get the macroscopic polarization we start from the microscopic polarization density in the crystal

$$\boldsymbol{P}^{\mathrm{micr}}(\boldsymbol{r}) = \sum_{l\varkappa} \boldsymbol{p}(l\varkappa)\, \delta\big(\boldsymbol{r} - \boldsymbol{x}(l\varkappa)\big). \tag{1.5.19}$$

Using (1.5.12) and the Fourier expansion

$$\sum_l \delta\big(\boldsymbol{r} - \boldsymbol{x}(l\varkappa)\big) = \sum_{\boldsymbol{K}} C(\boldsymbol{K})\, \mathrm{e}^{i\boldsymbol{K}\boldsymbol{r}} \tag{1.5.20}$$

with

$$C(\boldsymbol{K}) = \frac{1}{v_a} \int_{v_a} \mathrm{d}\boldsymbol{r}\, \mathrm{e}^{-i\boldsymbol{K}\boldsymbol{r}} \sum_l \delta\big(\boldsymbol{r} - \boldsymbol{x}(l\varkappa)\big) = \frac{1}{v_a} \exp\big(-i\boldsymbol{K}\boldsymbol{x}(\varkappa)\big), \tag{1.5.21}$$

where $\boldsymbol{K}$ is a vector of the reciprocal lattice and $\int_{v_a} \mathrm{d}\boldsymbol{r}$ denotes the integral over the volume of a primitive unit cell, (1.5.19) can be rewritten as

$$\boldsymbol{P}^{\mathrm{micr}}(\boldsymbol{r}) = \sum_{\boldsymbol{K}} \boldsymbol{P}(\boldsymbol{K})\, \mathrm{e}^{i(\boldsymbol{k}+\boldsymbol{K})\boldsymbol{r}} \tag{1.5.22}$$

with

$$\boldsymbol{P}(\boldsymbol{K}) = \frac{1}{v_a} \sum_{\varkappa} \boldsymbol{p}(\varkappa)\, \mathrm{e}^{-i\boldsymbol{K}\boldsymbol{x}(\varkappa)}. \tag{1.5.23}$$

As usual in electrodynamics in media, we obtain a macroscopic quantity by averaging the corresponding microscopic quantity over the volume of a primitive unit cell, which, in essence, is equivalent to a restriction to the Fourier coefficient $\boldsymbol{K} = 0$ in an expansion like (1.5.20). Thus, we have for the macroscopic polarization

$$\boldsymbol{P}(\boldsymbol{r}) = \mathrm{e}^{i\boldsymbol{kr}} \boldsymbol{P} \tag{1.5.24}$$

with

$$\boldsymbol{P} = \frac{1}{v_a} \sum_{\varkappa} \boldsymbol{p}(\varkappa). \tag{1.5.25}$$

If we now insert (1.5.15) into (1.5.25) we obtain for the amplitude of the macroscopic polarization (1.5.25)

$$P_\alpha = \frac{1}{v_a} \sum_{\varkappa'\alpha'} Z_{\alpha\alpha'}(\varkappa' \mid \boldsymbol{k})\, v_{\alpha'}(\varkappa')/\sqrt{M_{\varkappa'}} + \sum_{\alpha'} \chi^0_{\alpha\alpha'}(\boldsymbol{k})\, E_{\alpha'}, \tag{1.5.26}$$

where

$$Z_{\alpha\alpha'}(\varkappa \mid \boldsymbol{k}) = \sum_{\varkappa'} M_{\alpha\alpha'}(\varkappa'\varkappa \mid \boldsymbol{k}) \tag{1.5.27}$$

defines the $\boldsymbol{k}$-dependent tensor of the effective charge for an ion of type $\varkappa$ and

$$\chi^0_{\alpha\alpha'}(\boldsymbol{k}) = \frac{1}{v_a} \sum_{\varkappa\varkappa'} P_{\alpha\alpha'}(\varkappa\varkappa' \mid \boldsymbol{k}) \tag{1.5.28}$$

defines the $\boldsymbol{k}$-dependent optical frequency (static electronic) susceptibility.

To get a relation between the macroscopic polarization $\boldsymbol{P}$ and the macroscopic electric field $\boldsymbol{E}$ we use the Maxwell equations, which in absence of external charges, external currents, and magnetization read

$$\nabla \boldsymbol{D} = 0, \qquad \nabla \boldsymbol{B} = 0 \tag{1.5.29), (1.5.30}$$

$$\nabla \times \boldsymbol{E} = \frac{1}{c} \dot{\boldsymbol{B}}, \qquad \nabla \times \boldsymbol{B} = \frac{1}{c} \dot{\boldsymbol{D}}, \tag{1.5.31), (1.5.32}$$

where c is the speed of light, and $\boldsymbol{B}$ and $\boldsymbol{D}$ are the magnetic induction and the dielectric displacement, respectively. Taking the curl of (1.5.31) and combining the resulting expression with the time derivative of (1.5.32) we obtain

$$\nabla \times (\nabla \times \boldsymbol{E}) = -\frac{1}{c^2} \frac{\partial^2 \boldsymbol{D}}{\partial t^2}. \tag{1.5.33}$$

Using in (1.5.33)

$$\boldsymbol{D} = \boldsymbol{E} + 4\pi \boldsymbol{P} \tag{1.5.34}$$

the relation between $\boldsymbol{E}$ and $\boldsymbol{P}$ becomes

$$\nabla \times (\nabla \times \boldsymbol{E}) + \frac{1}{c^2} \frac{\partial^2 \boldsymbol{E}}{\partial t^2} = -\frac{4\pi}{c^2} \frac{\partial^2 \boldsymbol{P}}{\partial t^2}. \tag{1.5.35}$$

Writing now

$$\boldsymbol{E}(\boldsymbol{r}, t) = \boldsymbol{E}\, \mathrm{e}^{i\boldsymbol{kr} - i\omega t}, \quad \boldsymbol{P}(\boldsymbol{r}, t) = \boldsymbol{P}\, \mathrm{e}^{i\boldsymbol{kr} - i\omega t} \tag{1.5.36), (1.5.37}$$

we obtain from (1.5.35)

$$\left[1 - \frac{k^2c^2}{\omega^2}\,\mathbf{1}_t(\hat{\boldsymbol{k}})\right]\boldsymbol{E} = -4\pi\boldsymbol{P} \tag{1.5.38}$$

or

$$\boldsymbol{E} = -4\pi\boldsymbol{K}(\boldsymbol{k},\omega)\,\boldsymbol{P} \tag{1.5.39}$$

with the tensor

$$\boldsymbol{K}(\boldsymbol{k},\omega) = \left[1 - \frac{k^2c^2}{\omega^2}\,\mathbf{1}_t(\hat{\boldsymbol{k}})\right]^{-1} = \mathbf{1}_l(\hat{\boldsymbol{k}}) + \mathbf{1}_t(\hat{\boldsymbol{k}})\,\frac{1}{1-\dfrac{c^2k^2}{\omega^2}}. \tag{1.5.40}$$

Here $k = |\boldsymbol{k}|$, $\hat{\boldsymbol{k}} = \boldsymbol{k}/|\boldsymbol{k}|$, $\mathbf{1}$ is the unit tensor, and $\mathbf{1}_l(\hat{\boldsymbol{k}})$ and $\mathbf{1}_t(\hat{\boldsymbol{k}})$ are the longitudinal and transverse unit tensors, respectively, defined by

$$1_{l,\alpha\alpha'}(\hat{\boldsymbol{k}}) = \hat{k}_\alpha\hat{k}_{\alpha'} \tag{1.5.41}$$

and

$$\mathbf{1}_t(\hat{\boldsymbol{k}}) = \mathbf{1} - \mathbf{1}_l(\hat{\boldsymbol{k}}). \tag{1.5.42}$$

The tensors $\mathbf{1}_l(\hat{\boldsymbol{k}})$ and $\mathbf{1}_t(\hat{\boldsymbol{k}})$ have the property that they can decompose any vector $\boldsymbol{a}$ uniquely into a longitudinal and a transverse component with respect to the vector $\boldsymbol{k}$:

$$\mathbf{1}_l(\hat{\boldsymbol{k}})\,\boldsymbol{a} = \hat{\boldsymbol{k}}(\hat{\boldsymbol{k}}\boldsymbol{a}) \tag{1.5.43}$$

$$\mathbf{1}_t(\hat{\boldsymbol{k}})\,\boldsymbol{a} = -\hat{\boldsymbol{k}}\times(\hat{\boldsymbol{k}}\times\boldsymbol{a}). \tag{1.5.44}$$

Note that $\mathbf{1}_i(\hat{\boldsymbol{k}})\,\mathbf{1}_j(\hat{\boldsymbol{k}}) = \delta_{ij}\mathbf{1}_i(\hat{\boldsymbol{k}}) \qquad (i, j = \mathrm{l}, \mathrm{t})$.

Making use now of (1.5.38) we obtain from (1.5.26) the following relation between the macroscopic electric field and the displacements of the ions:

$$\boldsymbol{E} = -\frac{4\pi}{v_a}\sum_{\varkappa'}\boldsymbol{T}^{-1}(\boldsymbol{k},\omega)\,\boldsymbol{Z}(\varkappa'\mid\boldsymbol{k})\,\boldsymbol{v}(\varkappa')/\sqrt{M_{\varkappa'}}, \tag{1.5.45}$$

where $\boldsymbol{T}^{-1}(\boldsymbol{k},\omega)$ is the reciprocal of the tensor

$$\boldsymbol{T}(\boldsymbol{k},\omega) = \boldsymbol{\varepsilon}^0(\boldsymbol{k}) - \frac{k^2c^2}{\omega^2}\,\mathbf{1}_t(\boldsymbol{k}), \tag{1.5.46}$$

which involves the $\boldsymbol{k}$-dependent optical frequency dielectric tensor

$$\boldsymbol{\varepsilon}^0(\boldsymbol{k}) = \mathbf{1} + 4\pi\boldsymbol{\chi}^0(\boldsymbol{k}). \tag{1.5.47}$$

When (1.5.45) is substituted into (1.5.14), and we take account of (1.5.27) and label the $3r$ solutions of the resulting set of equations by $j = 1, 2, \ldots, 3r$, we obtain

$$\omega_j^2(\boldsymbol{k})\,\boldsymbol{v}(\varkappa\mid\boldsymbol{k}j) = \sum_{\varkappa'}\boldsymbol{D}(\varkappa\varkappa'\mid\boldsymbol{k})\,\boldsymbol{v}(\varkappa'\mid\boldsymbol{k}j), \tag{1.5.48}$$

where the modified dynamical matrix $\boldsymbol{D}(\varkappa\varkappa' \mid \boldsymbol{k})$ is defined by

$$\boldsymbol{D}(\varkappa\varkappa' \mid \boldsymbol{k}) = \boldsymbol{C}(\varkappa\varkappa' \mid \boldsymbol{k}) + \frac{4\pi}{v_a} \frac{1}{(M_\varkappa M_{\varkappa'})^{1/2}} \boldsymbol{Z}^+(\varkappa \mid \boldsymbol{k})\, \boldsymbol{T}^{-1}(\boldsymbol{k}, \omega)\, \boldsymbol{Z}(\varkappa' \mid \boldsymbol{k}) \tag{1.5.49}$$

(with $\omega = \omega_j(\boldsymbol{k})$).

The first term on the right hand side of (1.5.49) describes the effects of the short range forces and the second term the effects of the macroscopic electric field in the crystal.

In the limiting case $kc/\omega \gg 1$, i.e. in the electrostatic approximation, we get from (1.5.40) and (1.5.46)

$$\boldsymbol{K}(\boldsymbol{k}, \omega) \simeq \mathbf{1}_l(\hat{\boldsymbol{k}}), \tag{1.5.50}$$

$$\boldsymbol{T}(\boldsymbol{k}, \omega) \simeq \mathbf{1}_l(\hat{\boldsymbol{k}})\, \epsilon^0(\boldsymbol{k}) - \frac{k^2c^2}{\omega^2} \mathbf{1}_t(\hat{\boldsymbol{k}}) \tag{1.5.51}$$

with the $\boldsymbol{k}$-dependent optical frequency dielectric constant

$$\epsilon^0(\boldsymbol{k}) = \sum_{\alpha\alpha'} \hat{k}_\alpha \varepsilon^0_{\alpha\alpha'}(\boldsymbol{k})\, \hat{k}_{\alpha'}. \tag{1.5.52}$$

It can be readily verified that the reciprocal of the tensor (1.5.51) reads

$$\boldsymbol{T}^{-1}(\boldsymbol{k}, \omega) = \left(\epsilon^0(\boldsymbol{k})\right)^{-1} \mathbf{1}_l(\hat{\boldsymbol{k}}) + O\left(\frac{\omega^2}{k^2c^2}\right). \tag{1.5.53}$$

In the limiting case $kc/\omega \gg 1$ the macroscopic electric field is purely longitudinal, as seen by inserting (1.5.50) into (1.5.39), and by (1.5.53) the dynamical matrix (1.5.49) takes the form

$$\boldsymbol{D}(\varkappa\varkappa' \mid \boldsymbol{k}) = \boldsymbol{C}(\varkappa\varkappa' \mid \boldsymbol{k}) + \frac{4\pi}{v_a\epsilon^0(\boldsymbol{k})} \frac{1}{(M_\varkappa M_{\varkappa'})^{1/2}} \boldsymbol{Z}^+(\varkappa \mid \boldsymbol{k})\, \mathbf{1}_l(\hat{\boldsymbol{k}})\, \boldsymbol{Z}(\varkappa' \mid \boldsymbol{k}). \tag{1.5.54}$$

To simplify the discussion of the equation of motion (1.5.48) with the dynamical matrices (1.5.49) or (1.5.54), we consider in the following the long wavelength vibrations in a cubic crystal containing two nonequivalent ions in a primitive unit cell, e.g. in a crystal with the rock salt, zinc blende or cesium chloride structure.

The restriction to the limit $\boldsymbol{k} \to 0$ means that in the potential (1.5.1) $\boldsymbol{E}$ is constant (cf. (1.5.13)) and $\boldsymbol{u}(l\varkappa)$ does not depend on l (cf. (1.5.11)). Thus, in (1.5.1) the summations over the cell indices can be completely performed and over the indices of the kind of ions partially performed. From equations (1.5.2) and (1.5.3), which hold now for the remaining dependences of the latter type of indices, we obtain that the nondiagonal force constants equal each other in magnitude and sign, that the nondiagonal and diagonal force constants are equal in magnitude but different in sign, and that the ions in a cell are oppositely charged. Further, from the symmetry relations (1.5.6) to (1.5.8) we see that for a cubic crystal the force constants, charge tensor, and polarizability tensor are diagonal in the indices α and α' and do not depend on α. Consequently, in the long wavelength limit we have for the crystal at hand

$$C_{\alpha\alpha'}(\varkappa\varkappa' \mid 0) = \delta_{\alpha\alpha'}\mu(M_\varkappa M_{\varkappa'})^{-1/2} \operatorname{sgn} \varkappa \operatorname{sgn} \varkappa' \omega_{\mathrm{T}}^2, \tag{1.5.55}$$

$$Z_{\alpha\alpha'}(\varkappa \mid 0) = \delta_{\alpha\alpha'} \operatorname{sgn} \varkappa\, e_{\mathrm{T}}^*, \tag{1.5.56}$$

$$\varepsilon_{\alpha\alpha'}(0) = \delta_{\alpha\alpha'}\varepsilon_0, \tag{1.5.57}$$

where we have labelled the two ions in each primitive unit cell by $\varkappa = +, -$, and $\operatorname{sgn} \varkappa = +1, -1$, if $\varkappa = +, -$, respectively; μ is the reduced mass of the two ions in a unit cell, ω_{T} is a characteristic frequency and e_{T}^* a characteristic effective charge, ε_0 is the optical frequency dielectric constant.

First, let us now consider the long wavelength vibrations in the electrostatic approximation. From (1.5.48), (1.5.54), and (1.5.55) to (1.5.57) we obtain the following equation of motion:

$$\omega_j^2(0)\, v_\alpha(\varkappa \mid 0j) = \sum_{\varkappa'\alpha'} \Big\{\delta_{\alpha\alpha'}\mu(M_\varkappa M_{\varkappa'})^{-1/2} \operatorname{sgn} \varkappa \operatorname{sgn} \varkappa' \omega_{\mathrm{T}}^2 + \frac{4\pi(e_{\mathrm{T}}^*)^2}{v_a\varepsilon_0}(M_\varkappa M_{\varkappa'})^{-1/2} \operatorname{sgn} \varkappa \operatorname{sgn} \varkappa'\, \hat{k}_\alpha \hat{k}_{\alpha'}\Big\}\, v_{\alpha'}(\varkappa' \mid 0j). \tag{1.5.58}$$

There are two types of solutions of this equation. The first type is given by

$$\omega_j^2(0) = 0, \quad v_\alpha(\varkappa \mid 0j) = (M_\varkappa/M_{\mathrm{T}})^{1/2}\, u_\alpha(j), \quad j = 1, 2, 3, \tag{1.5.59}$$

where $M_{\mathrm{T}} = M_+ + M_-$ is the total mass of atoms in a primitive unit cell and $\boldsymbol{u}(j)$ ($j = 1, 2, 3$) are three mutually perpendicular unit vectors. Denoting by $\boldsymbol{u}(l\varkappa \mid \boldsymbol{k}j)$ the displacement of the ion $(l\varkappa)$ in the mode $(\boldsymbol{k}j)$, then we see from (1.5.59) and (1.5.11) that

$$\boldsymbol{u}(l + \mid 0j) = \boldsymbol{u}(l - \mid 0j), \qquad j = 1, 2, 3, \tag{1.5.60}$$

i.e. the modes (1.5.59) are acoustic.

The second type of solution of (1.5.58) is given by

$$v_\alpha(\varkappa \mid 0j) = (\mu/M_\varkappa)^{1/2} \operatorname{sgn} \varkappa\, w_\alpha(j), \qquad j = 4, 5, 6, \tag{1.5.61}$$

where $\boldsymbol{w}(j)$ ($j = 4, 5, 6$) are also any three mutually perpendicular unit vectors. Note that for these modes we have

$$M_+\boldsymbol{u}(l + \mid 0j) + M_-\boldsymbol{u}(l - \mid 0j) = 0, \tag{1.5.62}$$

i.e. the two ions in a primitive unit cell vibrate in anti-phase and the center of mass of the ions remains fixed. Cleary, such a motion of charged ions gives rise to a fluctuating dipole moment by means of which the crystal can interact with electromagnetic radiation. As mentioned above, from this property the name optical vibration arises for vibrations of type (1.5.61). When we insert (1.5.61) into (1.5.58) we obtain

$$\left(\omega_j^2(0) - \omega_{\mathrm{T}}^2\right) w_\alpha(j) = \frac{4\pi(e_{\mathrm{T}}^*)^2}{\mu v_a \varepsilon_0} \sum_{\alpha'} \hat{k}_\alpha \hat{k}_{\alpha'} w_{\alpha'}(j). \tag{1.5.63}$$

This equation in turn has solutions of two types. If $w_\alpha(j)$ is assumed to be parallel to $\boldsymbol{k}$ we have

$$\omega_4^2(0) = \omega_T^2 + \frac{4\pi(e_T^*)^2}{\mu v_a \varepsilon_0} \equiv \omega_L^2, \tag{1.5.64}$$

and if $w_\alpha(j)$ is assumed to be perpendicular to $\boldsymbol{k}$ we have

$$\omega_j^2(0) = \omega_T^2, \qquad j = 5, 6. \tag{1.5.65}$$

Clearly, the mode (1.5.64) is longitudinal, whereas the modes (1.5.65) are transverse. The former is called longitudinal optical (LO) mode and the latter ones transverse optical (TO) modes. We see from (1.5.64) and (1.5.65) that the longitudinal macroscopic electric field, which is caused by the LO mode (cf. (1.5.45) and (1.5.53)), acts back on the LO mode. It enhances the force constant governing the LO mode and thus raises the frequency of this mode with respect to the frequency of the TO modes, which in the electrostatic approximation have no macroscopic electric field.

In the limiting case $kc/\omega \gg 1$, which we have considered here so far, the time needed to transverse the distance between two lattice sites in the crystal at the speed of light is very much less than the vibrational period of an optical mode with frequency ω. In this case the retardation of the Coulomb interaction can be neglected, i.e. the Coulomb interaction may be assumed to interact instantaneously between the ions (electrostatic approximation). However, in the case that $kc/\omega \lessapprox 1$ the retardation of the Coulomb interaction can no longer be neglected. In the following we investigate this latter limiting case.

Because of $kc/\omega \lessapprox 1$ we may assume that the dynamical matrix (1.5.49) depends on $\boldsymbol{k}$ only through the tensor $\boldsymbol{T}^{-1}(\boldsymbol{k}, \omega)$:

$$\boldsymbol{D}(\varkappa\varkappa' \mid \boldsymbol{k}) = \boldsymbol{C}(\varkappa\varkappa' \mid 0) + \frac{4\pi}{v_a} \frac{1}{(M_\varkappa M_{\varkappa'})^{1/2}} \boldsymbol{Z}^+(\varkappa \mid 0)\, \boldsymbol{T}^{-1}(\boldsymbol{k}, \omega)\, \boldsymbol{Z}(\varkappa' \mid 0). \tag{1.5.66}$$

When $\boldsymbol{\varepsilon}^0(\boldsymbol{k})$ is replaced in $\boldsymbol{T}(\boldsymbol{k}, \omega)$ (cf. (1.5.46)) by $\mathbf{1}\varepsilon_0$, since $kc/\omega \lessapprox 1$, then $\boldsymbol{T}^{-1}(\boldsymbol{k}, \omega)$ reads

$$\boldsymbol{T}^{-1}(\boldsymbol{k}, \omega) = \left[\varepsilon_0 \mathbf{1} - \frac{c^2k^2}{\omega^2}\, \mathbf{1}_l(\hat{\boldsymbol{k}})\right] \Big/ \varepsilon_0 \left(\varepsilon_0 - \frac{c^2k^2}{\omega^2}\right), \tag{1.5.67}$$

as can be readily verified. Thus, in the limiting case $\dfrac{kc}{\omega} \lessapprox 1$ the lattice vibrations in the model at two ions in a primitive unit cell are described by the equation of motion (1.5.48) with the dynamical matrix (1.5.66), for which we need (1.5.55), (1.5.56) and (1.5.67). To solve the resulting equation we make the ansatz

$$v_\alpha(\varkappa \mid \boldsymbol{k}) = \sum_{j=1}^{6} v_\alpha^{(0)}(\varkappa \mid 0j)\, Q_j(\boldsymbol{k}), \tag{1.5.68}$$

where $v_\alpha^{(0)}(\varkappa \mid 0j)$'s $(j = 1, 2, \ldots, 6)$ are the solutions of the eigenvalue problem with the short range part of the dynamical matrix alone

$$\left(\omega_j^{(0)}(0)\right)^2 v_\alpha^{(0)}(\varkappa \mid 0j) = \sum_{\varkappa'\alpha'} C_{\alpha\alpha'}(\varkappa\varkappa' \mid 0)\, v_{\alpha'}^{(0)}(\varkappa' \mid 0j) \tag{1.5.69}$$

with $C(\varkappa\varkappa' \mid 0)$ from (1.5.55), and the $Q_j(\boldsymbol{k})$'s are coefficients which must be determined. Obviously, the eigensolutions of (1.5.69) are given by (1.5.59), (1.5.61), (1.5.64) and (1.5.65), if we put there formally $e_T^* = 0$. In writing (1.5.68) we have accounted for the fact that the $v_\alpha^{(0)}(\varkappa \mid 0j)$'s form a complete set of eigenvectors. When we now insert the ansatz (1.5.68) into the equation of motion and use the orthonormality of the $v_\alpha^{(0)}(\varkappa \mid 0j)$'s we obtain the following equation:

$$\begin{aligned} \omega^2 Q_j(\boldsymbol{k}) = \left(\omega_j^{(0)}(0)\right)^2 Q_j(\boldsymbol{k}) + \frac{4\pi}{v_a} \sum_{j'} \sum_{\varkappa'\alpha'} \sum_{\varkappa\alpha} (M_\varkappa M_{\varkappa'})^{-1/2}\, (e_T^*)^2 \\ \times \operatorname{sgn} \varkappa \operatorname{sgn} \varkappa' v_\alpha^{(0)}(\varkappa \mid 0j)\, T_{\alpha\alpha'}^{-1}(\boldsymbol{k}, \omega)\, v_{\alpha'}^{(0)}(\varkappa' \mid 0j')\, Q_j(\boldsymbol{k}). \end{aligned} \tag{1.5.70}$$

On account of (1.5.67), (1.5.59) and (1.5.61) the second term on the right hand side of (1.5.70) reads

$$\frac{4\pi}{v_a} \frac{(e_T)^{*2}}{\mu} \begin{cases} \dfrac{1}{\varepsilon_0} & \text{for} \quad j = 4 \text{ (LO)}, \\[2ex] \dfrac{1}{\varepsilon_0 - \dfrac{k^2c^2}{\omega^2}} & \text{for} \quad j = 5 \text{ (TO)}, \end{cases} \tag{1.5.71}$$

so that we obtain from (1.5.70) for the LO mode

$$\omega_L^2 = \omega_T^2 + \frac{4\pi(e_T^*)^2}{\mu v_a \varepsilon_0} \tag{1.5.72}$$

and for the TO modes

$$\omega_T^2(\boldsymbol{k}) = \omega_T^2 + 4\pi(e_T^*)^2/\mu v_a \left(\varepsilon_0 - \frac{k^2c^2}{\omega_T^2(\boldsymbol{k})}\right). \tag{1.5.73}$$

The latter equation can be rewritten as

$$\begin{aligned} \left(\omega_T^\pm(\boldsymbol{k})\right)^2 = \frac{1}{2} \{\omega_L^2 + (k^2c^2/\varepsilon_0) \\ \pm [(\omega_L^2 + k^2c^2/\varepsilon_0)^2 - 4\omega_T^2 k^2c^2/\varepsilon_0]^{1/2}\}. \end{aligned} \tag{1.5.74}$$

According to (1.5.73) for $k^2c^2/\varepsilon_0\omega_T^2(\boldsymbol{k}) \gg 1$ it holds that

$$\omega_T^2(\boldsymbol{k}) \to \omega_T^2. \tag{1.5.75}$$

Let us now discuss the results obtained on account of the retardation of the Coulomb interaction. When we account for retardation we have a longitudinal as well as a transverse macroscopic electromagnetic field in the crystal.

This can be seen from (1.5.45) and (1.5.67). The longitudinal field affects the LO mode in the same manner as in the electrostatic approximation, which results from a comparison of (1.5.64) and (1.5.72). Just as the longitudinal field is caused by the LO mode and acts back on it, the transverse field acts back on the TO modes which are its source. In this manner the LO and TO modes alter their frequencies from what they are in absence of the longitudinal and transverse fields, respectively. At the same time, the dispersion relation of the transverse electromagnetic field propagating through the crystal is affected

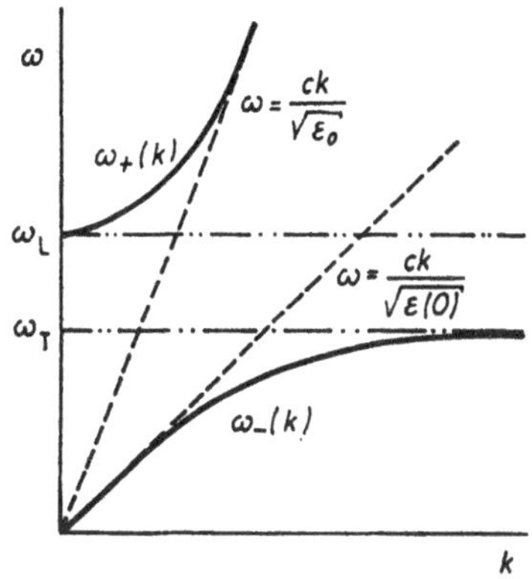

Fig. 1.10. Polariton dispersion curves for a cubic ionic crystal with two atoms per unit cell (see text)

by its coupling to the TO modes of the crystal. This strong interaction between electromagnetic waves and mechanical vibrations gives rise to mixed normal modes, called polaritons, which consist of admixtures of electromagnetic and mechanical vibrations. The weighting of each of these contributions depends on the wave vector $\boldsymbol{k}$. This can be seen from figure 1.10, where the functions $\omega_T^{\pm}(\boldsymbol{k})$ (1.5.74) are plotted. The lower branch $\omega_T^{-}(\boldsymbol{k})$ corresponds to a purely electromagnetic wave propagating through a medium with a dielectric constant $\varepsilon(0) = \varepsilon_0\omega_L^2/\omega_T^2$ in the limit $k \to 0$, and to a purely mechanical transverse optical mode in absence of retardation as $k \to \infty$. The upper branch $\omega_T^{+}(\boldsymbol{k})$ corresponds to a purely mechanical vibration with frequency ω_L as $k \to 0$ and to a purely electromagnetic wave propagating through a medium with a dielectric constant ε_0 as $k \to \infty$. However, for $k \approx \sqrt{\varepsilon_0}\,\omega/c$ the branches $\omega_T^{\pm}(\boldsymbol{k})$ have a more complicated nature. We see that retardation effects are important only for very small wave vectors, i.e. for $k < \omega_T/c$.

Concluding this section we now derive the expression for the lattice contribution to the dielectric tensor. To do so, we consider the first term on the right hand side of equation (1.5.26), which represents the macroscopic polarization in the crystal on neglection of the polarizability of the ions. In the model at hand with two ions in a primitive unit cell the displacements of the ions involved in this term are governed by the following equation:

$$\omega^2 v_\alpha(\varkappa) = \sum_{\varkappa'\alpha'} \{\delta_{\alpha\alpha'}\mu(M_\varkappa M_{\varkappa'})^{-1/2} \operatorname{sgn}\varkappa \operatorname{sgn}\varkappa' \, \omega_T^2\} \, v_{\alpha'}(\varkappa')$$

$$- \frac{1}{\sqrt{M_\varkappa}} \operatorname{sgn}\varkappa \, e_T^* E_\alpha, \tag{1.5.76}$$

which we have obtained from (1.5.14), (1.5.27), (1.5.55) and (1.5.56). As can be readily verified, the solution of (1.5.76) can be written as

$$v_\alpha(\varkappa) = \sum_{\varkappa'\alpha'} \sum_{j=1}^{6} \frac{v_\alpha^{(0)}(\varkappa \mid 0j) \, v_{\alpha'}^{(0)}(\varkappa' \mid 0j)}{(\omega_j^{(0)})^2 - \omega^2} \frac{\operatorname{sgn}\varkappa' e_T^*}{\sqrt{M_{\varkappa'}}} E_{\alpha'}, \tag{1.5.77}$$

where the $v_\alpha^{(0)}(\varkappa \mid 0j)$'s and the $\omega_j^{(0)}$'s are respectively the eigenvectors and eigenvalues of the equation of motion (1.5.69), which involves the short range part of the dynamical matrix only. Substituting (1.5.77) for (1.5.26) we obtain for the lattice part of the polarization

$$P_\alpha = \sum_{\alpha'} \chi_{\alpha\alpha'}(\omega) \, E_{\alpha'} \tag{1.5.78}$$

with the tensor of the dielectric susceptibility of the lattice

$$\chi_{\alpha\alpha'}(\omega) = \frac{1}{v_a} \sum_{j=1}^{6} \sum_{\varkappa\varkappa'} \frac{e_T^* \operatorname{sgn}\varkappa}{\sqrt{M_\varkappa}} \frac{v_\alpha^{(0)}(\varkappa \mid 0j) \, v_{\alpha'}^{(0)}(\varkappa' \mid 0j)}{(\omega_j^{(0)})^2 - \omega^2} \frac{e_T^* \operatorname{sgn}\varkappa'}{\sqrt{M_{\varkappa'}}}. \tag{1.5.79}$$

Since for acoustic modes $v_\alpha^{(0)}(\varkappa \mid 0j)/\sqrt{M_\varkappa}$ ($j = 1, 2, 3$) does not depend on $\varkappa$, these modes do not contribute to the susceptibility (1.5.79). When we account for the fact that for $v_\alpha^{(0)}(\varkappa \mid 0j)$ ($j = 4, 5, 6$) an expression like (1.5.61) holds and that the corresponding eigenfrequency $\omega_j^{(0)}$ ($j = 4, 5, 6$) equals ω_T we can rewrite (1.5.79) as

$$\chi_{\alpha\alpha'}(\omega) = \delta_{\alpha\alpha'} \frac{(e_T^*)^2}{\mu v_a} \frac{1}{\omega_T^2 - \omega^2}. \tag{1.5.80}$$

The total dielectric tensor is the sum of the lattice susceptibility (1.5.80) and the static electronic tensor (1.5.57). Thus we have

$$\varepsilon_{\alpha\alpha'}(\omega) = \delta_{\alpha\alpha'}\varepsilon(\omega) \tag{1.5.81}$$

with

$$\varepsilon(\omega) = \varepsilon_0 + \frac{4\pi(e_T^*)^2}{\mu v_a} \frac{1}{\omega_T^2 - \omega^2} \tag{1.5.82}$$

or equivalently

$$\varepsilon(\omega) = \varepsilon_0(\omega_L^2 - \omega^2)/(\omega_T^2 - \omega^2), \tag{1.5.83}$$

where we have used (1.5.72). From (1.5.83) we see that the ratio of ω_L^2 and ω_T^2 is given by

$$\omega_L^2/\omega_T^2 = \varepsilon(0)/\varepsilon_0, \tag{1.5.84}$$

where $\varepsilon(0)$ is the static dielectric constant. This relation is known as the Lyddane-Sachs-Teller (LST) relation (LYDDANE et al., 1941). The LST relation is experimentally well established. It is very useful in spectroscopy for determining one unknown parameter such as a characteristic frequency, $\varepsilon(0)$ or ε_0. For a cubic crystal with s branches of infrared active TO modes which can interact with electromagnetic radiation of given polarization the LST relation reads (KUROSAWA, 1961)

$$\prod_{j=1}^{s} \frac{\omega_{Lj}^2}{\omega_{Tj}^2} = \frac{\varepsilon(0)}{\varepsilon_0}. \tag{1.5.85}$$

Note that the LST relation was also extended to anharmonic systems (BARKER, 1975), systems with permanent dipoles (CHANG, 1976), and finite wave vectors (BURSTEIN et al., 1971). BARKER (1975) pointed out that the LST relation can be obtained from general principles of statistical mechanics. For a review of lattice dynamics of binary ionic solids see SINGH, 1982.

1.6. Microscopic theory of lattice dynamics

1.6.1. Microscopic theory versus phenomenological approaches

In a phenomenological approach to lattice dynamics the force constants are viewed as parameters to be determined from experimental data. To obtain a meaningful fit, the underlying model must not involve too many parameters and these must have a definite physical meaning. Different models have been developed for the various systems.

Especially the lattice dynamics of insulators have attracted much interest in recent years. In these materials, an account of the polarization of the ions and an explanation of the effective ionic charges observed are very challenging problems. In the preceding section we presented a phenomenological approach to lattice dynamics of insulators (ionic crystals). Alternative approaches are the shell model, the deformation dipole model, and the bond charge model. The three approaches are concisely described in the following (for more details see COCHRAN, 1971, SINHA, 1973, HARDY, 1974, BRÜESCH, 1982 (chapter 4), SINGH, 1982, and references therein).

In the shell model the ions are regarded as cores, to which massless shells, representing the outer valence electrons are bound by isotropic springs. In this model there is only one type of polarization: the relative displacement of the core and the shell, irrespective of whether this is brought about by an electric field or by the influence of short-range forces (fig. 1.11). By allowing for additional radial symmetric deformability of the ions, i.e. a so-called breathing deformability, the fit of experimental phonon dispersion curves by the shell model can be substantially improved.

In the deformation dipole model there are two possible types of polarization, which are assumed to be independent: (i) electric field induced and (ii) overlap induced polarization. The latter polarization originates from the

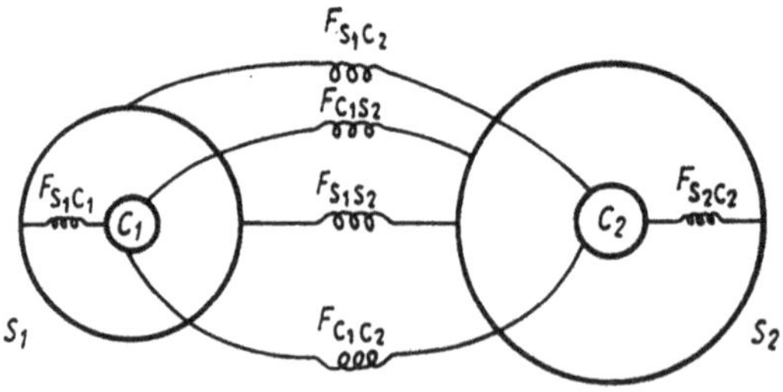

Fig. 1.11. Illustration of the short-range force constants in the shell model. $F_{S_2C_1}$ denotes the force constant between the core 1 and the shell 2, etc.

distortion of the special symmetry of the charge clouds around the ions by the overlap with the charge distributions of its neighbouring ions (fig. 1.12). Accordingly, a lattice wave produces a distortion of the charge cloud about a given ion which is equivalent to a point dipole at the undisplaced position of the ion.

The applicability of the shell model to covalent materials has been called in question because it is unphysical to divide the electrons in a bond between

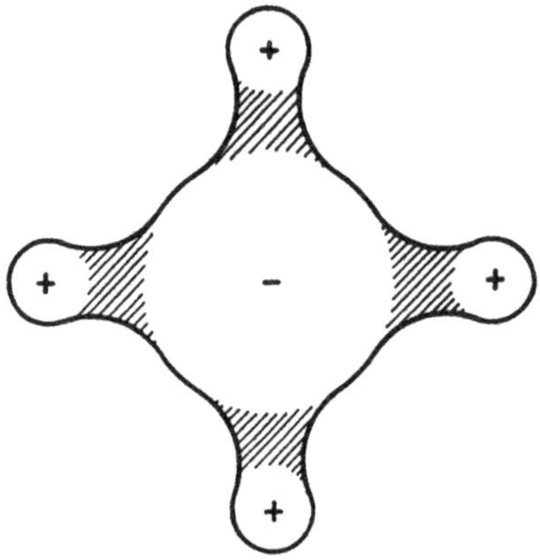

Fig. 1.12. Schematic representation of the distortion of the charge distribution around a negative ion due to the overlap with charge distributions of its neighbouring positive ions

two atoms, when they are shared between the two. The bond charge model seems to provide a more adequate description of these materials. In this model the incomplete screening of an ion in a covalent crystal by a diagonal dielectric matrix is compensated by bond charges to be placed mid-way between neighbouring ions (fig. 1.13). The bond charges yield effective noncentral forces between the ions and thus produce the stability of a diamond-type lattice against shear. Accordingly, they represent the covalent character of the bond-

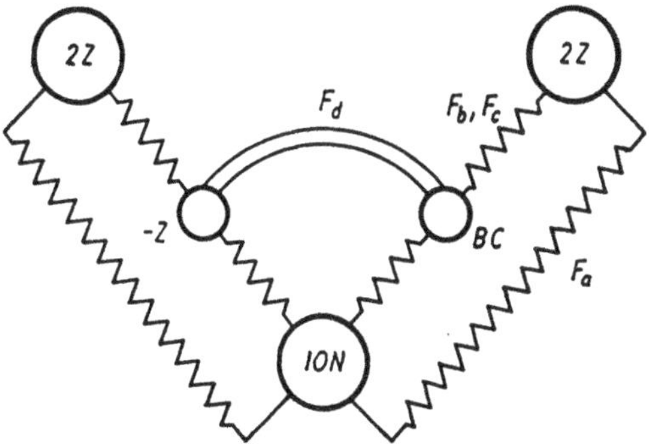

Fig. 1.13. Schematic representation of the bond charge model. $2Z$ is the remaining ion charge due to incomplete screening by a diagonal dielectric matrix. $-Z$ is the bond charge (BC). There are short-range central forces between the ions (F_a) and Coulomb forces between the ions and the BC's (F_b). If the BC's are not held fixed midway between the atoms but allowed to move adiabatically (like the shells in the shell model), short-range ion-BC forces (F_c) are needed to stabilize the BC's on their sites. F_d describes bond-bond interaction (after WEBER, 1977)

ing. By allowing for adiabatic motion of the bond charges the agreement of the calculated phonon spectrum with the experimentally determined one can be improved.

At first sight there is no relationship between the different phenomenological models used for studying lattice vibrations in the various systems. The microscopic theory of phonons provides a tool for examining a possible such relationship between the different models, because it represents a unified treatment of lattice vibrations in conducting as well as in non-conducting crystals.

In the microscopic theory of phonons a general expression for the force constants is derived from a microscopic examination of the forces between the ions, which consist of (i) direct ion-ion interaction and (ii) interaction via mediating electrons. The expression for the force acting on an ion is derived by studying the response of the electrons of the crystal to the field arising from the vibrating ions. Adopting the Born-Oppenheimer approximation and the harmonic approximation one gets the forces and thus the dynamical matrix in terms of the inverse dielectric matrix $\epsilon^{-1}(\boldsymbol{k} + \boldsymbol{K}, \boldsymbol{k} + \boldsymbol{K}')$ ($\boldsymbol{k}$ — wave vector reduced to the first Brillouin zone; $\boldsymbol{K}$, $\boldsymbol{K}'$ — reciprocal lattice vectors). (PICK et al., 1970, for reviews see SHAM, 1974, and BILZ et al., 1974).

A study of the analytic properties of ϵ^{-1} as a function of $\boldsymbol{k}$ near $\boldsymbol{k} = 0$ turns out to be important for establishing the connection between a microscopic and a phenomenological approach. In this way the dynamical matrix (1.5.54) derived for insulators (ionic crystals) from a purely phenomenological point of view can be rederived microscopically. In particular, a microscopic expression for the effective charge tensor can be found. For metals, on the other hand, the analytical behaviour of ϵ^{-1} shows that there are no such effective charge terms.

From an examination of the long wavelength acoustic modes it follows that for an insulator all acoustic-mode frequencies vanish as $\boldsymbol{k} \to \boldsymbol{0}$, if the neutrality condition of the effective charges is satisfied. The latter condition implies that off-diagonal elements of ϵ^{-1}, which determine the local field corrections, must be included. This manifests the importance of local field effects in insulators (including semiconductors like Ge or Si, too). In metals on the other hand there is no such condition. This fact suggests (as crudest approximation) complete omission of the nondiagonal elements of ϵ^{-1}.

Thus, with the possible exception of simple metals, a practical microscopic treatment of lattice dynamics is impossible unless the problem of inversion of a nondiagonal dielectric matrix has been solved. It should be noted that recent theoretical investigations suggest that local field corrections may be important also in simple metals due to strong deviations from the free-electron behaviour of the conduction electrons especially near the cores (Weilacher and Bross, 1977).

The Wannier representation makes practical an inversion of the dielectric matrix. In this way one can derive a microscopic shell model, which provides a quantum-mechanical justification for the macroscopic shell model (Sinha, 1973, Hanke, 1973; for a review see Hanke, 1978).

Generally, the use of the microscopic theory for explicit phonon calculations requires more or less rigorous approximations to reduce the numerical effort. While microscopic calculations have shown promising results in recent years, the use of phenomenological models is still an indispensable tool in the majority of phonon investigations, especially when experimental data are to be explained.

1.6.2. Dielectric matrix and lattice dynamics[1]

To obtain the microscopic expression of the dynamical matrix either all electrons are treated equivalently or the electrons are divided into electrons responsible for the 'dynamical' interatomic forces (valence electrons) and electrons moving bodily with the nuclei (core electrons). The latter splitting is suggested by the fact that the contribution of the valence electrons to the dielectric matrix is large compared to the remaining contributions. A theory which emphasizes the special role of the valence electrons is the pseudopotential theory, where the distortion of the core electrons by the nuclear displacements is completely neglected and the Pauli principle is accounted for by the requirement that the wave functions of the valence electrons must be orthogonal to the core electron wave functions. The effects resulting from the latter requirement can be simulated by a repulsive potential which largely

[1] To a certain extent the formulation of the microscopic theory of lattice dynamics presented in sections 1.6.2. and 1.6.3. parallels that by Brauer and Streitwolf (1977).

cancels the attractive core potential. In microscopic phonon theory the real pseudopotential, which is generally nonlocal, is in most cases replaced by a local model potential. This will be treated in this section, too.

Further assumptions and approximations generally used in microscopic phonon theory are: (i) each ion has a classical equilibrium position, which means that we adopt the Born-Oppenheimer approximation; (ii) an ion displaced from its equilibrium position interacts with the other ions and the valence electrons by the classical electromagnetic field they create; (iii) this field can be expanded with respect to the displacements of the ions and the equations can be linearized, which means that we adopt the harmonic approximation.

When we account for electron-electron interaction in the Hartree approximation, the potential at site $\boldsymbol{r}$ in a crystal is given by

$$V_0(\boldsymbol{r}) = V_{\mathrm{i}}(\boldsymbol{r}) + \int \mathrm{d}\boldsymbol{r}' \frac{e\langle\varrho(\boldsymbol{r}')\rangle}{|\boldsymbol{r} - \boldsymbol{r}'|}, \tag{1.6.1}$$

where $\langle\varrho(\boldsymbol{r})\rangle$ is the mean electron density and $V_{\mathrm{i}}(\boldsymbol{r})$ is the ionic potential which is assumed here to be additively composed,

$$V_{\mathrm{i}}(\boldsymbol{r}) = \sum_{l\varkappa} v_\varkappa\big(\boldsymbol{r} - \boldsymbol{R}(l\varkappa)\big), \tag{1.6.2}$$

where $v_\varkappa\big(\boldsymbol{r} - \boldsymbol{R}(l\varkappa)\big)$ represents the local pseudopotential of the $\varkappa$th ion in the lth primitive unit cell situated at $\boldsymbol{R}(l\varkappa)$. By the Poisson equation we obtain a rigid charge distribution $\varrho_\varkappa\big(\boldsymbol{r} - \boldsymbol{R}(l\varkappa)\big)$ centred round $\boldsymbol{R}(l\varkappa)$:

$$\varrho_\varkappa\big(\boldsymbol{r} - \boldsymbol{R}(l\varkappa)\big) = -\frac{1}{4\pi e}\,\Delta v_\varkappa\big(\boldsymbol{r} - \boldsymbol{R}(l\varkappa)\big) \tag{1.6.3}$$

or Fourier transformed (cf. (A2.9), (A2.10))

$$\varrho_\varkappa(\boldsymbol{q}) = \frac{v_\varkappa(\boldsymbol{q})\, q^2}{4\pi e}. \tag{1.6.4}$$

The effective charge $[\varrho_\varkappa(\boldsymbol{q})]_{\boldsymbol{q}=0}$ of this local pseudopotential is by definition $Z_\varkappa$, the nuclear charge minus that of the core electrons.

The equilibrium positions $\boldsymbol{x}(l\varkappa)$ of the ions are determined by the requirement of vanishing Lorentz force acting on each ion. Assuming permanent magnetic moments to be absent, the equilibrium condition for the ion $(l\varkappa)$ reads

$$0 = \int \mathrm{d}\boldsymbol{r}\varrho_\varkappa\big(\boldsymbol{r} - \boldsymbol{x}(l\varkappa)\big)\left[\boldsymbol{E}^{\mathrm{eq}}(\boldsymbol{r}) - \boldsymbol{E}_{l\varkappa}\big(\boldsymbol{r} - \boldsymbol{x}(l\varkappa)\big)\right], \tag{1.6.5}$$

where $\boldsymbol{E}^{\mathrm{eq}}(\boldsymbol{r})$ is the equilibrium electric field (cf. appendix 2):

$$\boldsymbol{E}^{\mathrm{eq}}(\boldsymbol{r}) = -\frac{1}{e}\,\nabla V_0(\boldsymbol{r}) \tag{1.6.6}$$

($V_0(\boldsymbol{r})$ to be taken with ions placed at $\boldsymbol{x}(l\varkappa)$) or Fourier transformed

$$\boldsymbol{E}^{\mathrm{eq}}(\boldsymbol{K}) = -\frac{i}{e}\,\boldsymbol{K}V_0(\boldsymbol{K}) = -\frac{i}{e}\,\boldsymbol{K}\Big[V_{\mathrm{i}}(\boldsymbol{K}) + \frac{4\pi e}{K^2}\,\langle\varrho(\boldsymbol{K})\rangle\Big]. \tag{1.6.7}$$

By the last term in (1.6.5) the contribution of the ion $(l\varkappa)$ to $\boldsymbol{E}^{\mathrm{eq}}$ is substracted, since an ion does not act on itself.

Writing

$$\boldsymbol{R}(l\varkappa) = \boldsymbol{x}(l\varkappa) + \boldsymbol{u}(l\varkappa) \tag{1.6.8}$$

we shall now obtain an equation of motion for the displacement $\boldsymbol{u}(l\varkappa)$,

$$M_\varkappa\,\frac{\partial^2\boldsymbol{u}(l\varkappa, t)}{\partial t^2} = \boldsymbol{F}(l\varkappa, t), \tag{1.6.9}$$

where $\boldsymbol{F}(l\varkappa, t)$ is the force acting on the ion $(l\varkappa)$ and $M_\varkappa$ is the mass of the $\varkappa$th ion.

The vibrating ions give rise to a current density

$$\boldsymbol{j}(\boldsymbol{r}, t) = \sum_{l\varkappa} \varrho_\varkappa\big(\boldsymbol{r} - \boldsymbol{x}(l\varkappa)\big)\,\frac{\partial\boldsymbol{u}(l\varkappa, t)}{\partial t} \tag{1.6.10}$$

or Fourier transformed

$$\boldsymbol{j}(\boldsymbol{q}, \omega) = -i\omega\,\frac{q^2}{4\pi e}\sum_{l\varkappa} \mathrm{e}^{-i\boldsymbol{q}\boldsymbol{x}(l\varkappa)}v_\varkappa(\boldsymbol{q})\;\boldsymbol{u}(l\varkappa, \omega). \tag{1.6.11}$$

This current density is now considered as an external current density in the sense of appendix 2 which causes an electromagnetic field (cf. (A2.21)) acting on the electrons in the crystal thus inducing a new field. The resulting internal field, which is the sum of the external and the induced field, is given by

$$\boldsymbol{E}^{\mathrm{int}}(\boldsymbol{q}, \omega) = -\frac{4\pi i}{\omega}\sum_{\boldsymbol{q}'} \boldsymbol{T}^{-1}(\boldsymbol{q}, \boldsymbol{q}', \omega)\,\boldsymbol{j}(\boldsymbol{q}', \omega), \tag{1.6.12}$$

where $\boldsymbol{T}^{-1}$ is the inverse of the tensor $\boldsymbol{T}$ (A2.28). The total microscopic field is given by $\boldsymbol{E}^{\mathrm{micr}} = \boldsymbol{E}^{\mathrm{eq}} + \boldsymbol{E}^{\mathrm{int}}$ and the Lorentz force on the ion $(l\varkappa)$ caused by this field reads

$$\boldsymbol{F}(l\varkappa, \omega) = \int \mathrm{d}\boldsymbol{r}\varrho_\varkappa\big(\boldsymbol{r} - \boldsymbol{R}(l\varkappa)\big)\big[\boldsymbol{E}^{\mathrm{micr}}(\boldsymbol{r}, \omega) - \boldsymbol{E}_{l\varkappa}\big(\boldsymbol{r} - \boldsymbol{R}(l\varkappa)\big)\big]. \tag{1.6.13}$$

Here we have not accounted for contributions from the magnetic field, since these are of second order in the displacements, and we are interested in first order contributions only. $\boldsymbol{E}_{l\varkappa}$ does not contribute to the first order term, since it is of zeroth order in the displacements. This can be seen by replacing $\boldsymbol{r} \to \boldsymbol{r} + \boldsymbol{R}(l\varkappa)$ in (1.6.13). Accounting for the equilibrium condition (1.6.5) we thus obtain for the first order expression of the force

$$\boldsymbol{F}(l\varkappa, \omega) = \frac{1}{V}\sum_{\boldsymbol{q}} \varrho_\varkappa(-\boldsymbol{q})\,\big[\boldsymbol{E}^{\mathrm{eq}}(\boldsymbol{q})\,i\big(\boldsymbol{q}\boldsymbol{u}(l\varkappa, \omega)\big) + \boldsymbol{E}^{\mathrm{int}}(\boldsymbol{q}, \omega)\big]\,\mathrm{e}^{i\boldsymbol{q}\boldsymbol{x}(l\varkappa)}. \tag{1.6.14}$$

Inserting now (1.6.4), (1.6.7), (1.6.12) and (1.6.11) into (1.6.14) and using (1.2.17), the equation of motion (1.6.9) takes the form

$$-M_\varkappa\omega^2u_\alpha(l\varkappa,\omega) = -\frac{v_a}{V}\sum_{\boldsymbol{k}}\sum_{l'\varkappa'\alpha'} e^{i\boldsymbol{k}(\boldsymbol{x}(l\varkappa)-\boldsymbol{x}(l'\varkappa'))}C_{\alpha\alpha'}(\varkappa\varkappa' \mid \boldsymbol{k}\omega)\, u_{\alpha'}(l'\varkappa',\omega)$$
$$+\frac{1}{V}\sum_{\boldsymbol{K}\alpha'} K_\alpha K_{\alpha'}\, e^{i\boldsymbol{K}\boldsymbol{x}(\varkappa)}v_\varkappa^*(\boldsymbol{K})\,\frac{K^2}{4\pi^2e^2}\,V_0(\boldsymbol{K})\,u_{\alpha'}(l\varkappa,\omega), \qquad (1.6.15)$$

where

$$C_{\alpha\alpha'}(\varkappa\varkappa' \mid \boldsymbol{k}\omega) = \frac{1}{v_a}\frac{1}{4\pi e^2}\sum_{\boldsymbol{K}\boldsymbol{K}'} q^2v_\varkappa^*(\boldsymbol{q})\, e^{i\boldsymbol{K}\boldsymbol{x}(\varkappa)}$$
$$\times\, T^{-1}_{\alpha\alpha'}(\boldsymbol{k}+\boldsymbol{K},\boldsymbol{k}+\boldsymbol{K}',\omega)\, e^{-i\boldsymbol{K}'\boldsymbol{x}(\varkappa')}v_{\varkappa'}(\boldsymbol{q}')\,q'^2. \qquad (1.6.16)$$

Here $\boldsymbol{q} = \boldsymbol{k}+\boldsymbol{K}$, $\boldsymbol{q}' = \boldsymbol{k}+\boldsymbol{K}'$. This labeling will be used in the following, too.

Now let us examine the analytic behaviour of the matrix $\boldsymbol{C}$ near $\boldsymbol{k} = 0$. The existence of long-range Coulomb forces in insulators gives rise to a non-analytic part of $C_{\alpha\alpha'}(\varkappa\varkappa' \mid \boldsymbol{k}\omega)$ for $\boldsymbol{k} = 0$. To separate this part we invert $\boldsymbol{T}(\boldsymbol{k}+\boldsymbol{K},\boldsymbol{k}+\boldsymbol{K}',\omega)$, which is an $(\infty\times\infty)$ square matrix with respect to $\boldsymbol{K},\boldsymbol{K}'$, by making use of the following formula for the inversion of a square matrix $\boldsymbol{M}$:

$$\boldsymbol{M} = \begin{bmatrix}\boldsymbol{P} & \boldsymbol{Q}\\ \boldsymbol{R} & \boldsymbol{S}\end{bmatrix}, \qquad \boldsymbol{M}^{-1} = \begin{bmatrix}\boldsymbol{W} & \boldsymbol{X}\\ \boldsymbol{Y} & \boldsymbol{Z}\end{bmatrix}. \qquad (1.6.17)$$

Here $\boldsymbol{P}$ and $\boldsymbol{S}$ are square matrices and it is easy to verify that

$$\left.\begin{aligned} \boldsymbol{X} &= -\boldsymbol{W}\boldsymbol{Q}\boldsymbol{S}^{-1},\\ \boldsymbol{Y} &= -\boldsymbol{S}^{-1}\boldsymbol{R}\boldsymbol{W},\\ \boldsymbol{W} &= (\boldsymbol{P}-\boldsymbol{Q}\boldsymbol{S}^{-1}\boldsymbol{R})^{-1},\\ \boldsymbol{Z} &= \boldsymbol{S}^{-1}+\boldsymbol{Y}\boldsymbol{W}^{-1}\boldsymbol{X}. \end{aligned}\right\} \qquad (1.6.18)$$

We now identify $\boldsymbol{P}$ with the $\boldsymbol{K} = \boldsymbol{K}' = 0$ element of $\boldsymbol{T}$ and $\boldsymbol{W}$ with the analogous element of $\boldsymbol{T}^{-1}$. $\boldsymbol{S}$ is then the matrix $\boldsymbol{T}$ reduced by the rows and columns with $\boldsymbol{K} = 0$, which is denoted in the following by $\mathring{\boldsymbol{T}}$. Writing with the aid of (A2.28), (1.5.40), and (A2.54)

$$\boldsymbol{T}(\boldsymbol{k}+\boldsymbol{K},\boldsymbol{k}+\boldsymbol{K}',\omega)$$
$$= \varepsilon(\boldsymbol{k}+\boldsymbol{K},\boldsymbol{k}+\boldsymbol{K}',\omega) - \delta_{\boldsymbol{K}\boldsymbol{K}'}\frac{q^2c^2}{\omega^2}\,1_t(\hat{\boldsymbol{q}}), \qquad (1.6.19)$$

where ε is the dielectric tensor, we see that for $\boldsymbol{K},\boldsymbol{K}' \neq 0$ the second term on the right hand side of (1.6.19) dominates. Thus $\mathring{\boldsymbol{T}}$ is nearly a purely transverse matrix and can be approximately inverted (cf. (1.5.53)):

$$\mathring{T}^{-1}_{\alpha\alpha'}(\boldsymbol{k}+\boldsymbol{K},\boldsymbol{k}+\boldsymbol{K}',\omega)$$
$$= \hat{q}_\alpha\mathring{\varepsilon}^{-1}(\boldsymbol{k}+\boldsymbol{K},\boldsymbol{k}+\boldsymbol{K}',\omega)\hat{q}'_{\alpha'} + O\left(\frac{\omega^2}{c^2q^2}\right). \qquad (1.6.20)$$

Here $\mathring{\epsilon}^{-1}$ is the matrix obtained from the dielectric matrix (cf. (A2.55), (1.6.19))

$$\epsilon(\boldsymbol{k}+\boldsymbol{K}, \boldsymbol{k}+\boldsymbol{K}', \omega) = \sum_{\alpha\alpha'} \hat{q}_\alpha T_{\alpha\alpha'}(\boldsymbol{k}+\boldsymbol{K}, \boldsymbol{k}+\boldsymbol{K}', \omega)\, \hat{q}'_{\alpha'} \tag{1.6.21}$$

by reducing ϵ to $\mathring{\epsilon}$ (neglection of the rows and columns with $\boldsymbol{K} = \boldsymbol{0}$) and inverting the resulting matrix.

By inverting $\boldsymbol{T}^{-1}$ in (1.6.16) in the manner described above, $\boldsymbol{C}$ from (1.6.16) becomes a sum of two parts $\boldsymbol{C}^{(1)}$ and $\boldsymbol{C}^{(2)}$, which in general have different analytic behaviour, and which are given by

$$\left.\begin{aligned} C^{(1)}_{\alpha\alpha'}(\varkappa\varkappa' \mid \boldsymbol{k}\omega) = {} & \frac{1}{v_a}\frac{1}{4\pi e^2} \sum_{\boldsymbol{K},\boldsymbol{K}'\neq 0} q^2 v_\varkappa{}^*(\boldsymbol{q})\, \mathrm{e}^{i\boldsymbol{K}\boldsymbol{x}(\varkappa)} \hat{q}_\alpha \\ & \times \mathring{\epsilon}^{-1}(\boldsymbol{k}+\boldsymbol{K}, \boldsymbol{k}+\boldsymbol{K}', \omega)\, \hat{q}'_{\alpha'}\, \mathrm{e}^{-i\boldsymbol{K}'\boldsymbol{x}(\varkappa')}\, v_{\varkappa'}(\boldsymbol{q}')\, q'^2, \end{aligned}\right\} \tag{1.6.22}$$

$$C^{(2)}_{\alpha\alpha'}(\varkappa\varkappa' \mid \boldsymbol{k}\omega) = \frac{4\pi}{v_a} \sum_{\beta\beta'} \boldsymbol{Z}^*_{\beta\alpha}(\varkappa \mid \boldsymbol{k}, -\omega)\, T^{-1}_{\beta\beta'}(\boldsymbol{k}, \omega)\, Z_{\beta'\alpha'}(\varkappa' \mid \boldsymbol{k}, \omega) \tag{1.6.23}$$

with the $\boldsymbol{k}$-dependent dynamical effective charge tensor

$$\begin{aligned} Z_{\alpha\alpha'}(\varkappa \mid \boldsymbol{k}\omega) = {} & \frac{1}{4\pi e} \sum_{\boldsymbol{K}} q^2 v_\varkappa(\boldsymbol{q})\, \mathrm{e}^{-i\boldsymbol{K}\boldsymbol{x}(\varkappa)} \\ & \times \Big[\delta_{\alpha\alpha'}\delta_{\boldsymbol{K}\boldsymbol{0}} - (1 - \delta_{\boldsymbol{K}\boldsymbol{0}}) \sum_{\boldsymbol{K}'\neq 0} \sum_{\beta} \varepsilon_{\alpha\beta}(\boldsymbol{k}, \boldsymbol{k}+\boldsymbol{K}', \omega) \\ & \times \hat{q}_\beta{}' \mathring{\epsilon}^{-1}(\boldsymbol{k}+\boldsymbol{K}', \boldsymbol{k}+\boldsymbol{K}, \omega)\, \hat{q}_{\alpha'}\Big]. \end{aligned} \tag{1.6.24}$$

$\boldsymbol{T}^{-1}$ in (1.6.23) is the inverse of the tensor

$$\boldsymbol{T}(\boldsymbol{k}, \omega) \equiv [\boldsymbol{T}^{-1}(\boldsymbol{k}, \boldsymbol{k}, \omega)]^{-1}, \tag{1.6.25}$$

whose physical meaning will be discussed later.

The adiabatic approximation consists now in approximating the dynamic response by a static one, i.e. we account for the fact that due to the large mass difference between electrons and ions the valence electrons instantaneously follow the motion of the ions. Thus we put $\omega = 0$ in $\boldsymbol{C}^{(1)}$ (1.6.22) and $\boldsymbol{Z}$ (1.6.23). By retaining the ω-dependence of $\boldsymbol{T}(\boldsymbol{k}, \omega)$ the retardation of the Coulomb interaction between the ions, and thus polariton effects, can be accounted for.

When we neglect polariton effects (i.e. we put $\omega = 0$ in $\boldsymbol{T}(\boldsymbol{k}, \omega)$) the total $\boldsymbol{T}(\boldsymbol{q}, \boldsymbol{q}', \omega)$ (cf. (1.6.19)) is nearly a purely transverse matrix and in analogy to (1.6.20) we have

$$T^{-1}_{\alpha\alpha'}(\boldsymbol{k}+\boldsymbol{K}, \boldsymbol{k}+\boldsymbol{K}', 0) = \hat{q}_\alpha \epsilon^{-1}(\boldsymbol{k}+\boldsymbol{K}, \boldsymbol{k}+\boldsymbol{K}', 0)\, \hat{q}'_{\alpha'}. \tag{1.6.26}$$

Inserting (1.6.26) into (1.6.25) we obtain for $\boldsymbol{C}^{(2)}$ (1.6.23)

$$C^{(2)}_{\alpha\alpha'}(\varkappa\varkappa' \mid \boldsymbol{k}0) = \frac{1}{\epsilon^0(\boldsymbol{k})} \frac{4\pi}{v_a} \sum_{\beta\beta'} [\hat{k}_\beta Z_{\beta\alpha}(\varkappa \mid \boldsymbol{k}, 0)]^* [\hat{k}_{\beta'} Z_{\beta'\alpha'}(\varkappa' \mid \boldsymbol{k}, 0)], \tag{1.6.27}$$

where (cf. (A2.69))

$$\epsilon^0(\boldsymbol{k}) \equiv [\epsilon^{-1}(\boldsymbol{k}, \boldsymbol{k}, \omega = 0)]^{-1} \tag{1.6.28}$$

is the $\boldsymbol{k}$-dependent optical frequency dielectric constant, and it is

$$\left.\begin{aligned} &\sum_\beta \hat{k}_\beta Z_{\beta\alpha}(\varkappa \mid \boldsymbol{k}, 0) \\ &= \frac{\epsilon^0(\boldsymbol{k})}{4\pi e} \sum_{\boldsymbol{K}} q^2 v_\varkappa(\boldsymbol{q})\, \mathrm{e}^{-i\boldsymbol{K}\boldsymbol{x}(\varkappa)} \hat{q}_\alpha \epsilon^{-1}(\boldsymbol{k}, \boldsymbol{k} + \boldsymbol{K}, 0) \equiv Z_\alpha(\varkappa \mid \boldsymbol{k}). \end{aligned}\right\} \tag{1.6.29}$$

To investigate now the analytic behaviour of $\boldsymbol{C}^{(1)}$ and $\boldsymbol{C}^{(2)}$ for $\boldsymbol{k}$ near zero, we examine the structure of $\epsilon(\boldsymbol{k} + \boldsymbol{K}, \boldsymbol{k} + \boldsymbol{K}') \equiv \epsilon(k + \boldsymbol{K}, \boldsymbol{k} + \boldsymbol{K}', \omega)$ and its inverse at $\boldsymbol{k} \to \boldsymbol{0}$ by making use of the expression for ϵ obtained in the self-consistent field approximation (SCF). It can be shown that the analytic properties derived in the SCF are rigorous. In the SCF we have (see (A2.56), (A2.77))

$$\begin{aligned} \epsilon(\boldsymbol{q}, \boldsymbol{q}') = \delta_{\boldsymbol{q}\boldsymbol{q}'} + \frac{4\pi}{qq'} \left(-\frac{e^2}{\boldsymbol{R}}\right) \sum_{\boldsymbol{k}\nu \boldsymbol{k}'\nu'} \langle \boldsymbol{k}\nu|\, \mathrm{e}^{i\boldsymbol{q}'\boldsymbol{r}}\, |\boldsymbol{k}'\nu'\rangle \langle \boldsymbol{k}'\nu'|\, \mathrm{e}^{-i\boldsymbol{q}\boldsymbol{r}}\, |\boldsymbol{k}\nu\rangle \\ \times \frac{f_0(E_{\boldsymbol{k}'\nu'}) - f_0(E_{\boldsymbol{k}\nu})}{E_{\boldsymbol{k}'\nu'} - E_{\boldsymbol{k}\nu} + \hbar(\omega + i\varepsilon)}, \qquad \varepsilon \to +0, \end{aligned} \tag{1.6.30}$$

where $|\boldsymbol{k}\nu\rangle$ denotes the Bloch function with wave vector $\boldsymbol{k}$ and band index ν (cf. (A2.72)), $E_{\boldsymbol{k}\nu}$ is the corresponding energy, and $f_0(E_{\boldsymbol{k}\nu})$ is the Fermi distribution function (A2.75). As seen from an inspection of (1.6.30), in an insulator the only allowed transitions are interband. Making use of the identity

$$[\mathrm{e}^{i\boldsymbol{q}\boldsymbol{r}}, H_0] = -\frac{\hbar}{2m} \sum_\alpha (\mathrm{e}^{i\boldsymbol{q}\boldsymbol{r}} p_\alpha + p_\alpha\, \mathrm{e}^{i\boldsymbol{q}\boldsymbol{r}}), \tag{1.6.31}$$

where H_0 is the one-particle Hamiltonian in SCF (A2.70) and $\boldsymbol{p}$ is the momentum operator, we have

$$\langle \boldsymbol{k}\nu|\, \mathrm{e}^{i\boldsymbol{q}\boldsymbol{r}}\, |\boldsymbol{k}'\nu'\rangle = \frac{\hbar}{m} \sum_\alpha q_\alpha \frac{\langle \boldsymbol{k}\nu|\, p_\alpha\, \mathrm{e}^{i\boldsymbol{q}\boldsymbol{r}}\, |\boldsymbol{k}'\nu'\rangle}{E_{\boldsymbol{k}\nu} - E_{\boldsymbol{k}'\nu'} + \dfrac{\hbar^2}{2m} q^2}, \tag{1.6.32}$$

i.e. for interband transitions each of the matrix elements in (1.6.30) goes to zero linearly with $\boldsymbol{q}$. We see that $\epsilon(\boldsymbol{k} + \boldsymbol{K}, \boldsymbol{k} + \boldsymbol{K}')$ has a well-defined limit for $\boldsymbol{K}, \boldsymbol{K}' \neq \boldsymbol{0}$, i.e. $\mathring{\epsilon}$ and thus $\mathring{\epsilon}^{-1}$ are always analytic functions of $\boldsymbol{k}$ at $\boldsymbol{k} \to \boldsymbol{0}$. Consequently, $\boldsymbol{C}^{(1)}$ (1.6.22) is an analytic function of $\boldsymbol{k}$ as $\boldsymbol{k} \to \boldsymbol{0}$. It describes short-range forces only. Further, from (1.6.30) and (1.6.32) we see that as $\boldsymbol{k} \to 0$

$$\begin{aligned} &\epsilon(\boldsymbol{k}, \boldsymbol{k}) \to \boldsymbol{k}\boldsymbol{Q}\boldsymbol{k}/k^2, \\ &\epsilon(\boldsymbol{k} + \boldsymbol{K}, \boldsymbol{k}) \to \boldsymbol{k}\boldsymbol{P}^*(\boldsymbol{K})/k, \\ &\epsilon(\boldsymbol{k}, \boldsymbol{k} + \boldsymbol{K}) \to \boldsymbol{k}\boldsymbol{P}(\boldsymbol{K})/k, \end{aligned} \tag{1.6.33}$$

where $\boldsymbol{Q}$ is a tensor and $\boldsymbol{P}(\boldsymbol{K})$ are vectors all independent of $\boldsymbol{k}$. Using (1.6.33), (1.6.17) and (1.6.18) we find that as $\boldsymbol{k} \to 0$

$$\begin{aligned} &1/\epsilon^{-1}(\boldsymbol{k}, \boldsymbol{k}) \to \boldsymbol{k}\boldsymbol{T}\boldsymbol{k}/k^2, \\ &\epsilon^{-1}(\boldsymbol{k}, \boldsymbol{k}+\boldsymbol{K})/\epsilon^{-1}(\boldsymbol{k}, \boldsymbol{k}) \to \boldsymbol{k}\boldsymbol{U}(\boldsymbol{K})/k, \\ &\epsilon^{-1}(\boldsymbol{k}+\boldsymbol{K}, \boldsymbol{k})/\epsilon^{-1}(\boldsymbol{k}, \boldsymbol{k}) \to \boldsymbol{k}\boldsymbol{U}^*(\boldsymbol{K})/k, \end{aligned} \tag{1.6.34}$$

where $\boldsymbol{T}$ is a tensor and $\boldsymbol{U}(\boldsymbol{K})$ are vectors all well defined as $\boldsymbol{k} \to 0$. Using (1.6.27) to (1.6.29) and (1.6.34) we get for $\boldsymbol{k} \to 0$

$$C^{(2)}_{\alpha\alpha'}(\varkappa\varkappa' \mid \boldsymbol{k}0) = \frac{4\pi}{v_a}\frac{1}{\epsilon^0(\hat{\boldsymbol{k}})} \sum_{\beta\beta'} [\hat{k}_\beta Z_{\beta\alpha}(\varkappa)]^* [\hat{k}_{\beta'} Z_{\beta'\alpha'}(\varkappa)], \tag{1.6.35}$$

where

$$Z_{\alpha\beta}(\varkappa) = Z_{\alpha\beta}(\varkappa \mid 0, 0) \tag{1.6.36}$$

is the dynamical effective charge tensor. We see from (1.6.35) that in contrast to $\boldsymbol{C}^{(1)}$ (1.6.22), $\boldsymbol{C}^{(2)}$ is discontinuous at $\boldsymbol{k} = 0$. This nonanalytic behaviour arises from the long-range Coulomb forces between the ions.

Before we proceed to discuss lattice vibrations in nonmetallic materials we consider the analytic behaviour of $\boldsymbol{C}$ for metals as $\boldsymbol{k} \to 0$. On examining (1.6.30), and accounting for the fact that in a metal both the interband and the intraband contributions are present, we see that as $\boldsymbol{k} \to 0$

$$\begin{aligned} &\epsilon(\boldsymbol{k}, \boldsymbol{k}) \to A/k^2, \\ &\epsilon(\boldsymbol{k}+\boldsymbol{K}, \boldsymbol{k}) \to B(\boldsymbol{K})/k, \\ &\epsilon(\boldsymbol{k}, \boldsymbol{k}+\boldsymbol{K}) \to B^*(\boldsymbol{K})/k, \\ &\epsilon(\boldsymbol{k}+\boldsymbol{K}, \boldsymbol{k}+\boldsymbol{K}') \to C(\boldsymbol{K}, \boldsymbol{K}'), \end{aligned} \tag{1.6.37}$$

and thus with (1.6.17) and (1.6.18) that as $\boldsymbol{k} \to 0$

$$\begin{aligned} &1/\epsilon^{-1}(\boldsymbol{k}, \boldsymbol{k}) \to D(\boldsymbol{K})/k^2, \\ &\epsilon^{-1}(\boldsymbol{k}+\boldsymbol{K}, \boldsymbol{k})/\epsilon^{-1}(\boldsymbol{k}, \boldsymbol{k}) \to E(\boldsymbol{K})/k, \\ &\epsilon^{-1}(\boldsymbol{k}, \boldsymbol{k}+\boldsymbol{K})/\epsilon^{-1}(\boldsymbol{k}, \boldsymbol{k}) \to E^*(\boldsymbol{K})/k, \\ &\epsilon^{-1}(\boldsymbol{k}+\boldsymbol{K}, \boldsymbol{k}+\boldsymbol{K}') \to F(\boldsymbol{K}, \boldsymbol{K}'), \end{aligned} \tag{1.6.38}$$

where A, B, C, D, E and F are functions well defined as $\boldsymbol{k} \to 0$. Using (1.6.37) and (1.6.38) we see that $\boldsymbol{C}^{(1)}$ as well as $\boldsymbol{C}^{(2)}$ are analytic functions as $\boldsymbol{k} \to 0$. Thus, for metals the separation of $\boldsymbol{C}$ into $\boldsymbol{C}^{(1)}$ and $\boldsymbol{C}^{(2)}$ according to (1.6.22) and (1.6.23) is not of importance, and with (1.6.16) and (1.6.26) we have in adiabatic approximation

$$\begin{aligned} C_{\alpha\alpha'}(\varkappa\varkappa' \mid \boldsymbol{k}0) = \frac{1}{v_a}\frac{1}{4\pi e^2} \sum_{\boldsymbol{K}\boldsymbol{K}'} q^2 v_\varkappa^*(\boldsymbol{q})\, e^{i\boldsymbol{K}\boldsymbol{x}(\varkappa)} \hat{q}_\alpha \\ \times\, \epsilon^{-1}(\boldsymbol{k}+\boldsymbol{K}, \boldsymbol{k}+\boldsymbol{K}', 0)\, \hat{q}'_{\alpha'} e^{-i\boldsymbol{K}'\boldsymbol{x}(\varkappa')} v_{\varkappa'}(\boldsymbol{q}')\, q'^2, \end{aligned} \tag{1.6.39}$$

where for metals ϵ^{-1} does not contain any non-analytic terms as $\boldsymbol{k} \to 0$.

1.6.3. Effective charge neutrality condition, acoustic modes, and dynamical matrix

On replacing in (1.6.29) the inverse dielectric matrix by the charge density response function by means of (A2.62), and making use of the acoustic sum rules (A2.88) we obtain

$$\lim_{k\to 0} \sum_{\varkappa} Z_\alpha(\varkappa \mid \boldsymbol{k}) = \lim_{k\to 0} \hat{k}_\alpha \epsilon^0(\boldsymbol{k}) \left[\sum_{\varkappa} \frac{k^2 v_\varkappa(\boldsymbol{k})}{4\pi e} + \frac{v_a}{V} \langle \varrho(0) \rangle\right] = 0. \quad (1.6.40)$$

Here the first term in the brackets is the total charge of the ions and the second term the total charge of the valence electrons (screening electrons) in a primitive unit cell. The right hand side of (1.6.40) vanishes because of the charge neutrality of the crystal. For a cubic crystal the charge tensor (1.6.36) is diagonal. Accordingly, the transverse effective charge $e^*(\varkappa)$ of an ion of type $\varkappa$ is given by

$$\begin{aligned} e^*(\varkappa) &= \lim_{k\to 0} \sum_{\alpha\beta} \hat{k}_\alpha Z_{\alpha\beta}(\varkappa \mid \boldsymbol{k}, 0)\, \hat{k}_\beta \\ &= \lim_{k\to 0} \frac{\epsilon^0(\boldsymbol{k})}{4\pi e} \sum_{\boldsymbol{K}} \frac{q}{k}\, v_\varkappa(\boldsymbol{q})\, \mathrm{e}^{-i\boldsymbol{K}\boldsymbol{x}(\varkappa)} \boldsymbol{q}\hat{\boldsymbol{k}} \epsilon^{-1}(\boldsymbol{k}, \boldsymbol{k}+\boldsymbol{K}, 0) \\ &= Z_\varkappa + \lim_{k\to 0} \frac{\epsilon^0(\boldsymbol{k})}{4\pi e} \sum_{\boldsymbol{K}}{}' \frac{q}{k}\, v_\varkappa(\boldsymbol{q})\, \mathrm{e}^{-i\boldsymbol{K}\boldsymbol{x}(\varkappa)} \boldsymbol{q}\hat{\boldsymbol{k}} \epsilon^{-1}(\boldsymbol{k}, \boldsymbol{k}+\boldsymbol{K}, 0), \end{aligned} \quad (1.6.41)$$

where the prime in the last line means that in the sum the term $\boldsymbol{K} = \boldsymbol{0}$ is missing. $Z_\varkappa$ is the nuclear charge minus that of the core electrons (cf. (1.6.4)). From (1.6.40) and (1.6.41) we obtain the following effective charge neutrality condition:

$$\sum_{\varkappa} e^*(\varkappa) = 0. \quad (1.6.42)$$

In particular, in the case of elemental covalent crystals, such as group IV semiconductors, all sublattices are identical and we have from (1.6.42) that $e^*(\varkappa) = 0$ for all $\varkappa$. Since the sum of the bare charges $Z_\varkappa$ equals the total charge of the screening electrons, it must hold that $\sum_{\varkappa} Z_\varkappa \neq 0$. Therefore, because of (1.6.42) at least for one type of ions the second term on the right hand side in the last line of (1.6.41) cannot disappear, which means that off-diagonal elements of the inverse dielectric matrix ϵ^{-1} are essential in an insulator to ensure the validity of the charge neutrality condition (1.6.40) or (1.6.42), respectively, and thus to ensure the existence of acoustic modes, as will be shown below. Consequently, in a non-metallic system ϵ^{-1} cannot be a diagonal matrix. In a metal on the other hand, there is no such requirement, which greatly facilitates any first-principle phonon calculation. For more details on effective charge calculations the reader is referred to Littlewood (1979) and references therein.

Now we turn to the problem of the existence of acoustic modes. On investigating long wavelength acoustic modes we can neglect retardation effects in (1.6.15), since for such modes we have $k = \omega/v_s \gg \omega/c$ (v_s — appropriate velocity of sound), i.e. the condition for absent retardation effects $kc/\omega > 1$ (cf. section 1.5.) is well satisfied. Inserting (cf. (1.5.11))

$$u_\alpha(l\varkappa, \omega) = \frac{1}{\sqrt{M_\varkappa}} v_\alpha(l\varkappa, \omega) \, e^{i\boldsymbol{k}\boldsymbol{x}(l\varkappa)} \tag{1.6.43}$$

into (1.6.15) and making use of (1.6.26) we obtain for the dynamical matrix (cf. (1.3.37), (1.5.48))

$$D_{\alpha\alpha'}(\varkappa\varkappa' \mid \boldsymbol{k}) = \frac{1}{\sqrt{M_\varkappa M_{\varkappa'}}} \Big[C_{\alpha\alpha'}(\varkappa\varkappa' \mid \boldsymbol{k}) - \delta_{\varkappa\varkappa'} \times \frac{1}{V} \frac{1}{4\pi e^2} \sum_{\boldsymbol{K}} K_\alpha K_{\alpha'} \, e^{i\boldsymbol{K}\boldsymbol{x}(\varkappa)} v_\varkappa^*(\boldsymbol{K}) \, K^2 V_0(\boldsymbol{K}) \Big], \tag{1.6.44}$$

where $C_{\alpha\alpha'}(\varkappa\varkappa' \mid \boldsymbol{k}) \equiv C_{\alpha\alpha'}(\varkappa\varkappa' \mid \boldsymbol{k}, 0)$ is given by (1.6.39). Using (A2.62) we get from (1.6.44) with (1.6.7) and the sum rules (A2.87)

$$\sum_{\varkappa'} C_{\alpha\alpha'}(\varkappa\varkappa' \mid 0) = \frac{1}{V} \frac{1}{4\pi e^2} \sum_{\boldsymbol{K}} K_\alpha K_{\alpha'} \, e^{i\boldsymbol{K}\boldsymbol{x}(\varkappa)} \, v_\varkappa^*(\boldsymbol{K}) \, K^2 V_0(\boldsymbol{K}) \,. \tag{1.6.45}$$

Thus with (1.6.45) we obtain for the dynamical matrix (1.6.44)

$$D_{\alpha\alpha'}(\varkappa\varkappa' \mid \boldsymbol{k}) = \frac{1}{\sqrt{M_\varkappa M_{\varkappa'}}} \Big[C_{\alpha\alpha'}(\varkappa\varkappa' \mid \boldsymbol{k}) - \delta_{\varkappa\varkappa'} \sum_{\varkappa''} C_{\alpha\alpha'}(\varkappa\varkappa'' \mid 0) \Big]. \tag{1.6.46}$$

According to (1.3.52) and (1.3.38) the existence of acoustic modes (the frequency of which tends to zero as $\boldsymbol{k} \to 0$ and whose amplitude for $\boldsymbol{k} \to 0$ is given by (1.3.53)) requires

$$\lim_{\boldsymbol{k} \to 0} \sum_{\varkappa'} \sqrt{M_{\varkappa'}} \, D_{\alpha\alpha'}(\varkappa\varkappa' \mid \boldsymbol{k}) = 0 \,. \tag{1.6.47}$$

We see from (1.6.46) that in the case of metals the condition (1.6.47) is trivially satisfied, because $\boldsymbol{C}$ is a continuous function of $\boldsymbol{k}$ as $\boldsymbol{k} \to 0$. However, in the case of an insulator, because of the discontinuity of $\boldsymbol{C}^{(2)}$ (1.6.35) at $\boldsymbol{k} = 0$, the charge neutrality condition (1.6.40), i.e. the existence of non-zero off-diagonal elements of the dielectric matrix, is a necessary and sufficient condition for (1.6.47).

To demonstrate the important role of the nondiagonal elements of the dielectric matrix (i.e. of local field effects) on phonons in non-metallic materials we present in figures 1.14 and 1.15 the phonon spectrum of silicon calculated without and with the nondiagonal elements by using a simplified pseudopotential technique. We see that the main shortcomings of the results without local field consist in imaginary TA (transverse acoustic) frequencies leaving the crystal unstable against shear. This can be understood in terms of absent

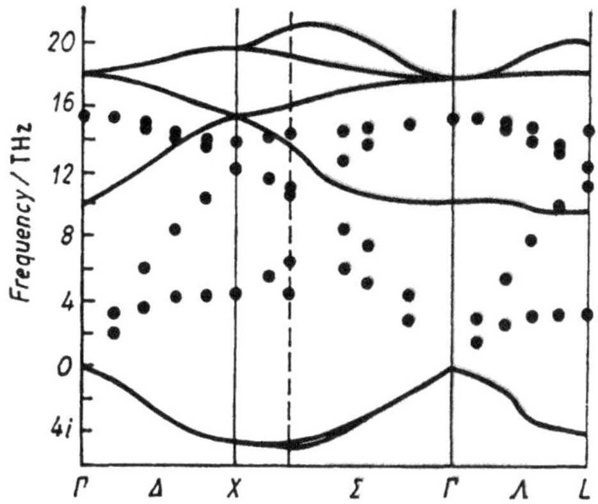

Fig. 1.14. Phonon dispersion curves for silicon. The full lines are theoretical results obtained by neglection of the non-diagonal elements of the dielectric matrix. Note that this neglecting leads to imaginary TA branches. The points are experimental results (after BERTONI et al., 1974, experimental points from DOLLING, 1963)

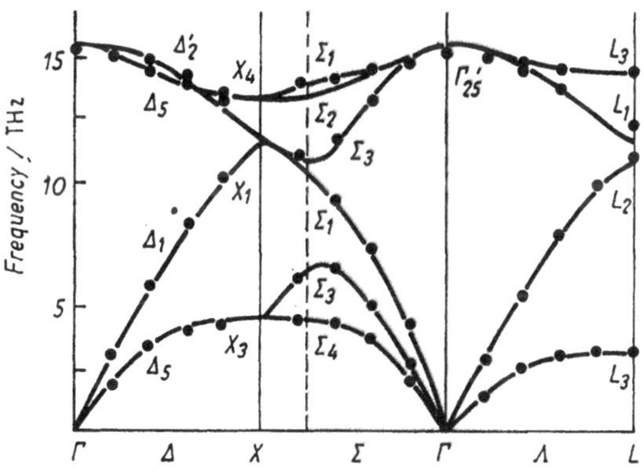

Fig. 1.15. Phonon dispersion curves for silicon. The full lines are theoretical results obtained on account of non-diagonal elements of the dielectric matrix. The points are experimental results (after TURNER and INKSON, 1978, experimental points from DOLLING, 1963)

bond charges and thus of absent noncentral forces which arise from the off-diagonal screening. Moreover, because of the incomplete screening of the ionic charges by the diagonal dielectric matrix assumed, the LA (longitudinal acoustic) frequencies go to a finite limit as $\boldsymbol{k} \to 0$. That is, the acoustic sum rule is not fulfilled. An account of nondiagonal elements of the dielectronic matrix, however, provides real TA branches, which compare satisfactorily with experimental data.

The microscopic expression for the dynamical matrix of metals is given by (1.6.46) with $\boldsymbol{C}$ from (1.6.39). In metals in most cases the pseudo-potentials may be assumed to be weak enough to perturb only very slightly the electronic functions of a free electron gas. It is then possible to take for ϵ the expression of a free electron gas, i.e. the Lindhard function, which is readily obtained from the SCF expression (A2.77):

$$\epsilon(\boldsymbol{k}+\boldsymbol{K}, \boldsymbol{k}+\boldsymbol{K}') = \delta_{\boldsymbol{K}\boldsymbol{K}'} \left\{1 + \frac{2me^2k_F}{\hbar^2\pi\,|\boldsymbol{k}+\boldsymbol{K}|^2} \times \left[1 + \frac{1-x^2}{2x} \ln \frac{1+x}{1-x}\right]\right\}, \tag{1.6.48}$$

where $x = |\boldsymbol{k}+\boldsymbol{K}|/2k_F$ with $\hbar k_F$ — the Fermi momentum of the free electron gas. The singularity at $x = 1$ in ϵ from (1.6.48) can give rise to an interesting feature in the phonon spectrum of metals, which consists in a reduction of the vibrational frequency (soft mode) of modes with wave vector $2k_F$ (Kohn anomaly). Whereas in three-dimensional systems this effect is weak, for one-dimensional systems, a giant Kohn anomaly is predicted, which has also been observed in certain quasi one-dimensional systems (fig. 1.16).

The microscopic expression for the dynamical matrix of a non-metallic material reads

$$D_{\alpha\alpha'}(\varkappa\varkappa' \mid \boldsymbol{k}) = \frac{1}{\sqrt{M_\varkappa M_{\varkappa'}}} \left[C^{(1)}_{\alpha\alpha'}(\varkappa\varkappa' \mid \boldsymbol{k}, 0) - \delta_{\varkappa\varkappa'} \sum_{\varkappa''} C^{(1)}_{\alpha\alpha'}(\varkappa\varkappa'' \mid 0, 0)\right] + \frac{1}{\sqrt{M_\varkappa M_{\varkappa'}}} C^{(2)}_{\alpha\alpha'}(\varkappa\varkappa' \mid \boldsymbol{k}, \omega), \tag{1.6.49}$$

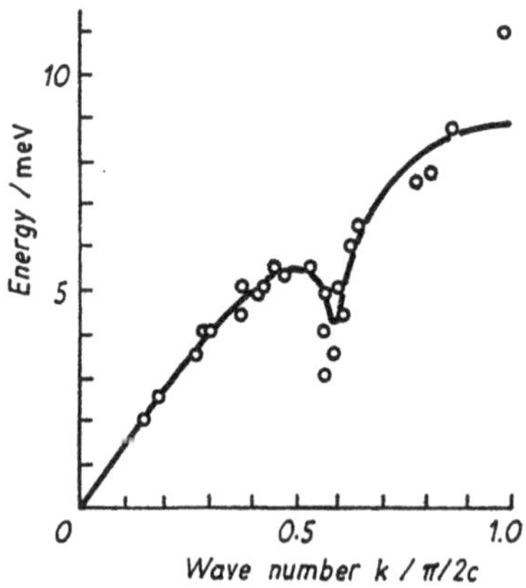

Fig. 1.16. Effect of the Kohn anomaly in the longitudinal acoustic phonon branch in the chain direction of the quasi one-dimensional system $K_2Pt(CN)_4 \cdot Br_{0,3} \cdot 3H_2O$. The solid line is a fit of the theoretical results obtained by a model treating the electrons as a one-dimensional free electron gas. The points arè experimental results. c is the lattice constant in chain direction (after GLÄSER, 1974, experimental points from RENKER et al., 1973)

where $\boldsymbol{C}^{(1)}$ is given by (1.6.22) with $\omega = 0$, and on account of polariton effects (cf. (1.6.23)) $\boldsymbol{C}^{(2)}$ has the form

$$C^{(2)}_{\alpha\alpha'}(\varkappa\varkappa' \mid \boldsymbol{k}, \omega) = \frac{4\pi}{v_a} \sum_{\beta\beta'} Z^*_{\beta\alpha}(\varkappa \mid \boldsymbol{k}, 0)\, T^{-1}_{\beta\beta'}(\boldsymbol{k}, \omega)\, Z_{\beta'\alpha'}(\varkappa' \mid \boldsymbol{k}, 0), \qquad (1.6.50)$$

or upon neglecting polariton effects $\boldsymbol{C}^{(2)}$ is given by (1.6.27). On comparing (1.6.49), (1.6.50) and (1.5.49), (1.6.49) turns out to be the microscopic version of the phenomenological expression (1.5.49). To see this, one should notice that the tensor $\boldsymbol{T}$ occurring in (1.6.50) has the meaning of the macroscopic $\boldsymbol{T}$-tensor which, by definition, connects the total field in a crystal with the external current density by an equation analogous to (A2.27). This fact results from (1.6.12) (with $\boldsymbol{q} = \boldsymbol{q}' = \boldsymbol{k}$) and (1.6.25). With the aid of the equation (being the analogue to (1.6.19)) for the macroscopic $\boldsymbol{T}$-tensor, the equivalence of the $\boldsymbol{T}$-tensors in (1.6.50) and (1.5.46) can be proved.

Note that from the connection between (1.6.49), (1.6.50) and (1.5.49) it follows that $e^*(\varkappa)$ in (1.6.41) represents the effective Born charge, or transverse effective charge (cf. section 1.5.).

1.6.4. Electron response in Wannier representation

Any first principle calculation of phonons on the basis of equations (1.6.46), (1.6.39) requires that we invert the dielectric matrix $\epsilon(\boldsymbol{k} + \boldsymbol{K}, \boldsymbol{k} + \boldsymbol{K}', \omega)$. In the free electron limit ϵ is a diagonal matrix in $\boldsymbol{K}$ and $\boldsymbol{K}'$ (cf. (1.6.48)) and can be inverted as a scalar. In general, however, where the lattice periodicity is important (for example in transition metals, semiconductors, ionic crystals), a direct inversion of the infinite-dimensional matrix ϵ is impossible.

An inversion of ϵ can be achieved using the Wannier representation in the SCF expression for ϵ (Pick, 1971, Hanke, 1973, and see the reviews by Hanke, 1978, and Sinha, 1978, and references therein). To show this we write ϵ (1.6.30) as

$$\epsilon(\boldsymbol{k} + \boldsymbol{K}, \boldsymbol{k} + \boldsymbol{K}', \omega) = \delta_{\boldsymbol{K}\boldsymbol{K}'} + \frac{4\pi}{|\boldsymbol{k} + \boldsymbol{K}|\,|\boldsymbol{k} + \boldsymbol{K}'|} \left(-\frac{e^2}{V}\right)$$
$$\times \sum_{\tilde{\boldsymbol{k}}\nu\nu'} \frac{f_0(E_{\tilde{\boldsymbol{k}}\nu}) - f_0(E_{\tilde{\boldsymbol{k}}+\boldsymbol{k}\nu'})}{E_{\tilde{\boldsymbol{k}}\nu} - E_{\tilde{\boldsymbol{k}}+\boldsymbol{k}\nu'} + \hbar(\omega + i\varepsilon)} \langle \tilde{\boldsymbol{k}}\nu |\, e^{-i(\boldsymbol{k}+\boldsymbol{K})\boldsymbol{r}}\, | \tilde{\boldsymbol{k}} + \boldsymbol{k}\nu' \rangle$$
$$\times \langle \tilde{\boldsymbol{k}} + \boldsymbol{k}\nu' |\, e^{i(\boldsymbol{k}+\boldsymbol{K}')\boldsymbol{r}}\, | \tilde{\boldsymbol{k}}\nu \rangle, \quad \varepsilon \to +0, \qquad (1.6.51)$$

where the tilde denotes electron wave vectors. Let us consider a number of electron bands which may have some contact with each other but which are well separated from the other bands. Then it is possible to construct Wannier functions $|nl\rangle$ $\big(\langle \boldsymbol{r} \mid nl \rangle = \varphi_n(\boldsymbol{r} - \boldsymbol{x}(l))\big)$ with exponential decay at infinity and with the property of orthonormality

$$\langle nl \mid n'l' \rangle = \delta_{nn'}, \delta_{ll'}. \qquad (1.6.52)$$

The Bloch waves are obtained by a unitary transformation

$$|\tilde{k}\nu\rangle = N^{-1/2} \sum_n c_{\nu n}(\tilde{k}) \sum_l e^{ikx(l)} |n\rangle l, \tag{1.6.53}$$

where N is the number of lattice vectors and $c_{\nu n}(\tilde{k})$, the so-called eigenvectors of the band complex, is a unitary matrix in indices n, ν. The number of Wannier functions associated with a unit cell is equal to the number of connected bands. Note that in materials with more than one atom per unit cell, the Wannier functions involve more than one atom, in general, they contain contributions from all atoms in a cell.

In terms of these Wannier functions we can ϵ from (1.6.51) rewrite as

$$\begin{aligned}\epsilon(\boldsymbol{k}+\boldsymbol{K}, \boldsymbol{k}+\boldsymbol{K}', \omega) &= \delta_{\boldsymbol{K}\boldsymbol{K}'} - \sum_{ss'} A_s(\boldsymbol{k}+\boldsymbol{K})\, N_{ss'}(\boldsymbol{k}, \omega)\, A_{s'}^*(\boldsymbol{k}+\boldsymbol{K}') \\ &\equiv 1 - \boldsymbol{A}\boldsymbol{N}\boldsymbol{A}^+,\end{aligned} \tag{1.6.54}$$

where

$$\begin{aligned}A_s(\boldsymbol{k}+\boldsymbol{K}) = \frac{2e\sqrt{\pi}}{|\boldsymbol{k}+\boldsymbol{K}|} \int \mathrm{d}\boldsymbol{r}\, \varphi_{n_1}(\boldsymbol{r}) \\ \times \exp\big(-i(\boldsymbol{k}+\boldsymbol{K})\,\boldsymbol{r}\big)\, \varphi_{n_2}\big(\boldsymbol{r}-\boldsymbol{x}(l)\big)\end{aligned} \tag{1.6.55}$$

and

$$\begin{aligned}N_{ss'}(\boldsymbol{k}, \omega) = \frac{1}{V} \sum_{\nu\nu'\tilde{\boldsymbol{k}}} \exp\left[i(\boldsymbol{k}+\tilde{\boldsymbol{k}})\,\big(\boldsymbol{x}(l)-\boldsymbol{x}(l')\big)\right] \\ \times c_{\nu n_1}^*(\tilde{\boldsymbol{k}})\, c_{\nu' n_2}(\tilde{\boldsymbol{k}}+\boldsymbol{k})\, c_{\nu' n_2'}^*(\tilde{\boldsymbol{k}}+\boldsymbol{k})\, c_{\nu n_1'}(\tilde{\boldsymbol{k}}) \\ \times \frac{f_0(E_{\tilde{\boldsymbol{k}}\nu}) - f_0(E_{\tilde{\boldsymbol{k}}+\boldsymbol{k}\nu'})}{E_{\tilde{\boldsymbol{k}}\nu} - E_{\tilde{\boldsymbol{k}}+\boldsymbol{k}\nu'} + \hbar(\omega+i\varepsilon)}, \quad \varepsilon \to +0.\end{aligned} \tag{1.6.56}$$

Here the subscript s stands for the set of indices l, n_1, n_2 occurring in (1.6.55). Because of its separable form the dielectric matrix (1.6.54) can be formally readily inverted, which yields

$$\left.\begin{aligned}\epsilon^{-1}(\boldsymbol{k}+\boldsymbol{K}, \boldsymbol{k}+\boldsymbol{K}', \omega) &= \delta_{\boldsymbol{K}\boldsymbol{K}'} + \sum_{ss'} A_s(\boldsymbol{k}+\boldsymbol{K})\, S_{ss'}^{-1}(\boldsymbol{k}, \omega)\, A_{s'}^*(\boldsymbol{k}+\boldsymbol{K}') \\ &= 1 + \boldsymbol{A}\boldsymbol{S}^{-1}\boldsymbol{A}^+,\end{aligned}\right\} \tag{1.6.57}$$

where the matrix $\boldsymbol{S}$ is given by

$$\boldsymbol{S} = \boldsymbol{N}^{-1} - \boldsymbol{V} \tag{1.6.58}$$

with

$$V_{ss'} = \sum_{\boldsymbol{K}} A_s^*(\boldsymbol{k}+\boldsymbol{K})\, A_{s'}(\boldsymbol{k}+\boldsymbol{K}). \tag{1.6.59}$$

For practical purposes the dimension of the matrix $\boldsymbol{N}$ involved in $\boldsymbol{S}^{-1}$ in (1.6.57) may not be too large, i.e. the Wannier functions must be well localized and the number of electronic bands to be accounted for must be manageable. Further, for want of quantitative expressions for Wannier functions, in practice, instead of such functions linear-combinations-of-atomic-orbitals

(LCAO) functions are often used for calculating ϵ^{-1}. In the case of metals, specifically in the case of transition metals, the dimension of $\boldsymbol{N}$ can be reduced by using a combined LCAO-OPW (orthogonal plane wave) description, which splits the dielectric matrix into two parts: a diagonal part connected with intra-band transitions in the s-type conduction band (the s-electrons are described by OPW functions) and a nondiagonal part connected with the contribution from the well localized d-electrons. Thus we have

$$\epsilon(\boldsymbol{k}+\boldsymbol{K}, \boldsymbol{k}+\boldsymbol{K}', \omega) = \epsilon_0(\boldsymbol{k}+\boldsymbol{K}, \omega)\, \delta_{\boldsymbol{K}\boldsymbol{K}'} - \sum_{ss'} A_s(\boldsymbol{k}+\boldsymbol{K})\, N_{ss'}(\boldsymbol{k}, \omega)\, A_{s'}^*(\boldsymbol{k}+\boldsymbol{K}') \tag{1.6.60}$$

and the inverse of the dielectric matrix is given by

$$\begin{aligned}\epsilon^{-1}(\boldsymbol{k}+\boldsymbol{K}, \boldsymbol{k}+\boldsymbol{K}', \omega) &= \epsilon_0^{-1}(\boldsymbol{k}+\boldsymbol{K}, \omega)\, \delta_{\boldsymbol{K}\boldsymbol{K}'} \\ &\quad + \epsilon_0^{-1}(\boldsymbol{k}+\boldsymbol{K}, \omega) \sum_{ss'} A_s(\boldsymbol{k}+\boldsymbol{K})\, S_{ss'}^{-1}(\boldsymbol{k}, \omega) \\ &\quad \times A_{s'}^*(\boldsymbol{k}+\boldsymbol{K}')\, \epsilon_0^{-1}(\boldsymbol{k}+\boldsymbol{K}'),\end{aligned} \tag{1.6.61}$$

where $\boldsymbol{S}$ is given by (1.6.58), but now with

$$V_{ss'} = \sum_{\boldsymbol{K}} A_s^*(\boldsymbol{k}+\boldsymbol{K})\, A_{s'}(\boldsymbol{k}+\boldsymbol{K})\, \epsilon_0^{-1}(\boldsymbol{k}+\boldsymbol{K}, \omega). \tag{1.6.62}$$

Inserting (1.6.61) into (1.6.39) the matrix $\boldsymbol{C}$, which determines the dynamical matrix (1.6.46), reads

$$\left.\begin{aligned}C_{\alpha\alpha'}(\varkappa\varkappa' \mid \boldsymbol{k}) = {} & \frac{1}{v_a} \frac{1}{4\pi e^2} \sum_{\boldsymbol{K}} q^2 v_\varkappa^*(q)\, \hat{q}_\alpha \epsilon_0^{-1}(\boldsymbol{q})\, \hat{q}_{\alpha'} v_{\varkappa'}(\boldsymbol{q})\, q^2 \\ & \times e^{i\boldsymbol{K}(\boldsymbol{x}(\varkappa)-\boldsymbol{x}(\varkappa'))} \\ & + \frac{1}{v_a} \sum_{ss'} U_\alpha^s(\boldsymbol{k}, \varkappa)\, S_{ss'}^{-1} U_{\alpha'}^{s'*}(\boldsymbol{k}, \varkappa'),\end{aligned}\right\} \tag{1.6.63}$$

where

$$U_\alpha^s(\boldsymbol{k}, \varkappa) = \frac{1}{2e\sqrt{\pi}} \sum_{\boldsymbol{K}} e^{i\boldsymbol{K}\boldsymbol{x}(\varkappa)} \hat{q}_\alpha q^2 \frac{v_\varkappa^*(\boldsymbol{q})}{\epsilon_0(\boldsymbol{q})} A_s(\boldsymbol{q}) \tag{1.6.64}$$

with $\boldsymbol{q} = \boldsymbol{k} + \boldsymbol{K}$ and $\epsilon_0(\boldsymbol{q}) \equiv \epsilon_0(\boldsymbol{q}, \omega = 0)$. Note that in the case of a crystal with more than one atom per unit cell we can express the Wannier functions by linear combinations of orbitals localized at the sites of the atoms of the corresponding cells. In the formulas given above $A_s(\boldsymbol{k}+\boldsymbol{K})$ has to be replaced then by $A_s(\boldsymbol{k}, \boldsymbol{K}) = A_s(\boldsymbol{k}+\boldsymbol{K})\, S(\boldsymbol{K})$, where

$$S(\boldsymbol{K}) = \frac{1}{r} \sum_{\varkappa} e^{-i\boldsymbol{x}(\varkappa)\boldsymbol{K}} \tag{1.6.65}$$

(r — number of atoms per cell) is the structure factor and $A_s(\boldsymbol{k}+\boldsymbol{K})$ now involves the orbitals localized at the atomic sites.

The first term in (1.6.63) corresponds to the electrostatic coupling between the ions which is screened by the dielectric function $\epsilon_0(\boldsymbol{q})$. The second term can be viewed as arising from a distribution of dipoles which are connected with the well localized electrons from the occupied bands. In the case of a simple metal the second term is usually neglected, i.e. ϵ is assumed to be diagonal. In insulators we have $\epsilon_0(\boldsymbol{q}) = 1$, which leads to dipolar models. In covalent materials, but also in noble metals and transition metals, both the $\epsilon_0(\boldsymbol{q})$ and the $N_{ss'}$-terms have to be taken into account in calculating the dielectric matrix. This leads to a model with screened dipoles.

By a suitable parametrization the expression (1.6.63), which represents a general screened dipole model, can be shown to reproduce also either the shell model or the bond charge model according to the values of the parameters involved. This makes this expression useful for studying the validity of various methods and models developed to describe lattice vibrations in different types of solids (including simple metals, transition metals, covalent materials and ionic crystals) and to examine their relationship to each other.

Phonon calculations based on a dielectric matrix of type (1.6.60) have been performed with a fair degree of success for different systems such as ionic crystals, transition metals and transition metal compounds (see Hanke, 1978, and references therein). In particular, phonon anomalies (phonon softening) observed in some transition metals and transition-metal compounds (see fig. 1.17) could be reproduced by such calculations. These anomalies have been ascribed to the second term on the right hand side of (1.6.63), i.e. to the quasi-localized nature of electronic states present in such materials. In the special case of transition metal compounds, owing to hybridization of metal d states and

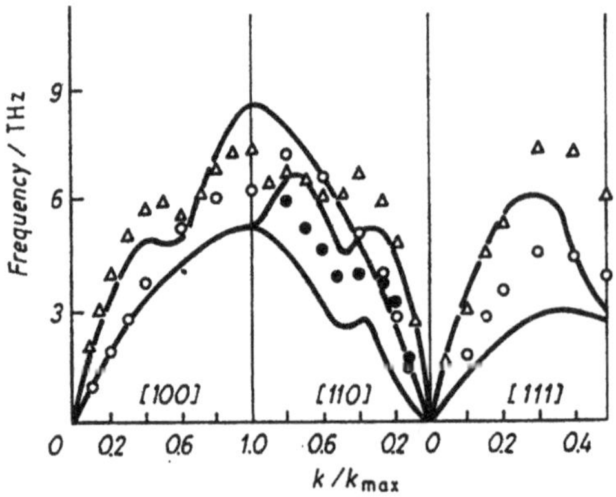

Fig. 1.17. Dispersion curves for acoustic phonon branches in NbC (having NaCl structure). The full lines are results of theoretical calculations based on a dielectric matrix of type (1.6.60). Circles and triangles are experimental results. Dips in the dispersion curves (soft modes) occur also in some other transition-metal compounds (after Hanke, 1978, experimental results from Smith and Gläser, 1970)

non-metal p states near the Fermi level, the denominator of the matrix $\boldsymbol{S}^{-1} = (\boldsymbol{N}^{-1} - \boldsymbol{V})^{-1}$ can become small for certain $\boldsymbol{k}$-values. This leads to a resonance-like increase in the density response, which is described by ϵ^{-1} (see appendix 2) or by $\boldsymbol{AS}^{-1}\boldsymbol{A}^{+}$. The occurring charge fluctuations are strongly coupled to the lattice (by the term $\boldsymbol{US}^{-1}\boldsymbol{U}^{+}$ in the dynamical matrix) and thus give rise to the phonon anomalies and, eventually, even to a structural phase transition. It is interesting to note that the electron phonon parameter also involves the matrix $\boldsymbol{S}^{-1}$, which explains the interrelation between phonon softening and high superconducting transition temperatures observed in transition metals and their compounds (see Hanke, 1978, and Sinha, 1980). Note that the phonon anomalies in transition-metal compounds have alternatively been explained by pecularities of the topology of the electronic bands near the Fermi energy (Varma and Weber, 1979, Weber et al., 1979, and see the reviews by Sinha, 1980, and Allen, 1980), which emphasizes the important role of the metallic part of the bonding. Furthermore, Falter and Selmke (1981) pointed out that such phonon anomalies can arise from pecularities of the Fermi surface giving rise to changes in the polarizability. In the latter paper the problem of inverting the dielectric matrix ϵ is managed by decomposing ϵ into a suitable approximate part $\bar{\epsilon}$ and a remainder, so that all interactions in the electron-ion system are renormalized by $\bar{\epsilon}^{-1}$, and perturbation theory can be used for calculating the dynamical matrix (cf. Falter and Selmke, 1980). Such a procedure is earsier to handle than the dielectric inversion approach in a localized basis described above.

Let us draw attention to an alternative microscopic approach to lattice dynamics. In this approach one attempts to calculate and use the adiabatic wave functions also for the nonequilibrium positions of the atoms by making use of localized wave functions, which are calculated by minimizing the Hartree-Fock expression for the total energy for a general but fixed atomic configuration. To this end the non-equilibrium orbitals are decomposed into two parts: (i) one part follows the cores if these are displaced, (ii) the other part describes the deformation of the orbitals if the atoms are displaced. In this way the energy splits into a rigid ion part and into a deformation part, which yields an expression for the dynamical matrix slightly more general than that of the deformation dipole model. Such an approach has been performed by Zeyher (1978) for non-metals, and a similar one by Varma and Weber (1979) for transition metals and compounds. The calculation of force constants, phonon dispersion curves, effective charges etc. for GaAs by Kunc and Martin (1982) belongs to this category of first principle phonon theories, too. Historically, the first such approach was proposed by Tolpygo and coworkers in the fifties (Tolpygo, 1950, 1959).

CHAPTER 2

LATTICE DYNAMICS OF CRYSTALS WITH COMPOSITIONAL DISORDER

2.1. General features of disordered systems

2.1.1. Introduction

In a crystalline material the lattice periodicity considerably simplifies any theoretical study of physical properties. In such a material the dynamical matrix has a simple recurrent pattern, which allows the eigenvalue problem to be reduce from one of order dN (d — dimension of the system, N — total number of atoms in it) to one associated with a set of matrices of order dr (r — number of atoms in a primitive unit cell). However, almost all real materials are not periodic and even carefully prepared crystals contain lattice defects. Because of the lack of periodicity the dynamical problem of such a system cannot be solved exactly except for a few special cases.

The simplest case of an aperiodic crystal is that of a crystal with a low concentration of defects (impurities) which do not interact with each other (section 2.2.). In this case the dynamical problem can be reduced to that of a single isolated defect. The latter problem can be solved exactly. An important tool for doing this is the Green function method. Especially double-time temperature-dependent Green functions turn out to be very suitable, because they are directly related to time-dependent correlation functions, by which experimental quantities can be expressed (section 2.1.3., appendix 3).

Defects modify both the vibrational eigenfrequencies and eigenvectors of the normal modes. The eigenfrequencies inside the bands of allowed frequencies are shifted by a defect only a small amount to $O(1/N)$. The eigenfrequencies near the band edges, however, can merge with the regions of forbidden frequencies. If the perturbation of the lattice by the defect is sufficiently large, the frequency shift of the latter normal modes with respect to the band edges can be rather large. The resulting modes are called localized modes, because in such a mode the amplitude of the atoms decreases faster than exponentially with increasing distance from the defect. Even though a defect shifts the frequency of modes inside a band only to $O(1/N)$ the amplitude of these modes can be strongly affected in a region around the defect of size $(1/N)$th of the crystal, especially when the defect has a characteristic frequency in the range of the band modes. In the latter case a resonance (quasilocalized) mode occurs, which is characterized by a large amplitude of atoms around the defect in a narrow frequency range about the characteristic defect frequency. Another effect caused by defects is the relaxation of selection rules that depends upon the conservation of the crystal momentum $\boldsymbol{k}$. This enables modes which in perfect crystals are not allowed to take part in infrared absorption or Raman scattering (such as acoustic modes in one-phonon processes) to contribute to the spectra in the presence of defects. The spectra can also be affected by modifications of the point symmetry of the crystal due to defects. Accordingly, optical methods turn out to be very suitable for studying the effects of defects on lattice dynamics.

The approximation of isolated defects is a good one only if the defect concentration is low, although some properties of lattices with moderate defect concentrations can be understood fairly well, at least qualitatively, on the basis of this approximation, too. With increasing defect concentration the defects can no longer be treated as being independent each other, and thus we have to deal with a disordered solid as such. Besides phenomena, similar to those found in systems with single defects, in systems with larger defect concentrations new phenomena occur, such as localized states due to clusters of defects or, more generally, localization of states due to the absence of long-range order (Anderson localization). The basic problem in dealing with disordered solids is that the microscopic configuration of a given sample is unknown. Moreover, given that the microscopic configuration is known one could not really solve the corresponding eigenvalue problem. Therefore, the conception of the ensemble average (configuration average) of physically observable quantities was developed, where an ensemble consists of elements corresponding to all possible atomic configurations which cannot be distinguished macroscopically. Thus, assuming that the experiment generally provides ensemble averaged quantities, from the very beginning the theoretican asks only for ensemble averages of observables.

Numerical as well as analytical methods have been used for investigating vibrational properties of disordered solids. The starting point of a calculation by a numerical method is the equation of motion of the atoms for a particular disordered system consisting of N atoms. For large N, one assumes that properties computed from this particular system relate not only to this, but to a whole ensemble of systems of which it is representative, i.e. the system at hand is assumed to be self-averaging. The essence of the application of analytical methods is the concept of expressing physical properties of a disordered system as ensemble averages of the corresponding properties of systems with well-defined configurations. The disordered system under consideration is generally assumed to be ergodic, i.e. the distribution of the stochastic quantities in a particular system is assumed to be the same as that over an ensemble of systems with the same statistical structure.

The so-called negative eigenvalue theorem (section 2.1.2.) is one of the major sources of the great progress reached in the last two decades in calculating vibrational properties of disordered solids by means of numerical methods. For certain types of dynamical matrices this theorem enables an accurate estimate of eigenvalue distribution by considerably fewer operations than would be needed in a full diagonalization procedure. Selected eigenvectors of normal modes and accordingly spectra, such as infrared, Raman and neutron scattering spectra, can also be calculated in this way, even for rather realistic three-dimensional disordered systems. Numerical results are of great importance also as criterion for the validity of approximations used in applying analytical methods. Among the analytical methods for studying disordered systems the Green function method is the most important one, because is quite compatible with the conception of an ensemble avergae: one has to calculate the ensemble average of an appropriate Green function and in many cases the experimental quantity of interest can then be readily expressed by it. In the last decade great progress has been made in elaborating standard approximations for calculating ensemble averages of Green functions. Both the Green function method and the negative eigenvalue theorem have very much stimulated the investigation of vibrational properties of disordered solids.

Roughly speaking there are two broad classes of disordered systems: (i) systems with substitutional, cellular or compositional disorder, (ii) systems with positional, topological or structural disorder. Compositionally disordered alloys and mixed crystals, but also crystals with substitutional impurities are among the systems of the first class. Glassy and amorphous materials, and superionic conductors are among the systems of the second class. In this chapter we deal with systems of the first class only. Chapter 3 will be devoted to systems of the second class. In the latter chapter we shall consider also the general problem of localization of states due to the absence of long-range order.

The most important approximation for treating elementary excitations in compositionally disordered systems by means of Green functions is the so-called coherent potential approximation (CPA) (SOVEN, 1967, TAYLOR, 1967). In this approximation the ensemble average of a Green function is calculated by embedding the atoms into an effective medium, which is self-consistently determined by the requirement of vanishing scattering (by the atoms) of the excitations governed by the effective medium. The CPA is a kind of mean-field approximation analogous to the molecular field theory in magnetism. By the CPA and especially by its extensions for also including clusters of atoms, rather good agreement with numerical and experimental results has been achieved in recent years (see section 2.3.).

For the subject considered in this chapter the reader is referred to the following reviews: for impurities in crystals see MARADUDIN (1966a, b), LUDWIG (1967), MARADUDIN et al. (1971), TAYLOR (1975) and BARKER and SIEVERS (1975); for mixed crystals and alloys see ELLIOTT et al. (1974) and BARKER and SIEVERS (1975); for the CPA see YONEZAWA and MORIGAKI (1973), for numerical methods see DEAN (1972), BELL (1976) and VISSCHER and GUBERNATIS (1980) and, in particular, for application of the Green function method and numerical methods to one-dimensional disordered systems see WU et al. (1980).

2.1.2. Numerical determination of frequency spectra and eigenvectors

Consider a system with the dynamical matrix $D_{\alpha\alpha'}(ll') = \Phi_{\alpha\alpha'}(ll')(M_l M_{l'})^{-1/2}$ (cf. (1.3.10)), where the Φ's are the harmonic force constants and M_l is the mass of the lth atom. Note that $\boldsymbol{D}$ is a real, symmetric matrix and that therefore its eigenvalues are real. In the case of short range interactions $\boldsymbol{D}$ has a simple band form with all elements beyond a certain distance from the main diagonal being zero. First we deal with the problem of calculating the frequency spectrum of $\boldsymbol{D}$. To do this we denote by $\eta(\boldsymbol{D}_1(\omega^2))$ the number of negative eigenvalues of $\boldsymbol{D}_1(\omega^2) \equiv \boldsymbol{D} - \omega^2\mathbf{1}$, which equals the number of eigenvalues of $\boldsymbol{D}$ smaller than ω^2. The distribution function of the squares of the eigenfrequencies $G(\omega^2)$ (cf. section 1.4.) can be directly related to $\eta(\boldsymbol{D}_1(\omega^2))$ as

$$\eta(\boldsymbol{D}_1(\omega^2)) = dN \int_0^{\omega^2} \mathrm{d}(\omega'^2)\, G(\omega'^2), \tag{2.1.1}$$

where d is the dimension of the system and N is the number of atoms in it. Making use of this equation, the squared frequency distribution of the system can be derived in the form of a histogram. The negative eigenvalue theorem provides an efficient method for evaluation of the quantities $\eta(\boldsymbol{D}_1(\omega^2))$ for the large-order band matrices $\boldsymbol{D}$ occurring in lattice dynamics.

Partitioning the matrix $\boldsymbol{D}_1(\omega^2)$ as

$$\boldsymbol{D}_1(\omega^2) = \begin{bmatrix} \boldsymbol{X}_1 & \boldsymbol{Y}_1 \\ \boldsymbol{Y}_1^{\mathrm{T}} & \boldsymbol{Z}_1 \end{bmatrix}, \tag{2.1.2}$$

where $\boldsymbol{X}_1$, $\boldsymbol{Z}_1$ are square matrices and $\boldsymbol{Y}_1^{\mathrm{T}}$ denotes the transpose of the matrix $\boldsymbol{Y}_1$, the negative eigenvalue theorem implies

$$\eta(\boldsymbol{D}_1(\omega^2)) = \eta(\boldsymbol{X}_1) + \eta(\boldsymbol{D}_2(\omega^2)), \tag{2.1.3}$$

where $\boldsymbol{D}_2(\omega^2)$ is defined in terms of the blocks of $\boldsymbol{D}_1(\omega^2)$ as

$$\boldsymbol{D}_2(\omega^2) = \boldsymbol{Z}_1 - \boldsymbol{Y}_1^T\boldsymbol{X}_1^{-1}\boldsymbol{Y}_1. \tag{2.1.4}$$

If $\boldsymbol{D}_2(\omega^2)$ is partitioned analogously as $\boldsymbol{D}_1(\omega^2)$ in (2.1.2) and if we pursue this procedure, we finally obtain

$$\eta(\boldsymbol{D}_1(\omega^2)) = \sum_{i=1}^{dN} \eta(\boldsymbol{X}_i) \tag{2.1.5}$$

with

$$\left.\begin{aligned} \boldsymbol{D}_i(\omega^2) &= \begin{bmatrix} \boldsymbol{X}_i & \boldsymbol{Y}_i \\ \boldsymbol{Y}_i^T & \boldsymbol{Z}_i \end{bmatrix}, \\ \boldsymbol{D}_{i+1}(\omega^2) &\doteq \boldsymbol{Z}_i - \boldsymbol{Y}_i^T\boldsymbol{X}_i^{-1}\boldsymbol{Y}_i. \end{aligned}\right\} \tag{2.1.6}$$

In this way $\eta(\boldsymbol{D}_1(\omega^2))$, for the full order matrix $\boldsymbol{D}_1(\omega^2)$, is related to a set of quantities $\eta(\boldsymbol{X}_i)$ for smaller order matrices $\boldsymbol{X}_i$. For convenience one generally uses for $\boldsymbol{X}_i$ the 11-element of the matrix $\boldsymbol{D}_i(\omega^2)$, i.e. a scalar. Thus, for the real, symmetric matrix $\boldsymbol{D}$ the number of eigenvalues which are smaller than ω^2 simply becomes the number of negative scalar quantities of the set $\{\boldsymbol{X}_i(\omega^2)\}$.

The proof of the negative eigenvalue theorem is based on the fact that the number of negative eigenvalues of a symmetric matrix is equal to the number of changes in the sign between consecutive leading principal minors, if one starts with the zeroth minor as positive. Note that the ith leading principal minor of a matrix $\boldsymbol{A}$ is the determinant arising from the first i rows and columns of $\boldsymbol{A}$. To verify (2.1.5) we write

$$\begin{bmatrix} \boldsymbol{X}_1 & \boldsymbol{Y}_1 \\ \boldsymbol{Y}_1^T & \boldsymbol{Z}_1 \end{bmatrix} = \begin{bmatrix} \mathbf{1}^{(x)} & 0 \\ \boldsymbol{Y}_1^T \boldsymbol{X}_1^{-1} & \mathbf{1}^{(z)} \end{bmatrix} \begin{bmatrix} \boldsymbol{X}_1 & \boldsymbol{Y}_1 \\ 0 & (\boldsymbol{Z}_1 - \boldsymbol{Y}_1^T \boldsymbol{X}_1^{-1} \boldsymbol{Y}_1 \end{bmatrix}, \tag{2.1.7}$$

where $\boldsymbol{X}_1^{-1}$ has been assumed to be not singular and where $\mathbf{1}^{(x)}$ and $\mathbf{1}^{(z)}$ are unit matrices of the same order as $\boldsymbol{X}_1$ and $\boldsymbol{Z}_1$, respectively. From the structure of the first factor on the right hand side of (2.1.7) follows that the leading principal minors of the second factor are identical to those of $\boldsymbol{D}_1(\omega^2)$. Assuming for simplicity $\boldsymbol{X}_1$ to be a scalar and accounting for the decomposition procedure (2.1.6), the result (2.1.5) becomes evident. For a rigorous proof of (2.1.5) the reader is referred to DEAN (1972).

A calculation of selected eigenvectors of normal modes requires an accurate evaluation of the corresponding eigenvalues. Such an eigenvalue can be obtained by means of the method described above by using the property that $\eta(\boldsymbol{D}_1(\omega^2))$ changes by 1 as ω^2 passes through an exact eigenfrequency ω_s^2. Starting with a suitable frequency interval around ω_s one can bracket ω_s by successively bisecting this interval and by investigating the change of $\eta(\boldsymbol{D}_1(\omega^2))$. Let ω' be the computed approximation to ω_s, the eigenvalue whose vector we wish to determine. This eigenvector can be obtained by the method of inverse iteration which consists in an iteration of the equation

$$(\boldsymbol{D} - \omega'^2 1)\, \boldsymbol{v}_k = \tilde{\boldsymbol{v}}_{k-1}, \qquad k = 1, 2, \ldots. \tag{2.1.8}$$

Here

$$\tilde{\boldsymbol{v}}_k = \boldsymbol{v}_k / \max(\boldsymbol{v}_k), \tag{2.1.9}$$

where max $(\boldsymbol{v}_k)$ denotes the element of the largest modulus in $\boldsymbol{v}_k$. The initial trial vector $\tilde{\boldsymbol{v}}_0$ can be chosen arbitrarily or by physical intuition. On representing $\tilde{\boldsymbol{v}}_0$ by the true eigenvectors $\boldsymbol{x}_l$ $(l = 1, \ldots, dN)$ of $\boldsymbol{D}$ as

$$\tilde{\boldsymbol{v}}_0 = \sum_{l=1}^{dN} b_l \boldsymbol{x}_l, \tag{2.1.10}$$

then, up to a normalization factor, we have after k iterations

$$\tilde{\boldsymbol{v}}_k = \sum_{l=1}^{dN} b_s (\omega_l^2 - \omega'^2)^{-k}\, \boldsymbol{x}_l. \tag{2.1.11}$$

Thus, if ω' is much closer to one eigenvalue ω_s than to any other, the component of $\boldsymbol{x}_i$ $(i \neq s)$ in $\tilde{\boldsymbol{v}}_k$ decreases rapidly with increasing k and a rapid convergence to $\boldsymbol{x}_l$ occurs.

Numerically calculated vibrational squared spectra for two component mass disordered linear chains with nearest-neighbour interactions are shown in figures 2.1 and 2.2. To graphical accuracy, at any rate, the curve for the 32000-atom chain (fig. 2.1a)) agrees with that for a 250000-atom chain (fig. 2.1b)) and it may therefore be regarded as having converged to the form of a

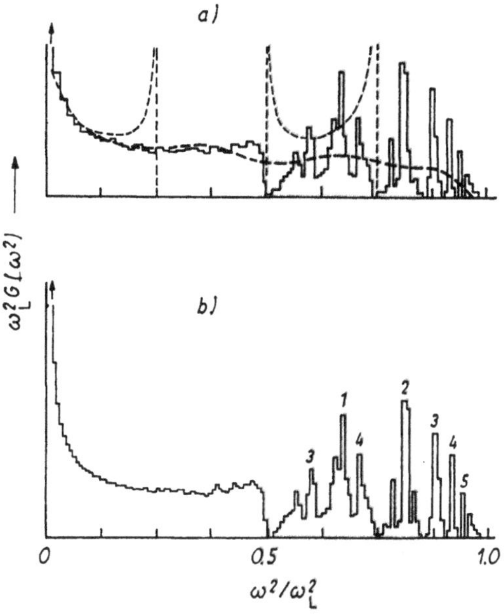

Fig. 2.1. Computed squared frequency spectrum for a mass disordered two-component chain with an equal number of light and heavy atoms, a mass ratio between the atomic species 2 : 1 and a length of 32000 atoms in a) and 250000 atoms in b). In a) the broken curve gives the result of a moment trace calculation and the dotted curve gives the squared frequency spectrum of an ordered chain. In b) above the individual peaks, arising from clusters of light atoms, the sizes of these clusters are designated. ω_L is the maximum frequency of a lattice containing light atoms only (after DEAN, 1972)

chain with an infinite number of atoms. On comparing these curves with the corresponding one for a 8000-atom chain (fig. 2.2) we see that even the latter reproduces all the essential spectral features quite well. In figure 2.1a) we have also depicted the results of a moment-trace calculation of the state density which is based on the first seven moments obtained by (1.4.28) from the dynamical matrix $\boldsymbol{D}$. We see that such a calculation cannot reproduce the true frequency spectra because of the low number of polynomials accounted for. From figures 2.1 and 2.2 we see that the squared frequency spectrum of a disordered chain differs considerably from the simple U-shaped spectrum of the regulár monatomic chain or the double U-shaped spectrum of regular diatomic chain, the latter is shown for comparison in figure 2.1a). From an inspection of numerically calculated eigenvectors the peaks in the high frequency region can be shown to be due to highly localized vibrations of clusters of one, two, three, ... light atoms bounded by heavy atoms. The corresponding peaks are labeled in figure 2.1b) by '1', '2', '3', ..., respectively. The modes in the low and medium frequency region turn out to be much more extended

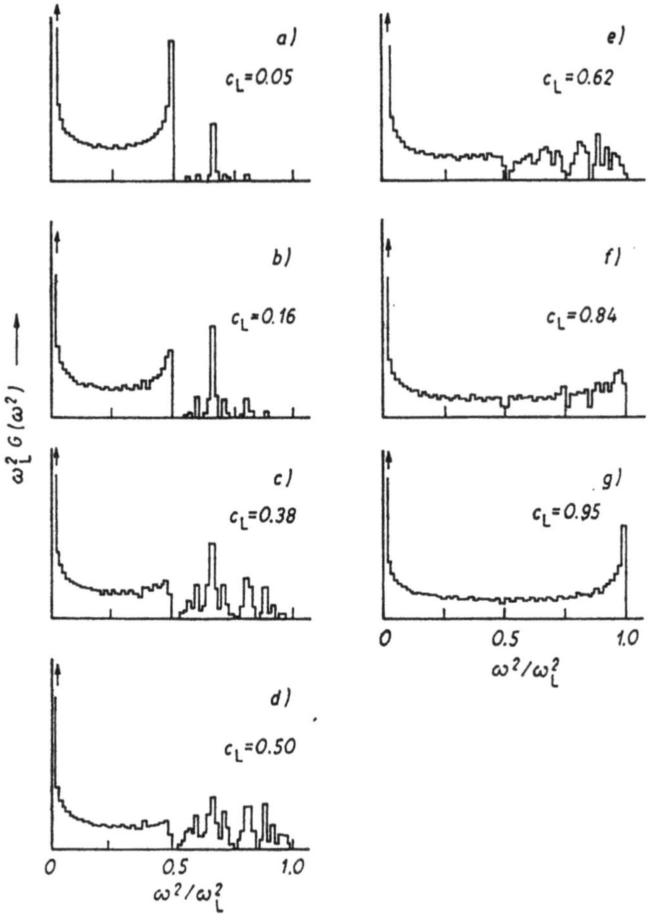

Fig. 2.2. Computed squared frequency spectrum for a mass disordered two-component chain with a mass ratio between the atomic species 2:1 and a length of 8000 atoms. c_L denotes the concentration of the light atoms (after DEAN, 1972)

than their high frequency counterparts. Further numerical results, in particular in the context with the localization problem, are described and discussed in section 3.

Another important method for studying lattice dynamics is the molecular-dynamics technique (RAHMAN, 1964, VERLET, 1967). In this technique one solves the Newton equations of motion using a computer for a finite number of atoms, with periodic boundary conditions, for a given interaction potential between the atoms. Assuming the system is ergodic, estimates for microcanonical ensemble averages may be obtained from time averages. A canonical ensemble can be simulated by considering, in place of the Newton equations, a coupled set of Langevin equations. In this approach one assumes that the atoms suffer collisions with much lighter particles which represent the heat bath defining the temperature (SCHNEIDER and STOLL, 1978). The molecular-dynamics technique has the advantage that anharmonicity can be included exactly. For an application of this technique to lattice dynamics of a disordered system see, for instance, JACUCCI et al. (1978).

The equation-of-motion method is related to the molecular-dynamics technique (BEEMAN and ALBEN, 1977, and references therein). It is a very effective method for studying harmonic disordered systems (see section 3.3.1.).

2.1.3. Green functions and multiple scattering theory in lattice dynamics

In a compositionally disordered systems the atoms are generally assumed to be placed randomly at the sites of a regular lattice, i.e. at $\boldsymbol{x}(l\varkappa) = \boldsymbol{x}(l) + \boldsymbol{x}(\varkappa)$, where $\boldsymbol{x}(l)$ labels the position of the lth unit cell and $\boldsymbol{x}(\varkappa)$ the position of the $\varkappa$th site within a cell. In the harmonic approximation the Hamiltonian of such a lattice is given by (cf. (1.3.1))

$$H = \sum_{l\varkappa\alpha} \frac{p_\alpha^2(l\varkappa)}{2M(l\varkappa)} + \frac{1}{2} \sum_{ll'\varkappa\varkappa'\alpha\alpha'} \Phi_{\alpha\alpha'}(l\varkappa, l'\varkappa')\, u_\alpha(l\varkappa)\, u_{\alpha'}(l'\varkappa'), \tag{2.1.12}$$

where $p_\alpha(l\varkappa)$ and $u_\alpha(l\varkappa)$ are, respectively, the cartesian coordinates of the momentum and displacement operators of the atom at site $(l\varkappa)$, and the M's and Φ's denote the masses of the atoms and the force constants, respectively. Note that in a compositionally disordered system, in general, the mass depends on l, and that Φ is not translationally invariant, unlike the situation in an ordered reference system described by the Hamiltonian

$$H^0 = \sum_{l\varkappa\alpha} \frac{p_\alpha^2(l\varkappa)}{2M^0(\varkappa)} + \frac{1}{2} \sum_{ll'\varkappa\varkappa'\alpha\alpha'} \Phi^0_{\alpha\alpha'}(l\varkappa, l'\varkappa')\, u_\alpha(l\varkappa)\, u_{\alpha'}(l'\varkappa'), \tag{2.1.13}$$

with masses $M^0(\varkappa)$ and force constants $\Phi^0_{\alpha\alpha'}(l\varkappa, l'\varkappa') = \Phi^0_{\alpha\alpha'}(0\varkappa, l' - l\varkappa')$.

Temperature-dependent double time Green functions are very suitable for evaluating thermodynamically averaged physical quantities. Most of these quantities contain the displacement-displacement correlation function (cf. appendix 3)

$$J_{\alpha\alpha'}(l\varkappa, l'\varkappa', \omega) = \int_{-\infty}^{\infty} \mathrm{d}t\, \mathrm{e}^{i\omega t} \langle u_\alpha(l\varkappa, t)\, u_{\alpha'}(l'\varkappa', 0)\rangle, \tag{2.1.14}$$

where

$$u_\alpha(l\varkappa, t) = \mathrm{e}^{\frac{i}{\hbar}Ht}\, u_\alpha(l\varkappa)\, \mathrm{e}^{-\frac{i}{\hbar}Ht}, \tag{2.1.15}$$

and

$$\langle\ldots\rangle = \mathrm{Tr}\,\{\mathrm{e}^{-\beta H}\ldots\}/\mathrm{Tr}\,\{\mathrm{e}^{-\beta H}\}, \qquad \beta = 1/kT. \tag{2.1.16}$$

As will be shown in the following the displacement-displacement correlation function can be readily obtained from the displacement-displacement double time Green function which is defined as

$$G_{\alpha\alpha'}(l\varkappa, l'\varkappa', \omega) = \int_{-\infty}^{\infty} \mathrm{d}t \, \mathrm{e}^{i\omega t} \langle\langle u_\alpha(l\varkappa, t); u_{\alpha'}(l'\varkappa', 0)\rangle\rangle, \tag{2.1.17}$$

where

$$\langle\langle A(t); B(t')\rangle\rangle = \mp \frac{i}{\hbar} \, \Theta[\pm(t - t')] \, \langle [A(t), B(t')]\rangle. \tag{2.1.18}$$

Here the upper and lower signs respectively give the retarded and advanced Green function with $\Theta(t)$ being the Heaviside step function. A and B are arbitrary operators. Because of (2.1.18), ω may be replaced in (2.1.17) by a complex frequency z with $\mathrm{Im}\, z > 0$ ($\mathrm{Im}\, z < 0$) for the retarded (advanced) Green function, i.e. the retarded (advanced) Green function is analytic in the upper (lower) complex frequency half-plane. Of both these analytic parts of the Green functions a new Green function can be composed which is analytic in the whole complex frequency plane, except on the real axis. This Green function is related to the correlation function (2.1.14) via the spectral representation

$$G_{\alpha\alpha'}(l\varkappa, l'\varkappa', z) = \frac{1}{2\pi\hbar} \int_{-\infty}^{\infty} \mathrm{d}\omega' \, \frac{(1 - \mathrm{e}^{-\beta\hbar\omega'}) \, J_{\alpha\alpha'}(l\varkappa, l'\varkappa', \omega')}{z - \omega'}, \tag{2.1.19}$$

as can be shown by using (2.1.17), (2.1.18), (2.1.14) and the property

$$\langle A(t) \, B(t')\rangle = \langle B(t') \, A(t + i\hbar\beta)\rangle \tag{2.1.20}$$

following from the cyclic property of the trace [$\mathrm{Tr}\,(ABC) = \mathrm{Tr}\,(BCA) = \mathrm{Tr}\,(CAB)$]. From (2.1.19) and by means of the identity

$$\frac{1}{\omega - \omega' \pm i\varepsilon} = \frac{P}{\omega - \omega'} \mp i\pi\delta(\omega - \omega') \tag{2.1.21}$$

(P — principal part, $\varepsilon \to +0$) the correlation function (2.1.14) can be found to be given by the discontinuity of the Green function across the real axis

$$\begin{aligned} &G_{\alpha\alpha'}(l\varkappa, l'\varkappa', \omega + i\varepsilon) - G_{\alpha\alpha'}(l\varkappa, l'\varkappa', \omega - i\varepsilon) \\ &= -\frac{i}{\hbar} (1 - \mathrm{e}^{-\beta\hbar\omega}) \, J_{\alpha\alpha'}(l\varkappa, l'\varkappa', \omega) \\ &= 2i \, \mathrm{Im} \, G_{\alpha\alpha'}(l\varkappa, l'\varkappa', \omega + i\varepsilon), \qquad \varepsilon \to +0. \end{aligned} \tag{2.1.22}$$

The last line of this equation will be proved somewhat later (using (2.1.43)).

When we differentiate equation (2.1.18) with respect to time, we obtain the equation of motion of the Green function

$$i\hbar \frac{\mathrm{d}}{\mathrm{d}t} \langle\langle A(t); B(t')\rangle\rangle = \delta(t - t') \, \langle [A(t), B(t)]\rangle + \langle\langle [A(t), H]; B(t')\rangle\rangle. \tag{2.1.23}$$

In order to get a closed equation for the displacement-displacement Green function we have to apply the equation of motion (2.1.23) twice and to account for the relations (cf. (1.3.5), (1.3.6))

$$[u_\alpha(l\varkappa), H] = i\hbar p_\alpha(l\varkappa)/M(l\varkappa) \tag{2.1.24}$$

and

$$[p_\alpha(l\varkappa), H] = -i\hbar \sum_{l'\varkappa'\alpha'} \Phi_{\alpha\alpha'}(l\varkappa, l'\varkappa')\, u_{\alpha'}(l'\varkappa'), \tag{2.1.25}$$

which can be obtained by using the Hamiltonian (2.1.12) and the commutation relations (cf. (1.3.7))

$$[u_\alpha(l\varkappa), p_{\alpha'}(l'\varkappa')] = i\hbar\delta_{ll'}\delta_{\varkappa\varkappa'}\delta_{\alpha\alpha'},\ [u_\alpha(l\varkappa), u_{\alpha'}(l'\varkappa')] = [p_\alpha(l\varkappa), p_{\alpha'}(l'\varkappa')] = 0. \tag{2.1.26}$$

On Fourier transforming the resulting equation for the Green function with respect to time we obtain

$$\sum_{l''\varkappa''\alpha''} [M(l\varkappa)\,\omega^2\delta_{ll''}\delta_{\varkappa\varkappa''}\delta_{\alpha\alpha''} - \Phi_{\alpha\alpha''}(l\varkappa, l''\varkappa'')] \times G_{\alpha''\alpha'}(l''\varkappa'', l'\varkappa', \omega) = \delta_{ll'}\delta_{\varkappa\varkappa'}\delta_{\alpha\alpha'}. \tag{2.1.27}$$

Condensed to matrix form this equation reads

$$[\boldsymbol{M}\omega^2 - \boldsymbol{\Phi}]\, \boldsymbol{G}(\omega) = \boldsymbol{1} \tag{2.1.28}$$

or

$$\boldsymbol{G}^{-1}(\omega) = \boldsymbol{M}\omega^2 - \boldsymbol{\Phi}. \tag{2.1.29}$$

For the Green function of a perfect crystal, described by the Hamiltonian H^0 from (2.1.13), a similar equation holds

$$\left(\boldsymbol{G}^0(\omega)\right)^{-1} = \boldsymbol{M}^0\omega^2 - \boldsymbol{\Phi}^0. \tag{2.1.30}$$

Introducing the perturbation matrix

$$\boldsymbol{V}(\omega) = (\boldsymbol{M}^0 - \boldsymbol{M})\,\omega^2 + \boldsymbol{\Phi} - \boldsymbol{\Phi}^0, \tag{2.1.31}$$

which in terms of components is suitably rewritten as

$$V_{\alpha\alpha'}(l\varkappa, l'\varkappa', \omega) = M^0(\varkappa)\,\omega^2\epsilon(l\varkappa)\,\delta_{ll'}\delta_{\varkappa\varkappa'}\delta_{\alpha\alpha'} + \Delta\Phi_{\alpha\alpha'}(l\varkappa, l'\varkappa') \tag{2.1.32}$$

with

$$\epsilon(l\varkappa) = \left(M^0(\varkappa) - M(l\varkappa)\right)/M^0(\varkappa) \tag{2.1.33}$$

and

$$\Delta\Phi_{\alpha\alpha'}(l\varkappa, l'\varkappa') = \Phi_{\alpha\alpha'}(l\varkappa, l'\varkappa') - \Phi^0_{\alpha\alpha'}(l\varkappa, l'\varkappa'), \tag{2.1.34}$$

we obtain from (2.1.29) to (2.1.31) the following Dyson equation relating $\boldsymbol{G}$ and $\boldsymbol{G}^0$:

$$\boldsymbol{G}(\omega) = \boldsymbol{G}^0(\omega) + \boldsymbol{G}^0(\omega)\,\boldsymbol{V}(\omega)\,\boldsymbol{G}(\omega). \tag{2.1.35}$$

To derive the spectral representation of $\boldsymbol{G}$ we write (2.1.28) as

$$[\boldsymbol{1}\omega^2 - \boldsymbol{D}]\,\boldsymbol{M}^{1/2}\boldsymbol{G}\boldsymbol{M}^{1/2} = \boldsymbol{1} \tag{2.1.36}$$

where $\boldsymbol{D}$ is the dynamical matrix (cf. eq. (1.3.10))

$$\boldsymbol{D} = \boldsymbol{M}^{-1/2}\boldsymbol{\Phi}\boldsymbol{M}^{-1/2}, \tag{2.1.37}$$

which can be diagonalized by a unitary matrix, whose sth row is given by $B_\alpha^{(s)}(l\varkappa)$ which is the sth eigenvector of $\boldsymbol{D}$ (cf. section 1.3.). By making use of this unitary matrix, from (2.1.36) we can readily obtain the following spectral representation:

$$G_{\alpha\alpha'}(l\varkappa, l'\varkappa', \omega) = \frac{1}{\sqrt{M(l\varkappa)\,M(l'\varkappa')}} \sum_s \frac{B_\alpha^{(s)}(l\varkappa)\,B_{\alpha'}^{(s)}(l'\varkappa')}{\omega^2 - \omega_s^2}, \tag{2.1.38}$$

where, as in section 1.3., the eigenvectors of $\boldsymbol{D}$ have been assumed to be real and ω_s^2 denotes the sth eigenvalue of $\boldsymbol{D}$. Quite similarly, we obtain also the spectral representation of $\boldsymbol{G}^0$. Note that the row $(\boldsymbol{k}j)$ ($\boldsymbol{k}$ — wave vector, j — branch index) of the unitary matrix diagonalizing the dynamical matrix of the perfect crystal $\boldsymbol{D}^0$ is given by $\frac{1}{\sqrt{N}}\, e_\alpha(\varkappa \mid \boldsymbol{k}j) \exp\big(i\boldsymbol{k}\boldsymbol{x}(l)\big)$ (cf. (1.3.31), (1.3.41)), which is the $(\boldsymbol{k}j)$th eigenvector of $\boldsymbol{D}^0$. Thus we obtain

$$G^0_{\alpha\alpha'}(l\varkappa, l'\varkappa', \omega) = \frac{1}{\sqrt{M^0(\varkappa)\,M^0(\varkappa')}}\,\frac{1}{N} \sum_{\boldsymbol{k}j} \frac{e_\alpha(\varkappa \mid \boldsymbol{k}j)\, e^*_{\alpha'}(\varkappa' \mid \boldsymbol{k}j)}{\omega^2 - \omega_j^2(\boldsymbol{k})}\, e^{i\boldsymbol{k}(\boldsymbol{x}(l) - \boldsymbol{x}(l'))}, \tag{2.1.39}$$

where $\omega_j^2(\boldsymbol{k})$ is the $(\boldsymbol{k}j)$th eigenvalue of $\boldsymbol{D}^0$.

According to the prescription given above, the retarded Green functions $\boldsymbol{G}$ and $\boldsymbol{G}^0$ are obtained by replacing ω by $\omega + i\varepsilon$, $\varepsilon \to +0$, in (2.1.38) and (2.1.39), respectively. As can be seen from the resulting expressions, the eigenfrequencies give rise to poles in the retarded Green function lying in the lower complex frequency half plane near the real axis.

The phonon density of states of a perfect crystal is given by (cf. (1.4.11), (1.4.13), (1.4.15))

$$g^0(\omega) = \frac{2\omega}{3rN} \sum_{\boldsymbol{k}j} \delta\big(\omega^2 - \omega_j^2(\boldsymbol{k})\big) \qquad (\omega > 0). \tag{2.1.40}$$

On account of (1.3.44) and (2.1.21) we obtain from (2.1.39) and (2.1.40)

$$\begin{aligned} g^0(\omega) &= -\frac{2\omega}{3r\pi N} \operatorname{Im} \operatorname{Tr} (\boldsymbol{M}^0)^{1/2}\, \boldsymbol{G}^0(\omega + i\varepsilon)\, (\boldsymbol{M}^0)^{1/2} \\ &= -\frac{2\omega}{3r\pi N} \operatorname{Im} \operatorname{Tr} \boldsymbol{M}^0\boldsymbol{G}^0(\omega + i\varepsilon), \qquad \varepsilon \to +0. \end{aligned} \tag{2.1.41}$$

Quite similarly, from (2.1.38) we get the phonon density of states of a disordered system

$$\begin{aligned} g(\omega) &= -\frac{2\omega}{3r\pi N} \operatorname{Im} \operatorname{Tr} \boldsymbol{M}^{1/2}\boldsymbol{G}(\omega + i\varepsilon)\, \boldsymbol{M}^{1/2} \\ &= -\frac{2\omega}{3r\pi N} \operatorname{Im} \operatorname{Tr} \boldsymbol{M}\boldsymbol{G}(\omega + i\varepsilon), \qquad \varepsilon \to +0. \end{aligned} \tag{2.1.42}$$

Thus, when the mass varies from site to site, the phonon density of states is not given directly by $\boldsymbol{G}$ or $\boldsymbol{G}^0$, respectively, but by the imaginary part of the corresponding mass weighted Green function.

In the following we summarize some general properties of the displacement-displacement Green function.

Since the displacement operators are even under time reversal, the displacement-displacement correlation function $\boldsymbol{J}(\omega)$ (2.1.14) is a real quantity. Hence from (2.1.19) we obtain

$$\boldsymbol{G}^*(\omega + i\varepsilon) = \boldsymbol{G}(\omega - i\varepsilon), \tag{2.1.43}$$

from which it follows that the discontinuity of the Green function across the real axis (2.1.22) is equal to $2i \operatorname{Im} \boldsymbol{G}(\omega + i\varepsilon)$.

On replacing z by $\omega \pm i\varepsilon$ in (2.1.19), and taking into account (2.1.21) and the fact that $\boldsymbol{J}(\omega)$ is a real quantity we obtain the following relation (Kramers-Kroning relation) between the real and the imaginary part of the Green function:

$$\left.\begin{aligned} \operatorname{Re} \boldsymbol{G}(\omega) &= \pm \frac{1}{\pi} P \int_{-\infty}^{\infty} \mathrm{d}\omega' \, \frac{\operatorname{Im} \boldsymbol{G}(\omega')}{\omega' - \omega} \\ &= \pm \frac{2}{\pi} P \int_{0}^{\infty} \mathrm{d}\omega' \, \frac{\omega' \operatorname{Im} \boldsymbol{G}(\omega')}{(\omega')^2 - \omega^2}, \end{aligned}\right\} \tag{2.1.44}$$

where the upper sign holds for the retarded and the lower sign for the advanced Green function. The last expression in (2.1.44) was obtained by using

$$\boldsymbol{J}(-\omega) = \mathrm{e}^{-\hbar\omega\beta} \boldsymbol{J}(\omega), \tag{2.1.45}$$

which holds for real $\boldsymbol{J}(\omega)$.

The Green function $\boldsymbol{G}$ can be shown to satisfy the following sum rule, which is of use in optical studies:

$$\left(\frac{2}{\pi}\right) \int_{0}^{\infty} \mathrm{d}\omega \, \omega \operatorname{Im} \boldsymbol{G}(\omega) \, \boldsymbol{M} = \mp 1, \tag{2.1.46}$$

where the upper sign holds for the retarded and the lower sign for the advanced Green function. This sum rule can be derived either by making use of (2.1.38) or of (2.1.17) (integrating the latter equation by parts with respect to time). By the latter method this sum rule can be shown to hold also for anharmonic systems as only the kinetic part of the Hamiltonian is used. Note that on deriving (2.1.46) we have taken into account that $\operatorname{Im} \boldsymbol{G}(\omega)$ is an odd function of ω, which follows from (2.1.45).

As pointed out above, the eigenfrequencies of the lattice are determined by the poles of the Green function $\boldsymbol{G}$, or correspondingly by the zeros of $\det |\boldsymbol{G}^{-1}|$. When we are interested only in the frequencies which are affected by defects (disorder), then we have to solve

$$\Delta(\omega) \equiv \det |\mathbf{1} - \boldsymbol{G}^0(\omega)\, \boldsymbol{V}| = 0, \tag{2.1.47}$$

where $\boldsymbol{G}^0$ is the Green function in the absence of disorder and $\boldsymbol{V}$ is the perturbation matrix (2.1.31). To see this, we rewrite (2.1.35) as

$$\boldsymbol{G}^{-1} = (\boldsymbol{G}^0)^{-1} [\mathbf{1} - \boldsymbol{G}^0\boldsymbol{V}]. \tag{2.1.48}$$

From this equation and (2.1.29), (2.1.30), (2.1.37), and (2.1.47) we find

$$\Delta(\omega) = \frac{\det |\boldsymbol{G}^{-1}|}{\det |(\boldsymbol{G}^0)^{-1}|} = \frac{\det |\boldsymbol{M}^{1/2}(\omega^2\mathbf{1} - \boldsymbol{D})\, \boldsymbol{M}^{1/2}|}{\det |(\boldsymbol{M}^0)^{1/2}\, (\omega^2\mathbf{1} - \boldsymbol{D}^0)\, (\boldsymbol{M}^0)^{1/2}|}$$

$$= \frac{\det |\boldsymbol{M}|}{\det |\boldsymbol{M}^0|} \prod_s \frac{\omega^2 - \omega_s{}^2}{\omega^2 - (\omega_s{}^0)^2}, \tag{2.1.49}$$

where ω_s and $\omega_s{}^0$ denote the eigenfrequencies of the crystal with and without disorder, respectively. From (2.1.49) we see that the normal modes which are unaffected by the disorder cancel in the numerator and the denominator of the quotient. Solely the modes affected by the disorder contribute to the zeros of the determinant $\Delta(\omega)$. The order of $\Delta(\omega)$ is governed by the order of $\boldsymbol{V}$. In some cases the order of $\boldsymbol{V}$ is small, for instance, in the case of a single defect, especially in the absence of force constant changes. A calculation of the perturbed frequencies by (2.1.47) then becomes a relatively simple task.

Now we introduce the t-matrix, $\boldsymbol{T}$, by

$$\boldsymbol{G}(\omega) = \boldsymbol{G}^0(\omega) + \boldsymbol{G}^0(\omega)\, \boldsymbol{T}(\omega)\, \boldsymbol{G}^0(\omega). \tag{2.1.50}$$

When we express $\boldsymbol{T}$ by $\boldsymbol{V}$ by means of (2.1.35) and (2.1.50),

$$\boldsymbol{T} = \boldsymbol{V}[\mathbf{1} - \boldsymbol{G}^0\boldsymbol{V}]^{-1}, \tag{2.1.51}$$

and compare (2.1.51) with (2.1.48), we see that the poles of $\boldsymbol{T}$ determine the eigenfrequencies which are affected by introducing disorder.

We calculate $\boldsymbol{T}$ using multiple scattering theory (cf. VELICKY et al., 1968). To do this we assume that the perturbation matrix $\boldsymbol{V}$ can be decomposed into a sum of individual perturbation matrices as

$$\boldsymbol{V} = \sum_i \boldsymbol{V}^i, \tag{2.1.52}$$

where $\boldsymbol{V}^i$ denotes the perturbation due to a defect at site $(l_i\varkappa_i)$. This decomposition will be discussed below in more detail. According to (2.1.50) the t-matrix $\boldsymbol{T}$ can be regarded as describing the scattering of an undistorted lattice wave, represented by $\boldsymbol{G}^0$, by defects with the perturbation matrix $\boldsymbol{V}$, so that a distorted lattice wave $\boldsymbol{G}$ results. When $\boldsymbol{V}$ has the form (2.1.52), we obtain with (2.1.51)

$$\boldsymbol{T} = [\mathbf{1} + \boldsymbol{T}\boldsymbol{G}^0]\, \boldsymbol{V} = \sum_i [\mathbf{1} + \boldsymbol{T}\boldsymbol{G}^0]\, \boldsymbol{V}^i \equiv \sum_i \boldsymbol{T}^i. \tag{2.1.53}$$

Thus, in this case $\boldsymbol{T}$ becomes a sum of contributions which can be ascribed to individual defects. On replacing $\boldsymbol{T}$ by $\sum_i \boldsymbol{T}^i$ in the third expression in (2.1.53) then the contribution of the ith defect, $\boldsymbol{T}^i$, takes the form

$$\boldsymbol{T}^i = \Big[1 + \sum_{i'(\neq i)} \boldsymbol{T}^{i'}\boldsymbol{G}^0\Big]\, \boldsymbol{t}^i, \tag{2.1.54}$$

where

$$\boldsymbol{t}^i \doteq \boldsymbol{V}^i[1 - \boldsymbol{G}^0\boldsymbol{V}^i]^{-1}. \tag{2.1.55}$$

The expression in the brackets in equation (2.1.54) describes the effect of all defects, besides the ith, on the lattice wave running to the ith defect. The atomic t-matrix, $\boldsymbol{t}^i$, describes the complete scattering of a wave by the ith defect.

As mentioned above, in the case of a disordered system we have to calculate the configuration average $\langle \boldsymbol{G}\rangle$ of the Green function $\boldsymbol{G}$. According to (2.1.50) this task is equivalent to calculating the average t-matrix $\langle \boldsymbol{T}\rangle$, because $\boldsymbol{G}^0$ is, by definition, unaffected by the averaging procedure. We have

$$\langle \boldsymbol{G}\rangle = \boldsymbol{G}^0 + \boldsymbol{G}^0\langle \boldsymbol{T}\rangle\, \boldsymbol{G}^0. \tag{2.1.56}$$

Furthermore, on account of (2.1.53) a calculation of $\langle \boldsymbol{T}\rangle$ requires that we calculate $\langle \boldsymbol{T}^i\rangle$, because

$$\langle \boldsymbol{T}\rangle = \sum_i \langle \boldsymbol{T}^i\rangle. \tag{2.1.57}$$

On averaging equation (2.1.54) we obtain

$$\begin{aligned}\langle \boldsymbol{T}^i\rangle &= \Big\langle\Big[1 + \sum_{i'(\neq i)} \boldsymbol{T}^{i'}\boldsymbol{G}^0\Big]\, \boldsymbol{t}^i\Big\rangle \\ &= \Big[1 + \sum_{i'(\neq i)} \langle \boldsymbol{T}^{i'}\rangle\, \boldsymbol{G}^0\Big] \langle \boldsymbol{t}^i\rangle + \Big\langle \sum_{i'(\neq i)} (\boldsymbol{T}^{i'} - \langle \boldsymbol{T}^{i'}\rangle)\, (\boldsymbol{t}^i - \langle \boldsymbol{t}^i\rangle)\Big\rangle. \end{aligned}\tag{2.1.58}$$

The first term in the last form of equation (2.1.58) describes the scattering of an average incident wave by an atom with an average atomic t-matrix. The second term describes the correlation between fluctuations in the incident wave and the atomic t-matrix.

Now we define a self-energy $\boldsymbol{\Sigma}$ by

$$\langle \boldsymbol{G}\rangle = \boldsymbol{G}^0 + \boldsymbol{G}^0\boldsymbol{\Sigma}\langle \boldsymbol{G}\rangle. \tag{2.1.59}$$

$\boldsymbol{\Sigma}$ has translational symmetry. In general $\boldsymbol{\Sigma}$ is complex and frequency dependent. It plays the role of an additional effective mass or force constant in the averaged system. Once $\boldsymbol{\Sigma}$ is known, the averaged Green function $\langle \boldsymbol{G}\rangle$ can be readily obtained from (2.1.59). Using (2.1.56) and (2.1.59) the following relation between $\boldsymbol{\Sigma}$ and $\langle \boldsymbol{T}\rangle$ results:

$$\boldsymbol{\Sigma} = [1 + \langle \boldsymbol{T}\rangle\, \boldsymbol{G}^0]^{-1}\, \langle \boldsymbol{T}\rangle. \tag{2.1.60}$$

A relatively simple approximation for calculating $\langle \boldsymbol{T}\rangle$ consists in a complete neglection of the correlation term in the last form of equation (2.1.58). This is a so-called single-site approximation (SSA). Using this approximation we obtain from (2.1.58) and (2.1.57)

$$\langle \boldsymbol{T} \rangle = [1 + \langle \boldsymbol{T} \rangle \, \boldsymbol{G}^0] \sum_i \langle \boldsymbol{t}^i \rangle \, [1 + \boldsymbol{G}^0 \langle \boldsymbol{t}^i \rangle]^{-1}. \tag{2.1.61}$$

Inserting (2.1.61) into (2.1.60) we get for $\boldsymbol{\Sigma}$ in SSA

$$\boldsymbol{\Sigma} = \sum_i \langle \boldsymbol{t}^i \rangle \, [1 + \boldsymbol{G}^0 \langle \boldsymbol{t}^i \rangle]^{-1}. \tag{2.1.62}$$

We discuss this expression somewhat later.

Now we turn to the decomposition (2.1.52) of the perturbation matrix. First we consider a cubic system with pure mass disorder. In this case, in (2.1.52) $\boldsymbol{V}^i$ reads

$$V_{\alpha\alpha'}^{(l_i\varkappa_i)}(l\varkappa, l'\varkappa', \omega) = M^0(\varkappa) \, \epsilon(l\varkappa) \, \omega^2 \delta_{l_i l} \delta_{\varkappa_i \varkappa} \delta_{ll'} \delta_{\varkappa\varkappa'} \delta_{\alpha\alpha'}, \tag{2.1.63}$$

where $\epsilon(l\varkappa)$ is defined by (2.1.33).

It is convenient to write $\epsilon(l\varkappa)$ as

$$\epsilon(l\varkappa) = \sum_j \eta^j_{(l\varkappa)} \epsilon^j(\varkappa), \tag{2.1.64}$$

where

$$\epsilon^j(\varkappa) = \left(M^0(\varkappa) - M^j\right) / M^0(\varkappa) \tag{2.1.65}$$

and

$$\eta^j_L = \begin{cases} 1 \text{ if an atom of kind } j \text{ is placed on site } L = (l\varkappa), \\ 0 \text{ otherwise.} \end{cases} \tag{2.1.66}$$

Here M^j is the mass of an atom of kind j and η^j_L is the occupation number of site L.

The occupation numbers have the properties

$$\eta^j_L \eta^{j'}_L = \eta^j_L \delta_{jj'}, \qquad \sum_j \eta^j_L = 1, \qquad \sum_l \eta^j_{(l\varkappa)} = c_\varkappa{}^j N, \tag{2.1.67}$$

where $c_\varkappa{}^j$ denotes the concentration of atoms in the $\varkappa$th sublattice and N is the number of atoms in a sublattice. N is assumed to be independent of $\varkappa$. The configuration averaging acts on the occupation numbers, because these contain all the statistics of the atoms. Thus the result of an averaging process contains averages such as $\langle \eta^j_L \rangle$, $\langle \eta^j_L \eta^{j'}_{L'} \rangle$, $\langle \eta^j_L \eta^{j'}_{L'} \eta^{j''}_{L''} \rangle$ etc., which characterize the occupation probabilities of one, two, three, etc. sites of the lattice (provided that in the latter expressions the site indices differ pairwise from each other). These quantities represent the one-, two-, three-particle correlation functions, respectively, and, at least in principle, they can be taken from experimental results. In particular, we have

$$\langle \eta^j_L \rangle = c_\varkappa{}^j, \tag{2.1.68}$$

and the two-particle correlation function can be determined by neutron, electron or, X-ray scattering experiments. At the end of this section we will make some further remarks on averages of occupation numbers.

On inserting (2.1.64) into (2.1.63) we obtain

$$V^{(l_t\varkappa_t)} = \sum_j \eta^j_{(l_t\varkappa_t)} V^{j(l_t\varkappa_t)}, \tag{2.1.69}$$

where

$$V^{j(l_t\varkappa_t)}_{\alpha\alpha'}(l\varkappa, l'\varkappa', \omega) = M^0(\varkappa)\, \epsilon^j(\varkappa)\, \omega^2 \delta_{l_t l} \delta_{\varkappa_t\varkappa} \delta_{ll'} \delta_{\varkappa\varkappa'} \delta_{\alpha\alpha'}. \tag{2.1.70}$$

Expanding the inverse matrix in (2.1.55) formally as a power series with respect to G^0V^j and using (2.1.69) and (2.1.70), the atomic t-matrix becomes

$$t^{(l_t\varkappa_t)} = \sum_j \eta^j_{(l_t\varkappa_t)} t^{j(l_t\varkappa_t)}, \tag{2.1.71}$$

where

$$t^{j(l_t\varkappa_t)}_{\alpha\alpha'}(l\varkappa, l'\varkappa', \omega) = \delta_{l_t l} \delta_{\varkappa_t\varkappa} \delta_{ll'} \delta_{\varkappa\varkappa'} \delta_{\alpha\alpha'} t^j(\varkappa, \omega) \tag{2.1.72}$$

with

$$t^j(\varkappa, \omega) = \frac{M^0(\varkappa)\, \epsilon^j(\varkappa)\, \omega^2}{1 - M^0(\varkappa)\, \epsilon^j(\varkappa)\, \omega^2 G^0(\varkappa, \omega)}, \tag{2.1.73}$$

where

$$G^0(\varkappa, \omega) \equiv G^0_{\alpha\alpha}(0\varkappa, 0\varkappa, \omega). \tag{2.1.74}$$

Note that cubic structure symmetry ensures that $G^0_{\alpha\alpha'}(0\varkappa, 0\varkappa, \omega) = \delta_{\alpha\alpha'} G^0(\varkappa, \omega)$. When we average equation (2.1.71) and use (2.1.68), we obtain the average atomic t-matrix

$$\langle t^{(l_t\varkappa_t)} \rangle = \sum_j c^j_{\varkappa_t} t^{j(l_t\varkappa_t)}. \tag{2.1.75}$$

Hence, according to (2.1.62), (2.1.75), (2.1.72), and (2.1.73) for mass disorder the SSA expression of the self-energy reads

$$\Sigma_{\alpha\alpha'}(l\varkappa, l'\varkappa', \omega) = \delta_{ll'}\, \delta_{\varkappa\varkappa'}\, \delta_{\alpha\alpha'} \Sigma(\varkappa, \omega), \tag{2.1.76}$$

where

$$\Sigma(\varkappa, \omega) = \sum_j c_\varkappa^j t^j(\varkappa, \omega) \Big/ \Big[1 + G^0(\varkappa, \omega) \sum_j c_\varkappa^j t^j(\varkappa, \omega)\Big]. \tag{2.1.77}$$

In presence of force constant changes the situation is more complicated than for pure mass disorder. In the former case, owing to translational invariance, i.e. owing to the relation (cf. (1.2.7))

$$\sum_{l'\varkappa'} \Phi_{\alpha\alpha'}(l\varkappa, l'\varkappa') = 0, \tag{2.1.78}$$

changes in the off-diagonal coupling constants (off-diagonal disorder) are accompanied by changes in the site-diagonal terms (on-diagonal disorder) of the perturbation matrix. Hence, in general, the decomposition (2.1.52) of the perturbation matrix is valid only if the defect concentration is sufficiently low, in order that diagonal contributions to V at nondefect sites arising from (2.1.78) are connected essentially with a single defect. In other words, we must demand that the defect spaces, i.e. the regions of defect-affected atoms around the actual defects, rarely overlap. Furthermore, as will be shown in the following, the decomposition (2.1.52) of the perturbation matrix is also valid for any system whose force constants superimpose linearly (Mostoller and Kaplan, 1977).

As an example of such a system we consider a monatomic AB alloy with the force constants

$$\boldsymbol{\Phi}^{\mathrm{AB}}(ll') = \frac{1}{2}\,[\boldsymbol{\Phi}^{\mathrm{AA}}(ll') + \boldsymbol{\Phi}^{\mathrm{BB}}(ll')]. \tag{2.1.79}$$

Here $\boldsymbol{\Phi}^{\mathrm{AB}}(ll')$ denotes the force constant (matrix) between an atom A at site l and an atom B at l'. The force constants in the alloy differ from those in the reference crystal (A crystal) by (cf. (2.1.34))

$$\Delta\boldsymbol{\Phi}(ll') = \frac{1}{2}\,[\boldsymbol{\Phi}^{\mathrm{BB}}(ll') - \boldsymbol{\Phi}^{\mathrm{AA}}(ll')] \times \begin{cases} 1 \text{ if a B atom is at site } l \text{ or } l', \\ 2 \text{ if a B atom is at site } l \text{ and } l'. \end{cases} \tag{2.1.80}$$

Hence we can write

$$\Delta\boldsymbol{\Phi}(ll') = \sum_{l_i j} \eta^j_{l_i}\,\Delta\boldsymbol{\Phi}^{j(l_i)}(ll'), \tag{2.1.81}$$

where $j = \mathrm{A}, \mathrm{B}$ and

$$\Delta\boldsymbol{\Phi}^{j(l_i)}(ll') = \frac{1}{2}\,[\boldsymbol{\Phi}^{\mathrm{BB}}(ll') - \boldsymbol{\Phi}^{\mathrm{AA}}(ll')]\,[\delta_{ll_i} + \delta_{l'l_i}]\,\delta_{j\mathrm{B}}. \tag{2.1.82}$$

The corresponding site-diagonal force constant changes the result from (2.1.78) and (2.1.80). To illustrate this, let us consider a single B atom embedded in an environment of A atoms. On account of nearest-neighbour force constant changes only, we obtain

$$\Delta\boldsymbol{\Phi}(l'l') = -\Delta\boldsymbol{\Phi}(ll') = -\frac{1}{2}\,[\boldsymbol{\Phi}^{\mathrm{BB}}(ll') - \boldsymbol{\Phi}^{\mathrm{AA}}(ll')], \tag{2.1.83}$$

$$\Delta\boldsymbol{\Phi}(ll) = -\sum_{l'} \Delta\boldsymbol{\Phi}(ll'), \tag{2.1.84}$$

where l denotes the site of the B atom and l' the nearest neighbour sites of l. Note that these site-diagonal changes are at all sites in the defect space, not just at the defect site.

Allowing also for mass disorder in the AB alloy at hand, then the contribution of site l_i to the defect matrix reads

$$\boldsymbol{V}^{(l_i)} = \sum_j \eta^j_{l_i}\,\boldsymbol{V}^{j(l_i)}, \tag{2.1.85}$$

where

$$V^{j(l_i)}_{\alpha\alpha'}(ll', \omega) = \left\{M_{\mathrm{A}}\omega^2\epsilon^{\mathrm{B}}\delta_{l_il}\delta_{ll'}\delta_{\alpha\alpha'} + \frac{1}{2}\,[\Phi^{\mathrm{BB}}_{\alpha\alpha'}(ll') - \Phi^{\mathrm{AA}}_{\alpha\alpha'}(ll')]\,[\delta_{ll_i} + \delta_{l'l_i}]\right\}\delta_{jB}. \tag{2.1.86}$$

We can write

$$V^{j(l_i)}_{\alpha\alpha'}(ll', \omega) = \sum_{ss'} V^{j(0)}_{\alpha\alpha'}(ss', \omega)\, \delta_{l,s+l_i}\delta_{l',s'+l_i}. \tag{2.1.87}$$

Here s and s' denote the sites of the defect space around a defect at site 0, i.e. s and s' are the site indices defined relative to l_i. Using (2.1.55) and (2.1.87) the atomic t-matrix at site l_i can also be expressed by that at site 0:

$$\boldsymbol{t}^{(l_i)} = \sum_j \eta^j_{l_i} \boldsymbol{t}^{j(l_i)}, \tag{2.1.88}$$

where

$$t^{j(l_i)}_{\alpha\alpha'}(ll', \omega) = \sum_{ss'} t^{j(0)}_{\alpha\alpha'}(ss', \omega)\, \delta_{l,s+l_i}\delta_{l',s'+l_i}, \tag{2.1.89}$$

$$t^{j(0)}_{\alpha\alpha'}(ss', \omega) = \sum_{s''\alpha''} V^{j(0)}_{\alpha\alpha''}(ss'', \omega)\, X^{-1j}_{\alpha''\alpha}(s''s', \omega), \tag{2.1.90}$$

$$X^j_{\alpha\alpha'}(ss', \omega) = \delta_{ss'}\delta_{\alpha\alpha'} - \sum_{s''\alpha''} G^0_{\alpha\alpha''}(ss'', \omega)\, V^{j(0)}_{\alpha''\alpha'}(s''s', \omega). \tag{2.1.91}$$

Thus, $\boldsymbol{t}^{(l_i)}$ is a matrix whose order is given by the size of the defect space.

Averaging equation (2.1.88) we obtain the average atomic t-matrix

$$\langle \boldsymbol{t}^{(l_i)} \rangle = \sum_j c^j \boldsymbol{t}^{j(l_i)}. \tag{2.1.92}$$

Inserting (2.1.92) into (2.1.62) and accounting for (2.1.89) the SSA expression of the self-energy for the system at hand becomes

$$\Sigma_{\alpha\alpha'}(ll', \omega) = \sum_{l_i} \sum_{ss'} \delta_{l,s+l_i}\delta_{l',s'+l_i} K_{\alpha\alpha'}(ss', \omega), \tag{2.1.93}$$

where

$$\boldsymbol{K} = \sum_j c^j \boldsymbol{t}^{j(0)} [1 + \boldsymbol{G}^0 \sum_j c^j \boldsymbol{t}^{j(0)}]^{-1}. \tag{2.1.94}$$

In (2.1.94) all matrices are matrices in the defect space. Equation (2.1.93) shows that $\boldsymbol{\Sigma}$ decomposes into an overlapping set of local self-energy matrices $\boldsymbol{K}$, the order of which is given by the size of the defect space. On account of (2.1.86) we obtain from (2.1.94), (2.1.90) and (2.1.91) the local self-energy matrix of the AB alloy

$$\boldsymbol{K} = c\boldsymbol{V}[1 - (1 - c)\, \boldsymbol{G}^0\boldsymbol{V}]^{-1}, \tag{2.1.95}$$

where $c = c^{\mathrm{B}}$ and $V = V^{\mathrm{B}(0)}$.

Concluding this section we consider configuration averages of products of occupation numbers.

If the system is completely disordered, the occupation probabilities of different lattice points are independent of each other and thus we have

$$\left.\begin{aligned} &\langle \eta^j_L \eta^{j'}_{L'} \rangle = c^j_\varkappa c^{j'}_{\varkappa'} \qquad (L \neq L';\ L \equiv (l, \varkappa)), \\ &\langle \eta^j_L \eta^{j'}_{L'} \eta^{j''}_{L''} \rangle = c^j_\varkappa c^{j'}_{\varkappa'} c^{j''}_{\varkappa''} \qquad (L, L', L'' \text{ pairwise different}) \text{ etc.} \end{aligned}\right\} \tag{2.1.96}$$

In this case the configuration of atoms is completely characterized by the concentrations of the various kinds of atoms.

If the system is partially ordered, we can write

$$\left.\begin{aligned}\langle\eta_L^j\eta_{L'}^{j'}\rangle &= c_x^j c_{x'}^{j'} + \alpha_{LL'}^{jj'},\\ \langle\eta_L^j\eta_{L'}^{j'}\eta_{L''}^{j''}\rangle &= c_x^j c_{x'}^{j'} c_{x''}^{j''} + \alpha_{LL'L''}^{jj'j''} \quad \text{etc.}\end{aligned}\right\} \tag{2.1.97}$$

Here the α's are the parameters describing the order. These parameters are not independent of each other, because they satisfy sum rules resulting from the second form of (2.1.67):

$$\left.\begin{aligned}&\sum_j \alpha_{LL'}^{jj'} = \sum_{j'} \alpha_{LL'}^{jj'} = 0,\\ &\sum_j \alpha_{LL'L''}^{jj'j''} = \alpha_{L'L''}^{j'j''} \qquad \begin{aligned}&\text{(and similiar expressions}\\ &\text{from the sum over } j' \text{ and } j'') \text{ etc.}\end{aligned}\end{aligned}\right\} \tag{2.1.98}$$

Let us consider, in particular, the two-particle correlation function in a monatomic AB alloy. In this case only one usually writes

$$\alpha_{ll'}^{jj'} = c^{\mathrm{A}}c^{\mathrm{B}}z^j z^{j'}\alpha_{l-l'}, \tag{2.1.99}$$

where $z^{\mathrm{A}} = -1$, $z^{\mathrm{B}} = 1$ and $\alpha_{l-l'}$ are the so-called Warren parameters which depend only on the spacing between the atoms. Since $\langle\eta_l{}^j\eta_{l'}^{j'}\rangle \geqq 0$, we have

$$-c^{\mathrm{B}}/c^{\mathrm{A}} \leqq \alpha_{l-l'} \leqq 1 \qquad (c^{\mathrm{B}} \leqq c^{\mathrm{A}}). \tag{2.1.100}$$

A positive nearest-neighbour Warren parameter means that there is a tendency to undergo phase separation into an A und a B crystal, because atoms of the same kind are preferred as nearest neighbours. A negative nearest neighbour Warren parameter, on the other hand, means that there is a tendency to formation of an AB crystal, because in this case different atoms are preferred as nearest neighbours. If $\alpha_{l-l'}$ vanishes with increasing spacing between the atoms, long range order is absent. In presence of long range order, $\alpha_{l-l'}$ alters periodically with the spacing $|\boldsymbol{x}(l) - \boldsymbol{x}(l')|$.

We close this section by summarizing some general properties (known as Herglotz property) holding for the average Green function and the self-energy as functions of the complex frequency z. From (2.1.28) we obtain

$$[\boldsymbol{M}(z^*)^2 - \boldsymbol{\Phi}]\, \boldsymbol{G}(z^*) = 1 \tag{2.1.101}$$

and

$$[\boldsymbol{M}(z^*)^2 - \boldsymbol{\Phi}]\, \boldsymbol{G}^+(z) = 1, \tag{2.1.102}$$

where $\boldsymbol{G}^+$ is the hermitian adjoint of $\boldsymbol{G}$. On comparing (2.1.101) and (2.1.102) we find

$$\boldsymbol{G}(z^*) = \boldsymbol{G}^+(z), \tag{2.1.103}$$

which must hold also for the configuration-averaged Green function

$$\langle\boldsymbol{G}(z^*)\rangle = \langle\boldsymbol{G}(z)\rangle^+. \tag{2.1.104}$$

By (2.1.59) we obtain

$$\left[\left(\boldsymbol{G}^0(z^*)\right)^{-1} - \boldsymbol{\Sigma}(z^*)\right] \langle \boldsymbol{G}(z^*)\rangle = \boldsymbol{1} \tag{2.1.105}$$

and

$$\left[\left(\left(\boldsymbol{G}^0(z)\right)^{-1}\right)^+ - \boldsymbol{\Sigma}^+(z)\right] \langle \boldsymbol{G}(z)\rangle^+ = \boldsymbol{1}. \tag{2.1.106}$$

On rewriting (2.1.106) with the aid of (2.1.103) and (2.1.104) and comparing the resulting equation with (2.1.105) we find that

$$\boldsymbol{\Sigma}(z^*) = \boldsymbol{\Sigma}^+(z). \tag{2.1.107}$$

The properties (2.1.103) and (2.1.107) are called reality condition.

Another important property is that Im $\boldsymbol{G}$ and Im $\boldsymbol{\Sigma}$ are negative definite when Im $z^2 > 0$. Note that an operator A is positive definite, if $\langle\psi| A |\psi\rangle > 0$ for any state $|\psi\rangle$ of the Hilbert space. Im A denotes the anti-hermitian part (or "imaginary part") of A:

$$\text{Im}\, A = (A - A^+)/2i. \tag{2.1.108}$$

By (2.1.108), (2.1.101) and (2.1.102) we find that

$$\text{Im}\, \boldsymbol{G} = -\boldsymbol{G}^+\boldsymbol{M}\boldsymbol{G}\, \text{Im}\, z^2, \tag{2.1.109}$$

and averaging this equation we obtain

$$\text{Im}\, \langle\boldsymbol{G}\rangle = -\langle\boldsymbol{G}^+\boldsymbol{M}\boldsymbol{G}\rangle\, \text{Im}\, z^2. \tag{2.1.110}$$

Hence, for Im $z^2 > 0$, Im $\boldsymbol{G}$ and Im $\langle\boldsymbol{G}\rangle$ are negative definite. This so-called property of definiteness guarantees a positive state density (cf. (2.1.42)). It can also be proved for $\boldsymbol{\Sigma}(z)$ (see Ducastelle, 1974).

Furthermore, $\langle\boldsymbol{G}(z)\rangle$ and $\boldsymbol{\Sigma}(z)$ lie analytically off the real z-axis (property of analyticity), and it holds that

$$\langle\boldsymbol{G}(z)\rangle \sim \frac{1}{z^2}\,\boldsymbol{1} \qquad (|z| \to \infty) \tag{2.1.111}$$

and

$$\frac{\boldsymbol{\Sigma}(z)}{z^2} \sim 0 \qquad (|z| \to \infty), \tag{2.1.112}$$

which are the so-called boundary conditions.

A failure of the Herglotz property in some approximations for calculating $\langle\boldsymbol{G}\rangle$ gives rise to non-physical behaviour of the state density (cf. section 2.3.4.).

2.2. Localized and resonant modes in systems with low defect concentrations

2.2.1. The isolated defect approximation

If in a crystal the defect concentration is sufficiently low, the defects do not interact with each other. In this case we may expect that dynamical properties can be adequately described by expanding the quantities of interest in powers of the defect concentration. By truncating this expansion at the term linear in the defect concentration, we reduce the dynamical problem to that of isolated defects. To see this, let us consider the low concentration limit of the t-matrix and of the self-energy.

Since at very low defect concentrations the interaction between the defects is negligible, we may start from the average t-matrix, $\langle \boldsymbol{T} \rangle$, in single-site approximation (2.1.61). Due to the fact that the average atomic t-matrix, $\langle \boldsymbol{t}^i \rangle$, is proportional to the defect concentration (cf. (2.1.75), (2.1.92)), from (2.1.61) $\langle \boldsymbol{T} \rangle$ can be readily found to be

$$\langle \boldsymbol{T} \rangle = c \sum_i \boldsymbol{t}^{\mathrm{d}(i)} + O(c^2). \tag{2.2.1}$$

Here we have assumed, for simplicity, that there is only one kind of defect in the crystal. Its concentration is c: $\boldsymbol{t}^{\mathrm{d}(i)}$ denotes the atomic t-matrix of a defect at site i (cf. (2.1.72), (2.1.89)). The sum in (2.2.1) runs over all lattice sites.

In the preceding section the poles of $\langle \boldsymbol{T} \rangle$ were shown to determine the frequencies of the defect-affected vibrational modes. We see from (2.2.1) that in linear approximation with respect to c the poles of $\langle \boldsymbol{T} \rangle$ agree with those of $\boldsymbol{t}^{\mathrm{d}(i)}$. Hence, in this approximation the defect-affected frequencies can be obtained by studying the dynamics of an isolated defect.

Various experimental quantities can be expressed by the displacement-displacement correlation function and thus, using (2.1.22), by the imaginary part of the displacement-displacement Green function (cf. appendix 3 and see below). At low defect concentrations the defect induced contribution to such a quantity becomes proportional to the imaginary part of the self-energy $\boldsymbol{\Sigma}$ (e.g. the infrared absorption, the Raman and the neutron-scattering cross section). According to (2.1.62), in linear approximation $\boldsymbol{\Sigma}$ with respect to c is

$$\boldsymbol{\Sigma} = c \sum_i{}' \boldsymbol{t}^{\mathrm{d}(i)}. \tag{2.2.2}$$

Hence, at low defect concentrations a calculation of such a quantity can be reduced to a calculation of the corresponding quantity for a system with a single defect, followed by a multiplication of the result thus obtained by the number of defects in the system.

However, an expansion with respect to c is not appropriate for studying all vibrational effects at low defect concentrations. For instance, it fails on investigations of the vibrational frequency spectrum in the vicinity of frequencies at which van Hove singularities (cf. section 1.4.) occur in the crystal without defects. Using such an expansion, the van Hove singularities are preserved at any concentrations, i.e. the sharpness of these singularities is not, as expected, eroded with increasing defect concentration. This is a consequence of the nonanalytic behaviour of the frequency spectrum at the van Hove singularities, which makes any expansion with respect to c incorrect. Note that the erosion of the singularities can be described by some form of self-consistent theory (see section 2.3.4.).

We now compile some formulas needed in the following sections for investigating vibrational properties of crystals with isolated defects. For isolated defects we use a notation analogous to that adopted in the preceding section for the many-defect problem. The perturbation matrix $\boldsymbol{V}$ is given by (2.1.31) with mass and force constant changes due to a single defect, which is assumed to be placed at the origin of the lattice. The order of the matrix $\boldsymbol{V}$ is determined by the size of the defect space. If we denote by $\boldsymbol{G}$ and $\boldsymbol{G}^0$ the Green functions of a crystal with and without a defect, respectively, the following Dyson equation holds (cf. (2.1.35))

$$\boldsymbol{G} = \boldsymbol{G}^0 + \boldsymbol{G}^0\boldsymbol{V}\boldsymbol{G}. \tag{2.2.3}$$

Note that for studying effects of defects on a variety of physical phenomena only elements of $\boldsymbol{G}$ in the defect space, denoted here by $\hat{\boldsymbol{G}}$, are needed. From (2.2.3) $\hat{\boldsymbol{G}}$ can be readily seen to satisfy the equation

$$\hat{\boldsymbol{G}} = \hat{\boldsymbol{G}}^0 + \hat{\boldsymbol{G}}^0\boldsymbol{V}\hat{\boldsymbol{G}}, \tag{2.2.4}$$

where $\hat{\boldsymbol{G}}^0$ denotes the elements of $\boldsymbol{G}^0$ in the defect space.

For a single defect the t-matrix, $\boldsymbol{t}$, is defined by analogy to (2.1.50) as

$$\boldsymbol{G} = \boldsymbol{G}^0 + \boldsymbol{G}^0\boldsymbol{t}\boldsymbol{G}^0. \tag{2.2.5}$$

On comparing this equation with (2.2.4) we obtain (cf. (2.1.55))

$$\boldsymbol{t} = \boldsymbol{V}[1 - \boldsymbol{G}^0\boldsymbol{V}]^{-1}. \tag{2.2.6}$$

Equation (2.2.6) shows that $\boldsymbol{t}$ has nonzero elements in the defect space only. Thus we may also write

$$\boldsymbol{t} = \boldsymbol{V}[1 - \hat{\boldsymbol{G}}^0\boldsymbol{V}]^{-1}. \tag{2.2.7}$$

As mentioned above, the poles of $\boldsymbol{t}$ determine the defect-affected vibrational frequencies. Rewriting (2.2.4) as

$$\hat{\boldsymbol{G}} = [1 - \hat{\boldsymbol{G}}^0\boldsymbol{V}]^{-1}\,\hat{\boldsymbol{G}}^0, \tag{2.2.8}$$

and comparing (2.2.7) and (2.2.8), we see that these frequencies are also given by the poles of $\hat{\boldsymbol{G}}$.

Besides defect-affected frequencies, the disturbance of the vibrational amplitudes of the atoms around a given defect is of great interest. From the range of this disturbance some information on the validity of the isolated defect approximation may be expected. Using (2.1.22) and the property $\operatorname{Im}\boldsymbol{G}(-\omega + i\varepsilon) = -\operatorname{Im}\boldsymbol{G}(\omega + i\varepsilon)$ (which results from (2.1.22) and (2.1.45)), the thermal displacement squared is readily found to be given by

$$\langle \boldsymbol{u}^2(l\varkappa)\rangle = -\frac{\hbar}{\pi}\int_0^\infty d\omega \coth(\beta\hbar\omega/2)\sum_\alpha \operatorname{Im} G_{\alpha\alpha}(l\varkappa, l\varkappa, \omega). \tag{2.2.9}$$

As an example for an experimental quantity we consider below the one-phonon infrared absorption. According to (A3.20), (A3.19), (A.3.22), and (2.1.17) for a cubic crystal the absorption coefficient is given by (neglecting Lorentz field corrections)

$$\alpha(\omega) = -\frac{4\pi\omega n}{c'}\frac{1}{V}\sum_{ll'\varkappa\varkappa'\beta\beta'} M_{\alpha,\beta}(l\varkappa)\, M_{\alpha,\beta}(l'\varkappa')\operatorname{Im} G_{\beta\beta'}(l\varkappa, l'\varkappa', \omega), \tag{2.2.10}$$

where c' is the velocity of light in the medium, V is the volume of the crystal, $M_{\alpha,\beta}(l\varkappa)$ is the αth component of the first order expansion coefficient of the electric dipole moment with respect to the displacement $u_\beta(l\varkappa)$ (i.e. the effective charge tensor), and n is the number of defects (see also PLUMELLE et al., 1979).

2.2.2. The mass defect approximation

As already mentioned in section 2.1.2., defects can give rise to localized and resonant vibrational modes, and they can relax selection rules arising from a conservation of the crystal momentum. Here we will study such effects by considering a single mass defect placed at the origin of a cubic Bravais lattice. Using (2.1.71) to (2.1.73), the t-matrix (2.2.6) is readily found to read in this case as

$$t_{\alpha\alpha'}(l, l', \omega) = \delta_{l0}\delta_{ll'}\delta_{\alpha\alpha'}t(\omega), \tag{2.2.11}$$

where

$$t(\omega) = \frac{M^0\epsilon\omega^2}{1 - M^0\epsilon\omega^2 G^0(\omega)}, \tag{2.2.12}$$

$$\epsilon = (M^0 - M)/M^0. \tag{2.2.13}$$

Here M^0 is the mass of a host lattice atom, M is the mass of the defect atom and $G^0(\omega)$ is the diagonal element of the Green function of the host lattice (cf. (2.1.74)).

For $\omega > \omega_L$ (ω_L — maximum frequency of the host lattice) $G^0(\omega)$ is real and positive, which results from (2.1.14) by using (2.1.39) (replacing there ω by $\omega + i\varepsilon$) and (2.1.21). Hence, according to (2.2.12), $t(\omega)$ may exhibit a pole at $\omega > \omega_L$ for $\epsilon > 0$, i.e. for a light defect. The corresponding normal mode is called localized mode, because the displacement amplitudes of the atoms vibrating in such a mode decay faster than exponentially with increasing distance from the defect.

From (2.2.12) we see that the eigenfrequency of the localized mode is the solution of the equation

$$1 = M^0\epsilon\omega^2 G^0(\omega). \tag{2.2.14}$$

Using $G^0(\omega) = \frac{1}{3} \sum_\alpha G^0_{\alpha\alpha}(0, 0, \omega)$ (valid for a cubic crystal), (2.1.39) and the orthonormality condition (1.3.44), the eigenfrequency equation (2.2.14) takes the form

$$1 = \frac{\epsilon\omega^2}{3N} \sum_{kj} \frac{1}{\omega^2 - \omega_j^2(\boldsymbol{k})} \tag{2.2.15}$$

(N — number of sites in the Bravais lattice). With the aid of (1.4.13) and (1.4.15), (2.2.15) can be rewritten as

$$1 = \epsilon\omega^2 \int_0^{\omega_L} d\omega' \frac{g^0(\omega')}{\omega^2 - \omega'^2}, \tag{2.2.16}$$

where $g^0(\omega')$ is the vibrational density of states of the host lattice.

In general, $g^0(\omega) \to 0$ like $(\omega_L^2 - \omega^2)^{1/2}$ as $\omega \to \omega_L$ (cf. (1.4.25)) and the integral in (2.2.16) does not diverge as $\omega \to \omega_L$. Hence, in this case we have no solution for $\epsilon \to 0$. ϵ must exceed some critical value in order that a localized mode occurs. In fig. 2.3 the ϵ-dependence of a localized mode frequency, ω_0,

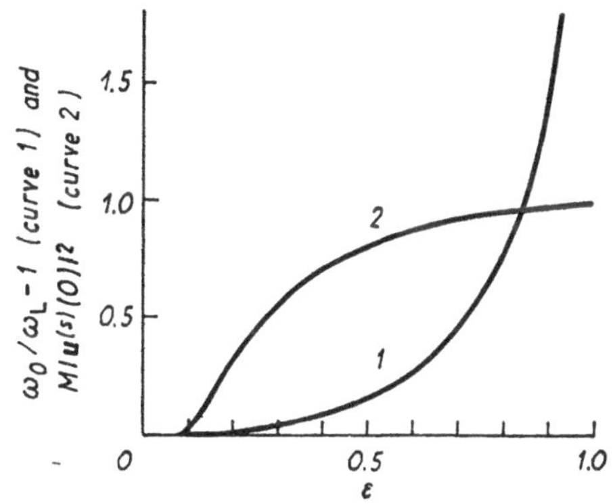

Fig. 2.3. The relative frequency ω_0/ω_L (curve 1) (ω_L — maximum frequency of the perfect lattice) and the contribution of the defect atom at the origin to the normalized mean square amplitude (curve 2) for a localized mode as a function of the parameter $\epsilon = (M^0 - M)/M^0$ (M^0 mass of a host lattice atom, M — mass of the defect atom) for a Debye spectrum (after DAWBER and ELLIOTT, 1963a)

is shown for a defect in a cubic lattice whose frequency spectrum is approximated by a Debye spectrum (cf. (1.4.38)). Note that for such a spectrum the integral in (2.2.16) is singular for $\omega \to \omega_L$ so that a (threefold degenerate) localized mode occurs for all $\epsilon > 0$.

Some information on the amplitude of the atoms in a localized mode can be obtained by examining the amount of amplitude on the defect atom. To do this, we introduce the displacement amplitude of the lth atom in the sth normal mode by (cf. (1.3.8))

$$u_\alpha^{(s)}(l) = B_\alpha^{(s)}(l)/\sqrt{M_l}. \tag{2.2.17}$$

Here $B_\alpha^{(s)}(l)$ is the corresponding eigenvector of the secular equation (1.3.11) and M_l is the mass of the lth atom. From the orthonormality condition (1.3.12) we obtain for the defect model at hand

$$M \sum_\alpha [u_\alpha^{(s)}(0)]^2 = 1 - M^0 \sum_{\alpha, l(\neq 0)} [u_\alpha^{(s)}(l)]^2. \tag{2.2.18}$$

On the other hand, according to (2.1.38), we have

$$\operatorname{Im} \sum_\alpha G_{\alpha\alpha}(0, 0, \omega + i\varepsilon) = -\pi \sum_s \sum_\alpha [u_\alpha^{(s)}(0)]^2 \, \delta(\omega^2 - \omega_s^2). \tag{2.2.19}$$

In particular, this equation can be used to calculate $\sum_\alpha [u_\alpha^{(s)}(0)]^2$ for the localized mode. Then, by means of (2.2.18), information on the amplitude of the host lattice atoms can be obtained.

From (2.2.4) we get (for $\omega > \omega_L$)

$$\begin{aligned}\operatorname{Im} G_{\alpha\alpha}(0, 0, \omega + i\varepsilon) &= \operatorname{Im} \{G^0_{\alpha\alpha}(0, 0, \omega + i\varepsilon)/[1 - M^0 \epsilon(\omega + i\varepsilon)^2 \, G^0_{\alpha\alpha}(0, 0, \omega + i\varepsilon)]\} \\ &= \pi \frac{G^0_{\alpha\alpha}(0, 0, \omega)}{\dfrac{\mathrm{d}}{\mathrm{d}\omega} [M^0 \epsilon \omega^2 G^0_{\alpha\alpha}(0, 0, \omega)]} \, \delta(\omega - \omega_0).\end{aligned} \tag{2.2.20}$$

On deriving the second line of this equation, we have taken into account that $G^0(\omega)$ is real and positive for $\omega > \omega_L$. Using (2.2.14) and (2.2.15), (2.2.20) can be rewritten as

$$\begin{aligned}&\operatorname{Im} G_{\alpha\alpha}(0, 0, \omega + i\varepsilon) \\ &= -\frac{\pi}{2\omega_0} \left[\frac{\epsilon^2 M^0 \omega_0^4}{3N} \sum_{kj} \frac{1}{\left(\omega^2 - \omega_j^2(\boldsymbol{k})\right)^2} - \epsilon M^0\right]^{-1} \delta(\omega - \omega_0).\end{aligned} \tag{2.2.21}$$

Thus, from (2.2.21) and (2.2.19), by (1.4.13) and (1.4.15), we finally find for the squared amplitude of the defect atom vibrating with the frequency of the localized mode ω_0

$$\sum_\alpha [u_\alpha^{(s)}(0)]^2 = \frac{1}{\epsilon M^0} \left[\epsilon \omega_0^4 \int_0^{\omega_L} \mathrm{d}\omega' \, \frac{g^0(\omega')}{(\omega_0^2 - \omega'^2)^2} - 1\right]^{-1}. \tag{2.2.22}$$

In figure 2.3 $M \sum_\alpha [u_\alpha^{(s)}(0)]^2$ is shown in dependence of ϵ for the Debye model. We see that this quantity approaches 1 as $\epsilon \to 1$. That is, according to (2.2.18), the lighter the defect atom the stronger the spatial extent of the localized mode decreases.

Note that owing to the cubic symmetry assumed the frequency of the localized mode is three-fold degenerate. By reducing the symmetry, this degeneracy may be lifted, so that for a mass defect in a Bravais lattice up to three roots of (cf. (2.2.7))

$$\det |\mathbf{1} - \hat{G}^0 \boldsymbol{V}| = 0 \tag{2.2.23}$$

can occur.

According to (2.2.23) for a crystal with a gap in the vibrational state density, localized modes may also occur in the gap (gap modes). In figure 2.4 the occurrence of localized ($\omega > \omega_L$) and gap modes is illustrated for a cubic two atomic lattice with a defect on the sublattice $\varkappa = 1$. We see, for example,

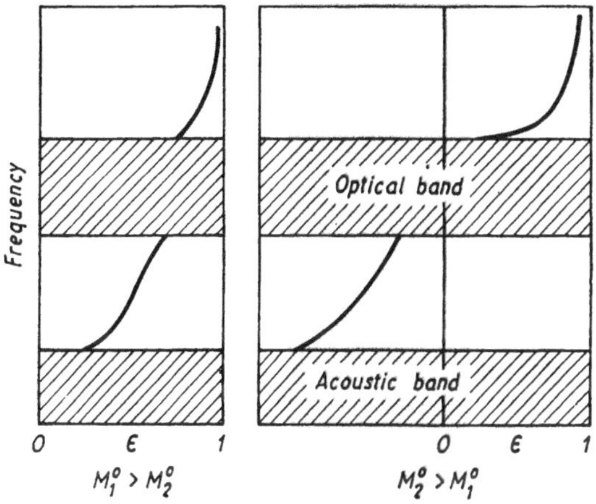

Fig. 2.4. The occurrence of gap and localized modes in dependence of $\epsilon = (M_1^0 - M_1)/M_1^0$. M_1, M_2 respectively denote the masses of atoms on sublattices $\varkappa = 1, 2$. The upperscript 0 denotes host lattice atoms (after Jaswal, 1965)

that a gap mode splits off the bottom of the optical band, if the defect atom is heavier than the lighter host atom. In general, critical defect masses exist also for the occurrence of localized gap modes.

Let us now consider modes with eigenfrequencies in the band region of the host crystal. These modes are perturbed always in the vicinity of the defect atom. Information on this perturbation can be obtained again from a study of the squared amplitude of the defect atom in such a mode, i.e. from a study of $\operatorname{Im} G_{\alpha\alpha}(0, 0, \omega + i\varepsilon)$ for $\omega < \omega_L$. For the latter quantity the first line of equation (2.2.20) is further valid. However, we must now take into account that at $\omega < \omega_L$ $G^0_{\alpha\alpha}(0, 0, \omega + i\varepsilon)$ is in general a complex function, i.e.

$$\left.\begin{aligned} G^0_{\alpha\alpha}(0, 0, \omega + i\varepsilon) &= \operatorname{Re} G^0(\omega) + i \operatorname{Im} G^0(\omega) \\ &= \frac{1}{M^0} P \int\limits_0^{\omega_L} d\omega' \frac{g^0(\omega')}{\omega^2 - \omega'^2} - i \frac{\pi}{M^0} \frac{g^0(\omega)}{2\omega}. \end{aligned}\right\} \tag{2.2.24}$$

The second line of this equation was obtained by making use of the fact that for a cubic crystal $G^0_{\alpha\alpha}(0, 0, \omega)$ is independent of α, and by (2.1.39), (1.3.44), (2.1.21), (1.4.13) and (1.4.15).

Inserting (2.2.24) into (2.2.20) and comparing the resulting expression with (2.2.19), we find the following expression for the squared amplitude of the defect atom:

$$\sum_\alpha [u_\alpha^{(s)}(0)]^2 = \frac{1}{NM^0}$$

$$\times\left\{\left[1 - \epsilon\omega_s^2 P \int_0^{\omega_L} \mathrm{d}\omega' g^0(\omega')/(\omega_s^2 - \omega'^2)\right]^2 + [\pi\epsilon\omega_s g^0(\omega_s)/2]^2\right\}^{-1}, \tag{2.2.25}$$

which holds for $\omega < \omega_L$. We see that the amplitude of such a mode is of $O(N^{-1/2})$ (as for a lattice wave (cf. (1.4.58)), whereas the amplitude of a localized mode is of $O(1)$ (cf. (2.2.22)). Moreover, we see from (2.2.25) that the amplitude becomes very large for frequencies in the immediate neighbourhood of the solution of the equation

$$1 = \epsilon\omega_R^2 P \int_0^{\omega_L} \mathrm{d}\omega' \frac{g^0(\omega')}{\omega_R^2 - \omega'^2}. \tag{2.2.26}$$

In this case the denominator of (2.2.25) takes on a resonant form. The mode defined by (2.2.26) is called resonant mode. For such a mode the t-matrix and the Green function also take on a resonant form, i.e. their imaginary parts exhibit a lorentzian behaviour, provided that

$$\left[\frac{\mathrm{d}\omega^2 \operatorname{Re} G^0(\omega)}{\mathrm{d}\omega}\right]_{\omega=\omega_R} < 0, \tag{2.2.27}$$

where ω_R is given by (2.2.26). To see that this relation must be fulfilled, we use (2.2.3) and expand $G^0(\omega)$ in the denominator of the expression for $G(\omega) \equiv G_{\alpha\alpha}(0, 0, \omega + i\varepsilon)$ about ω_R. Assuming for simplicity

$$\frac{\mathrm{d}\omega^2 \operatorname{Im} G^0(\omega)}{\mathrm{d}\omega} \ll \frac{\mathrm{d}\omega^2 \operatorname{Re} G^0(\omega)}{\mathrm{d}\omega} \tag{2.2.28}$$

we obtain approximately

$$G(\omega) = \frac{-G^0(\omega_R)\,\gamma(\omega_R)/M^0\epsilon}{\omega - \omega_R + i\omega_R^2\gamma(\omega_R) \operatorname{Im} G^0(\omega_R)}, \tag{2.2.29}$$

where

$$\gamma^{-1}(\omega_R) = \left[\frac{\mathrm{d}\omega^2 \operatorname{Re} G^0(\omega)}{\mathrm{d}\omega}\right]_{\omega=\omega_R}. \tag{2.2.30}$$

As the retarded Green function must be analytic in the upper complex frequency half plane and as $\operatorname{Im} G^0(\omega + i\varepsilon) < 0$ (cf. (2.2.24)), $\gamma(\omega)$ must be negative to guarantee the correct analytic behaviour.

Thus, the large enhancement of the amplitude of the defect atom (cf. (2.2.25)) as well as a peak in the frequency spectrum (arising from (2.2.29)) may be used to characterize a resonant mode, whose frequency is determined by (2.2.26). We see that the damping of such a mode is proportional to the vibrational density of states of the host lattice (in (2.2.9) via Im $G^0(\omega_R)$), i.e. it is proportional to the number of modes into which the resonant mode can decay. For a heavy defect atom (large negative ϵ) the resonance occurs at

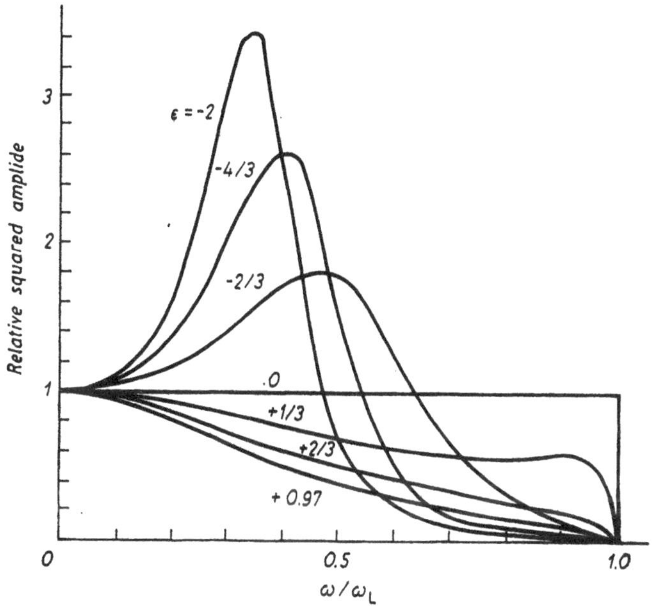

Fig. 2.5. The squared amplitude of a defect atom, relative to that of an atom in a perfect crystal, as a function frequency for the case of a Debye spectrum for various values $\epsilon = (M^0 - M)/M^0$ (M^0 — mass of a host atom, M — mass of the defect atom) (after Dawber and Elliott, 1963a)

low frequencies and the width is narrow. This tendency is illustrated in figure 2.5, where the squared amplitude of the defect atom is plotted for the case of a Debye spectrum. For $\epsilon > 0$ in a band mode the squared amplitude of the defect becomes smaller than that of an atom in a perfect lattice.

Valuable information on the vibrational amplitude near a defect atom can be obtained from a calculation of $\langle \boldsymbol{u}^2(l) \rangle$ by using (2.2.9). The quantity Im $G_{\alpha\alpha}(l, l, \omega)$ involved in this equation can be calculated along the lines described above. Results of such a calculation for a defect atom ($l = 0$) and a nearest neighbour atom ($l = 1$) in a fcc crystal for $\epsilon = -2$ (Cu: Au alloy) and $\epsilon = 0.57$ (Cu:Al alloy) are plotted in figures 2.6 and 2.7, respectively. According to these figures the disturbance of the vibrational amplitude due to a defect falls off very rapidly with increasing distance from the defect.

Let us now turn to the infrared absorption $\alpha(\omega)$ in the presence of a defect. We see from (2.2.10) that $\alpha(\omega)$ is governed by the imaginary part of the Green function. The simplest case is a homopolar crystal with a simple charged defect. With $M_{\alpha,\beta}(l) = \delta_{l0}\delta_{\alpha\beta}e^*$ (e^* — effective charge of the defect), in this case we have $\alpha(\omega) \sim \text{Im}\, G_{\alpha\alpha}(0, 0, \omega)$. Hence, by (2.2.10), (2.2.19), (2.2.22) and

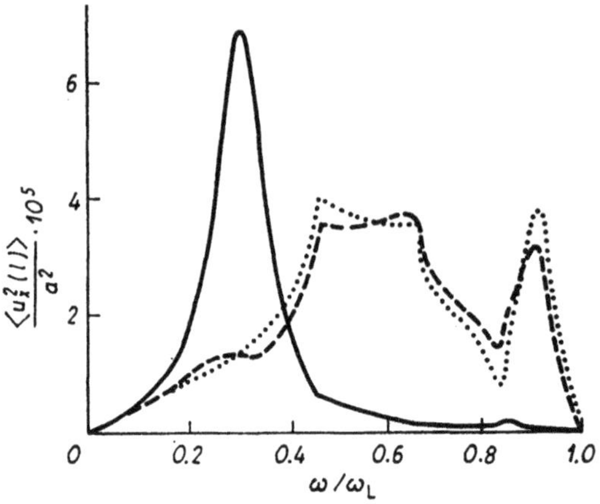

Fig. 2.6. Thermal average of the mean square amplitude $\langle u_x^2(l)\rangle/a^2$ (a — lattice constant) for an Au defect atom in a Cu host as a function of ω/ω_L (ω_L — maximum frequency of the host crystal) for $\epsilon = (M^0 - M)/M^0 = -2$ (M^0 — mass of a Cu atom, M — mass of an Au atom); —— defect atom, - - - - nearest neighbour atom, ······ unperturbed atom; $T = 50$ K (after Taylor, 1975)

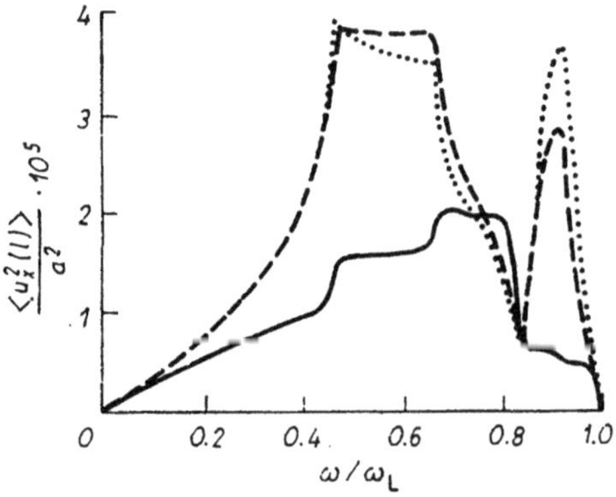

Fig. 2.7. Thermal average of the mean square amplitude $\langle u_x^2(l)\rangle/a^2$ (a — lattice constant) for an Al defect atom in a Cu host as a function of ω/ω_L (ω_L — maximum frequency of the host crystal) for $\epsilon = (M^0 - M)/M^0 = 0.57$ (M^0 — mass of a Cu atom, M — mass of an Au atom); —— defect atom, - - - - nearest neighbour atom, ····· unperturbed atom; $T = 50$ K (after Taylor, 1975)

(2.2.25) we obtain for charged isotopic defects

$$\alpha(\omega) = \frac{2\pi^2 n e^{*2}}{c' V M^0} \times \begin{cases} g^0(\omega) \left\{ \left[1 - \epsilon\omega^2 P \int\limits_0^{\omega_L} d\omega' \, \frac{g^0(\omega')}{\omega^2 - \omega'^2} \right]^2 + [\pi\epsilon\omega g^0(\omega)/2]^2 \right\}^{-1} & \text{for} \quad \omega < \omega_L, \\ \frac{\delta(\omega - \omega_0)}{\epsilon} \left\{ \epsilon\omega_0^4 \int\limits_0^{\omega_L} d\omega' \, \frac{g^0(\omega')}{(\omega_0^2 - \omega'^2)^2} - 1 \right\}^{-1} & \text{for} \quad \omega > \omega_L, \end{cases} \tag{2.2.31}$$

where n is the number of defect atoms in the crystal. Equation (2.2.31) predicts absorption in the whole frequency range of the band modes, which is proportional to the vibrational density of states modified by the denominator, which may exhibit resonances. Furthermore, if a localized mode exists a δ-peak is predicted to appear at the frequency of this mode.

Equation (2.2.31) has been applied to various systems, in particular to group III and V substitutional impurities in Si. For instance, rather good

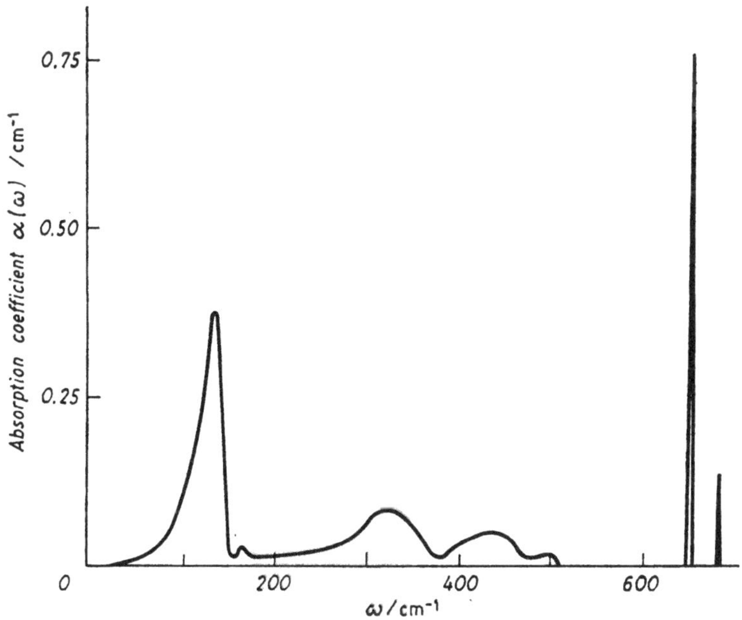

Fig. 2.8. Absorption spectrum predicted by (2.2.31) for Si doped with B^{10} and B^{11} in natural abundance. The peaks of the localized modes are reduced by about a factor 10 (after DAWBER and ELLIOTT, 1963b)

agreement between theoretical and experimental results has been obtained for Si:B. The theoretical absorption curve for Si:B is shown in figure 2.8. The inband absorption reflects the major peaks of the vibrational density of states of Si. The sharp local mode peak contains about 75 per cent of the B absorption. The calculated local mode frequency (at 664 cm^{-1} for B^{11}) turns out to be substantially higher than the observed value (at 620 cm^{-1} for B^{11}). This suggests that a B-defect also causes an appreciable force constant change for a B—Si bond (by about 14 per cent).

The defect activated absorption in the whole band of vibrational modes, which is predicted by (2.2.31), is of particular interest. Such behaviour is characteristic for disordered systems. Note that a perfect homopolar crystal has no one-phonon absorption and that a polar crystal has only one-phonon reststrahl lines arising from $\boldsymbol{k} = \mathbf{0}$ optical modes. The broad inband absorption in disordered systems reflects the relaxation of the $\boldsymbol{k} = \mathbf{0}$ selection rule due to the lack of lattice periodicity (cf. section 2.3.3.).

In figure 2.9 theoretical and experimental curves are given for the one-phonon absorption in dilute Ge:Si alloys. The theoretical curves were obtained

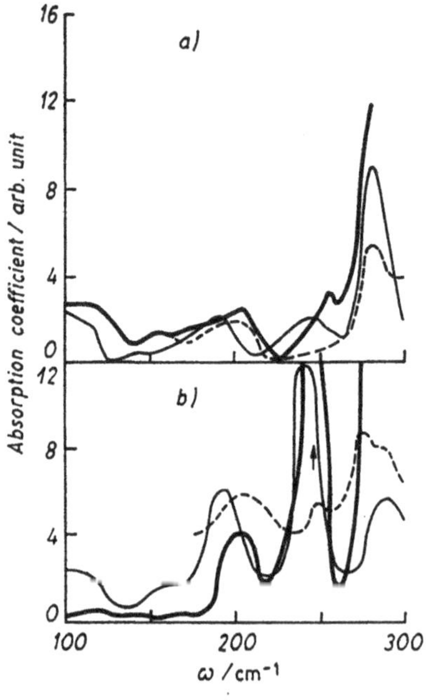

Fig. 2.9. Infrared absorption in Ge:Si alloys, a) - - - - - experimental results on $Ge_{0.95}$ $Si_{0.05}$, —— one phonon density of states for Ge, —— absorption predicted by mass defect approximation for isolated single Si atoms,
b) - - - - - experimental results on $Ge_{0.88}Si_{0.12}$, —— and —— absorption predicted for different modes (having E_u and A_{1u} symmetry, respectively) of paired Si atoms (after AGRAWAL et al., 1979)

by using the mass defect approximation ($M_{\mathrm{Ge}}/M_{\mathrm{Si}} = 2.6$). Note that in the case of an uncharged defect atom in a homopolar crystal, effective charges on the defect and its nearest neighbours arise from the wave function overlap between the defect atom and its neighbour atoms: $\alpha(\omega)$ is evaluated similarly as previously described. We see from figure 2.9 that for low defect concentrations the absorption bears resemblance to the vibrational density of states (fig. 2.9a)). At higher defect concentration, however, in the absorption spectrum new features appear, which may be ascribed to nearest-neighbour defect pairs.

Effects of force constant changes and defect pairs on vibrational properties cannot be described here in detail. Note that such effects can be treated quite similarly as effects of mass defects. However, the defect matrices are in the former case larger than for a simple mass defect. By applying group theory the problems of force constant changes and defect pairs can be simplified. For more details on this subject the reader is referred to MARADUDIN et al. (1971), chapter 8. Reviews of the extensive literature on experimental investigations of defect affected vibrational properties were given by MARADUDIN et al. (1971), BARKER and SIEVERS (1975) and TAYLOR (1975).

2.2.3. Description of defect vibrations by an Einstein oscillator

Following KRUMHANSL and MATTEW (1968) and DEDERICHS and ZELLER (1976) we give here a simple and physically evident description of the local vibrational behaviour of an isolated defect. The description is based on the Green function technique. It leads to an Einstein oscillator. This oscillator is characterized by an effective force constant describing the static response, by an effective mass consisting of the defect mass plus additional contributions from the other atoms near the defect, and by a damping constant, which may be expressed by the effective force constant and the effective mass. The latter quantities can be calculated by means of the Green function of the perfect lattice, but they can also be approximated by expressions which do not require a knowledge of the perfect lattice Green function. The Einstein oscillator description can be applied to resonant and localized defect vibrations.

According to (2.1.28), the Green function of an imperfect crystal is given by

$$[\boldsymbol{M}\omega^2 - \boldsymbol{\Phi}]\, \boldsymbol{G}(\omega) = \boldsymbol{1}, \tag{2.2.32}$$

where $\boldsymbol{M}$ and $\boldsymbol{\Phi}$, respectively, denote here the mass and force constant matrix in the presence of an isolated defect. The matrices $\boldsymbol{M}$, $\boldsymbol{\Phi}$ and $\boldsymbol{G}$ can be suitably partitioned as

$$\boldsymbol{M} = \begin{bmatrix} \boldsymbol{M}_{\mathrm{CC}} & 0 \\ 0 & \boldsymbol{M}_{\mathrm{RR}} \end{bmatrix}, \quad \boldsymbol{\Phi} = \begin{bmatrix} \boldsymbol{\Phi}_{\mathrm{CC}} & \boldsymbol{\Phi}_{\mathrm{CR}} \\ \boldsymbol{\Phi}_{\mathrm{RC}} & \boldsymbol{\Phi}_{\mathrm{RR}} \end{bmatrix}, \quad \boldsymbol{G} = \begin{bmatrix} \boldsymbol{G}_{\mathrm{CC}} & \boldsymbol{G}_{\mathrm{CR}} \\ \boldsymbol{G}_{\mathrm{RC}} & \boldsymbol{G}_{\mathrm{RR}} \end{bmatrix}, \tag{2.2.33}$$

where the subscript C denotes the "central" space, which in most cases contains the defect atom only, and R denotes the "remaining" space. Inserting

(2.2.33) into (2.2.32) we obtain

$$\left.\begin{aligned}
&(\boldsymbol{M}_{\mathrm{CC}}\omega^2 - \boldsymbol{\Phi}_{\mathrm{CC}})\, \boldsymbol{G}_{\mathrm{CC}} - \boldsymbol{\Phi}_{\mathrm{CR}}\boldsymbol{G}_{\mathrm{RC}} = \boldsymbol{1},\\
&-\boldsymbol{\Phi}_{\mathrm{RC}}\boldsymbol{G}_{\mathrm{CC}} + (\boldsymbol{M}_{\mathrm{RR}}\omega^2 - \boldsymbol{\Phi}_{\mathrm{RR}})\, \boldsymbol{G}_{\mathrm{RC}} = \boldsymbol{0},\\
&(\boldsymbol{M}_{\mathrm{CC}}\omega^2 - \boldsymbol{\Phi}_{\mathrm{CC}})\, \boldsymbol{G}_{\mathrm{CR}} - \boldsymbol{\Phi}_{\mathrm{CR}}\boldsymbol{G}_{\mathrm{RR}} = \boldsymbol{0},\\
&-\boldsymbol{\Phi}_{\mathrm{RC}}\boldsymbol{G}_{\mathrm{CR}} + (\boldsymbol{M}_{\mathrm{RR}}\omega^2 - \boldsymbol{\Phi}_{\mathrm{RR}})\, \boldsymbol{G}_{\mathrm{RR}} = \boldsymbol{1}.
\end{aligned}\right\} \tag{2.2.34}$$

As pointed out in the preceding section, the submatrix of $\boldsymbol{G}$ which is connected with the defect, $\boldsymbol{G}_{\mathrm{CC}}$, is of particular interest, because its poles or pseudopoles determine localized or resonant modes, respectively, and its imaginary part governs the squared amplitude of the defect atom and experimental quantities such as the infrared absorption. From the first equations (2.2.34) the following expression for $\boldsymbol{G}_{\mathrm{CC}}$ may be obtained:

$$\boldsymbol{G}_{\mathrm{CC}}(\omega) = [\boldsymbol{M}_{\mathrm{CC}}\omega^2 - \boldsymbol{\Phi}_{\mathrm{CC}} - \boldsymbol{\Phi}_{\mathrm{CR}}\hat{\boldsymbol{G}}_{\mathrm{RR}}\boldsymbol{\Phi}_{\mathrm{RC}}]^{-1}, \tag{2.2.35}$$

where $\hat{\boldsymbol{G}}_{\mathrm{RR}}$ is given by

$$[\boldsymbol{M}_{\mathrm{RR}}\omega^2 - \boldsymbol{\Phi}_{\mathrm{RR}}]\, \hat{\boldsymbol{G}}_{\mathrm{RR}}(\omega) = \boldsymbol{1}. \tag{2.2.36}$$

To obtain $\hat{\boldsymbol{G}}_{\mathrm{RR}}$ we proceed as on deriving (2.1.48). For a substitutional defect we find

$$\hat{\boldsymbol{G}}_{\mathrm{RR}} = [\boldsymbol{1} - \hat{\boldsymbol{G}}^0_{\mathrm{RR}}\boldsymbol{V}_{\mathrm{RR}}]^{-1}\, \hat{\boldsymbol{G}}^0_{\mathrm{RR}}, \tag{2.2.37}$$

where

$$[\boldsymbol{M}^0_{\mathrm{RR}}\omega^2 - \boldsymbol{\Phi}^0_{\mathrm{RR}}]\, \hat{\boldsymbol{G}}^0_{\mathrm{RR}} = \boldsymbol{1}. \tag{2.2.38}$$

Here $\boldsymbol{M}^0_{\mathrm{RR}}$ and $\boldsymbol{\Phi}^0_{\mathrm{RR}}$, respectively, denote the mass and force constant submatrices connected with the remaining space in a perfect crystal. $\boldsymbol{V}_{\mathrm{RR}}$ is the corresponding submatrix of the perturbation matrix (2.1.31) for an isolated defect. Using (2.2.38) and the last two equations (2.2.34) (but written down for the perfect crystal), the following expression for $\hat{\boldsymbol{G}}^0_{\mathrm{RR}}$ can be readily found

$$\hat{\boldsymbol{G}}^0_{\mathrm{RR}} = \boldsymbol{G}^0_{\mathrm{RR}} - \boldsymbol{G}^0_{\mathrm{RC}}[\boldsymbol{G}^0_{\mathrm{CC}}]^{-1}\, \boldsymbol{G}^0_{\mathrm{CR}}, \tag{2.2.39}$$

where $\boldsymbol{G}^0$ denotes the Green function of the perfect crystal.

Let us consider resonant modes. Such modes occur at frequencies ω_{R} defined by

$$\det |\mathrm{Re}\, \boldsymbol{G}^{-1}_{\mathrm{CC}}(\omega_{\mathrm{R}})| = 0, \tag{2.2.40}$$

provided that the imaginary part $\boldsymbol{G}^0(\omega_{\mathrm{R}})$ is small, i.e. if $\boldsymbol{G}_{\mathrm{CC}}(\omega)$ has a quasi pole. This definition is in accordance with that given in the preceding section. Note that Im $\boldsymbol{G}^0$ is small near the band edges, i.e. in a monatomic simple cubic lattice near $\omega \approx 0$ and $\omega \approx \omega_{\mathrm{L}}$, where ω_{L} is the maximum frequency of the perfect lattice. Restricting ourselves here to low frequency resonant modes we expand $\hat{\boldsymbol{G}}^{-1}_{\mathrm{CC}}(\omega)$ for small ω. Thus, according to (2.2.35), we have to expand $\hat{\boldsymbol{G}}_{\mathrm{RR}}(\omega)$ in powers of ω. If $\hat{\boldsymbol{G}}_{\mathrm{RR}}(\omega)$ does not show any resonance behaviour itself, the properties of $\hat{\boldsymbol{G}}_{\mathrm{RR}}(\omega)$ agree with those of $\boldsymbol{G}^0(\omega)$ and we may examine the low frequency expansion of $\boldsymbol{G}^0(\omega)$ instead of that of $\hat{\boldsymbol{G}}_{\mathrm{RR}}(\omega)$. To guarantee such a behaviour of $G_{RR}(\omega)$ we must possibly somewhat enlarge the defect space in some cases.

Consider first the imaginary part of $G^0(\omega)$ at low frequencies. According to (2.1.39), for a Bravais lattice $G^0(\omega)$ takes the form

$$G^0_{\alpha\alpha'}(l, l', \omega) = \frac{v_a}{(2\pi)^3} \frac{1}{M^0} \sum_j \int_{1.BZ} d\boldsymbol{k}\, e_\alpha(\boldsymbol{k}j)\, e_{\alpha'}(\boldsymbol{k}j) \frac{\exp\left[i\boldsymbol{k}(\boldsymbol{x}(l) - \boldsymbol{x}(l'))\right]}{(\omega + i\varepsilon)^2 - \omega_j^2(\boldsymbol{k})}, \quad \varepsilon \to +0 \tag{2.2.41}$$

(v_a — volume of a primitive unit cell), where we have used (1.4.18) and the fact that for a Bravais lattice the eigenvectors $\boldsymbol{e}(\boldsymbol{k}j)$ are real (cf. section 1.3.). In (2.2.41) $\exp\left[i\boldsymbol{k}(\boldsymbol{x}(l) - \boldsymbol{x}(l'))\right]$ may be replaced by $\cos \boldsymbol{k}(\boldsymbol{x}(l) - \boldsymbol{x}(l'))$. Using (2.1.21) we obtain from (2.2.41)

$$\operatorname{Im} G^0_{\alpha\alpha'}(l, l', \omega) = -\frac{\pi(\operatorname{sgn}\omega)\, v_a}{M^0(2\pi)^3} \times \sum_j \int_{1.BZ} d\boldsymbol{k}\, e_\alpha(\boldsymbol{k}j)\, e_{\alpha'}(\boldsymbol{k}j) \cos \boldsymbol{k}(\boldsymbol{x}(l) - \boldsymbol{x}(l'))\, \delta(\omega^2 - \omega_j^2(\boldsymbol{k})). \tag{2.2.42}$$

For low frequencies, only $\boldsymbol{k}$ vectors near $\boldsymbol{k} = 0$ contribute to the integral in (2.2.42). Here polarization vectors depend solely on $\hat{\boldsymbol{k}} = \boldsymbol{k}/|\boldsymbol{k}|$ and according to (1.3.54)

$$\omega_j(\boldsymbol{k}) = c_j(\hat{\boldsymbol{k}})\, |\boldsymbol{k}|, \tag{2.2.43}$$

where c_j is the velocity of sound connected with the jth branch. Expanding the cosine and integrating over $|\boldsymbol{k}|$, we obtain from (2.2.42) and (2.2.43)

$$\operatorname{Im} G^0_{\alpha\alpha'}(l, l', \omega) = -\frac{\pi v_a}{3M^0(2\pi)^3} \sum_j \int d\Omega_{\hat{k}} e_\alpha(\hat{\boldsymbol{k}}j)\, e_{\alpha'}(\hat{\boldsymbol{k}}j)\, \omega/c_j^3(\hat{\boldsymbol{k}}) + O(\omega^3). \tag{2.2.44}$$

For a cubic crystal, using $G^0_{\alpha\alpha'} = \delta_{\alpha\alpha'}G^0$ and the orthonormality condition (1.3.44), (2.2.44) can be rewritten as

$$\operatorname{Im} G^0_{\alpha\alpha'}(l, l', \omega) = -\frac{2v_a}{3M^0(2\pi)^3}\left(\frac{1}{c_l^3} + \frac{2}{c_t^3}\right)\omega\delta_{\alpha\alpha'} + O(\omega^3), \tag{2.2.45}$$

where c_l and c_t are the longitudinal and transverse velocity of sound, respectively. We see from (2.2.45) that at low frequencies $\operatorname{Im} G^0_{\alpha\alpha'}(l, l', \omega)$ is independent of indices l and l'.

According to (2.1.44), $\operatorname{Re} G^0(\omega)$ and $\operatorname{Im} G^0(\omega)$ are connected with each other by a Kramers-Kronig relation. Differentiating the relation (2.1.44) with respect to ω^2 we see that a Kramers-Kronig relation also connects $\frac{d}{d\omega^2} \operatorname{Re} G^0(\omega)$ and $\frac{d}{d\omega^2} \operatorname{Im} G^0(\omega)$. On account of these relations and of (2.2.45), we obtain the following low frequency expansion for Re $\boldsymbol{G}^0(\omega)$:

$$\operatorname{Re} \boldsymbol{G}^0(\omega) = \boldsymbol{G}^0(0) + \omega^2 \left[\frac{\mathrm{d}}{\mathrm{d}\omega^2} \operatorname{Re} \boldsymbol{G}^0(\omega)\right]_{\omega^2 \to +0} + \cdots. \tag{2.2.46}$$

Note that in (2.2.46) the limit $\omega^2 \to +0$ is due to the singularity of $\frac{\mathrm{d}}{\mathrm{d}\omega^2} \operatorname{Re} \boldsymbol{G}^0(\omega)$ at $\omega^2 = -0$, which results from the singularity of $\frac{\mathrm{d}}{\mathrm{d}\omega^2} \operatorname{Im} \boldsymbol{G}^0(\omega)$ by using the Kramers-Kronig relation.

Now we proceed to study low frequency resonant modes by examining the low frequency behaviour of $\boldsymbol{G}_{\mathrm{CC}}(\omega)$ from (2.2.35). On account of (2.2.46) we obtain for Re $\boldsymbol{G}_{\mathrm{RR}}(\omega)$

$$\operatorname{Re} \boldsymbol{G}_{\mathrm{RR}}(\omega) = -\boldsymbol{\Phi}_{\mathrm{RR}}^{-1} + \omega^2 \left[\frac{\mathrm{d}}{\mathrm{d}\omega^2} \operatorname{Re} \hat{\boldsymbol{G}}_{\mathrm{RR}}(\omega)\right]_{\omega^2 \to +0} + \cdots, \tag{2.2.47}$$

where we have used $\hat{\boldsymbol{G}}_{\mathrm{RR}}(0) = -\boldsymbol{\Phi}_{\mathrm{RR}}^{-1}$ (cf. (2.2.36)). The first term of (2.2.47) contributes to an effective force constant for the defect

$$\boldsymbol{f}_{\mathrm{CC}}^{\mathrm{eff}} = \boldsymbol{\Phi}_{\mathrm{CC}} - \boldsymbol{\Phi}_{\mathrm{CR}} \boldsymbol{\Phi}_{\mathrm{RR}}^{-1} \boldsymbol{\Phi}_{\mathrm{RC}} \tag{2.2.48}$$

and the second term to an effective mass

$$\boldsymbol{M}_{\mathrm{CC}}^{\mathrm{eff}} = \boldsymbol{M}_{\mathrm{CC}} - \boldsymbol{\Phi}_{\mathrm{CR}} \left\{\left[\frac{\mathrm{d}}{\mathrm{d}\omega^2} \operatorname{Re} \hat{\boldsymbol{G}}_{\mathrm{RR}}(\omega)\right]_{\omega^2 \to +0}\right\} \boldsymbol{\Phi}_{\mathrm{RC}}. \tag{2.2.49}$$

The imaginary part of $\hat{\boldsymbol{G}}_{\mathrm{RR}}(\omega)$, which behaves as described by (2.2.44), gives rise to a damping constant γ_{CC} as

$$\boldsymbol{M}_{\mathrm{CC}}^{\mathrm{eff}} \gamma_{\mathrm{CC}} \omega = -\boldsymbol{\Phi}_{\mathrm{CR}} [\operatorname{Im} \hat{\boldsymbol{G}}_{\mathrm{RR}}(\omega)] \, \boldsymbol{\Phi}_{\mathrm{RC}}. \tag{2.2.50}$$

Thus, from (2.2.35), (2.2.48), (2.2.49), and (2.2.50) we see that the low frequency behaviour of $\boldsymbol{G}_{\mathrm{CC}}(\omega)$ is given by

$$\boldsymbol{G}_{\mathrm{CC}}(\omega) = [\boldsymbol{M}_{\mathrm{CC}}^{\mathrm{eff}} \omega^2 - \boldsymbol{f}_{\mathrm{CC}}^{\mathrm{eff}} + i \boldsymbol{M}^{\mathrm{eff}} \gamma_{\mathrm{CC}} \omega]^{-1}. \tag{2.2.51}$$

This expression is similar to the Einstein approximation $\boldsymbol{G}_{\mathrm{CC}}(\omega) \approx [\boldsymbol{M}_{\mathrm{CC}} \omega^2 - \boldsymbol{\Phi}_{\mathrm{CC}}]^{-1}$ describing the defect atom vibrating in the field of the rest of the crystal whose atoms are held fixed. However, (2.2.51) takes into account the motion of all other atoms. It represents the Green function of a damped Einstein oscillator with effective mass $\boldsymbol{M}_{\mathrm{CC}}^{\mathrm{eff}}$, effective force constant $\boldsymbol{f}_{\mathrm{CC}}^{\mathrm{eff}}$ and damping constant γ_{CC}.

For simplicity, now we assume the central subspace C to contain the defect atom only. For cubic symmetry, the quantities $\boldsymbol{f}_{\mathrm{CC}}^{\mathrm{eff}}$, $\boldsymbol{M}_{\mathrm{CC}}^{\mathrm{eff}}$ and γ_{CC} become diagonal 3×3 matrices, i.e. $(\boldsymbol{f}_{\mathrm{CC}}^{\mathrm{eff}})_{\alpha\alpha'} = \delta_{\alpha\alpha'} f^{\mathrm{eff}}$ etc., and $\boldsymbol{G}_{\mathrm{CC}}(\omega)$ reads

$$[G_{\mathrm{CC}}(\omega)]_{\alpha\alpha'} = \delta_{\alpha\alpha'} [M^{\mathrm{eff}} \omega^2 - f^{\mathrm{eff}} + i M^{\mathrm{eff}} \gamma \omega]^{-1}, \tag{2.2.52}$$

which represents the Green function for an isotropic oscillator with velocity-dependent damping.

According to (2.2.9) and (2.2.19), the squared vibrational amplitude of the defect atom is governed by Im $[G_{CC}(\omega)]_{\alpha\alpha'}$, which on account of (2.2.52) is given by

$$\mathrm{Im}\,[G_{CC}(\omega)]_{\alpha\alpha'} = -\delta_{\alpha\alpha'} \frac{1}{M^{\mathrm{eff}}} \frac{\gamma\omega}{(\omega^2 - \omega_R{}^2)^2 + \gamma^2\omega^2}. \tag{2.2.53}$$

Here

$$\omega_R = \sqrt{f^{\mathrm{eff}}/M^{\mathrm{eff}}} \tag{2.2.54}$$

is the resonance frequency determined by the effective force constant and the effective mass. Before discussing (2.2.53) let us express γ in terms of f^{eff} and M^{eff}. To this end, first we point out that at low frequencies Im $\boldsymbol{G} \approx$ Im $\boldsymbol{G}^0$. Using (2.2.3) we obtain for Im $\boldsymbol{G}$

$$\begin{aligned}\mathrm{Im}\,\boldsymbol{G} &= \frac{1}{2i}\,(\boldsymbol{G} - \boldsymbol{G}^+)\\ &= \frac{1}{2i}\left[(1 - \boldsymbol{G}^0\boldsymbol{V})^{-1}\,\boldsymbol{G}^0 - (\boldsymbol{G}^0)^+\left(1 - \boldsymbol{V}^+(\boldsymbol{G}^0)^+\right)^{-1}\right]\\ &= (1 - \boldsymbol{G}^0\boldsymbol{V})^{-1}\,\mathrm{Im}\,\boldsymbol{G}^0\left(1 - \boldsymbol{V}^+(\boldsymbol{G}^0)^+\right)^{-1}\\ &= (1 + \boldsymbol{G}\boldsymbol{V})\,\mathrm{Im}\,\boldsymbol{G}^0(1 + \boldsymbol{V}^+\boldsymbol{G}^+),\end{aligned} \tag{2.2.55}$$

where the cross denotes the hermitian adjunct (note $\boldsymbol{G}^+ = \boldsymbol{G}^*$ and $\boldsymbol{V}^+ = \boldsymbol{V}$). Expanding Im $\boldsymbol{G}$ up to terms linear in ω we may replace $\boldsymbol{V}(\omega)$ by $\boldsymbol{V}(0) = \boldsymbol{\Phi} - \boldsymbol{\Phi}^0$, $\boldsymbol{G}(\omega)$ by $\boldsymbol{G}(0)$, and Im $\boldsymbol{G}^0$ by the expression (2.2.45). Note that the latter expression is independent of the site indices. Hence, using $\sum_{l'} \Phi_{\alpha\alpha'}(ll') = 0$ (1.3.52), for small ω (2.2.55) reduces to

$$\mathrm{Im}\,G_{\alpha\alpha'}(l, l', \omega) \approx \mathrm{Im}\,G^0_{\alpha\alpha'}(l, l', \omega) \approx -\delta_{\alpha\alpha'} \frac{3}{2\pi} \frac{1}{M^0\omega_D{}^3}\,\omega, \tag{2.2.56}$$

where the Debye frequency, ω_D, is given by (1.4.40). Comparing now (2.2.56) with the low frequency limit of (2.2.53), we find for the damping constant

$$\gamma = \frac{3}{2\pi} \frac{1}{M^0\omega_D{}^3} \frac{(f^{\mathrm{eff}})^2}{M^{\mathrm{eff}}}. \tag{2.2.57}$$

We can distinguish two kinds of resonances, "spring resonances" and "mass resonances". The former resonances arise from small f^{eff} and the latter ones from large M^{eff}. According to (2.2.57), spring resonances are much more weakly damped than mass resonances. Spring resonances give a large static response $G_{CC}(0) = 1/f^{\mathrm{eff}}$, i.e. for $f^{\mathrm{eff}} \to 0$ the defect configuration becomes unstable.

To proceed in discussing resonant modes, we now introduce the local frequency spectrum

$$g_\alpha{}^l(\omega) = -(2\omega M(l)/\pi)\,\mathrm{Im}\,G_{\alpha\alpha}(l, l, \omega) \tag{2.2.58}$$

being normalized as

$$\int_0^\infty \mathrm{d}\omega g_\alpha{}^l(\omega) = 1, \tag{2.2.59}$$

which can be shown by using (2.1.38) and (2.1.21). When we replace in (2.2.58) Im $G_{\alpha\alpha}(l, l, \omega)$ by Im $[G_{CC}(\omega)]_{\alpha\alpha}$ from (2.2.53), we obtain the following expression for the local spectrum of the defect d with mass M:

$$g_\alpha{}^{d}(\omega) = \frac{M}{M^{\mathrm{eff}}} \frac{2}{\pi} \frac{\gamma\omega^2}{(\omega^2 - \omega_R{}^2)^2 + \gamma^2\omega^2} \tag{2.2.60}$$

representing an asymmetric lorentzian, which is normalized as

$$\int_0^\infty d\omega \, g_\alpha{}^{d}(\omega) = M/M^{\mathrm{eff}}. \tag{2.2.61}$$

The normalization M/M^{eff} of $g_\alpha{}^{d}(\omega)$ represents the fraction of the total local spectrum of the defect which is contained in the resonant mode. In figure 2.10 $g_\alpha{}^{d}(\omega)$ is plotted for three different resonant frequencies. We see that the half-width, which is governed by γ from (2.2.57), goes to zero as $\omega_R \to 0$, and the local defect spectrum becomes $g_\alpha{}^{d}(\omega) = (M/M^{\mathrm{eff}})\, \delta(\omega)$.

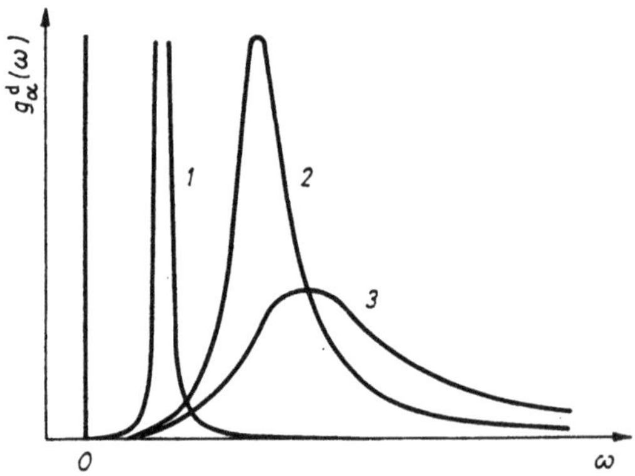

Fig. 2.10. Fraction of the total local defect spectrum being contained in a low frequency resonant mode, $g_\alpha{}^{d}(\omega)$, for three different resonant frequencies $0 < \omega_R^{(1)} < \omega_R^{(2)} < \omega_R^{(3)}$. At $\omega_R \to 0$ the spectrum becomes a δ-function (after DEDERICHS and ZELLER, 1976)

According to (2.2.48), (2.2.49), (2.2.37), and (2.2.39) f^{eff} and M^{eff} can be calculated by means of the Green function $\boldsymbol{G}^0$ of the perfect lattice. Since, in general, only numerical values are available for $\boldsymbol{G}^0$, approximations for calculating f^{eff} and M^{eff} without explicitly using $\boldsymbol{G}^0$ are valuable. For such approximations the reader is referred to DEDERICHS and ZELLER (1976). In this reference the method described above is applied to resonant modes near the upper band edge and to localized modes, too.

2.2.4. Small finite defect concentrations

As discussed in the preceding two sections, isolated defects can give rise to localized and resonant modes. Localized modes are characterized by frequencies outside the vibrational spectrum of the perfect crystal and by attenuation of the band modes near the defect. Resonant modes are characterized by a considerable enhancement of the vibrational amplitude near the defect for band modes in a narrow frequency range. Following ELLIOTT and TAYLOR (1967), here we show that small finite defect concentrations can give rise to a band of localized modes, instead of a single localized mode frequency, and to an appreciable enhancement of the vibrational density of states near a resonance.

We start from the Dyson equation (2.1.59). Introducing the $(\boldsymbol{k}, j)$ representation by

$$\langle G_{\alpha\alpha'}(l\varkappa, l'\varkappa', \omega)\rangle = [M^0(\varkappa)\, M^0(\varkappa')]^{1/2}\, N^{-1} \sum_{jj'\boldsymbol{k}} e_\alpha(\varkappa \mid \boldsymbol{k}j)\, e^*_{\alpha'}(\varkappa' \mid \boldsymbol{k}j') \times \exp\left[i\boldsymbol{k}\left(\boldsymbol{x}(l) - \boldsymbol{x}(l')\right)\right] \langle G_{jj'}(\boldsymbol{k}, \omega)\rangle, \tag{2.2.62}$$

$$\langle G_{jj'}(\boldsymbol{k}, \omega)\rangle = N^{-1} \sum_{\substack{l\varkappa\alpha \\ l'\varkappa'\alpha'}} [M^0(\varkappa)\, M^0(\varkappa')]^{1/2}\, e_\alpha{}^*(\varkappa \mid \boldsymbol{k}j)\, e_{\alpha'}(\varkappa' \mid \boldsymbol{k}j') \times \exp\left[-i\boldsymbol{k}\left(\boldsymbol{x}(l) - \boldsymbol{x}(l')\right)\right] \langle G_{\alpha\alpha'}(l\varkappa, l'\varkappa', \omega)\rangle, \tag{2.2.63}$$

$$\Sigma_{\alpha\alpha'}(l\varkappa, l'\varkappa', \omega) = [M^0(\varkappa) M^0(\varkappa')]^{1/2}\, N^{-1} \sum_{jj'\boldsymbol{k}} e_\alpha(\varkappa \mid \boldsymbol{k}j)\, e^*_{\alpha'}(\varkappa' \mid \boldsymbol{k}j') \times \exp\left[i\boldsymbol{k}\left(\boldsymbol{x}(l) - \boldsymbol{x}(l')\right)\right] \Sigma_{jj'}(\boldsymbol{k}, \omega) \tag{2.2.64}$$

and accounting for

$$G^0_{jj'}(\boldsymbol{k}, \omega) = \delta_{jj'}[\omega^2 - \omega_j{}^2(\boldsymbol{k})]^{-1}, \tag{2.2.65}$$

which results from a comparison of (2.2.62) and (2.1.39), the Dyson equation takes the form

$$\langle G_{jj'}(\boldsymbol{k}, \omega)\rangle = \delta_{jj'}[\omega^2 - \omega_j{}^2(\boldsymbol{k})]^{-1} + \sum_{j''} [\omega^2 - \omega_j{}^2(\boldsymbol{k})]^{-1}\, \Sigma_{jj''}(\boldsymbol{k}, \omega)\, \langle G_{j''j'}(\boldsymbol{k}, \omega)\rangle. \tag{2.2.66}$$

In the following we restrict ourselves to pure mass disorder in a monatomic cubic system and use the self-energy expression in single-site approximation given by (2.1.76) and (2.1.77). Then we have

$$\Sigma_{jj'}(\boldsymbol{k}, \omega) = \delta_{jj'}\Sigma(\omega), \tag{2.2.67}$$

where

$$\Sigma(\omega) = \frac{c\epsilon\omega^2}{1 - M^0\epsilon\omega^2(1 - c)\, G^0(\omega)}. \tag{2.2.68}$$

Here $\epsilon = (M^0 - M)/M^0$ (M^0 — mass of a host atom, M — mass of a defect atom), $G^0(\omega) = G^0_{\alpha\alpha}(0, 0, \omega)$ (cf. (2.2.24)) and c is the defect concentration.

From (2.2.66) to (2.2.68) we obtain

$$\langle G_{jj'}(\boldsymbol{k}, \omega)\rangle = \delta_{jj'}\langle G_j(\boldsymbol{k}, \omega)\rangle, \tag{2.2.69}$$

where

$$\langle G_j(\boldsymbol{k}, \omega)\rangle = [\omega^2 - \omega_j^2(\boldsymbol{k}) - \Sigma(\omega)]^{-1}. \tag{2.2.70}$$

In the following we examine Im $\langle G_j(\boldsymbol{k}, \omega)\rangle$. This quantity is of particular interest, because it governs inelastic neutron scattering (cf. (A3.95) and accounts for (2.1.22); for more details see ELLIOTT and TAYLOR, 1967), approximates the vibrational state density (cf. (2.1.42)) and via Im $\langle G(ll', \omega)\rangle$ (from (2.2.62) by noting that the $\boldsymbol{e}(\boldsymbol{k}j)$'s are real for a Bravais lattice) the infrared absorption spectrum, and other experimental quantities.

Using $G^0(\omega)$ from (2.2.24) we obtain for the self-energy

$$\left.\begin{aligned} \lim_{\varepsilon\to 0} \Sigma\,(\omega + i\varepsilon) &= \omega^2[\Delta(\omega) - i\Gamma(\omega)] \\ &= \frac{c\epsilon\omega^2[\alpha(\omega) - i\beta(\omega)]}{\alpha^2(\omega) + \beta^2(\omega)}, \end{aligned}\right\} \tag{2.2.71}$$

where

$$\alpha(\omega) = 1 - (1 - c)\,\epsilon\omega^2 \int\limits_0^{\omega_L} \frac{d\omega' g^0(\omega')}{\omega^2 - \omega'^2}, \tag{2.2.72}$$

$$\beta(\omega) = (1 - c)\,\epsilon\omega^2\,\frac{\pi g^0(\omega)}{2\omega}. \tag{2.2.73}$$

Here $g^0(\omega)$ is the vibrational state density of the perfect crystal. Inserting the self-energy (2.2.71) into (2.2.70), we find

$$\text{Im}\,\langle G_j(\boldsymbol{k}, \omega)\rangle = -\frac{\omega^2\Gamma(\omega)}{\left[\omega^2\big(1 - \Delta(\omega)\big) - \omega_j^2(\boldsymbol{k})\right]^2 + [\omega^2\Gamma(\omega)]^2}, \tag{2.2.74}$$

which holds for frequencies inside the range of $g^0(\omega)$ when $\Gamma(\omega) \neq 0$. According to (2.2.74), Im $\langle G_j(\boldsymbol{k}, \omega)\rangle$ should be peaked at ω given by

$$\omega^2\big(1 - \Delta(\omega)\big) - \omega_j^2(\boldsymbol{k}) = 0. \tag{2.2.75}$$

For heavy defects, an additional peak is expected at the resonant frequency ω_R being defined by the vanishing of $\Delta(\omega)$ or respectively $\alpha(\omega)$ (the latter condition corresponds to the resonance condition (2.2.26) modified by the finite defect concentration). For $\omega_j(\boldsymbol{k})$ very close to the resonant frequency, both the peaks should be of the same order of magnitude and close together. Note that a resonance occurs at low $g^0(\omega)$ and gives rise to a peak in $\Gamma(\omega)$. Since the vibrational state density of the crystal, $g\langle(\omega)\rangle$, is roughly speaking proportional to $\Gamma(\omega)$, we see that near a resonant frequency $g\langle(\omega)\rangle$ is appreciably enhanced with respect to $g^0(\omega)$. In figure 2.11 the relations between the functions $g^0(\omega)$, ω^2 Re $G^0(\omega)$, $\Delta(\omega)$, and $\Gamma(\omega)$ are shown for a resonant mode of defects with $\epsilon = -2$ in a copper host.

Consider now the imaginary part of the Green function outside the range of $g^0(\omega)$ when $\Gamma(\omega) = 0$. In this case we obtain from (2.2.70) with $\omega \to \omega + i\varepsilon$

$$
\begin{aligned}
-\mathrm{Im}\,\langle G_j(\boldsymbol{k}, \omega)\rangle &= \pi\delta\left[\omega^2\left(1 - \Delta(\omega)\right) - \omega_j^2(\boldsymbol{k})\right] \\
&= \pi\delta(\omega - \omega_l) \Big/ \left\{\frac{\mathrm{d}}{\mathrm{d}\omega}\left[\omega^2\left(1 - \Delta(\omega)\right) - \omega_j^2(\boldsymbol{k})\right]_{\omega=\omega_l}\right\} \\
&= \pi\delta(\omega - \omega_l) \Big/ \left\{2\omega_l\left[1 - \frac{\Delta(\omega_l)}{\omega_l^2}\left(1 - \frac{(1-c)\,\Delta(\omega_l)}{c}\,B(\omega_l)\right)\right]\right\} \\
&\approx \frac{c\pi\omega_l}{2\left(\omega_l^2 - \omega_j^2(\boldsymbol{k})\right)^2 B(\omega_l)}\,\delta(\omega - \omega_l),
\end{aligned}
\tag{2.2.76}
$$

where ω_l is the solution of

$$\omega_l^2\left(1 - \Delta(\omega_l)\right) - \omega_j^2(\boldsymbol{k}) = 0 \tag{2.2.77}$$

and

$$B(\omega) = \int\limits_0^{\omega_L} \mathrm{d}\omega'\omega'^2 g^0(\omega')/[\omega^2 - \omega'^2]^2. \tag{2.2.78}$$

The last form in (2.2.76) was obtained by assuming very small c.

The extra poles (2.2.76) are related to the existence of localized modes. Note that for an isolated defect the condition for a localized mode with frequency ω_0 reads (2.2.16)

$$M^0\epsilon\omega_0^2 G^0(\omega_0) = 1, \tag{2.2.79}$$

while (2.2.77) may be rewritten as

$$M^0\epsilon\omega_l^2 G^0(\omega_l) = \left[1 - \frac{c\epsilon\omega_l^2}{\omega_l^2 - \omega_j^2(\boldsymbol{k})}\right]\frac{1}{1-c}. \tag{2.2.80}$$

Here $G^0(\omega)$ is given by $\mathrm{Re}\,G^0(\omega)$ from (2.2.24). On expanding $\omega_l^2 G^0(\omega_l)$ in powers of $(\omega_l - \omega_0)$ and accounting for (2.2.79) and (2.2.80), we find for the frequencies ω_l, relative to ω_0, in lowest order with respect to c

$$\omega_l - \omega_0 = \frac{c[\omega_j^2(\boldsymbol{k}) - (1 - \epsilon)\,\omega_0^2]}{2\epsilon\omega_0 B(\omega_0)\,[\omega_0^2 - \omega_j^2(\boldsymbol{k})]}. \tag{2.2.81}$$

Equation (2.2.81) shows that for low finite defect concentrations instead of a single localized mode frequency a band of localized modes occurs, whose width is proportional to the defect concentration. However, the zero width of $\mathrm{Im}\,\langle G_j(\boldsymbol{k}, \omega)\rangle$ from (2.2.76) is not a reasonable result, because it means that plane-wave excitations of wave vector $\boldsymbol{k}$ are not damped in the disordered system, in contrast to our expectation. As will be shown in the following sections, this deficiency can be overcome by a self-consistent theory.

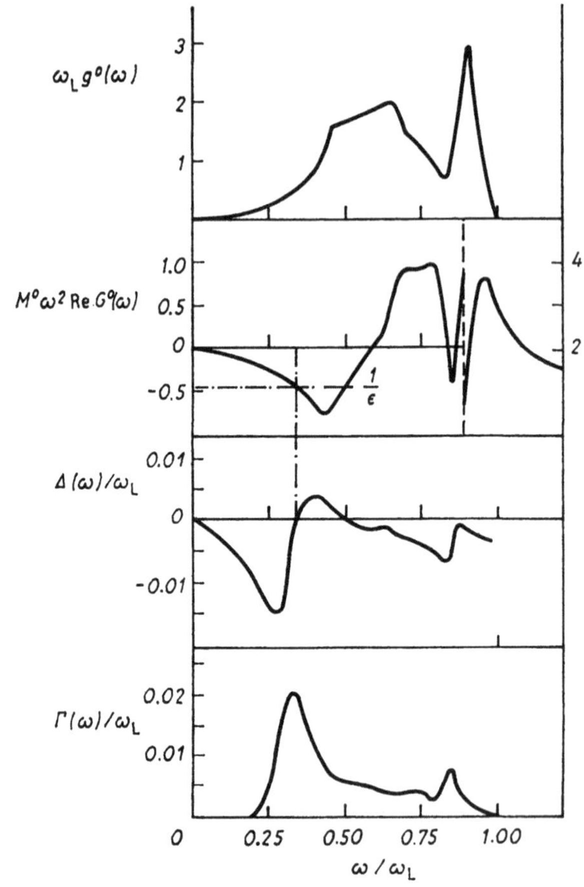

Fig. 2.11. Relations between the functions $g^0(\omega)$, $M^0\omega^2 \,\mathrm{Re}\, G^0(\omega)$, $\Delta(\omega)$, and $\Gamma(\omega)$ for defects with $\epsilon = -2$ in a Cu host. Note that beside a resonant mode at $\omega = 0.35\omega_L$ (ω_L — maximum frequency of a Cu host) there is an incipient resonance at $\omega = 0.86\ \omega_L$. At $\omega \geqq 0.9\omega_L$ the scale is changed for $M^0\omega^2 \,\mathrm{Re}\, G^0(\omega)$ (after TAYLOR, 1975)

2.3. Lattice vibrations in mixed crystals and alloys

2.3.1. Experimental findings

The main features of one-phonon infrared and Raman spectra of mixed crystals arise from long wavelength optical phonons. Most of the mixed crystals studied so far are of the $AB_{1-c}C_c$ type, where the end member crystals AB and AC are simple structures with two atoms per unit cell. With respect to the concentration dependence of the long wavelength modes, one can distinguish, in effect, two types of such mixed crystals: one-mode and two-mode crystals (see for example, the reviews by CHANG and MITRA, 1971, and by BARKER and SIEVERS, 1975).

$KCl_{1-c}Br$ and $GaAs_cP_{1-c}$, respectively, are good examples of one- and two-mode crystals. In figures 2.12 and 2.13 reflectivity spectra of these crystals

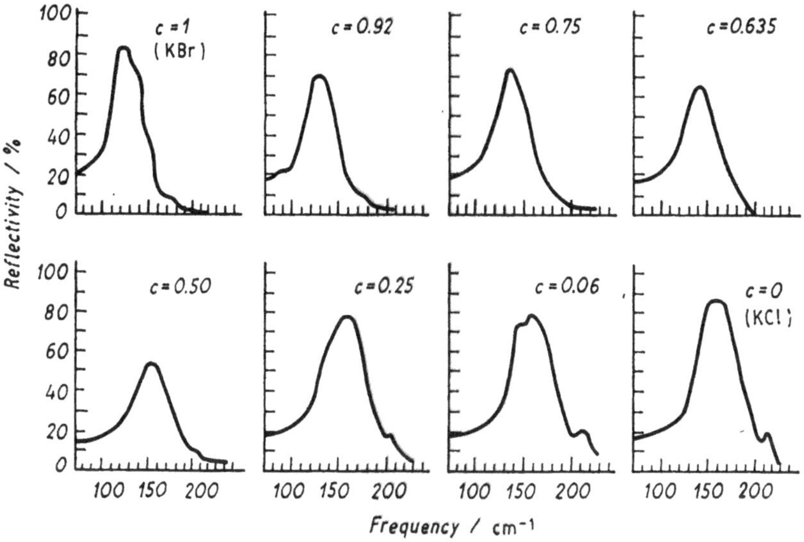

Fig. 2.12. Reflectivity spectra of $KCl_{1-c}Br_c$, which is a good example of one-mode behaviour (after FERTEL and PERRY, 1969)

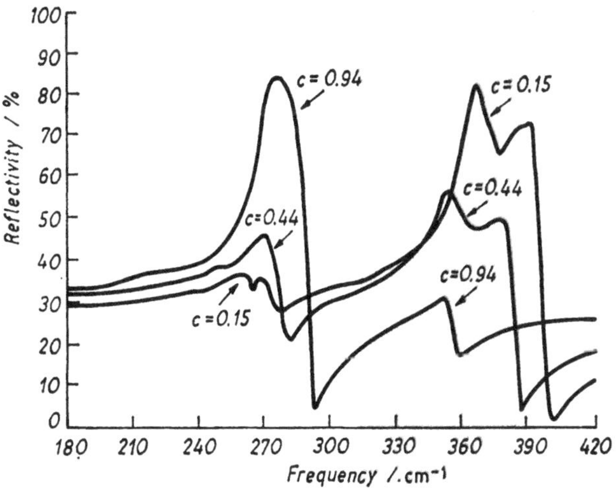

Fig. 2.13. Reflectivity spectra of $GaAs_cP_{1-c}$, which is a good example of two-mode behaviour (after VERLEUR and BARKER, 1966)

are shown. We see that $KCl_{1-c}Br_c$ has a single reststrahlen peak, whose position varies continuously with concentration from the reststrahlen frequency of one end-member to that of the other end-member, and which appears with approximately constant strength. On the other hand, for an intermediate mixing ratio, $GaAs_cP_{1-c}$ exhibits two reststrahlen peaks at frequencies close to those of the end members. The strength of each peak is approximately proportional to the mole fraction of the corresponding component.

Beside unambiguous one- or two-mode crystals, there are also mixed-mode crystals showing over a certain composition range a two-mode behaviour and over the remaining range a one-mode behaviour. Furthermore, a similar mode behaviour as observed for long-wave-length phonons exists also for phonons on zone boundaries (see, for example, GENZEL and BAUHOFER, 1976). The latter mode-behaviour can be studied experimentally by examining multi-phonon spectra.

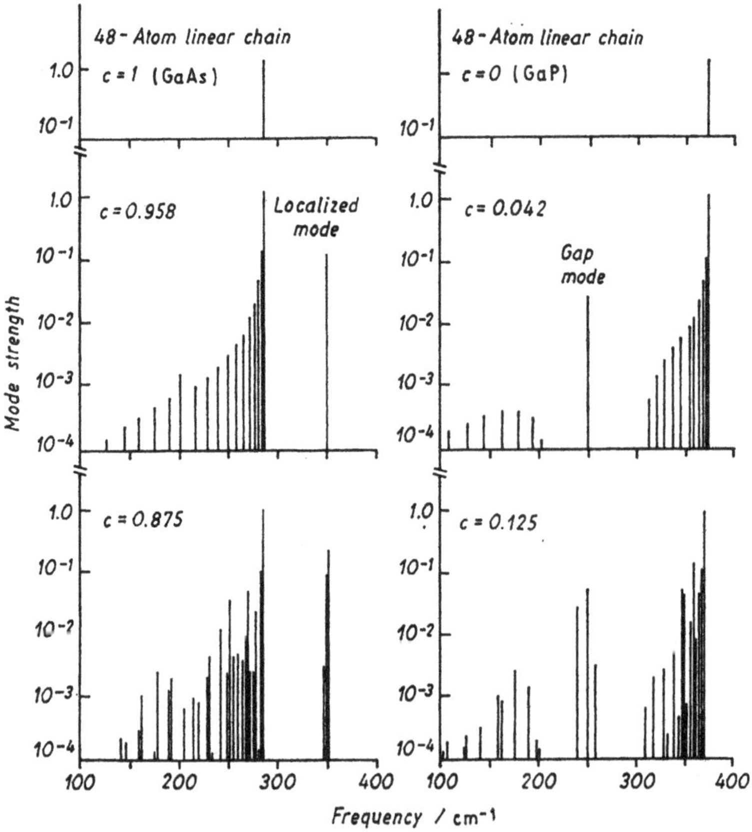

Fig. 2.14. Transverse mode strengths for a 48-atom linear chain model for $GaAs_cP_{1-c}$ (after BARKER and SIEVERS, 1975)

The problem now is how to understand these experimental features. The simplest attempt to establish criteria for predicting the mode behaviour is based on an examination of characteristic eigenmodes as a function of the mass ratios of the mixed crystal constituents. The restriction to an examination of the effect of mass changes is suggested by the fact that mixed crystals are formed only if the force and lattice constants of the end member crystals are similar. For long wavelength phonons, such a mass criterion predicts a two-mode behaviour, if near the one end of the composition range light defect atoms give rise to localized modes and if near the other end gap modes occur due to heavy defect atoms. This situation is well illustrated by figures 2.14 and 2.15, where for a simple 48-atom linear chain model for $GaAs_c/P_{1-c}$ the mode strengths (i.e. the weight factors of the δ-functions occurring in the spectral density) and the resulting reflectivity spectra are plotted. However, note that in such a one-dimensional model longitudinal modes are absent and

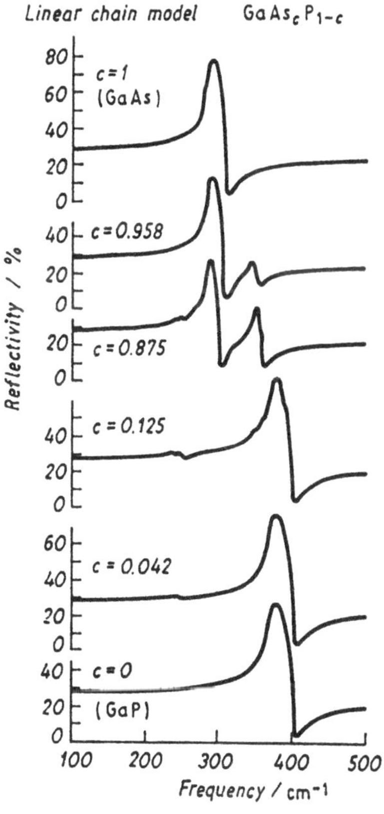

Fig. 2.15. Reflectivity for a 48-atom linear chain model for $GaAs_cP_{1-c}$ (cf. fig. 2.14) (after Barker and Sievers, 1975)

that in a three-dimensional crystal the reststrahlen band is placed in the frequency range between the LO and TO long wavelength modes. Hence, predictions on the existence of localized and gap modes, obtained by means of one-dimensional models, must be regarded with some caution. In figure 2.16 the different types of long wavelength mode behaviour are illustrated. The discussion of the mode behaviour will be continued in the following sections. Generally speaking, the main features of the mode behaviour are rather well understood at present, at least qualitatively. However, there are a lot of fine details in mixed crystal spectra which are still an interesting object of research.

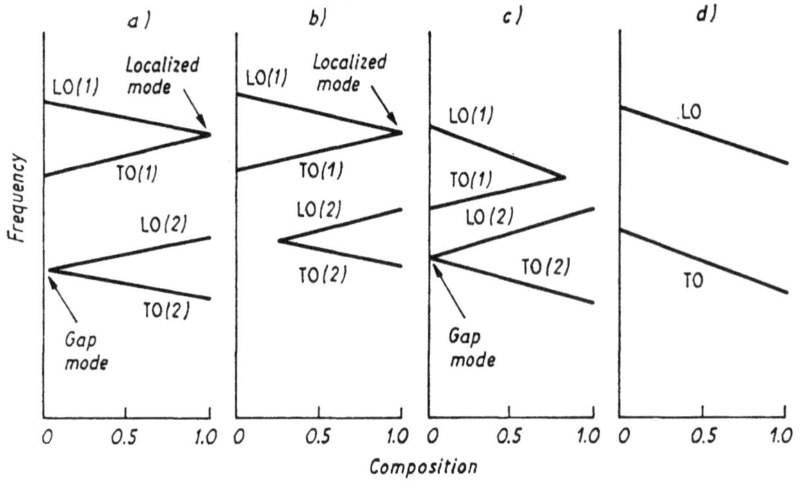

Fig. 2.16. Types of mixed crystal behaviour, a) gap mode and localized mode both allowed (two-mode behaviour), b) only localized mode allowed (mixed-mode behaviour), c) only gap allowed (mixed-mode behaviour), d) neither localized nor gap mode allowed (one-mode behaviour)

Now let us turn to the disorder induced density-of-states contribution to the one-phonon infrared and Raman spectra of mixed crystals, which arises from a relaxation of the crystal momentum ($\boldsymbol{k} = 0$) selection rule, quite analogously as in the case of single defects (cf. section 2.2.2.). It is often not very easy to identify experimentally such a contribution, because it can be masked by $\boldsymbol{k} \approx 0$ bands, resonant mode bands, and an overlap of one-phonon and multiphonon spectra. For illustration, in figure 2.17 the low frequency range of the Raman scattering spectrum of $Ge_{1-c}Si_c$ mixed crystals is shown, where the band below 120 cm^{-1} has been ascribed to disorder induced scattering of TA phonons of Ge and the band below 170 cm^{-1} to TA phonons of Si. Note that the remaining peaks in this figure have been identified as two-phonon peaks. The density-of-state contribution to the spectra will be discussed in more detail in the following sections, too.

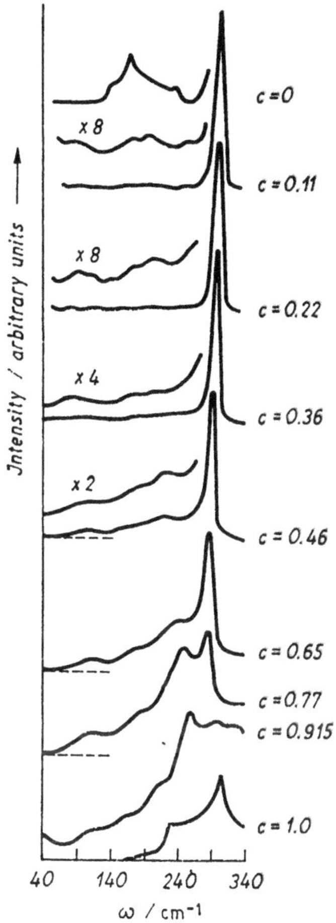

Fig. 2.17. First and second order Raman spectra of crystalline $Ge_{1-c}Si_c$ alloys at $T \approx 330$ K for $\omega \lessapprox 340$ cm^{-1} (after LANNIN, 1977)

2.3.2. The virtual crystal approximation and the random element isodisplacement model

As discussed in the preceding section, mixed crystals may be classified with respect to the concentration dependence of their long wavelength optical phonons. We now present two simple approximations for describing such concentration dependences: the virtual crystal approximation (VCA) and the random element isodisplacement (REI) model. In the VCA the mixed crystal is replaced by an ordered crystal with configuration-averaged masses and force constants, while in the REI model the mixed crystal is replaced by an ordered crystal with a pseudo unit cell which is configuration averaged. The VCA turns out to be suitable for describing one-mode behaviour and the REI model for describing two-mode behaviour.

Let us consider a mixed crystal of the $AB_{1-c}C_c$ type and restrict ourselves, for simplicity, to a one-dimensional system. The procedure used in the following can be readily applied also to systems with higher dimension. According to (2.1.27) and (1.3.52), the Green function of such a system obeys the equation

$$\omega^2 M(l\varkappa)\, G(l\varkappa, l'\varkappa', \omega) = \delta_{ll'}\delta_{\varkappa\varkappa'} + \sum_{l''\varkappa''(\neq l\varkappa)} \Phi(l\varkappa, l''\varkappa'')\,[G(l''\varkappa'', l'\varkappa', \omega) - G(l\varkappa, l'\varkappa', \omega)], \tag{2.3.1}$$

where the sublattice $\varkappa = 1$ is assumed to be completely occupied by A-atoms and the sublattice $\varkappa = 2$ randomly by B-atoms with concentration $(1 - c)$ and C-atoms with concentration c.

Using the occupation numbers defined by (2.1.66), the masses and coupling coefficients involved in (2.3.1) may be expressed as follows:

$$M(l\varkappa) = \sum_j \eta^j_{(l\varkappa)} M^j(\varkappa), \tag{2.3.2}$$

$$\Phi(l\varkappa, l'\varkappa') = \sum_{jj'} \eta^j_{(l\varkappa)}\eta^{j'}_{(l'\varkappa')}\Phi^{jj'}(l\varkappa, l'\varkappa'). \tag{2.3.3}$$

Here the index j runs from 1 to 3 and we define $M^1(1) = M_A$, $M^2(1) = M^3(1) = M^1(2) = 0$, $M^2(2) = M_B$, $M^3(2) = M_C$, where M_A, M_B, and M_C denote the masses of the A-, B- and C-atoms, respectively. Equation (2.3.3) is based on the assumption that a coupling coefficient is governed only by the two atoms connected by it. It is reasonable to assume $\Phi^{jj'}(l\varkappa, l'\varkappa') = \Phi^{jj'}(l - l'\varkappa, 0\varkappa')$. We may regard the coupling coefficients as being composed of a short range repulsive part and a long range electrostatic part. In two- or three-dimensional systems the latter part gives rise to a splitting of the LO and TO modes (cf. section 1.5.). An account of this effect is important for interpreting optical spectra of such systems (see, for example, LACINA and PERSHAN, 1970).

On averaging (2.3.1), the following Green functions occur:

$$\mathscr{G}(l\varkappa, l'\varkappa', \omega) \equiv \langle G(l\varkappa, l'\varkappa', \omega)\rangle, \tag{2.3.4}$$

$$\mathscr{G}^{j''}_{\substack{l''\\ \varkappa''}}(l\varkappa, l'\varkappa', \omega) \equiv \langle \eta^{j''}_{(l''\varkappa'')} G(l\varkappa, l'\varkappa', \omega)\rangle / \langle \eta^{j''}_{(l''\varkappa'')}\rangle, \tag{2.3.5}$$

$$\mathscr{G}^{j''\,j'''}_{\substack{l''\,l'''\\ \varkappa''\varkappa'''}}(l\varkappa, l'\varkappa', \omega) \equiv \frac{\langle \eta^{j''}_{(l''\varkappa'')}\eta^{j'''}_{(l'''\varkappa''')} G(l\varkappa, l'\varkappa', \omega)\rangle}{\langle \eta^{j''}_{(l''\varkappa'')}\eta^{j'''}_{(l'''\varkappa''')}\rangle}. \tag{2.3.6}$$

Here the symbol $\langle\ldots\rangle$ denotes the configuration average. The Green functions (2.3.5) and (2.3.6) are called conditionally averaged Green functions, because certain atoms are held fixed on averaging. Inserting now (2.3.2) and (2.3.3) into (2.3.1) and averaging the resulting equation, we obtain

$$\omega^2 \sum_j M^j(\varkappa)\, c^j_\varkappa \mathscr{G}^{j}_{\substack{l\\ \varkappa}}(l\varkappa, l'\varkappa', \omega) = \delta_{ll'}\delta_{\varkappa\varkappa'} + \sum_{\substack{l''\varkappa''(\neq l\varkappa)\\ jj''}} \Phi^{jj''}(l\varkappa, l''\varkappa'')\, c^j_\varkappa c^{j''}_{\varkappa''} \left[\mathscr{G}^{jj''}_{\substack{l\,l''\\ \varkappa\varkappa''}}(l''\varkappa'', l'\varkappa', \omega) - \mathscr{G}^{jj''}_{\substack{l\,l''\\ \varkappa\varkappa''}}(l\varkappa, l'\varkappa', \omega)\right], \tag{2.3.7}$$

where we have also used (2.3.5), (2.3.6) and (2.1.96). For the mixed crystal at hand we have $c_1^1 = 1$, $c_1^2 = c_1^3 = c_2^1 = 0$, $c_2^2 = (1 - c)$, $c_2^3 = c$.

We see that equation (2.3.7) connects two different Green functions. Hence, this equation cannot be solved without using a suitable approximation. Note that a hierarchy of connected equations for conditionally averaged Green functions with one, two, ... atoms held fixed can be obtained from (2.3.1) by multiplying this equation by η and products of η's and averaging the resulting equations. To get a solution, a finite closed set of equations must be obtained from this hierarchy. To this end, in general, a conditionally averaged Green function with $n + 1$ atoms held fixed is replaced by one with n fixed atoms. The crudest such approximation is the VCA, which consists in a complete neglection of any difference between the Green functions (2.3.4) to (2.3.6).

In the VCA (2.3.7) reduces to

$$\omega^2 \overline{M}(\varkappa)\, \mathcal{G}(l\varkappa, l'\varkappa', \omega) = \delta_{ll'}\delta_{\varkappa\varkappa'} + \sum_{l''\varkappa''(\neq l\varkappa)} \overline{\Phi}(l\varkappa, l''\varkappa'')\, [\mathcal{G}(l''\varkappa'', l'\varkappa', \omega) - \mathcal{G}(l\varkappa, l'\varkappa', \omega)], \quad (2.3.8)$$

where

$$\overline{M}(\varkappa) = \sum_j c_\varkappa{}^j M^j(\varkappa), \quad (2.3.9)$$

$$\overline{\Phi}(l\varkappa, l'\varkappa') = \sum_{jj'} \Phi^{jj'}(l\varkappa, l'\varkappa')\, c_\varkappa{}^j c_{\varkappa'}^{j'}. \quad (2.3.10)$$

Comparing (2.3.8) and (2.3.1), we see that (2.3.8) describes a crystal with average masses (2.3.9) and average coupling constants (2.3.10).

Now we expand $\mathcal{G}$ in a Fourier series,

$$\mathcal{G}(l\varkappa, l'\varkappa', \omega) = \frac{1}{N} \sum_k \mathcal{G}(\varkappa\varkappa', k, \omega)\, e^{ik[x(l\varkappa) - x(l'\varkappa')]}, \quad (2.3.11)$$

where N denotes the number of atoms in a sublattice and $x(l\varkappa)$ describes the position of the atom $(l\varkappa)$. Inserting (2.3.11) into (2.3.8), we obtain

$$\sum_{\varkappa''} \left\{ \left[\omega^2 \overline{M}(\varkappa) + \sum_{l'''\varkappa'''} (1 - \delta_{ll'''}\delta_{\varkappa\varkappa'''})\, \overline{\Phi}(l\varkappa, l'''\varkappa''') \right] \delta_{\varkappa\varkappa''} - \sum_{l''} (1 - \delta_{ll''}\delta_{\varkappa\varkappa''})\, \overline{\Phi}(l\varkappa, l''\varkappa'')\, e^{ik[x(l''\varkappa'') - x(l\varkappa)]} \right\} \mathcal{G}(\varkappa''\varkappa', k, \omega) = \delta_{\varkappa\varkappa'}. \quad (2.3.12)$$

The poles of $\mathcal{G}$ or, accordingly, the zeros of det $|\mathcal{G}^{-1}|$ determine the eigenfrequencies of the system. Since the expression in the braces in (2.3.12) is a real 2×2 matrix, we obtain from det $|\mathcal{G}^{-1}| = 0$ two phonon branches: an acoustic and an optical branch. Thus, the VCA leads to plane-wave-like excitations which are not damped by disorder. This is in contradiction to the fact that such excitations cannot be eigenmodes of a disordered system, because lattice periodicity is lacking.

Now let us study the long wavelength modes in the VCA. To do this we put $k \to 0$ in (2.3.12) and assume, for simplicity, nearest neighbour interaction only. Defining force constants F^{AB} and F^{AC} by

$$\begin{aligned}\Phi^{Aj}(l1, l2) &= \Phi^{jA}(l2, l1) = \Phi^{Aj}\big(l1, (l-1)\,2\big)\\ &= \Phi^{jA}\big((l-1)\,2, l1\big) \equiv -F^{Aj}, \qquad j = \mathrm{B, C},\end{aligned} \tag{2.3.13}$$

the resulting secular equation for the long wavelength modes takes the form

$$\begin{vmatrix} \omega^2 M_A - (1-c)\,F^{AB} - cF^{AC} & (1-c)\,F^{AB} + cF^{AC} \\ (1-c)\,F^{AB} + cF^{AC} & \omega^2[(1-c)\,M_B + cM_C] - (1-c)\,F^{AB} - cF^{AC} \end{vmatrix} = 0. \tag{2.3.14}$$

This equation has the two roots

$$\left.\begin{aligned} &\omega_1 = 0 \qquad \text{(acoustic mode)},\\ &\omega_2 = \left[\left((1-c)\,F^{AB} + cF^{AC}\right)\left(\frac{1}{M_A} + \frac{1}{(1-c)\,M_B + cM_C}\right)\right]^{1/2}\\ &\text{(optical mode)}. \end{aligned}\right\} \tag{2.3.15}$$

Consequently, for any concentration, the VCA yields a single long wavelength optical mode frequency. Hence, it may be used only to fit (not to predict) one-mode behaviour. It represents the crudest interpolation scheme over the entire concentration range.

Let us now turn to the REI model. We start again from (2.3.1). Replacing in this equation the masses and coupling constants by means of (2.3.2) and (2.3.3), multiplying the resulting equation with $\eta^j_{(l\varkappa)}$ and averaging it, we obtain

$$\begin{aligned}\omega^2 M^j(\varkappa)\, \mathcal{G}^{j}_{\substack{l\\ \varkappa}}(l\varkappa, l'\varkappa', \omega) &= \delta_{ll'}\delta_{\varkappa\varkappa'} + \sum_{\substack{l''\varkappa''(\neq l\varkappa)\\ j''}} \Phi^{jj''}(l\varkappa, l''\varkappa'')\\ &\times \frac{\langle \eta^j_{(l\varkappa)}\eta^{j''}_{(l''\varkappa'')}\rangle}{\langle \eta^j_{(l\varkappa)}\rangle}\left[\mathcal{G}^{jj''}_{\substack{ll''\\ \varkappa\varkappa''}}(l''\varkappa'', l'\varkappa', \omega) - \mathcal{G}^{jj''}_{\substack{ll''\\ \varkappa\varkappa''}}(l\varkappa, l'\varkappa', \omega)\right].\end{aligned} \tag{2.3.16}$$

Here, the hierarchy mentioned above is decoupled by using the following approximation in (2.3.16):

$$\mathcal{G}^{jj''}_{\substack{ll''\\ \varkappa\varkappa''}}(l\varkappa, l'\varkappa', \omega) \approx \mathcal{G}^{j}_{\substack{l\\ \varkappa}}(l\varkappa, l'\varkappa', \omega). \tag{2.3.17}$$

This is a single-site approximation (cf. section 2.1.3.), because an atom held fixed on averaging does not "feel" whether or not another atom is held fixed, too. That is, corresponding correlations between two and more atoms are neglected. With the aid of (2.3.17), (2.3.16) becomes

$$\begin{aligned}\omega^2 M^j(\varkappa)\, \mathcal{G}^{j}_{\substack{l\\ \varkappa}}(l\varkappa, l'\varkappa', \omega) &= \delta_{ll'}\delta_{\varkappa\varkappa'} + \sum_{\substack{l''\varkappa''(\neq l\varkappa)\\ j''}} \Phi^{jj''}(l\varkappa, l''\varkappa'')\, c^{j''}_{\varkappa''}\\ &\times \left[\mathcal{G}^{j''}_{\substack{l''\\ \varkappa''}}(l''\varkappa'', l'\varkappa', \omega) - \mathcal{G}^{j}_{\substack{l\\ \varkappa}}(l\varkappa, l'\varkappa', \omega)\right],\end{aligned} \tag{2.3.18}$$

where we have used also (2.1.96).

Note that in (2.3.18) only such elements of the Green function $\mathcal{G}^{j''}_{\varkappa''}(l\varkappa, l'\varkappa', \omega)$ occur, where $l'' = l$ and $\varkappa'' = \varkappa$. On the other hand, on deriving the approximations ATA and CPA (see section 2.3.4.) beside such diagonal elements non-diagonal elements must be accounted for also (see BÖTTGER, 1976).

$\mathcal{G}^{j}_{\varkappa}(l\varkappa, l'\varkappa', \omega)$ may be expanded in a Fourier series like $\mathcal{G}(l\varkappa, l'\varkappa', \omega)$ in (2.3.11). The equation (resulting from (2.3.18)) for the corresponding Fourier coefficients $\mathcal{G}^{j}(\varkappa\varkappa', k, \omega)$ reads

$$\sum_{\varkappa''j''} \Big\{\Big[\omega^2 M^j(\varkappa) + \sum_{l'''\varkappa'''j'''} (1 - \delta_{ll'''}\delta_{\varkappa\varkappa'''})\, c^{j'''}_{\varkappa'''} \Phi^{jj'''}(l\varkappa, l'''\varkappa''')\Big]\, \delta_{\varkappa\varkappa''}\delta_{jj''}$$
$$- c^{j''}_{\varkappa''} \sum_{l''} (1 - \delta_{ll''}\delta_{\varkappa\varkappa''})\, \Phi^{jj''}(l\varkappa, l''\varkappa'')\, \mathrm{e}^{ik[x(l''\varkappa'') - (l\varkappa)]}\Big\}$$
$$\times\, \mathcal{G}^{j''}(\varkappa''\varkappa', k, \omega) = \delta_{\varkappa\varkappa'}. \qquad (2.3.19)$$

Introducing a Green function $g^{jj'}$ by

$$\sum_{\varkappa''j''} \{\ldots\}\, g^{j''j'}(\varkappa''\varkappa', k, \omega) = \delta_{\varkappa\varkappa'}\delta_{jj'}, \qquad (2.3.20)$$

where $\{\ldots\}$ denotes the expression in the braces in (2.3.19), then it holds

$$\mathcal{G}(\varkappa\varkappa', k, \omega) = \sum_{j} c_{\varkappa}{}^{j} \mathcal{G}^{j}(\varkappa\varkappa', k, \omega) = \sum_{jj'} c_{\varkappa}{}^{j} g^{jj'}(\varkappa\varkappa', k, \omega). \qquad (2.3.21)$$

That is, the eigenfrequencies of the system may be determined from det $|\{\ldots\}| = 0$, because the poles of $\mathcal{G}$ and $g^{jj'}$ coincide. Since $\{\ldots\}$ is a real 3×3 matrix, det $|\{\ldots\}| = 0$ yields three branches of undamped phonons: one acoustic and two optical branches.

The secular equation resulting from (2.3.19) agrees with that for a crystal with a pseudo unit cell formed by atoms A, $(1 - c)$ B and c C (see fig. 2.18). The atoms B and C are located at the same site in the pseudo unit cell. In writing down the equation of motion for such a system, the masses and force

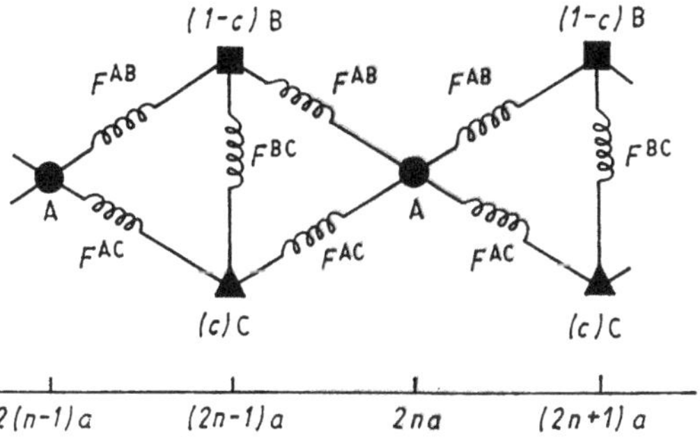

Fig. 2.18. Pseudo unit cell for a mixed diatomic chain, $AB_{1-c}C_c$ (after VARSHNEY et al., 1975)

constants are weighted by the concentration of the atoms involved. For more details on the formulation of this pseudo unit cell model and for explicit calculations of dispersion curves by means of it, the reader is referred to VARSHNEY et al. (1975).

Now let us investigate the long-wavelength limit of equation (2.3.19). We assume that the interatomic forces extend only up to the next nearest neighbours. With

$$\Phi^{jj'}\big(l2, (l+1)\,2\big) = \Phi^{j'j}\big((l+1)\,2, l2\big) = \Phi^{jj'}\big(l2, (l-1)\,2\big)$$
$$= \Phi^{j'j}\big((l-1)\,2, l2\big) \equiv -F^{jj'}/2, \qquad j, j' = \mathrm{B}, \mathrm{C}, \qquad (2.3.22)$$

and with (2.3.13) we obtain from (2.3.19) the following secular equation:

$$\begin{vmatrix} \omega^2 M_\mathrm{A} - (1-c)\,F^{\mathrm{AB}} - cF^{\mathrm{AC}} & (1-c)\,F^{\mathrm{AB}} & cF^{\mathrm{AC}} \\ F^{\mathrm{AB}} & \omega^2 M_\mathrm{B} - F^{\mathrm{AB}} - cF^{\mathrm{BC}} & cF^{\mathrm{BC}} \\ F^{\mathrm{AC}} & (1-c)\,F^{\mathrm{BC}} & \omega^2 M_\mathrm{C} - F^{\mathrm{AC}} - (1-c)\,F^{\mathrm{BC}} \end{vmatrix}$$
$$= 0. \qquad (2.3.23)$$

This equation is identical with the secular equation obtained by CHANG and MITRA (1968) for their REI model, which is the above pseudo unit cell model for $k = 0$. It is interesting to note that (2.3.23) does not contain the next nearest neighbour force constants F^{AA}, F^{BB} and F^{CC}. Contributions from these force constants, arising from the first and the second term in the braces in (2.3.19), cancel for $k = 0$. This reflects the fact that in the approximation at hand for $k = 0$ all atoms of a given type oscillate in phase, as in a perfect crystal. This isodisplacement is a basic assumption on deriving (2.3.23) by means of the REI model. Here it results from the decoupling (2.3.17).

Equation (2.3.23) has three roots, one of them being zero (acoustic mode). Near the end-member crystals, one of the remaining two modes is the defect mode, while the other mode is the optical mode. For a system exhibiting two-mode behaviour, the defect mode is a localized mode near the one end member and a gap mode near the other end member. Application of the REI model to one-mode and mixed-mode systems requires some additional assumptions, which in general are not very convincing. In the former case, for example, a resonant mode is thought of as being the defect mode, while in the latter case a combination of a resonant and a gap or localized mode seems to be appropriate. Benefits of the REI model are: (i) two-mode behaviour can be fitted, (ii) force constant changes can be included, (iii) interpolation over the whole concentration range is possible. Shortcomings of this model are: (i) all modes are described by undamped plane-wave excitations, irrespectivly whether a given mode is a defect or a band mode, (ii) comparison with experiment yields for the next-nearest neighbour force constant F^{BC} in general a value of the same order as that for nearest neighbour force constants. For a critical examination of the REI model the reader is referred, for example, to LUCOVSKY et al. (1976) and JAHNE and ULRICI (1980), see also the note on page 203.

2.3.3. Disorder activated modes in one-phonon infrared and Raman spectra

As already mentioned in section 2.3.1., one-phonon infrared and Raman spectra of disordered systems sometimes exhibit unambiguously disorder activated modes. In particular, these spectra may reflect certain features of the vibrational state density. Before considering this phenomenon in more detail, let us summarize some basic formulas for Raman scattering and infrared absorption.

Using (A3.56), (A3.57), (A3.61), (2.1.45) and (2.1.22) we obtain the following expression for the (non-resonant) one-phonon Raman scattering cross section:

$$\frac{d^2\sigma}{d\omega_s\, d\Omega} = \frac{\omega_s}{\omega_l}\left(\frac{e}{mc}\right)^4 \sum_{\alpha\alpha'\bar{\alpha}\bar{\alpha}'} e_\alpha{}^s e_{\alpha'}^{\,l} I_{\alpha\alpha'\bar{\alpha}\bar{\alpha}'}(\omega_s - \omega_l)\, e_{\bar{\alpha}}{}^s e_{\bar{\alpha}'}^{\,l}, \tag{2.3.24}$$

where

$$I_{\alpha\alpha'\bar{\alpha}\bar{\alpha}'}(\omega) = -\frac{\hbar}{\pi}\, n(\omega) \sum_{l\varkappa\beta} \sum_{\bar{l}\bar{\varkappa}\bar{\beta}} P^{(1)}_{\alpha\alpha',\beta}(l\varkappa)\, P^{(1)}_{\bar{\alpha}\bar{\alpha}',\bar{\beta}}(\bar{l}\bar{\varkappa})$$
$$\times \operatorname{Im} G_{\beta\bar{\beta}}(l\varkappa, \bar{l}\bar{\varkappa}, \omega + i\varepsilon), \qquad \varepsilon \to +0, \tag{2.3.25}$$

with

$$n(\omega) = [e^{\hbar\omega\beta} - 1]^{-1}. \tag{2.3.26}$$

Here $n(\omega)$ denotes the mean number of phonons with frequency ω (cf. (1.4.50)); $e_\alpha{}^l$ and $e_\alpha{}^s$ are, respectively, the αth cartesian components of the polarization vectors of the incident and the scattered light; ω_l and ω_s are, respectively, the frequencies of the incident and the scattered light, and $P^{(1)}_{\alpha\alpha',\beta}(l\varkappa)$ is the first order expansion coefficient (with respect to the displacement component $u_\beta(l\varkappa)$) of the $\alpha\alpha'$ component of the polarizability tensor. Thus, for $\omega = \omega_s - \omega_l > 0$, $I(\omega)$ from (2.3.25) is proportional to the number of phonons $n(\omega)$ and describes the Stokes component of the spectrum. If ω is negative, since $\operatorname{Im} G(-|\omega|) = -\operatorname{Im} G(|\omega|)$ (cf. (2.1.22) and (2.1.45)), the scattering cross section is proportional to $-n(-|\omega|) = n(|\omega|) + 1$ and describes the anti-Stokes component.

The one-phonon infrared absorption coefficient is given by (cf. (A3.20), (A3.21) or (2.2.10), respectively)

$$\alpha(\omega) = -\frac{4\pi\omega n}{c'}\frac{1}{V} \sum_{\substack{l\varkappa\beta \\ \bar{l}\bar{\varkappa}\bar{\beta}}} M^{(1)}_{\alpha,\beta}(l\varkappa)\, M^{(1)}_{\alpha,\bar{\beta}}(\bar{l}\bar{\varkappa})$$
$$\times \operatorname{Im} G_{\beta\bar{\beta}}(l\varkappa, \bar{l}\bar{\varkappa}, \omega + i\varepsilon), \qquad \varepsilon \to +0. \tag{2.3.27}$$

Here $M^{(1)}_{\alpha,\beta}(l\varkappa)$ is the first order expansion coefficient (with respect to $u_\beta(l\varkappa)$) of the αth component of the electric dipole moment (i.e. the effective charge).

Thus, we see that one-phonon Raman scattering and infrared absorption are both described by an expression of the type

$$O(\omega) \propto -\sum_{\substack{l\varkappa\alpha \\ l'\varkappa'\alpha'}} Q_\alpha(l\varkappa)\, Q_{\alpha'}(l'\varkappa') \operatorname{Im} G_{\alpha\alpha'}(l\varkappa, l'\varkappa', \omega), \tag{2.3.28}$$

where the quantities Q represent the first order expansion coefficients either of the polarizability or of the dipole moment. According to the familiar concept of configuration averaging, we must still average (2.3.28):

$$\langle O(\omega)\rangle \propto -\sum_{\substack{l\varkappa\alpha\\ l'\varkappa'\alpha'}} \langle Q_\alpha(l\varkappa)\, Q_{\alpha'}(l'\varkappa')\ \mathrm{Im}\ G_{\alpha\alpha'}(l\varkappa, l'\varkappa', \omega)\rangle, \tag{2.3.29}$$

where $\langle\ldots\rangle$ denotes the configuration average. Since, in general, disorder affects electronic as well as vibrational properties of a system, in (2.3.29) the averaging procedure acts simultaneously on the quantities Q, being dependent on the electronic structure, and on Im $\boldsymbol{G}$, which is governed by lattice dynamics. However, such a complex procedure cannot be accomplished at present. Therefore, here we restrict ourselves to a study of some simple special cases, where disorder affects either electronic or dynamical properties of a system.

First, let us assume that disorder affects the quantities Q only. In particular, we assume

$$Q_\alpha(l\varkappa) = \sum_j \eta^j_{(l\varkappa)} Q_\alpha{}^j(\varkappa), \tag{2.3.30}$$

where the $\eta^j_{(l\varkappa)}$'s are the occupation numbers defined in (2.1.66). Equation (2.3.30) is based on the assumption that $Q_\alpha(l\varkappa)$ depends only on the type of atom at site $(l\varkappa)$. Such an assumption implies that the wave functions governing $Q_\alpha(l\varkappa)$ are quite well localized at site $(l\varkappa)$.

Inserting (2.3.30) into (2.3.28) and accounting for

$$\langle \eta^j_{(l\varkappa)} \eta^{j'}_{(l'\varkappa')}\rangle = c_\varkappa{}^j c_{\varkappa'}^{j'}(1 - \delta_{ll'}\delta_{\varkappa\varkappa'}) + c_\varkappa{}^j \delta_{jj'}\delta_{ll'}\delta_{\varkappa\varkappa'} \tag{2.3.31}$$

we obtain

$$\begin{aligned}\langle O(\omega)\rangle \propto &-\sum_{jj'\varkappa\varkappa'\alpha\alpha'} c_\varkappa{}^j c_{\varkappa'}^{j'} Q_\alpha{}^j(\varkappa)\, Q_{\alpha'}^{j'}(\varkappa') \sum_{ll'} \mathrm{Im}\ G^0_{\alpha\alpha'}(l\varkappa, l'\varkappa', \omega)\\ &-\sum_{jj'\varkappa\alpha\alpha'} (c_\varkappa{}^j\delta_{jj'} - c_\varkappa{}^j c_\varkappa{}^{j'})\, Q_\alpha{}^j(\varkappa)\, Q_{\alpha'}^{j'}(\varkappa) \sum_l \mathrm{Im}\ G^0_{\alpha\alpha'}(l\varkappa, l\varkappa, \omega).\end{aligned} \tag{2.3.32}$$

Here $c_\varkappa{}^j$ is the concentration of the atoms of type j in the $\varkappa$th sublattice. Expressing $\boldsymbol{G}^0$ in (2.3.32) by its $(\boldsymbol{k}, \tilde{\jmath})$-representation (2.1.39) and using (2.1.21) and (1.3.65), then $\langle O(\omega)\rangle$ becomes

$$\begin{aligned}\langle O(\omega)\rangle \propto &\sum_{jj'\varkappa\varkappa'\alpha\alpha'} \frac{c_\varkappa{}^j c_{\varkappa'}^{j'} Q_\alpha{}^j(\varkappa)\, Q_{\alpha'}^{j'}(\varkappa')}{[M^0(\varkappa)\ M^0(\varkappa')]^{1/2}}\, N \sum_{\tilde{\jmath}} e_\alpha(\varkappa\mid 0\tilde{\jmath})\, e^*_{\alpha'}(\varkappa'\mid 0\tilde{\jmath})\, \delta\big(\omega^2 - \omega_{\tilde{\jmath}}{}^2(0)\big)\\ &+\sum_{jj'\varkappa\alpha\alpha'} (c_\varkappa{}^j\delta_{jj'} - c_\varkappa{}^j c_{\varkappa}^{j'})\, \frac{Q_\alpha{}^j(\varkappa)\, Q_{\alpha'}^{j'}(\varkappa')}{M^0(\varkappa)} \sum_{\boldsymbol{k}\tilde{\jmath}} e_\alpha(\varkappa\mid \boldsymbol{k}\tilde{\jmath})\, e^*_{\alpha'}(\varkappa\mid \boldsymbol{k}\tilde{\jmath})\, \delta\big(\omega^2 - \omega_{\tilde{\jmath}}{}^2(\boldsymbol{k})\big).\end{aligned} \tag{2.3.33}$$

On the right-hand side of (2.3.33), the first term describes the contribution of long wavelength optical phonons to the spectrum. Such a contribution is well known from perfect crystals. The second term is, in essence, proportional to the vibrational state density (cf. (1.4.13)). It describes the disorder activation of modes. Note that, as expected, this term vanishes, if $Q_\alpha^j(\varkappa)$ is independent of j, i.e. in the case of a perfect crystal.

Let us now study how Raman and infrared spectra are affected by disorder via lattice dynamics. To this end, we use the representation (2.1.38) for $\boldsymbol{G}$ shown in (2.3.28). Thus, we obtain

$$O(\omega) \propto \sum_{\substack{l\varkappa\alpha \\ l'\varkappa'\alpha}} \frac{Q_\alpha(l\varkappa)\, Q_{\alpha'}(l'\varkappa')}{\big(M(l\varkappa)\, M(l'\varkappa')\big)^{1/2}} \sum_s B_\alpha^{(s)}(l\varkappa)\, B_{\alpha'}^{(s)}(l'\varkappa')\, \delta(\omega^2 - \omega_s^2), \quad (2.3.34)$$

where $B^{(s)}(l\varkappa)$ is the sth (real) orthonormalized eigenvector of the dynamical matrix and ω_s is the corresponding eigenfrequency. Following GALEENER and SEN (1978), now we assume: (i) there are groups of eigenvectors related to each other by certain common features, for example, by similar motion of atoms, (ii) there is a one-to-one relationship between eigenfrequencies and eigenvectors. The latter assumption is based on theexpectation that in a disordered system there are no exact degeneracies of eigenvalues.

We label each group of eigenvectors by a band index $j = 1, \ldots, J$, and by N_j we denote the number of eigenvalues in each band. The frequencies in the jth band are listed in order of increasing value and then labelled ω_{ji}, where $i = 1, 2, \ldots, N_j$. According to the second assumption, we may write $\boldsymbol{B}^{(ji)}(l\varkappa) = B^j(l\varkappa, \omega_{ji}^2)$. Introducing the state density of the jth sub-band by

$$G_j(\omega^2) = \sum_{i=1}^{N_j}{}' \delta(\omega^2 - \omega_{ji}^2) \Big/ \sum_{j=1}^{J} N_j, \quad (2.3.35)$$

then (2.3.34) takes the form

$$O(\omega) \propto \sum_j D_j(\omega^2)\, G_j(\omega^2), \quad (2.3.36)$$

where

$$D_j(\omega^2) = \Big[\sum_{l\varkappa\alpha} Q_\alpha(l\varkappa)\, B_\alpha^j(l\varkappa, \omega^2)/\big(M(l\varkappa)\big)^{1/2}\Big]^2 \quad (2.3.37)$$

is a frequency dependent coupling coefficient. Thus, Raman and infrared spectra should exhibit a set of peaks each of which is proportional to a sub-band vibrational state density. However, the shape of such a peak may be governed much more by the frequency dependence of the coupling coefficient than by the sub-band state density (see also KLUG and WHALLEY, 1979, and references therein).

Consequently, disorder-activated features in one-phonon Raman and infrared spectra may be caused by disorder via the electronic structure as well as via the lattice dynamics. In general, one may expect that both these mechanisms are simultaneously effective.

It is noteworthy that a breakdown of the usual $\boldsymbol{k}$-selection rules of crystalline spectra has not only been observed in mixed crystals, but also in crystals with isolated defects (cf. section 2.2.), in amorphous materials (cf. section 3.3. and see BÖTTGER, 1974, and references therein), in superionic conductors (ALBEN and BURNS, 1977, BURNS et al., 1979), and in magnetic semiconductors, where it is induced by spin disorder (see Zeyher and Kress, 1979). The activation of modes by disorder is still a problem of great theoretical and experimental interest (cf. Bonneville, 1981, Carles et al., 1982).

We close this section with a study of this problem by means of a simple exactly solvable model of a disordered lattice. This model is a linear chain, in which the mass of one species of atom is infinite. For calculating the lattice Green function we need the eigenstates and eigenfrequencies of the following equation of motion for the atoms in an island of L atoms with finite mass, which is bordered by two infinitely heavy atoms (cf. (1.3.2)):

$$[M\omega^2 - 2\gamma]\, u(l) + \gamma u(l+1) + \gamma u(l-1) = 0 \tag{2.3.38}$$

with the boundary conditions

$$u(0) = u(L+1) = 0, \tag{2.3.39}$$

where $u(l)$ denotes the displacement of the atom at site la (a — lattice constant; $l = 0, 1, \ldots, L+1$), γ is the nearest neighbour force constant, and M the mass of an atom with finite mass. Introducing mass-dependent displacements (cf. (1.3.8)) and solving the eigenvalue problem we find the following eigenstates and eigenvalues:

$$B^{(s)}(l) = \sqrt{\frac{2}{L+1}}\, \sin(sla), \tag{2.3.40}$$

$$\omega_s^2 = \omega_M^2 \sin^2(sa/2), \tag{2.3.41}$$

where

$$s = \frac{m\pi}{(L+1)\,a}, \quad m = 1, 2, \ldots, L; \quad \text{and} \quad \omega_M = \sqrt{\frac{4\gamma}{M}}. \tag{2.3.42}$$

Inserting these solutions into the Green function (2.1.38), we obtain the lattice Green function $G_L(ll', \omega)$ for an island of L atoms with finite mass. If the disorder does not affect the electronic structure, then $Q_\alpha(l\varkappa)$ in (2.3.29) is independent of l, and we see that $O(\omega)$ is governed by the configuration average of $A_L(\omega^2) \equiv -\sum_{ll'} \operatorname{Im} G_L(ll', \omega)$, which represents, in essence, the so-called spectral density (cf. (2.3.86)) at $k = 0$. Furthermore, we need $G_L(\omega^2) = -\sum_l \operatorname{Im} G_L(ll, \omega)$, whose configuration average determines the vibrational state density. $A_L(\omega^2)$ and $G(\omega^2)$ are given by

$$A_L(\omega^2) = \frac{\pi}{M}\, \frac{2}{L+1} \sum_{\substack{s \\ (m \text{ odd})}} \cot^2(sa/2)\, \delta(\omega^2 - \omega_s^2), \tag{2.3.43}$$

$$G_L(\omega^2) = \frac{\pi}{M} \sum_s \delta(\omega^2 - \omega_s^2). \tag{2.3.44}$$

The averaging is performed by means of the weight function

$$f(L) = N(1-c)^2\, c^L, \tag{2.3.45}$$

which describes the probability of finding an island of L atoms with finite mass: $f(L)$ is normalized as

$$\sum_L Lf(L) = cN, \tag{2.3.46}$$

where c is the concentration of the finite atoms and N is the total number of atoms in the chain. By means of (2.3.45), we obtain for the averages of the quantities $A_L(\omega^2)$ and $G_L(\omega^2)$

$$\langle A(\omega^2)\rangle = \frac{2}{M}\frac{\omega_M}{\omega^3}\sqrt{1-\omega^2/\omega_M{}^2}\sum_{L=1}^{\infty}\sum_{n=1}^{\infty} f(L)\,\delta\big(2n-1-2(L+1)\,\lambda\big), \tag{2.3.47}$$

$$\langle G(\omega^2)\rangle = \frac{1}{M}\left[\omega_M\omega\sqrt{1-\omega^2/\omega_M{}^2}\right]^{-1}\sum_{L=1}^{\infty}\sum_{n=1}^{\infty} f(L)\,(L+1)\,\delta\big(n-2(L+1)\,\lambda\big), \tag{2.3.48}$$

where

$$\lambda = \frac{1}{\pi}\arcsin\frac{\omega}{\omega_M} \leqq 1/2. \tag{2.3.49}$$

In (2.3.47) and (2.3.48), for a given ω, different combinations of values n and L can provide nonzero contributions. To account for this fact, we insert

$$\sum_{pq} \delta_{\lambda,p/q} = 1 \tag{2.3.50}$$

into the right hand sides of these equations. In (2.3.50) p/q are fractions which cannot be reduced. In calculating $\langle A(\omega^2)\rangle$ we may then put $2n-1=(2k+1)p$, $2(L+1)=(2k+1)\,q$ with $k=0,1,\ldots$ and replace the summations over L and n by a summation over k. Thus, we obtain finally

$$\langle A(\omega^2)\rangle = \frac{N}{M}\frac{(1-c)^2}{c}\frac{\omega_M}{\omega^3}\sqrt{1-\omega^2/\omega_M{}^2}\sum_{\substack{p(\text{odd})\\ q(\text{even})}}\frac{1}{q}\ln\frac{1+c^{q/2}}{1-c^{q/2}}\,\delta(p/q-\lambda) \tag{2.3.51}$$

and, analogously, we get

$$\langle G(\omega^2)\rangle = \frac{N(1-c)^2}{M2c\omega_M\omega\sqrt{1-\omega^2/\omega_M{}^2}} \times\left\{\sum_{\substack{p(\text{odd})\\ q(\text{even})}}\frac{c^{q/2}}{1-c^{q/2}}\,\delta(p/q-\lambda)+\sum_{\substack{p(\text{arbitrary})\\ q(\text{odd})}}\frac{c^q}{1-c^q}\,\delta(p/q-\lambda)\right\}. \tag{2.3.52}$$

The procedure presented here can be readily extended also to a diatomic chain in which one species of atom is randomly replaced by atoms with infinite mass (Böttger and Kleinert, 1981). A diatomic chain has the advantage that the effect of disorder on an optical band can also be studied. Results of such a study are plotted in figure 2.19. For comparison, this figure also shows results of a CPA calculation and a cluster approximation (cf. section

2.3.4.). We see that disorder strongly broadens the δ-functions characterizing the spectral density of a perfect chain. There is some resemblance between the spectral density and the density of vibrational states, but the two quantities are far from being identical.

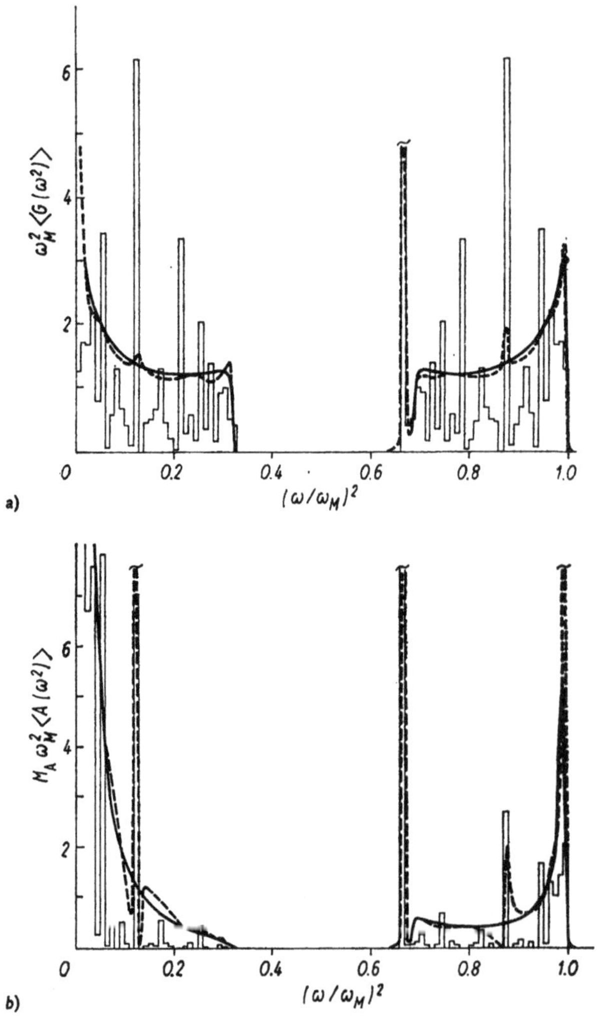

Fig. 2.19. a) Vibrational state density, $\langle G(\omega^2) \rangle$, and b) spectral density, $\langle A(\omega^2) \rangle$, for a one-dimensional $AB_{1-c}C_c$ ($c = 0.3$) alloy (mass ratio $M_A : M_B = 1:2$) calculated (for $M_c = \infty$) by formulas analogous to (2.3.47) and (2.3.48) (histogram) and (for $M_c = 50 M_B$) by the CPA (solid curve) and an embedded cluster method (dashed curve) (after Böttger and Kleinert, 1981)

2.3.4. The average *t*-matrix approximation and the coherent-potential approximation

Using multiple scattering theory, the conception of a single site approximation was introduced in section 2.1.3. In such an approximation any explicit influence of one scatterer on another one is neglected. The individual scattering atoms are viewed as being embedded in an effective medium whose choice is open. Here we consider the following two single site approximations: the average *t*-matrix approximation (ATA) and the coherent potential approximation (CPA). In the ATA the effective medium is determined by requiring that it reproduces the average *t*-matrix at each atom. In the CPA, on the other hand, the effective medium is made self-consistent by requiring that the *t*-matrix for scattering off a single scatterer is zero on average.

Let us consider a mass disordered $A_{1-c}B_c$ alloy with a cubic Bravais lattice. The equation for the Green function (2.1.18) of such a system can be written as

$$\boldsymbol{G}^{-1} = (\boldsymbol{G}^0)^{-1} - \boldsymbol{V}, \tag{2.3.53}$$

where

$$(\boldsymbol{G}^0)^{-1} = \boldsymbol{M}^0\omega^2 - \boldsymbol{\Phi}^0 \tag{2.3.54}$$

and

$$\boldsymbol{V} = (\boldsymbol{M}^0 - \boldsymbol{M})\,\omega^2. \tag{2.3.55}$$

Here, $\boldsymbol{G}^0$ is the Green function of a perfect reference crystal having a mass matrix $\boldsymbol{M}^0 = M^0\mathbf{1}$ and a force constant matrix $\boldsymbol{\Phi}^0$. Note that M^0 does not necessarily agree with the mass of an A or B atom.

We are interested in the self-energy $\boldsymbol{\Sigma}$ defined by (cf. (2.1.59))

$$\langle \boldsymbol{G}\rangle = \boldsymbol{G}^0 + \boldsymbol{G}^0\,\boldsymbol{\Sigma}\,\langle \boldsymbol{G}\rangle, \tag{2.3.56}$$

where $\langle \boldsymbol{G}\rangle$ denotes the configuration average of $\boldsymbol{G}$. $\boldsymbol{\Sigma}$ has the symmetry of the lattice. It describes the effective medium. In general, $\boldsymbol{\Sigma}$ will be complex and non-hermitian. On calculating $\boldsymbol{\Sigma}$, it is often convenient to introduce a diagonal matrix $\boldsymbol{\sigma} = \sigma\mathbf{1}$, with a σ whose choice is open. By means of $\boldsymbol{\sigma}$, (2.3.53) is rewritten as

$$\boldsymbol{G}^{-1} = \tilde{\boldsymbol{G}}^{-1} - \tilde{\boldsymbol{V}}, \tag{2.3.57}$$

where

$$\tilde{\boldsymbol{G}}^{-1} = (\boldsymbol{G}^0)^{-1} - \boldsymbol{\sigma}, \tag{2.3.58}$$

$$\tilde{\boldsymbol{V}} = \boldsymbol{V} - \boldsymbol{\sigma}. \tag{2.3.59}$$

Here, $\tilde{\boldsymbol{G}}$ may be viewed as the Green function of a new reference crystal and $\tilde{\boldsymbol{V}}$ as the corresponding perturbation matrix. The self-energy $\tilde{\boldsymbol{\Sigma}}$, defined by

$$\langle \boldsymbol{G}\rangle = \tilde{\boldsymbol{G}} + \tilde{\boldsymbol{G}}\tilde{\boldsymbol{\Sigma}}\langle \boldsymbol{G}\rangle, \tag{2.3.60}$$

is related to $\boldsymbol{\Sigma}$ by

$$\boldsymbol{\Sigma} = \tilde{\boldsymbol{\Sigma}} + \boldsymbol{\sigma}. \tag{2.3.61}$$

In calculating $\tilde{\Sigma}$ we make use of (2.1.52), (2.1.69), (2.1.70) and write

$$\tilde{V} = \sum_{\substack{l_t \\ j(\mathrm{A,B})}} \eta^j_{(l_t)} \tilde{V}^{j(l_t)}, \tag{2.3.62}$$

where

$$\tilde{V}^{j(l_t)}_{\alpha\alpha'}(ll', \omega) = (v^j - \sigma)\, \delta_{l_t l} \delta_{ll'} \delta_{\alpha\alpha'}, \tag{2.3.63}$$

$$v^j = (M^0 - M^j)\, \omega^2. \tag{2.3.64}$$

Furthermore using (2.3.61) to (2.3.63) and (2.1.71) to (2.1.73), the atomic t-matrix takes the form

$$t^{(l_t)} = \sum_j \eta^j_{(l_t)} t^{j(l_t)}, \tag{2.3.65}$$

where

$$t^{j(l_t)}_{\alpha\alpha'}(ll', \omega) = t^j(\omega)\, \delta_{ll_t} \delta_{ll'} \delta_{\alpha\alpha'}, \tag{2.3.66}$$

$$t^j(\omega) = (v^j - \sigma)/[1 - (v^j - \sigma)\, \tilde{G}(0)], \tag{2.3.67a}$$

$$\tilde{G}(0) \equiv \tilde{G}_{\alpha\alpha}(0, 0, \omega). \tag{2.3.67b}$$

Inserting the t-matrix (2.3.67a) into the single site expression for the self-energy given by (2.1.77), and accounting for (2.3.61), we finally obtain

$$\mathbf{\Sigma} = \Sigma \mathbf{1}, \tag{2.3.68}$$

where

$$\Sigma = \sigma + \langle t \rangle / [1 + \tilde{G}(0)\, \langle t \rangle] \tag{2.3.69}$$

and

$$\langle t \rangle = \sum_j c^j t^j, \tag{2.3.70}$$

with t^j given by (2.3.67).

Equation (2.3.69) can be used in two ways: either we insert into this equation the quantities $\langle t \rangle$ and $\tilde{G}(0)$ which correspond to a given choice of σ, or we determine σ self-consistently from the condition

$$\langle t \rangle = 0 \tag{2.3.71}$$

on account of (2.3.58). The former approach is called the average t-matrix approximation (ATA), because the second term on the right hand side of (2.3.69) represents at each site an effective potential determined by the average t-matrix. This approach is not self-consistent, and the resulting Σ will, in general, depend on the choice of σ. The latter self-consistent approach is usually referred to as the coherent potential approximation (CPA). In this approach we obtain, from (2.3.71) and (2.3.69), $\Sigma = \sigma$ and thus, from (2.3.61) and (2.3.60), $\langle \mathbf{G} \rangle = \tilde{\mathbf{G}}$. The CPA yields a self-energy, which is invariant with respect to the choice of the reference lattice (see below). It is the best single-site approximation. However, the CPA is much more difficult to implement numerically than the ATA.

Let us now draw attention to the ATA. Putting $\sigma = 0$ and choosing the perfect A-crystal as reference, then from (2.3.69), (2.3.70), (2.3.67) and (2.3.64) we obtain

$$\Sigma = \frac{c(M^{\mathrm{A}} - M^{\mathrm{B}})\,\omega^2}{1 - (1-c)\,(M^{\mathrm{A}} - M^{\mathrm{B}})\,\omega^2 G^0(0)}. \tag{2.3.72}$$

Here, M^{A} and M^{B}, respectively, denote the mass of an A- and a B-atom, $G^0(0)$ is the element (2.3.67b) of the Green function of the perfect A-crystal, and $c \equiv c^{\mathrm{B}}$. The self-energy (2.3.72) is identical with that given in (2.2.68). Note that this self-energy and the resulting Green function are not symmetric in the A- and B-atom types. In particular, an increase of the concentration of the B-atoms tends to cancel the resonant behaviour of (2.3.72), so that as $c \to 1$, the self-energy (2.3.72) tends to that resulting from the virtual crystal approximation (VCA).

The result of the ATA can be substantially improved with respect to (2.3.72) by using a VCA reference crystal. To do this, in (2.3.69) we put

$$\sigma = \sum_j c^j v^j \equiv \bar{v}, \tag{2.3.73}$$

which yields

$$\langle t\rangle = \frac{(1-c)\,(v^{\mathrm{A}} - \bar{v})}{1 - (v^{\mathrm{A}} - \bar{v})\,\tilde{G}(0)} + \frac{c(v^{\mathrm{B}} - \bar{v})}{1 - (v^{\mathrm{B}} - \bar{v})\,\tilde{G}(0)}. \tag{2.3.74}$$

Here, $\tilde{G}(0)$ denotes the element (2.3.67b) of the Green function of a crystal described in the VCA. The average t-matrix (2.3.74) is symmetric in the A- and B-atom types. It is exact at either limit of concentration, so it is useful as an interpolation formula.

Now let us turn to the CPA. Using (2.3.70), (2.3.67a) and $\sigma = \Sigma$, the CPA-equation (2.3.71) takes the form

$$0 = \frac{(1-c)\,(v^{\mathrm{A}} - \Sigma)}{1 - (v^{\mathrm{A}} - \Sigma)\,\tilde{G}(0)} + \frac{c(v^{\mathrm{B}} - \Sigma)}{1 - (v^{\mathrm{B}} - \Sigma)\,\tilde{G}(0)}. \tag{2.3.75}$$

This equation must be solved self-consistently for Σ and the element (2.3.67b) of the Green function

$$\tilde{\boldsymbol{G}} = [(\boldsymbol{G}^0)^{-1} - \boldsymbol{\Sigma}]^{-1}. \tag{2.3.76}$$

The resulting Green function $\tilde{\boldsymbol{G}} = \langle \boldsymbol{G}\rangle$ is invariant with respect to the choice of the Green function $\boldsymbol{G}^0$ of the reference crystal. To see this, in (2.3.54) we replace $\boldsymbol{M}^0$ by $\boldsymbol{M}'^0$ and introduce, correspondingly, a Green function $\boldsymbol{G}'^0$ and a self-energy $\boldsymbol{\Sigma}'$ related to $\boldsymbol{G}^0$ and $\boldsymbol{\Sigma}$ as

$$(\boldsymbol{G}'^0)^{-1} = (\boldsymbol{G}^0)^{-1} - \boldsymbol{\Delta}, \tag{2.3.77}$$

$$\boldsymbol{\Sigma}' = \boldsymbol{\Sigma} - \boldsymbol{\Delta}, \tag{2.3.78}$$

where

$$\boldsymbol{\Delta} = (\boldsymbol{M}^0 - \boldsymbol{M}'^0)\,\omega^2 \equiv \Delta \mathbf{1}. \tag{2.3.79}$$

Thus, with (2.3.71) we have

$$G^{-1} = (G'^0)^{-1} - \Sigma' \tag{2.3.80}$$

and

$$0 = \frac{(1-c)\,(v'^{\mathrm{A}} - \Sigma')}{1 - (v'^{\mathrm{A}} - \Sigma')\,\tilde{G}(0)} + \frac{c\,(v'^{\mathrm{B}} - \Sigma')}{1 - (v'^{\mathrm{B}} - \Sigma')\,\tilde{G}(0)}. \tag{2.3.81}$$

Inserting now (2.3.78) and $v'^j = v^j - \Delta$ into (2.3.81), we find that (2.3.81) agrees with (2.3.75), i.e. both these equations yield the same $\tilde{G}(0)$.

The CPA expression (2.3.75) is symmetric in the A- and B-atom types. It is exact at either limit of concentration, and so it is a useful interpolation formula.

Introducing $v^{\mathrm{A}} = \delta/2$, $v^{\mathrm{B}} = -\delta/2$ and $\bar{v} = (1-c)\,v^{\mathrm{A}} + cv^{\mathrm{B}}$, the CPA equation (2.3.75) may be written as

$$\Sigma = \bar{v} + \frac{c(1-c)\,\delta^2\tilde{G}(0)}{1 + [\bar{v} + \Sigma]\,\tilde{G}(0)}, \tag{2.3.82}$$

where $\tilde{G}(0)$ results from (2.3.76b). Similarly, using (2.3.73) and (2.3.74) the ATA expression (2.3.69) for the self-energy becomes

$$\Sigma = \bar{v} + \frac{c(1-c)\,\delta^2\tilde{G}(0)}{1 + 2\bar{v}\tilde{G}(0)}, \tag{2.3.83}$$

where $\tilde{G}(0)$ is given by the VCA Green function, i.e. by (2.3.76b) after replacing Σ by $\bar{v}$. Comparing (2.3.82) and (2.3.83), we see that the ATA may be viewed as the first iteration (i.e. replacing $\Sigma \to \bar{v}$) of (2.3.82) towards self-consistency.

The Green function elements $\tilde{G}(0)$ in (2.3.74) and (2.3.75) can be expressed by the frequency spectrum of the unperturbed reference crystal. To see this, we make use of (2.1.39), (1.3.44), and (1.4.13) and obtain for the Green function element $G^0(0, \omega) \equiv G^0_{\alpha\alpha}(0, 0, \omega)$ of the (simple cubic) reference crystal the following expression:

$$G^0(0, \omega) = \frac{1}{3} \sum_{\alpha} G^0_{\alpha\alpha}(0, 0, \omega) = \frac{1}{3M^0N} \sum_{kj} \frac{1}{\omega^2 - \omega_j^2(\boldsymbol{k})}$$

$$= \frac{1}{M^0} \int_0^\infty \mathrm{d}\omega'^2 \, \frac{G(\omega'^2)}{\omega^2 - \omega'^2}. \tag{2.3.84}$$

Here $G(\omega^2)$ denotes the distribution function of the squared eigenfrequencies of the reference crystal: ω is in general a complex quantity in (2.3.84). Since in the single site approximation the self-energy is a diagonal matrix (see (2.3.68)) $\tilde{G}(0)$ in (2.3.75) can be obtained from $G^0(0, \omega)$ by replacing ω^2 in (2.3.84) by $\omega^2 - \Sigma(\omega)/M^0$. Note that $\Sigma(\omega)$ is, in general, a complex quantity. Similarly, $\tilde{G}(0)$ in (2.3.74) results from $G^0(0, \omega)$ by replacing $\omega^2 \to \omega^2 - \bar{v}/M^0$.

Now let us compare some results obtained for frequency spectra of disordered alloys by the ATA, the CPA and machine calculations. For a three dimensional system with nearest neighbour interaction, results obtained by the ATA and by machine calculations are plotted in figure 2.20. We see from this figure that for a small amount of lighter defect atoms the ATA correctly produces a band of high-frequency localized states. With increasing defect concentration this band merges with the host band. A resonant defect band occurs when the concentration of the heavier atoms becomes small. It is obvious from figure 2.20 that the ATA does not correctly describe the band edges. The CPA, however, rather correctly reproduces the band edges. This can be seen from figure 2.21, where results of the CPA and machine calculations are presented for a simple cubic lattice with nearest neighbour interactions.

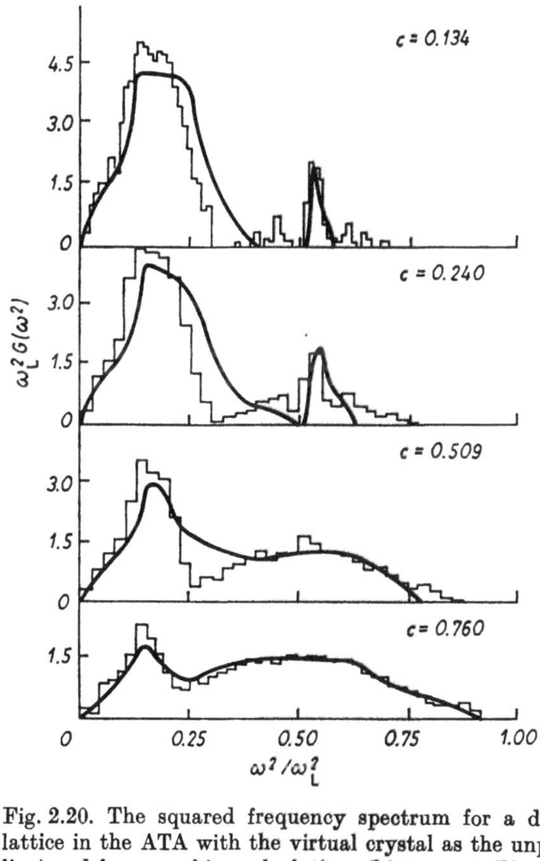

Fig. 2.20. The squared frequency spectrum for a disordered simple cubic lattice in the ATA with the virtual crystal as the unperturbed crystal (solid line) and from machine calculations (histogram). The mass ratio between the species of atoms is 3:1. c is the concentration of the light atoms. ω_L is the maximum frequency of the lattice containing light atoms only (after LEATH and GOODMAN, 1969)

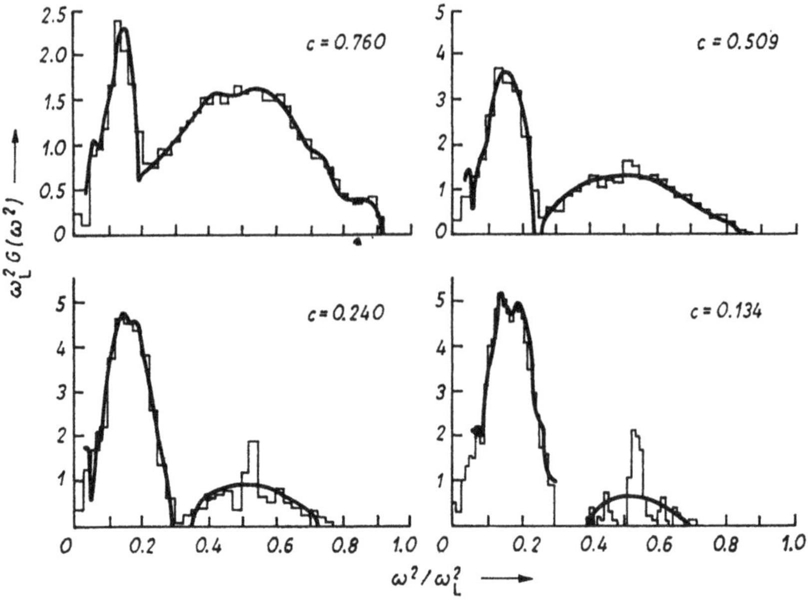

Fig. 2.21. The squared frequency spectrum for a disordered simple cubic lattice in the CPA (solid line) and from machine calculations (histogram). The mass ratio between the atomic species is 3:1. c is the concentration of the light atoms. ω_L is the maximum frequency of the lattice containing light atoms only (after TAYLOR, 1967)

Both the ATA and CPA fail to reproduce the spiky structure of the frequency spectra. This structure arises from clusters of light defect atoms. Clusters, however, are not accounted for in a single-site approximation. Note that the spiky structure of the spectra is especially pronounced in the case of disordered chains (cf. fig. 2.1), because in such a system fluctuations in the local composition play a dominant role. Such fluctuations are less important in a three-dimensional system, where a given site has a larger number of nearest neighbours and hence sees some average surroundings.

Note that differences between predictions of the ATA and the CPA may be masked by a featureless frequency spectrum of the perfect lattice. Such a spectrum has the lattice (simple cubic lattice with nearest neighbour interactions) underlying the calculations presented in figures 2.20 and 2.21. Substantial differences between ATA and CPA results may occur when a realistic frequency spectrum for the unperturbed lattice is used. This can be seen from figure 2.22, where frequency spectra calculated using the ATA and the CPA for a disordered fcc lattice are shown. It is obvious, that the ATA generally predicts spectra with much more structure than the CPA. This structure originates because the imaginary part of the self-energy in the ATA vanishes at frequencies above the maximum mean lattice frequency $\overline{\omega}_L$ (cf. section 2.2.4., where $\Gamma(\omega) = 0$ when $g^0(\omega) = 0$). The fact that Im Σ vanishes at $\omega > \overline{\omega}_L$ is very unsatisfactory in that it predicts undamped plane-wave excitations for a disordered system (cf. also (2.2.76)). In figure 2.22 for $\omega > \overline{\omega}_L$, the ATA-frequency spectra are shaded. The CPA provides Im $\Sigma \neq 0$ for all frequencies at which the frequency spectrum is nonzero. Hence the CPA predicts a finite lifetime for all plane-wave excitations. This is a consequence of its self-consistent nature.

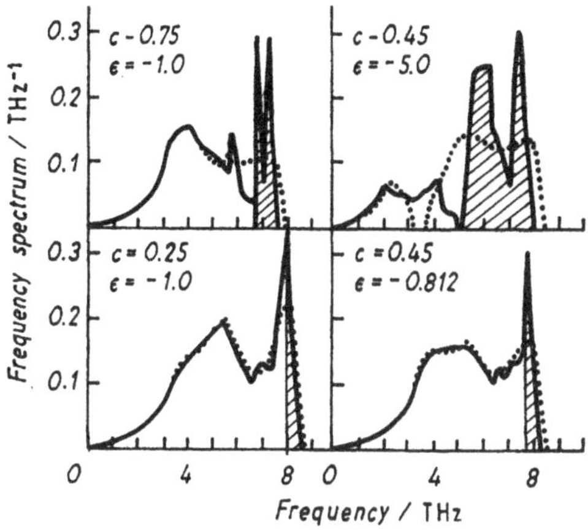

Fig. 2.22. ATA and CPA frequency spectra calculated for fcc random alloys: ——— ATA, - - - - - CPA. The regions, where the ATA yields a vanishing imaginary part of the self energy, are shaded. $\epsilon = (M^h - M^d)/M^h$ (M^h, M^d denote the mass of a host and a defect atom, respectively) and c is the concentration of the defect atoms. On the left ϵ is fixed and c varies, and on the right ϵ varies and c is fixed (after KAMITAKAHARA and TAYLOR, 1974)

Dispersion and damping of plane-wave excitations (phonons) in disordered alloys can be obtained from inelastic coherent neutron scattering experiments. Before comparing results of such experiments with results of CPA calculations, we discuss here some theoretical aspects of plane-wave excitations in disordered systems. As already mentioned in section 2.2.4., in a disordered alloy with a cubic Bravais lattice the one-phonon coherent neutron scattering cross section is proportional to Im $\langle G_j(\boldsymbol{k}, \omega)\rangle$ where, in the case of a diagonal self-energy matrix (cf. (2.3.68)), $\langle G_j(\boldsymbol{k}, \omega)\rangle$ is given by (cf. 2.2.70))

$$\langle G_j(\boldsymbol{k}, \omega)\rangle = [\omega^2 - \omega_j^2(\boldsymbol{k}) - \Sigma(\omega)]^{-1}. \tag{2.3.85}$$

Here $\Sigma(\omega)$ describes the self-energy in the ATA or the CPA, respectively, and $\omega_j(\boldsymbol{k})$ denotes the dispersion relation of the jth vibrational branch of the perfect reference crystal.

The quantity

$$\langle A_j(\boldsymbol{k}, \omega)\rangle = -\frac{1}{\pi} \operatorname{Im} \langle G_j(\boldsymbol{k}, \omega + i\varepsilon)\rangle, \quad \varepsilon \to +0 \ (\text{real } \omega), \tag{2.3.86}$$

is generally referred to as (configuration averaged) spectral density. By means of this quantity, the Green function $\langle G_j(\boldsymbol{k}, \omega)\rangle$, defined in the complex ω-plane with a cut along the real axis, can be expressed by a Cauchy integral as

$$\langle G_j(\boldsymbol{k}, \omega)\rangle = \int_{-\infty}^{\infty} \mathrm{d}\omega' \langle A_j(\boldsymbol{k}, \omega')\rangle/(\omega - \omega')$$

$$= \int_{0}^{\infty} \mathrm{d}\omega'^2 \langle A_j(\boldsymbol{k}, \omega')\rangle/(\omega^2 - \omega'^2). \tag{2.3.87}$$

Here, the second form has been obtained by using the fact that Im $\langle G_j(\boldsymbol{k}, \omega)\rangle$ is an odd function of ω (cf. section 2.3.1.). Similarly as in section 2.1.3., we may now define a retarted Green function by $\langle G_j(\boldsymbol{k}, \omega)\rangle^{\text{ret}} \equiv \langle G_j(\boldsymbol{k}, \omega + i\varepsilon)\rangle$ (real ω), which is an analytical function in the upper complex ω-half plane, but which has poles in the lower half plane. According to the general conception of elementary excitations, the real part of these poles describes the dispersion of plane-wave excitations and the imaginary part their damping. Note, however, that such an interpretation of the poles is reasonable only when their imaginary parts are smaller than the corresponding real parts, i.e. when there is no so-called overdamping of the excitations. Thus, replacing ω in (2.3.87) by $\omega + i\varepsilon$ and using (2.1.21), we find

$$\langle G_j(\boldsymbol{k}, \omega)\rangle^{\text{ret}} = P \int_{-\infty}^{\infty} \mathrm{d}\omega' \frac{\langle A_j(\boldsymbol{k}, \omega')\rangle}{\omega - \omega'} - i\pi \langle A_j(\boldsymbol{k}, \omega)\rangle. \tag{2.3.88}$$

This equation shows that in the lower complex ω-half plane the poles of $\langle G_j(\boldsymbol{k}, \omega)\rangle^{\text{ret}}$ agree with those of $\langle A_j(\boldsymbol{k}, \omega)\rangle$, because the first term on the right hand side of (2.3.88) is an analytical function. Consequently, a study of plane-wave excitations in disordered alloys requires that we investigate the poles of $\langle A_j(\boldsymbol{k}, \omega)\rangle$ in the lower complex ω-half plane. Such excitations will be well defined, when for real ω $\langle A_j(\boldsymbol{k}, \omega)\rangle$ is a lorentzian with a sufficiently small half-width.

For a disordered Ni_{1-c} Pt_c alloy, experimental data of one-phonon coherent neutron scattering are shown in figure 2.23. We see that the scattering intensity as a function of ω exhibits a double peak structure. This structure is due to resonant modes of the heavier Pt-atoms in the Ni-host ($M^{\text{Pt}}/M^{\text{Ni}} = 3.32$). The doubling of a phonon peak occurs at frequencies near the resonant frequency (see the discussion in section 2.2.4. after (2.2.75)). With increasing concentration of Pt-atoms the resonant behaviour is strongly enhanced, the phonon line shapes become well separated double peaks, and the corresponding dispersion curve splits, thereby producing a frequency gap (fig. 2.24). Figure 2.25 shows the scattering intensity for a $Rb_{1-c}K_c$ alloy as a function of the frequency. The double peak structure observed in this case is due to localized modes of the lighter K-atoms ($M^{\text{Rb}}/M^{\text{K}} = 2.18$). We see that with increasing concentration of K-atoms the localized mode peak becomes well resolved from

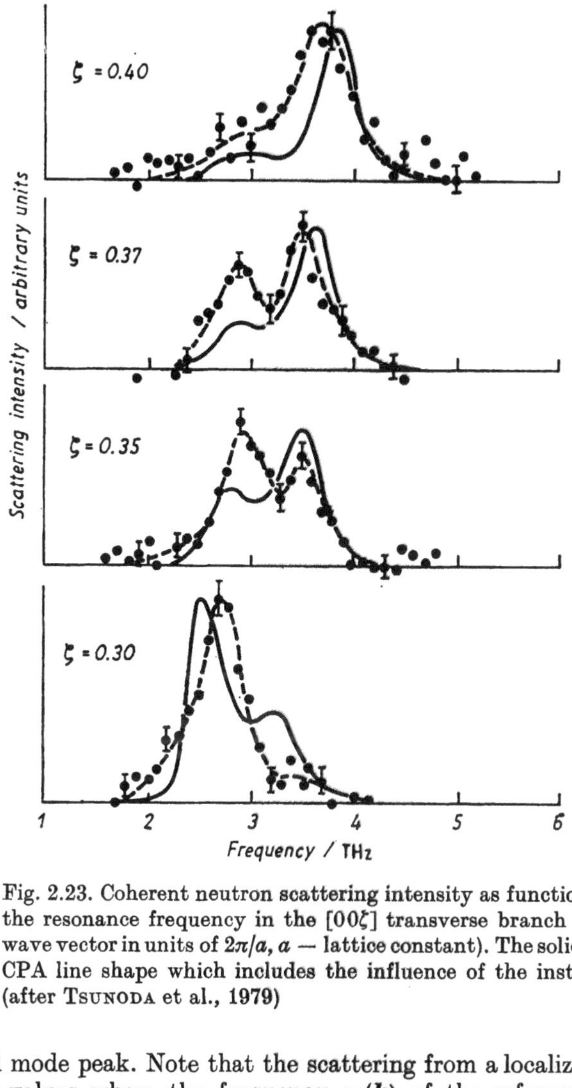

Fig. 2.23. Coherent neutron scattering intensity as function of frequency near the resonance frequency in the [00ζ] transverse branch of $Ni_{0.95}Pt_{0.05}$ (ζ — wave vector in units of $2\pi/a$, a — lattice constant). The solid curves indicate the CPA line shape which includes the influence of the instrumental resolution (after TSUNODA et al., 1979)

the band mode peak. Note that the scattering from a localized mode is strongest at $\boldsymbol{k}$-values where the frequency $\omega_j(\boldsymbol{k})$ of the reference crystal attains its maximum value (cf. (2.2.76)). Dispersion curves for phonons in $Rb_{1-c}K_c$ alloys are shown in figure 2.26. In this figure the dispersion curves of the localized mode are of particular interest. Note that in figures 2.23 and 2.25 the width of a phonon peak is generally some tenths of its mean frequency. Hence, the phonon conception may be used for interpreting the experimental results.

For comparison, figures 2.23 and 2.25 also show results of CPA calculations. We see that the CPA calculations are qualitatively in good agreement with the

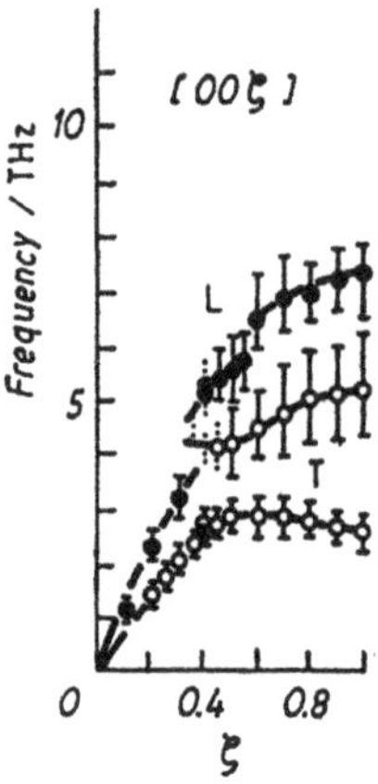

Fig. 2.24. Phonon dispersion curves for the [00ζ] direction for $Ni_{0.70}Pt_{0.30}$. ζ is the wave vector in units of $2\pi/a$ (a — lattice constant). The dotted lines show approximate positions of poorly defined scattering peaks (after TSUNODA et al., 1979)

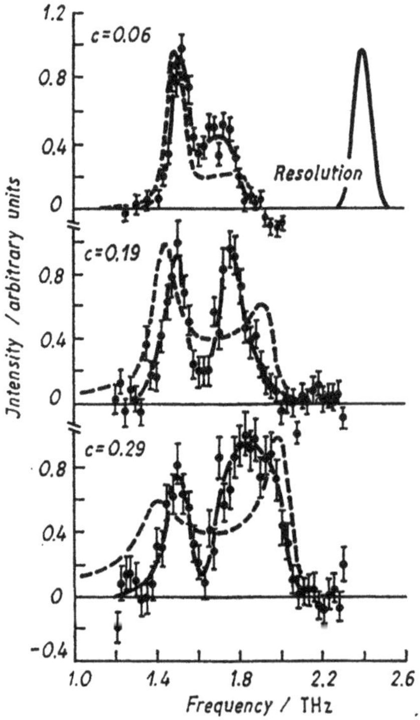

Fig. 2.25. Coherent neutron scattering intensity as a function of frequency for the [ζζ0] longitudinal zone boundary vibrations for $Rb_{1-c}K_c$ as a function of the concentration c ($\zeta = 0.5$; ζ — wave vector in units of $2\pi/a$, a — lattice constant). The high frequency peaks arise from localized modes of the lighter K-atoms. The dashed lines represent CPA results (after KAMITAKAHARA and COPLEY, 1978)

experiments, but there is much room for improvements in the theory. The remaining discrepancies between theory and experiment suggest the importance of force-constant changes and cluster effects in addition to single-site effects from mass disorder. In particular, neutron scattering experiments give some indications for a $\boldsymbol{k}$-dependence of the self-energy. Note that in the case of a pure mass disorder, the single site expression for Σ is site diagonal and thus $\boldsymbol{k}$-independent.

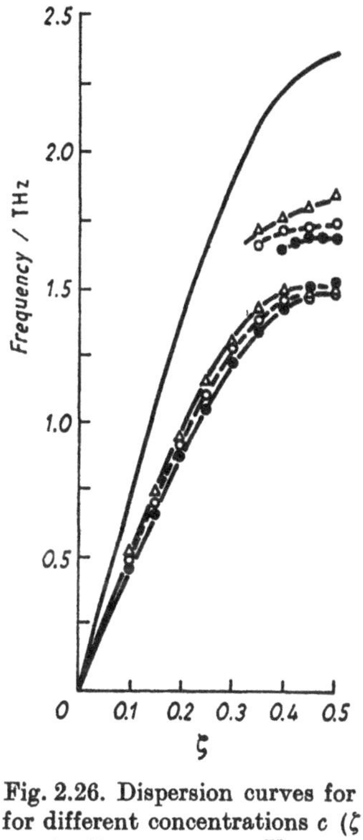

Fig. 2.26. Dispersion curves for the $[\zeta\zeta 0]$ longitudinal vibrations of $Rb_{1-c}K_c$ for different concentrations c (ζ — wave vector in units of $2\pi/a$, a — lattice constant); ——— pure K, —△— $c = 0.29$, - -○- - $c = 0{,}18$, —●— $c = 0{,}06$ (after Kamitakahara and Copley, 1978)

An account of off-diagonal disorder in alloys is a more intricate problem for phonons than for electrons, because in the former case off-diagonal disorder (changes in force constants) is accompanied by diagonal disorder due to the sum rule (2.1.78). An analogous constraint does not exist for electrons (cf. Whitelaw, 1981).

The assumption of a linear superposition of force constants, i.e. $\Phi^{AB} = (\Phi^{AA} + \Phi^{BB})/2$ (cf. section 2.1.3.), is at present the best approach for including off-diagonal disorder within the CPA. To obtain the CPA equations for this case, we use (2.1.85) to (2.1.95) and proceed as previously mentioned in this section. With the perfect A-crystal as reference crystal we find (using the notation of section 2.1.3.)

$$0 = (1 - c)(-\boldsymbol{K})[1 - \tilde{\boldsymbol{G}}(-\boldsymbol{K})]^{-1} + c(\boldsymbol{V} - \boldsymbol{K})[1 - \tilde{\boldsymbol{G}}(\boldsymbol{V} - \boldsymbol{K})]^{-1}, \quad (2.3.89)$$

where $\boldsymbol{V}$, $\boldsymbol{K}$ and $\tilde{\boldsymbol{G}}$ are matrices in the defect space, and

$$[(\boldsymbol{G}^0)^{-1} - \boldsymbol{\Sigma}]\,\tilde{\boldsymbol{G}} = \boldsymbol{1}, \quad (2.3.90\text{a})$$

$$\Sigma_{\alpha\alpha'}(ll', \omega) = \sum_{l_i} \sum_{ss'} \delta_{l,s+l_i}\delta_{l',s'+l_i} K_{\alpha\alpha'}(ss', \omega). \quad (2.3.90\text{b})$$

Equations (2.3.89) to (2.3.90b) form a closed set of complex matrix equations for $\boldsymbol{K}$, $\boldsymbol{\Sigma}$, and $\tilde{\boldsymbol{G}}$ which must be solved self-consistently.

The CPA, including force constant changes in the superposition limit, provides better overall agreement with neutron scattering data than the mass-defect CPA. But there remain marked discrepancies in the phonon line-shapes between theory and experiment (Mostoller and Kaplan, 1977), which suggests the need to include multiple scattering by clusters of atoms.

It was found that the approach which incorporates force-constant changes into the CPA by assuming Φ^{AB} to be the geometrical mean of Φ^{AA} and Φ^{BB} is more limited than the superposition approach. It does not yield a $\boldsymbol{k}$-dependent self-energy, and for certain sets of parameters it reduces to the VCA. Moreover, it often predicts no correct band edges (Mostoller and Kaplan, 1979). Both approaches cannot reproduce a spiky structure in the vibrational state density.

The lattice dynamics of mixed diatomic crystals have also been investigated by the ATA (Lacina and Pershan, 1970) and the CPA (Sen and Hartman, 1974, and Taylor, 1976). However, to date in CPA calculations solely short-range forces have been taken into account, because the inclusion of long-range forces is a cumbersome task. Hence, the results obtained cannot be compared with infrared and Raman spectra of crystals with long-range electrostatic forces.

For more details on the ATA and the CPA the reader is referred to the review by Elliott et al. (1974), where the reader will also find other methods for deriving the ATA and the CPA, such as diagram techniques and the method of the conditionally averaged Green functions.

Concluding this section, let us summarize some important properties of the CPA and the ATA. The CPA is a self-consistent single-site approximation. It is the best single-site approximation. It is symmetric with respect to interchange of host and defect atoms and gives the correct self-energy in both dilute limits $c \to 0$ and $(1 - c) \to 0$. It reproduces the exact frequency spectrum to order c or $(1 - c)$ in the concentration. Furthermore, the CPA is correct in both the weak and strong scattering limits, it preserves the translational invariance of the reference crystal (virtual crystal), and it exhibits the correct analytic properties of the self-energy. Note that the latter property is important for obtaining a single solution of the self-consistency equation and for obtaining a physical behaviour of the frequency spectrum (see Ducastelle, 1974). The failure of the CPA is in that it neglects multi-site scattering.

The ATA is a non-self-consistent single-site approximation. It is easier to handle than the CPA, but its results are of more limited validity. Like the CPA, it is symmetric in the host and defect atoms, it is correct in both the dilute limits $c \to 0$ and $(1 - c) \to 0$, and it reproduces the exact spectrum to c or $(1 - c)$. Note that the VCA fails already in c or $(1 - c)$ to reproduce single impurity modifications of the frequency spectrum.

2.3.5. Cluster approximations

In the preceding section effects of mass and force constant disorder have been investigated within the framework of a single site approximation. Certain effects of disordered systems cannot be reproduced by such an approximation, for instance, the spiky structure of the vibrational state density, which arises from clustering of atoms. The purpose of the present section is to describe some approaches allowing for cluster effects.

Let us begin with the so-called molecular coherent potential approximation (MCPA), which is sometimes also called cellular coherent potential approximation (CCPA). For convenience, we first reformulate the CPA as follows. Consider a system with the effective medium σ on all sites except the origin, where an atom of type j is assumed to be located. The Green function $\boldsymbol{G}_j$ of such a system obeys the equation (cf. (2.3.57))

$$\boldsymbol{G}_j^{-1} = \tilde{\boldsymbol{G}}^{-1} - \tilde{\boldsymbol{V}}^j, \tag{2.3.91}$$

where $\tilde{\boldsymbol{G}}$ and $\tilde{\boldsymbol{V}}^j \equiv V^{j(0)}_{\alpha\alpha'}(ll', \omega) = (v^j - \sigma)\,\delta_{l0}\delta_{ll'}\delta_{\alpha\alpha'}$, respectively, are defined by (2.3.58) and (2.3.63). From this equation we obtain

$$\boldsymbol{G}_j\tilde{\boldsymbol{V}}^j = \tilde{\boldsymbol{G}}[1 - \tilde{\boldsymbol{V}}^j\tilde{\boldsymbol{G}}]^{-1}\,\tilde{\boldsymbol{V}}^j \equiv \tilde{\boldsymbol{G}}\boldsymbol{t}^j. \tag{2.3.92}$$

Here $\boldsymbol{t}^j$ agrees with the t-matrix $\boldsymbol{t}^{j(0)}$ from (2.3.66). On replacing $\tilde{\boldsymbol{V}}^j$ on the left-hand side of (2.3.92) by

$$\tilde{\boldsymbol{V}}^j = \boldsymbol{t}^j[1 + \tilde{\boldsymbol{G}}\boldsymbol{t}^j]^{-1}, \tag{2.3.93}$$

we find

$$\boldsymbol{G}_j = \tilde{\boldsymbol{G}} + \tilde{\boldsymbol{G}}\boldsymbol{t}^j\tilde{\boldsymbol{G}}. \tag{2.3.94}$$

From this equation the CPA-condition (2.3.71)

$$\sum_j c^j\boldsymbol{t}^j = 0 \tag{2.3.95}$$

($\boldsymbol{t}^j$ is a 1×1 matrix in the CPA) can be written as

$$\langle G_j(0)\rangle \equiv \sum_j c^jG_j(0) = \tilde{G}(0), \tag{2.3.96}$$

where $\tilde{G}(0) \equiv G_{\alpha\alpha}(0, 0, \omega)$ aud $G_j(0) \equiv G_{j,\alpha\alpha}(0, 0, \omega)$. Equation (2.3.96) is the customary effective medium (mean field) condition: the effect of replacing the effective medium at a given site by the actual value of the random quantity (mass) must be zero on average.

In the MCPA the real disordered material outside some cluster C is replaced by an effective medium with a cluster-diagonal self-energy (TSUKADA, 1972, GONIS and GARLAND, 1978). Let $\boldsymbol{v}^j$ and $\boldsymbol{\sigma}$, respectively, be the defect matrix and the effective medium belonging to C and let $\tilde{\boldsymbol{V}}^j$ be a matrix containing solely a single nonzero submatrix, $\boldsymbol{v}^j - \boldsymbol{\sigma}$, which is connected with C. Inserting this matrix $\tilde{\boldsymbol{V}}^j$ into (2.3.91), then the Green function $\boldsymbol{G}_j$, defined by (2.3.91), describes a system composed of clusters, all of them, except the cluster C, containing the effective medium $\boldsymbol{\sigma}$. By (2.3.92) to (2.3.94) we find the following self-consistency condition (cf. (2.3.96))

$$\langle \hat{\boldsymbol{G}}_j \rangle = \hat{\tilde{\boldsymbol{G}}}, \tag{2.3.97}$$

or equivalently (cf. (2.3.95))

$$\langle [\boldsymbol{1} - (\boldsymbol{v}^j - \boldsymbol{\sigma})\, \hat{\tilde{\boldsymbol{G}}}]^{-1} (\boldsymbol{v}^j - \boldsymbol{\sigma}) \rangle = \boldsymbol{0}. \tag{2.3.98}$$

Here $\hat{\boldsymbol{G}}_j$ and $\hat{\tilde{\boldsymbol{G}}}$ respectively denote the submatrices of $\boldsymbol{G}_j$ and $\tilde{\boldsymbol{G}}$ belonging to the cluster C: $\langle \ldots \rangle$ denotes the averaging over all cluster configurations. Equation (2.3.97) or (2.3.98) have to be solved self-consistently together with

$$\tilde{\boldsymbol{G}}^{-1} = (\boldsymbol{G}^0)^{-1} - \boldsymbol{\Sigma}, \tag{2.3.99}$$

where $\boldsymbol{G}^0$ is the Green function of the reference crystal and $\boldsymbol{\Sigma}$ is the self-energy matrix having block diagonal form with the blocks $\boldsymbol{\sigma}$. Note that in the case of mass disorder for a n-site cluster the coherent potential $\boldsymbol{\sigma}$ is a $n \times n$ matrix. Equation (2.3.97) or, equivalently, (2.3.98) contains n^2 equations for determining the n^2 elements of $\boldsymbol{\sigma}$.

The MCPA yields an effective medium exhibiting an artifical superstructure, which depends on the cluster size. Hence, the MCPA has a drawback in that it does not yield an average Green function having the translational invariance of the reference crystal (or of the virtual crystal). However, the MCPA has the benefit of being a fully self-consistent cluster extension of the single-site CPA. It always yields a positive density of states and a Green function which is analytic throughout the upper (lower) complex frequency plane. It may also be extended to include off-diagonal disorder.

For a binary mass-disordered chain, in figure 2.27 the vibrational state density calculated in the MCPA is shown and compared with exact results. As can be seen, the MCPA reproduces much of the spiky structure of the state density. For a binary mass disordered chain, figure 2.28 shows the spectral density calculated by a Monte-Carlo method, by the CPA and the MCPA. We see that, unlike the CPA, the MCPA also reproduces the fine structure of the spectral function at high frequencies. Note that the effect of disorder increases with the frequency for mass disorder, because it is introduced through the quantity $M^j\omega^2$. Thus, according to figure 2.28, for larger disorder the CPA and the MCPA begin to differ appreciably. The artificial superstructure occurring in the MCPA does not appear to affect the good agreement between the MCPA and the numerical results. MCPA calculations have so far only been performed for relatively small clusters in linear chains.

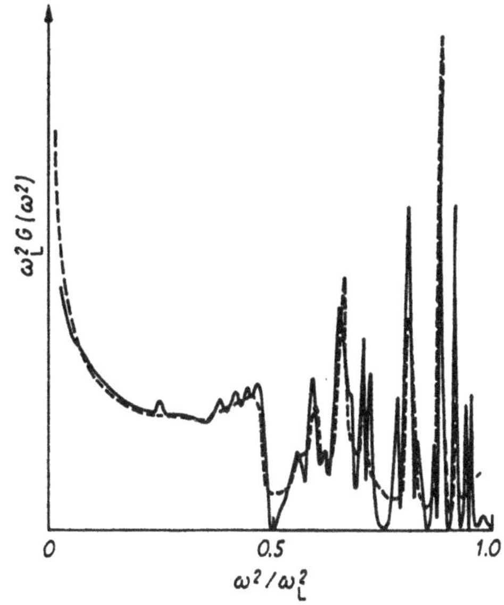

Fig. 2.27. Squared frequency spectrum for a mass-disordered two component chain with a mass ratio between the atomic species 2 : 1 and equal concentration of the species: ——— exact results, - - - - - 7-site MCPA. ω_L is the maximum frequency of the chain containing light atoms only (after HIRSCH, 1978)

The so-called homomorphic cluster coherent potential approximation (HCPA) (ODAGAKI and YONEZAWA, 1978, 1979) yields an analytic Green function, too. Moreover, unlike the MCPA, it avoids the loss of the translational invariance of the reference crystal. The HCPA is based on the assumption that the perturbation matrix (2.1.31) may be partitioned into "homomorphic" subunits as

$$V = \sum_C V^C, \tag{2.3.100}$$

where V^C is a matrix having only the single nonzero submatrix v^C which is connected with the sites of the cluster C. For n-site clusters v^C is a $n \times n$ matrix. In (2.3.100) the sum runs over all possible homomorphic clusters. To illustrate the decomposition (2.3.100) let us consider homomorphic two-site clusters (pairs of nearest neighbour atoms) on a square lattice with mass disorder. According to (2.1.52), (2.1.69) and (2.1.70), we have in this case

$$V = \sum_{l_1} V^{(l_1)}, \tag{2.3.101}$$

where

$$V^{(l_1)}_{\alpha\alpha'}(ll', \omega) = \sum_j \eta^j_{(l_1)} M^0 \epsilon^j \delta_{l_1 l} \delta_{ll'} \delta_{\alpha\alpha'}. \tag{2.3.102}$$

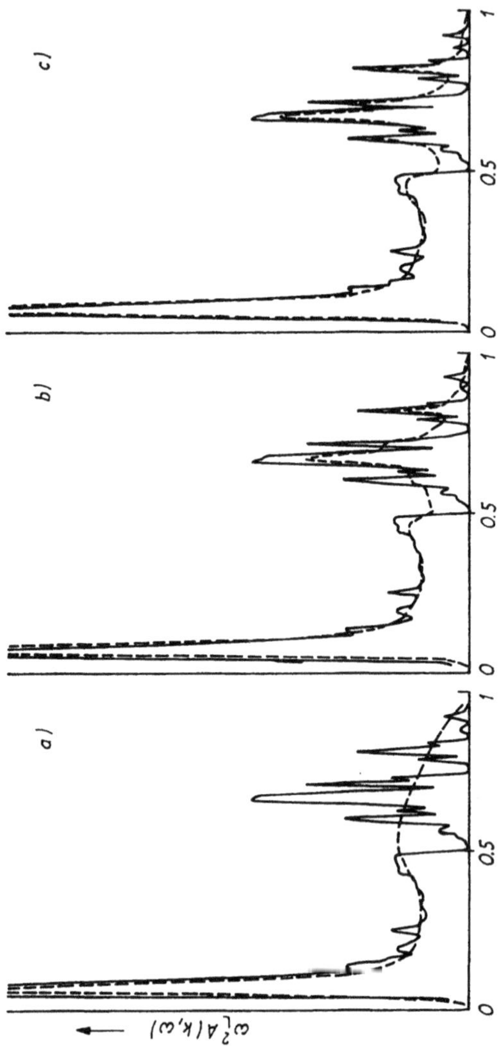

Fig. 2.28. Spectral function $A(k, \omega)$ for a mass-disordered two component chain with a mass ratio between the atomic species 2 : 1, equal concentration of the species, and k defined by $2(1 - \cos ka) = 0.4$ (a — lattice constant); ——— Monte-Carlo method; - - - -: a) CPA, b) 5-site MCPA and c) 7-site MCPA; ω_L is the maximum frequency of the chain containing light atoms only (after Hirsch, 1978)

We can share the relative mass change ϵ^j between the s nearest neighbour bonds connected with an atom by decomposing it into s equal portions (cf. fig. 2.29). In this way $\boldsymbol{V}$ can be rewritten as a sum of contributions from nearest neighbour pairs of atoms, which corresponds to the homomorphic partition (2.3.100). Note that for the HCPA the sum in (2.3.100) accounts for all nearest neighbour pairs, while for the MCPA the analogous sum includes only those nearest neighbour pairs which form molecules.

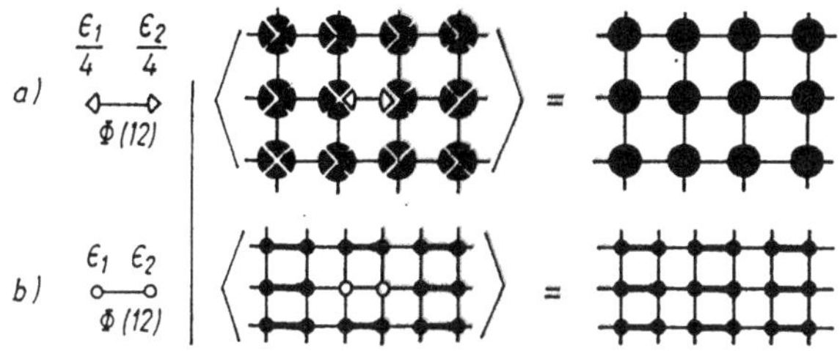

Fig. 2.29. The cluster CPA condition for a disordered square lattice for a) HCPA and b) MCPA. The cluster, the idea of the approximation and the effective medium are shown for both approximations (after ODAGAKI and YONEZAWA, 1979)

Now we continue to derive the HCPA. We express the self-energy as a sum of homomorphic parts:

$$\boldsymbol{\Sigma} = \sum_C \boldsymbol{\sigma}^C, \tag{2.3.103}$$

where $\boldsymbol{\sigma}^C$ is a matrix containing solely a single nonzero submatrix, $\boldsymbol{\sigma}$, which is connected with the sites of the cluster C, but which does not depend on C. Now we choose one of the homomorphic subunits of (2.3.100) as the disordered unit in a system having the effective medium $\boldsymbol{\sigma}$ in all other subunits. Quite analogous to the derivation of the MCPA, we find for the HCPA the following self-consistency condition:

$$\langle[\mathbf{1} - (\boldsymbol{v}^C - \boldsymbol{\sigma})\, \tilde{\boldsymbol{G}}^C]^{-1}\, (\boldsymbol{v}^C - \boldsymbol{\sigma})\rangle = \mathbf{0}, \tag{2.3.104}$$

where $\tilde{\boldsymbol{G}}^C$ denotes the submatrix of $\tilde{\boldsymbol{G}}$ being connected with the cluster C: The Green function $\tilde{\boldsymbol{G}}$ is given by (2.3.99) with the self-energy $\boldsymbol{\Sigma}$ from (2.3.103). $\langle\ldots\rangle$ denotes the averaging over all cluster configurations. As schematically shown in figure 2.29, the HCPA preserves the translational invariance of the reference crystal, whereas the MCPA produces a superstructure.

Besides the two-cluster approximations mentioned above, numerous other attempts have been made to go beyond the single-site CPA. Some of these attempts are very promising, including that based on the augmented-space formalism, which describes the disorder in terms of interactions with a fictitious field rather than by configurationally dependent quantities (KAPLAN and GRAY, 1976, DIEHL and LEATH, 1979, KAPLAN et al., 1980, GRAY and KAPLAN,

1981). However, many of these attempts suffer from certain limitations reducing their usefulness for studying disordered alloys. For instance, they do not give properly analytic Green functions, do not include the correct low (high) concentration limit, cannot account for off-diagonal disorder, are not self-consistent or do not preserve the translational invariance of the reference crystal (see Elliott et al., 1974, Gonis and Garland, 1978, and references therein).

Generally, cluster CPA theories are numerically cumbersome due to the self-consistent evaluation of the spatially varying self-energy. However, sometimes it suffices to calculate properties of disordered alloys by using a non-self-consistent cluster approximation. Such an approximation has the advantage that it requires much less numerical work than a self-consistent calculation. As an example, let us consider a state density calculation by the embedded-cluster method (Myles and Dow, 1979). This method is based on the idea that a small cluster of atoms embedded in an effective medium should produce the main features of the vibrational spectrum when the effective medium is suitable chosen. It uses an effective reference crystal, whose Green function $\tilde{\boldsymbol{G}}$ is defined by

$$\tilde{\boldsymbol{G}}^{-1} = (\boldsymbol{G}^0)^{-1} - \boldsymbol{\sigma}, \tag{2.3.105}$$

where $\boldsymbol{G}^0 = [\boldsymbol{M}^0\omega^2 - \boldsymbol{\Phi}^0]^{-1}$ denotes the Green function of a perfect reference crystal and $\boldsymbol{\sigma}$ is an effective medium. Now we replace in this effective reference crystal a cluster of n atoms by real atoms and consider the resulting system as a model of the alloy at hand. The Green function $\boldsymbol{G}$ of this system obeys the equation

$$\boldsymbol{G} = \tilde{\boldsymbol{G}} + \tilde{\boldsymbol{G}}\boldsymbol{V}\boldsymbol{G} \tag{2.3.106}$$

with the perturbation matrix

$$\boldsymbol{V} = (\boldsymbol{M}^0 - \boldsymbol{M})\,\omega^2 + \boldsymbol{\Phi} - (\boldsymbol{\Phi}^0 + \boldsymbol{\sigma}), \tag{2.3.107}$$

which is nonzero only inside the cluster. Here $\boldsymbol{M}$ and $\boldsymbol{\Phi}$ respectively denote the mass and force constant matrix of the atoms in the cluster. A calculation of the cluster vibrational state density $\langle G(\omega^2, n)\rangle$ requires the solution of (2.3.106) for the atoms within the cluster

$$\boldsymbol{G} = (1 - \tilde{\boldsymbol{G}}\boldsymbol{V})^{-1}\,\tilde{\boldsymbol{G}} \tag{2.3.108}$$

and the averaging of (cf. (2.1.42))

$$G(\omega^2, n) = -(1/\pi n)\,\mathrm{Im}\,\mathrm{Tr}_C\;\boldsymbol{M}\boldsymbol{G} \tag{2.3.109}$$

over all configurations of the n-atom cluster. Here Tr_C means the trace over the cluster sites only. As can be seen from figure 2.30, a CPA-effective medium $\boldsymbol{\sigma}$ yields a much better result than an ATA medium. With a CPA medium the embedded-cluster method reproduces all the major features of the numeri-

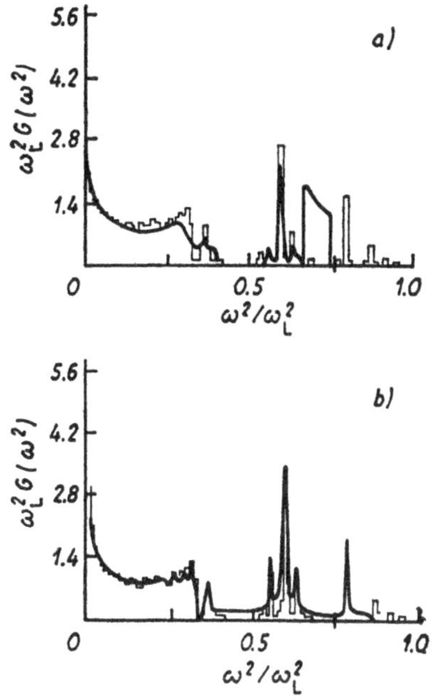

Fig. 2.30. Squared frequency spectrum for a mass disordered two component chain with a mass ratio between the atomic species 3:1 and a concentration of the light atoms $c = 0.25$: light line — computer results; heavy line — a) ATA + embedded 8-atom cluster, b) CPA + embedded 8-atom cluster (after Myles and Dow, 1979)

cally calculated spectrum for a disordered binary chain. Note that this method is especially applicable to concentrations $0.05 \lessapprox c \lessapprox 0.95$ (where cluster effects are important) and that it reduces to the single defect theory as $c \to 0$ and $(1 - c) \to 0$.

For an embedded cluster approach, which also allows the calculation of the spectral density, the reader is referred to Böttger and Kleinert (1981) (cf. also fig. 2.19).

Let us yet draw attention to the useful (ad hoc) self-consistent cluster theory of Butler (1973) which, in order to simplify MCPA, deals with a site diagonal effective medium instead of with a matrix one. Butler picks out a single diagonal element of the condition (2.3.97) and takes it as condition for the effective medium. Accordingly, the consistency between a cluster and its external medium is required only at a single site of the cluster: at the central site (self-consistent central-site approximation: SCCSA) or better at one of the equivalent boundary sites (self-consistent boundary-site approximation: SCBSA). Note that a boundary site feels better the external medium.

For one-dimensional systems ($d = 1$) the SCBSA was found to be identical with the MCPA, while for strong disorder the SCCSA provides unphysical results (caused by non-analycity). The SCBSA has been extended to $d = 3$ (KUMAR and JOSHI, 1978, NAKAO and DOYAMA, 1981). The resulting method is simpler to handle than MCPA, off-diagonal disorder can be included and the site diagonal effective medium does not violate the translational invariance, which the MCPA fails to give. Note that for $d = 3$, the equivalence of the SCBSA and MCPA does not hold, since for $d = 3$ one usually finds that the cells (reproducing the lattice when periodically continued) do not have all boundary sites equivalent.

Note that the HCPA was recently also used for studying stochastic charge carrier transport in random media (ODAGAKI and LAX, 1981) and localization by off-diagonal disorder (FERTIS et al., 1981, PURI and ODAGAKI, 1981; see below).

Cluster theories are still a very active field of research.

CHAPTER 3

VIBRATIONAL PROPERTIES OF SYSTEMS WITH STRUCTURAL DISORDER

3.1. Localized vibrational states

3.1.1. The spatial extent of modes in one-dimensional systems

As shown in section 2.2., a single defect can give rise to localized vibrational states if the mass and/or force constant changes due to the defect are sufficiently large. The question is now whether localized states may also occur when there is a finite defect concentration.

Localization of states due to disorder is one of the most attractive problems in the theory of disordered systems. It is of vital importance for an understanding of transport properties, but it is also an interesting problem as such. Localization due to the absence of long range order has been investigated for more than two decades. Nevertheless, so far it is only partly understood, even for the simplest systems.

First, let us consider some results of numerical investigations of disordered chains with a finite number N of atoms. Figure 3.1 shows the vibrational amplitude for a particular sample of a binary random chain with $N = 50$ containing 22 heavy and 28 light atoms. The mass ratio was taken to be 3. The modes are labelled in order of increasing eigenfrequency. We see that there are two quite distinct types of modes: extended modes at low frequencies and strongly localized modes at higher frequencies. It is obvious that the division between both types occurs between the 25th and 26th eigenmode. Similar localization also occurs in chains with glass-like (structural) disorder. Figure 3.2 shows the envelopes of some typical eigenmodes of a monatomic chain with $N = 512$ having a rectangular force-constant distribution function. We note again that the localization becomes stronger with increasing frequency. Topological disorder[1], no less than the kinds of disorder considered before, can lead to appreciable mode localization. Bell (1972) has simulated this kind of disorder in a chain by choosing on a random basis the atoms with which a given atom is permitted to interact. Numerical inspections of such chains show that mode localization increases drastically with increasing frequency.

Thus, most of the numerically calculated eigenstates of disordered chains are localized in a region which is small compared with the chain length. Localized states are obviously a characteristic property of disordered systems. Mott and Twose (1961) conjectured and Borland (1963) proved that all one-electron eigenstates of a disordered chain are localized. Dean (1964) presented

---

[1] The topological disorder is a structural disorder which cannot be removed by a deformation of the atomic array without cutting bonds between atoms.

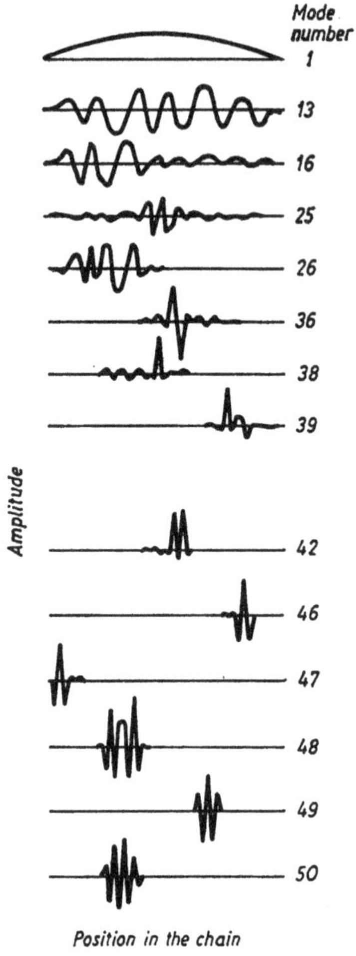

Fig. 3.1. Vibrational amplitude for a 50-atom mass disordered chain containing 22 heavy and 28 light atoms. The mass ratio equals 3. The modes are labelled in order of increasing eigenfrequency (after DEAN and BACON, 1963)

a proof similar to that of Borland for electrons that in a glass-like chain with nearest neighbour forces all eigenstates are localized. Borland's proof is based on the result that in a disordered system the envelope of a wave function grows exponentially with the distance from one end, at which the boundary conditions are given. From this he concluded that the eigenstates are localized in the sense that the envelope of an eigenfunction, on average, exhibits an exponential decay on either side of a given region.

In figure 3.3 we present schematically an (vibrational or electronic) eigenstate in a disordered chain. We see that the state may be appreciable in a given region of fluctuation of length L_f and that it decays exponentially with a

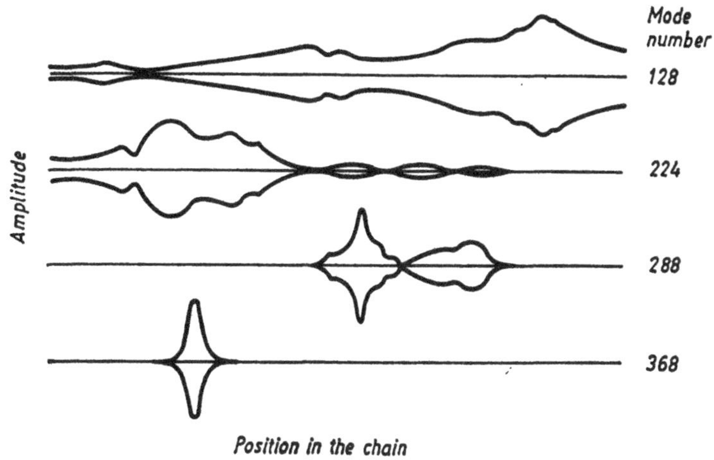

Fig. 3.2. Envelopes of normal modes in a monatomic 512-atom chain with force constant disorder. The modes are labelled in order of increasing eigenfrequency (after DEAN, 1964)

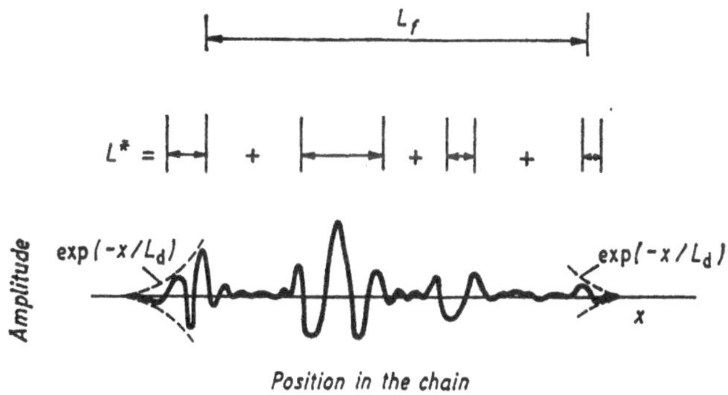

Fig. 3.3. Eigenstate of a one-dimensional disordered system. Outside a region of fluctuations of length L_f the state decays with a decay length L_d. The length L^* is defined by (3.1.1) (after PAPATRIANTAFILLOU et al., 1976)

decay length L_d outside this region. A further "length" characterizing a localized state is

$$L^*(\omega) = \left[\sum_l B^4(l, \omega) \Big/ \left(\sum_l B^2(l, \omega)\right)^2\right]^{-1} \tag{3.1.1}$$

giving the number of sites participating in an eigenstate of frequency ω. Here $B(l, \omega)$ is the mass-dependent normalized amplitude (cf. (1.3.8) and (1.3.12)) of this state (in higher dimensional systems $B(l, \omega) = |\boldsymbol{B}(l, \omega)|$). To illustrate L^* let us consider a monatomic system with N atoms, where N' atoms participate appreciably in the motion, with roughly equal amplitudes. In this case, we have $L^* = N'$. For the sinusoidal modes characteristic of periodic systems, it holds that $L^* \approx \frac{2}{3} N$. In figure 3.3 four regions mainly contribute to the length L^*. Note that the participation ratio

$$p(\omega) = L^*(\omega)/N \tag{3.1.2.}$$

is also often used for characterizing localized states instead of L^*.

Figure 3.4 shows computer results for the configuration-averaged decay length $\langle L_\mathrm{d}\rangle$ of vibrational eigenstates in a binary isotopically disordered chain

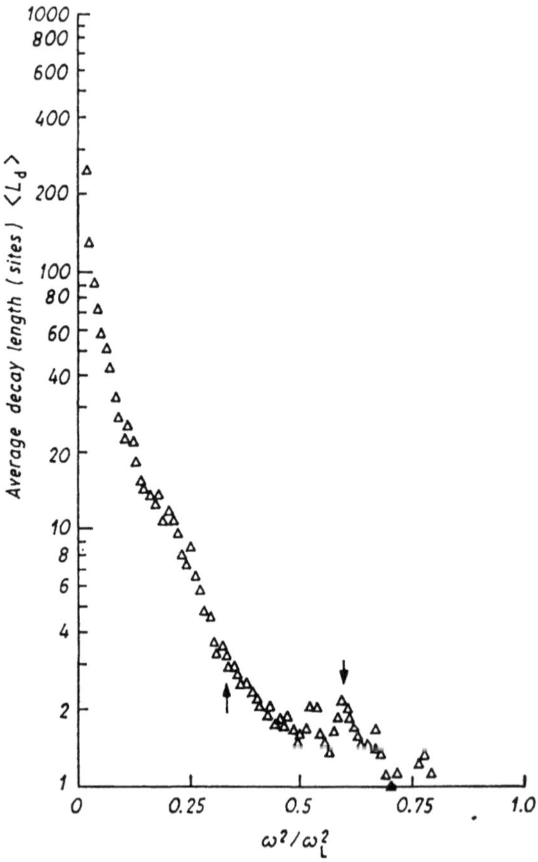

Fig. 3.4. Configuration-averaged decay length for eigenstates of a binary disordered semi-infinite chain with a concentration of light atoms $c = 0.5$. The mass ratio between the species of atoms is 3. ω_L is the maximum frequency of the chain composed of the lighter atoms only (after PAINTER and HARTMANN, 1976)

with $N = 1000$, concentration $c = 0.5$, and fixed initial boundary conditions at one end. We see that $\langle L_d \rangle$ tends to diverge as ω^2 tends to zero, which indicates a tendency to delocalization for the low frequency states. Near the band edge of the heavy atom host, marked by an upward arrow, a drastic drop in the decay length is observed. In $\langle L_d(\omega) \rangle$ peaks occur at the frequency of the single-defect localized mode, marked by a downward arrow, and at frequencies of isolated light-defect clusters. These modes obviously decay less rapidly than their nearby modes. An inspection of L^*, however, shows that L^* is smaller for these modes than for their nearby modes (PAINTER and HARTMANN, 1976).

Now let us turn to an analytical investigation of localization in a semi-infinite isotopically disordered harmonic chain. Such a chain is governed by the equation of motion (cf. (2.3.38))

$$[M(l)\,\omega^2 - 2\gamma]\,u(l) + \gamma u(l+1) + \gamma u(l-1) = 0 \tag{3.1.3}$$

with the initial boundary values $u(0)$ and $u(1)$ $\big(u^2(0) + u^2(1) \neq 0\big)$. Here $M(l)$ and $u(l)$ denote, respectively, the mass and the displacement of the lth atom, and γ is the nearest neighbour force constant. For an infinite or semi-infinite chain, an arbitrary value of ω within the region of allowed frequencies is, in general, arbitrarily close to an eigenvalue. Thus, the particular solution of (3.1.3) with frequency ω may be expressed as

$$\boldsymbol{X}_{l+1} = \boldsymbol{T}_l \boldsymbol{T}_{l-1} \dots \boldsymbol{T}_1 \boldsymbol{X}_1. \tag{3.1.4}$$

Here

$$\boldsymbol{X}_l = \begin{bmatrix} u(l) \\ u(l-1) \end{bmatrix} \quad \text{and} \quad \boldsymbol{T}_l = \begin{bmatrix} \alpha_l & -1 \\ 1 & 0 \end{bmatrix}, \tag{3.1.5}$$

where

$$\alpha_l = 2 - \frac{M(l)}{\gamma}\,\omega^2, \tag{3.1.6}$$

are called state vector and transfer matrix, respectively. The $M(l)$'s are mutually independent random variables. Therefore, information on the vibrational amplitude far from the fixed end of the chain may be obtained from a study of the limiting property of products of random matrices. By such a study, the exponential growth of the eigensolutions of (3.1.3) can be proved quite generally (see the review by ISHII, 1973, and see also STEPHEN, 1981).

Here we are interested in an explicit analytical expression for the decay length. To obtain this length, we introduce the so-called state ratio

$$Z_{l+1} = u(l+1)/u(l) \tag{3.1.7}$$

and rewrite (3.1.3) as

$$Z_{l+1} = \alpha_l - 1/Z_l. \tag{3.1.8}$$

Here Z_l and Z_{l+1} are random variables defined for the lth and $(l+1)$th site, respectively. The α_l's are the independent random variables, whose distribution is assumed to be described by the (normalized) function $P(\alpha)$. In the following we need the (normalized) probability density function $w_l(Z_l)$. According to standard mathematics, the function $w_{l+1}(Z_{l+1})$ is given in terms of the function $w_l(Z_l)$ by an integral equation of the form

$$w_{l+1}(Z_{l+1}) = \iint \mathrm{d}Z_l \, \mathrm{d}\alpha_l \, \delta\left(Z_{l+1} - \alpha_l + \frac{1}{Z_l}\right) w_l(Z_l) \, P(\alpha_l). \tag{3.1.9}$$

As $l \to \infty$, the quantities Z_{l+1} and Z_l become statistically at par, and $w_l(Z)$ converges uniformly to $w(Z)$. Dropping in (3.1.9) the subscripts at w_{l+1} and w_l, we obtain for $w(Z)$ the integral equation

$$w(Z) = \int \mathrm{d}\alpha \, w\left(\frac{1}{\alpha - Z}\right) \frac{P(\alpha)}{(\alpha - Z)^2}. \tag{3.1.10}$$

Following HIROTA (1973), we now assume that $P(\alpha)$ has a sharp maximum at the mean value α_0 and expand $w(1/(\alpha - Z))/(\alpha - Z)^2$ around α_0. Retaining only terms up to the second order in $(\alpha - \alpha_0)$, (3.1.10) becomes

$$w(Z) \approx w\left(\frac{1}{\alpha_0 - Z}\right) \frac{1}{(\alpha_0 - Z)^2} + \frac{\sigma^2}{2} \frac{\mathrm{d}^2}{\mathrm{d}Z^2} \left\{ w\left(\frac{1}{\alpha_0 - Z}\right) \frac{1}{(\alpha_0 - Z)^2} \right\}, \tag{3.1.11}$$

where $\sigma^2 = \langle(\alpha - \alpha_0)^2\rangle$ is the variance of α. Here, $\langle\ldots\rangle$ denotes the average performed with $P(\alpha)$. As can be readily verified, the expression

$$w(Z) = \frac{\sqrt{1 - \alpha_0^4/4}}{\pi} \frac{1}{Z^2 - \alpha_0 Z + 1} \left\{ 1 + \frac{\sigma^2}{\alpha_0} \frac{Z^3 - Z}{(Z^2 - \alpha_0 Z + 1)^2} \right\} \tag{3.1.12}$$

is a solution of (3.1.11), provided that $1 - \alpha_0^4/4 > 0$.

The ratio of lengths of successive state vectors is given by

$$\frac{|\boldsymbol{X}_{l+1}|}{|\boldsymbol{X}_l|} = \frac{(u^2(l+1) + u^2(l))^{1/2}}{(u^2(l) + u^2(l-1))^{1/2}} = \frac{(1 + Z_{l+1}^2)^{1/2}}{(1 + Z_l^2)^{1/2}} |Z_l|. \tag{3.1.13}$$

Averaging the logarithm of this ratio we obtain

$$\left\langle \ln \frac{|\boldsymbol{X}_{l+1}|}{|\boldsymbol{X}_l|} \right\rangle = \langle \ln |Z_l| \rangle + \left\langle \ln \frac{(1 + Z_{l+1}^2)^{1/2}}{(1 + Z_l^2)^{1/2}} \right\rangle. \tag{3.1.14}$$

For a stationary probability density the second term on the right hand side of (3.1.14) vanishes, and with (3.1.12) the first term becomes

$$\left.\begin{aligned} \gamma(\omega) &\equiv \lim_{l\to\infty} \langle \ln |Z_l| \rangle = \frac{1}{2} \int_{-\infty}^{\infty} \mathrm{d}Z \, w(Z) \ln Z^2 \\ &= \frac{\sigma^2}{8} \frac{1}{1 - \alpha_0^2/4} = \frac{(\langle M^2\rangle - \langle M\rangle^2)\,\omega^2}{8\langle M\rangle\,\gamma(1 - \langle M\rangle\,\omega^2/4\gamma)}, \\ &\left(1 - \frac{\langle M\rangle}{4\gamma}\,\omega^2 > 0\right). \end{aligned}\right\} \tag{3.1.15}$$

Here in the last form, $\langle \ldots \rangle$ denotes the average over the $M(l)$'s. As can be seen from (3.1.15), $\gamma(\omega)$ is a positive quantity. This means that the amplitude of a solution of (3.1.3) grows like $\exp\left(l\gamma(\omega)\right)$ on the average, and it is generally believed that this implies each eigenfunction is localized around a certain site l_0 like $\exp\left(-|l - l_0|\,\gamma(\omega)\right)$. In this sense $\gamma^{-1}(\omega)$ can be identified with the decay length $L_d(\omega)$. According to (3.1.15), at low frequencies $L \sim \omega^{-2}$ holds, i.e. the tendency to delocalization strongly increases with decreasing frequency.

With regard to the computer experiments discussed previously, note that localization of eigenstates in large finite systems is generally assumed to be closely correlated with the existence of exponentially localized eigenstates in infinite systems, although it is clear that in finite systems nonlocalized states are not precisely defined. In a chain with N atoms an eigenstate of frequency ω looks spatially well localized only if $\gamma(\omega)\, N/2 \gg 1$. From this condition and the low frequency limit of equation (3.1.15), the following condition for localized states results:

$$\omega \gg 4 \sqrt{\frac{\langle M \rangle \gamma}{N(\langle M^2 \rangle - \langle M \rangle^2)}} \equiv \omega_d. \tag{3.1.16}$$

Here ω_d is the so-called demarcation frequency. It tends to zero as $N \to \infty$.

It should be noted that in a disordered harmonic chain with fixed ends in the limit $N \to \infty$ the heat conduction vanishes, while it diverges for a periodic chain with fixed ends. The former fact is a consequence of the exponential growth of particular solutions, i.e. of the localization of the vibrational states (see the review by Ishii, 1973).

For more details on the possibility of calculating the various lengths characterizing a localized state in a disordered chain the reader is referred to Papatriantafillou et al. (1976).

3.1.2. Localization in the Anderson model

At present the localization problem in three-dimensional disordered systems is much better understood for electrons than for vibrational states. To a certain extent this fact is a consequence of the great importance of electron localization for electric properties of such materials, which has stimulated the investigations of the electron problem. But it is also due to specific computational problems encountered in studying localization of lattice vibrations, which arise from the matrix form of the force constants and from the sum rule (1.3.52) connecting diagonal and off-diagonal force constants.

Most of the studies of electron localization performed so far are based on the Anderson model (Anderson, 1958, 1978). This is a one-electron disordered tight-binding model. Using certain simplifying assumptions, a study of localization of vibrational states becomes mathematically quite similar to that of electron states in the Anderson model. Therefore, it seems reasonable to outline here the investigation of localization in the Anderson model and to point out in the next section, how to apply the results obtained here to the vibrational problem.

The Hamiltonian of the Anderson model reads

$$H = \sum_l |l\rangle \epsilon_l \langle l| + \sum_{ll'} |l\rangle V_{ll'} \langle l'| . \tag{3.1.17}$$

Here, $|l\rangle$ denotes the Wannier state connected with the lth site in a periodic lattice ($\langle l|l'\rangle = \delta_{ll'}, \sum_l |l\rangle \langle l| = 1$). Disorder is usually introduced by assuming that the ϵ_l's are mutually independent random variables with a common probability distribution $P(\epsilon_l)$ and that $V_{ll'} = V$ for nearest neighbours, otherwise zero (diagonal disorder). Anderson defines localization as follows: At time $t = 0$ an electron is put on some site, say site 0. If localized states overlap with site 0, the electron has a finite probability of being initially in each of these eigenstates, and there is a finite probability p_{00} of rediscovering the electron on this site at $t \to \infty$. On the other hand, if no localized states overlap with site 0, the electron will diffuse away and thus $p_{00} = 0$ results.

The time averaged probability of finding the electron in state $|0\rangle$ at $t \to \infty$, if it was at $t = 0$ in the same state $|0\rangle$, may be expressed as

$$p_{00} = \lim_{\varepsilon \to +0} \frac{\varepsilon}{\pi} \int_{-\infty}^{\infty} \mathrm{d}E G_0(E + i\varepsilon)\, G_0(E - i\varepsilon) , \tag{3.1.18}$$

where $G_0(E)$ is the zero diagonal element of the Green function

$$G(E) = [E - H]^{-1} , \tag{3.1.19}$$

i.e. $G_0(E) = \langle 0| [E - H]^{-1} |0\rangle$. The expression (3.1.18) can be derived as follows: If the electron is at $t = 0$ in the state $|\Psi(0)\rangle$ and at time t in the state $|\Psi(t)\rangle$, then the probability of finding the electron at t in state $|\Psi(0)\rangle$ is

$$\begin{aligned} |\langle \Psi(t) \mid \Psi(0)\rangle|^2 = &\sum_{\substack{nn' \\ (E_n \neq E_{n'})}} |\langle \Psi(0) \mid \Phi_n\rangle|^2 \, |\langle \Psi(0) \mid \Phi_{n'}\rangle|^2 \\ &\times \cos \frac{1}{\hbar} (E_n - E_{n'})\, t \\ &+ \sum_{\substack{nn' \\ (E_n = E_{n'})}} |\langle \Psi(0) \mid \Phi_n\rangle|^2 \, |\langle \Psi(0) \mid \Phi_{n'}\rangle|^2 , \end{aligned} \tag{3.1.20}$$

where $|\Phi_n\rangle$ is the eigenstate of H connected with the eigenvalue E_n. Taking the time average $\frac{1}{t} \int_0^t \mathrm{d}t \ldots$, the first term on the right-hand side of (3.1.20) vanishes for $t \to \infty$. The second term may be rewritten as

$$\lim_{\varepsilon \to 0} \frac{\varepsilon}{\pi} \int_{-\infty}^{\infty} \mathrm{d}E \, \langle \Psi(0)|\, G(E + i\varepsilon)\, |\Psi(0)\rangle \, \langle \Psi(0)|\, G(E - i\varepsilon)\, |\Psi(0)\rangle . \tag{3.1.21}$$

This can be readily varified by expanding $|\Psi(0)\rangle$ with respect to $|\Phi_n\rangle$. With $|\Psi(0)\rangle = |0\rangle$ the expression (3.1.21) agrees with the right-hand side of (3.1.18).

The expression p_{00} from (3.1.18) may be rewritten as

$$p_{00} = \int_{-\infty}^{\infty} \mathrm{d}E f_0(E), \tag{3.1.22}$$

where

$$f_0(E) = N_0(E)\, Z_0(E) \tag{3.1.23}$$

with

$$Z_0(E) = \lim_{\varepsilon\to 0} \left[1 - \frac{\Delta_0(E + i\varepsilon) - \Delta_0(E - i\varepsilon)}{2i\varepsilon}\right]^{-1} \tag{3.1.24}$$

and

$$\begin{aligned} N_0(E) &= \lim_{\varepsilon\to 0} \frac{i}{2\pi} \left[G_0(E + i\varepsilon) - G_0(E - i\varepsilon)\right] \\ &= \sum_n |\langle 0 \mid \Phi_n\rangle|^2\, \delta(E - E_n). \end{aligned} \tag{3.1.25}$$

Here $\Delta_0(E)$ is the self-energy defined by

$$G_0(E) = [E - \epsilon_0 - \Delta_0(E)]^{-1}, \tag{3.1.26}$$

and $N_0(E)$ is the (weighted) density of the states with energy E which overlap with the site 0. The second line in (3.1.25) was obtained by making use of

$$G(E) = \sum_n \frac{|\Phi_n\rangle\langle\Phi_n|}{E - E_n} \tag{3.1.27}$$

and of (2.1.21).

$f_0(E)$ is a positive quantity, since $G_0(E + i\varepsilon) = [G_0(E - i\varepsilon)]^*$. Hence, it may be viewed as the contribution to p_{00} of states with energy E. From (3.1.24) it follows that a branch cut in $\Delta_0(E)$ along the real axis corresponds to extended states, since in this case we have $Z_0(E) = 0$ and thus $f_0(E) = 0$ in spite of $N_0(E) \neq 0$. Localized states correspond to $Z_0(E) \neq 0$ and, of course, $N_0(E) \neq 0$. In the energy range of localized states the imaginary part of $\Delta_0(E)$ tends to zero as the imaginary part of E tends to zero, except at a finite number of points, where δ-function singularities appear. A vanishing of $Z_0(E)$ due to such a singularity implies nothing about localization or nonlocalization of states of energy E, because $N_0(E)$ also vanishes. Consequently, an eigenstate with energy E is localized only if $\Delta_0(E)$ is analytic at E. Note that the basic difference in the behaviour of the imaginary part of $\Delta_0(E)$ for localized and extended states vanishes if $\Delta_0(E)$ is configuration averaged, since then the δ-function singularities mentioned before will be broadened to continuous non-zero functions. Therefore, one has to evaluate the probability density function of $\Delta_0(E)$ instead of taking the configuration average of it.

In order to study the analytic properties of $\Delta_0(E)$ we expand $\Delta_0(E)$ in a suitable perturbation series and study its convergence. This procedure is based on the assumption that the states of energy E are localized, so that $\Delta_0(E)$ may be expanded in a series, since it is an analytic function. If the convergence of this series can be proved, the original assumption of localized states was correct, otherwise it was wrong and the states of energy E are extended.

To obtain a series expansion of $\Delta_0(E)$ in terms of known quantities we use the equation for the Green function

$$(E - \epsilon_l) G_{ll'} = \delta_{ll'} + \sum_{l''} V_{ll''} G_{l''l'} \quad (V_{ll} = 0), \tag{3.1.28}$$

where

$$G_{ll'} = \langle l| [E - H]^{-1} |l'\rangle, \tag{3.1.29}$$

and express by iteration on the right-hand side of this equation the nondiagonal elements of G by diagonal elements. In this manner we obtain by means of (3.1.26)

$$\Delta_0(E) = \sum_{l'} V_{0l'} g_{l'}(E) V_{l'l} + \sum_{ll'} V_{ll'} g_{l'}(E) V_{l'l''} g_{l''}(E) V_{l''0} + \cdots, \tag{3.1.30}$$

where

$$g_l(E) = [E - \epsilon_l]^{-1}. \tag{3.1.31}$$

The series (3.1.30) may be represented by all paths starting from site 0 and returning to it at the end only. This series has two essential shortcomings: (i) spurious divergencies may arise from a repeated occurrence of sites being connected with a small energy denominator; (ii) the series has not the correct analytic properties. With regard to the latter deficiency note that it holds

$$G^r_{ll'} = G_{ll'} - G_{lr} G_{rl'} / G_{rr}, \tag{3.1.32}$$

where G^r denotes the Green function of a system with $\epsilon_r \to \infty$. This relation can be obtained by replacing in (3.1.28) ϵ_l by $\epsilon_l' = \epsilon_l + (\epsilon_r' - \epsilon_r)\,\delta_{lr}$, by expressing $G(\ldots, \epsilon_r', \ldots)$ by $G(\ldots, \epsilon_r, \ldots)$ and finally by putting $\epsilon_r' \to \infty$. Equation (3.1.32) shows that the zeros of G_0 (poles of Δ_0) are given by the poles of G^0_{ll} (l is such that $G_{0l} \neq 0$). That is, the poles of Δ_0 do not coincide with the energies ϵ_l as suggested by (3.1.30).

Both the shortcomings mentioned above can be avoided by resumming the series (3.1.30) in such a way that no repitition of the same site on a path occurs. To realize this, let us rewrite (3.1.28) as

$$\sum_{ll'} S_{ll''} G_{l''l'} = \delta_{ll'}, \tag{3.1.33}$$

where

$$S_{ll'} = (E - \epsilon_l)\,\delta_{ll'} - V_{ll'}. \tag{3.1.34}$$

G_0 may be expressed as

$$G_0 = S^0/S, \tag{3.1.35}$$

where S is the determinant of the matrix $\mathbf{S}$, and S^0 is the principal minor which is obtained by deleting from $\mathbf{S}$ the 0th row and column.

From (3.1.35) and (3.1.26) we get

$$\Delta_0 = E - \epsilon_0 - \frac{S}{S^0}. \tag{3.1.36}$$

Analogously, we may define

$$\Delta_{l_i}^{0l_1\ldots l_{i-1}} = E - \epsilon_{l_i} - \frac{S^{0l_1\ldots l_{i-1}}}{S^{0l_1\ldots l_{i-1}l_i}} \tag{3.1.37}$$

and

$$G_{l_i}^{0l_1\ldots l_{i-1}} = [E - \epsilon_{l_i} - \Delta_{l_i}^{0l_1\ldots l_{i-1}}]^{-1}, \tag{3.1.38}$$

where $S^{0l_1\ldots l_i}$ is the principal minor obtained from S by deleting the 0th, l_1th, ..., and l_ith rows and columns. The determinant S may be written in terms of its principal minors (Fujita and Hori, 1972):

$$S = S_{l_1l_1}S^{l_1} + \sum_{m=2}^{\infty} \sum_{l_2l_3\ldots l_m}' (-1)^{m-1} S_{l_1l_2}S_{l_2l_3} \ldots S_{l_ml_1}S^{l_1l_2l_3\ldots l_m}, \tag{3.1.39}$$

where the prime means that the sum must be taken over all sets of distinct indices $l_2, l_3, \ldots, l_m$, all of which are distinct from l_1. Inserting now (3.1.39) into (3.1.36) and using (3.1.37) and (3.1.38) we obtain the so-called renormalized perturbation expansion (RPE)

$$\Delta_0(E) = \sum_{m=1}^{\infty} \sum_{l_1\ldots l_m}' V_{0l_1}G_{l_1}^0(E)\, V_{l_1l_2}G_{l_2}^{0l_1}(E) \ldots V_{l_{m-1}l_m}G_{l_m}^{0l_1\ldots l_{m-1}}(E)\, V_{l_m0}. \tag{3.1.40}$$

$(l_1, \ldots, l_m$ — distinct)

In analogy to (3.1.32) the poles of $G_{l_i}^{0l_1\ldots l_{i-1}}$ can be shown to agree with the zeros of $G_{l_{i+1}}^{0l_1\ldots l_i}$. Consequently, all poles in (3.1.40) cancel, except those of G_l^0. That is, $\Delta_0(E)$ from (3.1.40) exhibits the correct analytic behaviour. Repeated scattering from the same site, which leads in the series (3.1.30) to spurious divergencies, is eliminated. The price one pays is the introduction of an additive term $\Delta_{l_i}^{0l_1\ldots l_{i-1}}$ in each of the energy denominators. These terms may be expressed through relations similar to (3.1.40). Repeating this process of eliminating all the unknowns on the right-hand side of equation (3.1.40), we obtain an infinite continued fraction-like expression in each term of equation (3.1.40). The convergence of the whole RPE requires (i) the convergence of the latter iteration procedure for expressing the self-energies in the energy denominators and (ii) the convergence of the series (3.1.40) alone (that is, of the so-called renormalized perturbation series (RPS)). So far, studies of the convergence of the RPE are restricted to a study of either the convergence of the iteration procedure (Abou-Chacra et al., 1973, Heinrichs, 1977) or the RPS (Anderson, 1958, Economou and Cohen, 1972).

In investigating the iteration process, only the leading term on the right-hand side of equation (3.1.40) is retained. Furthermore, the fact is ignored that the self-energy on the right-hand side of the resulting equation differs from that on the left. Thus, the following equation is obtained:

$$\Delta_0(E) = \sum_l \frac{(V_{0l})^2}{E - \epsilon_l - \Delta_l(E)}. \tag{3.1.41}$$

Note that this equation is exact for a so-called Bethe lattice (cf. section 3.3.2.). The quantity of interest is the probability density distribution $w_{\Delta_0}(\Delta_0)$. Assuming for simplicity that the $V_{ll'}$'s are constant for nearest neighbour sites, and viewing the ϵ_l's as independent variables with the common probability distribution $P(\epsilon_l)$, then the probability density distribution is governed by

$$w_{\Delta_0}(\Delta_0) = \int \prod_{l=1}^{z} \mathrm{d}\epsilon_l \, \mathrm{d}\Delta_l \, \delta\left(\Delta_0 - \sum_l \frac{(V_{0l})^2}{E - \epsilon_l - \Delta_l(E)}\right) P(\epsilon_l) \, w_{\Delta_l}(\Delta_l). \tag{3.1.42}$$

Here z is the coordination number. Since in (3.1.41) Δ_0 and Δ_l are statistically at par, there is a self-consistency condition, namely that Δ_0 on the left should have the same statistical distribution as Δ_l on the right-hand side. This means that on both sides of equation (3.1.42) $w_{\Delta_i}(\Delta_i)$ may be replaced by $w(\Delta_i)$. The existence of localized states is studied by looking for a solution of the self-consistency equation, which yields Im $\Delta_0(E)$ proportional to Im E.

Examining the convergence of the RPS is based on the assumption that this convergence also implies the convergence of the iteration procedure. It reduces to the problem of finding the probability distribution of

$$\Delta_0^{(M)}(E) = V \sum' \frac{V}{e_{l_1}} \frac{V}{e_{l_2}} \cdots \frac{V}{e_{l_M}} \tag{3.1.43}$$

as $M \to \infty$. Here $e_{l_i} = E - \epsilon_{l_i} - \Delta_{l_i}^{0l_1 \ldots l_{i-1}}$ are random functions possessing a distribution function, and the sum runs over all M-step self-avoiding paths starting from and ending at site 0. At present the problem of finding the probability distribution of $\Delta_0^{(M)}(E)$ is far from being solved, since the various contributions to the sum (3.1.43) lack independence from each other, that is, one needs to know the degree of statistical independence among these contributions. Only the two limiting cases of no correlation or very strong correlation have been studied up to now. It turns out that in the latter case a so-called localization function $L(E)$ can be defined by (Economou and Cohen, 1972)

$$L(E) = \lim_{M\to\infty} \left(V^{M+1} \sum' \tilde{G}_{l_1}^{0} \tilde{G}_{l_2}^{0l_1} \ldots \tilde{G}_{l_M}^{0l_1 \ldots l_{M-1}}\right)^{1/M}, \tag{3.1.44}$$

where

$$\left.\begin{aligned} \ln \tilde{G}_{l_i}^{0l_1 \ldots l_{i-1}} &= \langle \ln |G_{l_i}^{0l_1 \ldots l_{i-1}}| \rangle \\ &= \langle \ln |[E - \epsilon_{l_i} - \Delta_{l_i}^{0 \ldots l_{i-1}}(E)]^{-1}| \rangle . \end{aligned}\right\} \tag{3.1.45}$$

Here the prime indicates summation over all M-step self-avoiding paths, and $\langle \ldots \rangle$ denotes the configuration average. $L(E)$ has the property that, if $L(E)$

$< 1 (> 1)$, the RPS converges (diverges) in probability, i.e. the corresponding states are localized (extended). The energies E_c, defined by $L(E_c) = 1$, separate the regions of localized and extended states. Note that the occurrence of configuration averages in (3.1.45) does by no means imply that the RPS has been configuration averaged. For more details on the latter method, often called $L(E)$-method, the reader is referred to the review by LICCIARDELLO and ECONOMOU (1975).

Among the main problems attacked in recent years by the above methods are (i) localization by off-diagonal disorder and (ii) localization in two-dimensional ($d = 2$) systems. It appeared that off-diagonal disorder gives rise to localization quite similarly as on-diagonal one (FERTIS et al., 1981, and PURI and ODAGAKI, 1981). However, in the former case for certain distributions of the off-diagonal elements, extended states may exist at the middle of the band no matter how large the degree of disorder is (ANTONIOU and ECONOMOU, 1977). In agreement with other methods (see the note added in proof on page 203) the present approach strongly indicates localization of all states for $d \leqq 2$ for nonzero (on-diagonal) disorder (SOUKOULIS and ECONOMOU, 1980), while for $d = 3$ complete localization requires a final amount of disorder (cf. SOUKOULIS and ECONOMOU, 1981).

3.1.3. Localization of vibrational states in three-dimensional systems

The condition for localization of vibrational states in disordered harmonic systems can be cast into a form which is quite analogous to that for localization of electron states in the Anderson model. For vibrational states we may define localization as follows. Let us assume that all atoms are at rest at time $t < 0$, and that at $t = 0$ an external force acts for a moment on some atom, say the atom at site 0, so that this atom starts to vibrate. If it still vibrates (with a finite amplitude) at $t \to \infty$, then localized vibrational states overlap with site 0, otherwise no such state exists.

The equation of motion for the system at hand reads (cf. (1.3.2))

$$M(l)\, \ddot{u}_\alpha(lt) = -\sum_{l'\alpha'} \Phi_{\alpha\alpha'}(ll')\, u_{\alpha'}(l't) + f_\alpha \delta_{l0} \delta(t), \tag{3.1.46}$$

where the second term on the right-hand side describes the external force acting on the system. From this equation we get for the Fourier transform of the displacements

$$u_\alpha(l\omega) = -\sum_{\alpha'} G_{\alpha\alpha'}(l0, \omega)\, f_{\alpha'}, \tag{3.1.47}$$

where G is the Green function given by (2.1.29) with $\omega \to \omega + i\varepsilon$. By (3.1.47) we obtain for the squared vibrational amplitude

$$|\boldsymbol{u}(0t)|^2 = \sum_{\alpha} [G_{\alpha\alpha'}(00, t)]^2\, f^2, \tag{3.1.48}$$

where we have assumed, for simplicity, $f_\alpha = \delta_{\alpha\alpha'} f$. Using (2.1.38) we may express $G_{\alpha\alpha'}(00, t)$ as

$$G_{\alpha\alpha'}(00, t) = -\frac{\Theta(t)}{M(0)} \sum_s \frac{B_\alpha^{(s)}(0)\, B_{\alpha'}^{(s)}(0)}{\omega_s} \sin \omega_s t. \tag{3.1.49}$$

When we now insert this expression into (3.1.48) and take the time average of the resulting equation, then we find

$$\lim_{t\to\infty} \frac{1}{t} \int_0^t \mathrm{d}t' \, |\boldsymbol{u}(0t')|^2 \, \frac{1}{f^2} = \frac{1}{[M(0)]^2} \sum_{\substack{\alpha s s' \\ (\omega_s = \omega_{s'})}} \frac{B_\alpha^{(s)}(0)\, B_{\alpha'}^{(s)}(0)\, B_\alpha^{(s')}(0)\, B_{\alpha'}^{(s')}(0)}{2\omega_s^2}. \tag{3.1.50}$$

By (2.1.38) the right-hand side of (3.1.50) can be readily shown to be equal to the sum over α of

$$p_{00}^{\alpha\alpha'} = \lim_{\varepsilon\to 0} \frac{\varepsilon}{\pi} \int_{-\infty}^{\infty} \mathrm{d}\omega G_{\alpha\alpha'}(00, \omega + i\varepsilon)\, G_{\alpha\alpha'}(00, \omega - i\varepsilon), \tag{3.1.51}$$

which is the analogue to the expression (3.1.18) governing electron localization.

Because of the matrix form of the force constants, an investigation of localization of vibrational states is, in general, much more complicated than an investigation of localization of electron states. For that reason, no general inspection of localization of vibrational states (in higher dimensional systems) on the basis of (3.1.51) has been performed so far. However, by some drastic assumptions the vibrational problem can be reduced to that of electrons, so that the results obtained for electrons can be used to get some qualitative insight into the problem of localization of vibrational states. To see this, let us consider the equation governing the Green function of a regular mass disordered system (cf. (2.1.27), (3.1.52))

$$\begin{aligned} &\sum_{\alpha''} \Big(M(l)\, \omega^2 \delta_{\alpha\alpha''} + \sum_{l''(\neq l)} \Phi_{\alpha\alpha''}(ll'')\Big)\, G_{\alpha''\alpha'}(ll') \\ &= \delta_{ll'}\delta_{\alpha\alpha'} + \sum_{\substack{l''(\neq l) \\ \alpha''}} \Phi_{\alpha\alpha''}(ll'')\, G_{\alpha''\alpha'}(l''l') \end{aligned} \tag{3.1.52}$$

and compare it with the equation (3.1.28) for the Green function of electrons in the Anderson model. Making the drastic assumption $\Phi_{\alpha\alpha'}(ll') = \delta_{\alpha\alpha'}\Phi(ll')$ and assuming further that $\Phi(ll') = -\gamma$ for nearest neighbours and zero otherwise, then we get for $G(ll') \equiv -G_{\alpha\alpha}(ll')$ (the minus sign is convenient for comparing the phonon and electron problems) the equation

$$\big(z\gamma - M(l)\, \omega^2\big)\, G(ll') = \delta_{ll'} + \gamma \sum_{l''(\neq l)} G(l''l'), \tag{3.1.53}$$

where z is the coordination number. With $V_{ll'} = V$ for nearest neighbours and zero otherwise (3.1.28) reads

$$(E - \epsilon_l)\, G_{ll'} = \delta_{ll'} + V \sum_{l''(\neq l)} G_{l''l'}. \tag{3.1.54}$$

We assume that the masses $M(l)$ (energies ϵ_l) possess a terminating distribution function centred and peaked at $M^0(\epsilon^0)$ with total width $\Delta M(W)$. By comparing (3.1.53) and (3.1.54), we see that the vibrational problem reduces to the electron problem if we make the substitutions

$$z\gamma = E,\ M(l)\,\omega^2 = \epsilon_l,\ M^0\omega^2 = \epsilon^0,\ \Delta M\omega^2 = W,\ \gamma = V. \tag{3.1.55}$$

A typical result of an investigation of electron localization by means of the $L(E)$-method (cf. section 3.1.2.) is shown in figure 3.5. The energies separating localized and extended states, defined by $L(E_c) = 1$, are represented by a dashed line in the $E' - W$ plane, where $E' = E - \epsilon^0$. The region of localized states ($L < 1$) is shaded and that of extended states ($L > 1$) unshaded. Using the third and fourth equation (3.1.55), then E', defined above, may be expressed as

$$E' = zV - \frac{M^0}{\Delta M}\,W. \tag{3.1.56}$$

Equation (3.1.56) represents a straight line in the $E' - W$ plane, on which ω^2 varies if one moves along it. This line has a slope $\tan\Phi = \dfrac{\Delta M}{M^0}$ and, since $M(l) \geqq 0$, the maximal slope is $\tan\Phi_{\max} = 2$. The frequency corresponding to

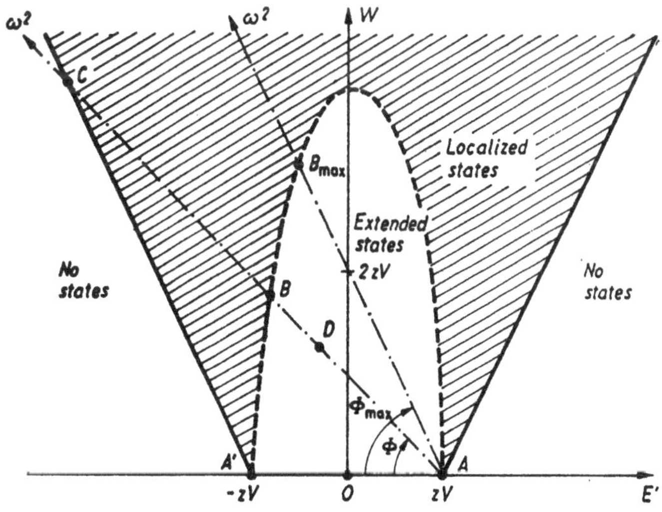

Fig. 3.5. Localization in the Anderson model. The solid lines represent the band edges, the dashed line separates extended and localized states. For electronic states, W is the total width of the distribution around the energy ϵ^0 (see text) and E' is the energy measured from ϵ^0. For vibrational states, the frequency axis is a straight line starting from the point A and making an angle Φ with the E'-axis (after Economou, 1971)

any point D in the plane is given by

$$\omega^2 = \frac{(AD)}{\left((M^0)^2 + (\Delta M)^2\right)^{1/2}}, \tag{3.1.57}$$

where (AD) denotes the distance between points D and A. The latter is the starting point of the line (3.1.56). Hence, for the vibrational problem the line (3.1.57) corresponds to the ω^2 axis with a scale proportional to $\left((M^0)^2 + (\Delta M)^2\right)^{1/2}$. We see from figure 3.4 that up to the frequency

$$\omega_B{}^2 = \frac{(AB)}{\left((M^0)^2 + (\Delta M)^2\right)^{1/2}} \tag{3.1.58}$$

(B lies on the curve $L(E_c) = 1$) all modes are extended. For small disorder, i.e. small Φ, only the high frequency modes are localized.

Figure 3.6 shows results of an investigation of mode localization in a simple isotopically disordered system having an acoustic and an optical band. In

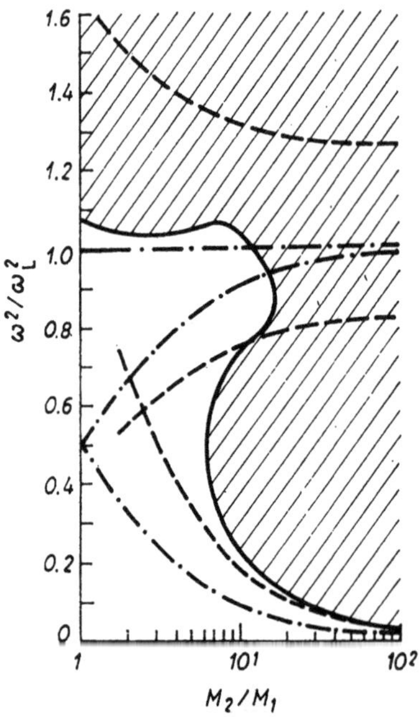

Fig. 3.6. Regions of localized (shaded) and extended (unshaded) vibrational states of a disordered simple cubic system with nearest neighbour forces and two species of atoms with masses described by a common distribution function peaked at M_1 and M_2 and having a total width $\Delta W_1 = M_1$ and $\Delta W_2 = 0.4 M_2$, respectively. -·-·-·-· designates the band edges for absent disorder; ω_L is the corresponding maximum frequency of the system, – – – – characterizes the changes of the band edges by disorder (after Böttger, 1972)

this system two species of atoms are alternately placed at the sites of a simple cubic lattice. The masses of the first (second) species are assumed to possess a common distribution, centred and peaked at $M_1(M_2)$, and to have a total width $\Delta W_1(\Delta W_2)$. The force constants are assumed to have the form described above in context with (3.1.53). The localization function was calculated by completely ignoring the self-energies in the last form in (3.1.45), which is a simple approximation. The results obtained indicate localization of the high frequency optical modes. For a sufficiently large disorder, localization of optical modes also occurs near the lower edge of the optical band and near the upper edge of the acoustic band.

It is worth noticing that, in contrast to the electron problem, in the vibrational problem off-diagonal disorder is always connected with on-diagonal disorder. Therefore, in the presence of off-diagonal disorder the results obtained for electrons may not be applied to vibrations. Note that in spin systems off- and on-diagonal disorders are correlated with each other, too.

Up to now the character of vibrational states in models of real structurally disordered materials has been investigated by numerical methods only (see section 3.3.1.). The general features obtained by these methods agree qualitatively with those described above.

3.2. Plane-wave-like excitations

3.2.1. Experimental and numerical results

Inelastic cold-neutron scattering experiments have indicated marked coherence effects in non-crystalline materials. These effects suggest an approximate description of vibrational excitations in such materials by plane-wave-like, i.e. phonon-like, excitations. Since plane waves cannot be exact eigenstates of a disordered system, these excitations are dampened by scattering on structural irregularities and the phonon conception is meaningful only if their lifetimes are greater than about one vibrational period. It is worth noticing that similar coherence effects have also been observed in simple liquids and amorphous ferromagnetic materials, which suggests the existence of phonon-like and magnon-like collective motions in simple liquids and amorphous ferromagnets, respectively.

Before discussing some characteristic experimental and numerical results, let us collect here some formulae for describing collective motions. The coherent neutron scattering cross section is the experimental quantity indicating such motions. It is proportional to the scattering function (dynamic structure factor) defined by (see (A3.67), (A3.74))

$$S(\boldsymbol{k}, \omega) = \frac{1}{2\pi N} \sum_{ll'} \int_{-\infty}^{\infty} \mathrm{d}t \, \mathrm{e}^{i\omega t} \langle \mathrm{e}^{-i\boldsymbol{k}\boldsymbol{R}(l,t)} \, \mathrm{e}^{i\boldsymbol{k}\boldsymbol{R}(l',0)} \rangle, \qquad (3.2.1)$$

where $\hbar \boldsymbol{k}$ and $\hbar\omega$ respectively are the momentum and the energy transfer in the scattering process; $\boldsymbol{R}(l, t)$ denotes the position of the lth atom at time t; $\langle\ldots\rangle$ represents the thermal and, in the case of need (disordered solids), the configuration average; and N is the number of atoms in the system. Note that in the literature the scattering function $S(\boldsymbol{k}, \omega)$ is often defined by the Fourier transform of the correlation function of the particle density fluctuations instead of the particle densities. In this case in (A3.72) the particle densities $n(\boldsymbol{x}, t)$ must be replaced by $\left(n(\boldsymbol{x}, t) - \varrho\right)$ (ϱ denotes the mean particle density), which gives rise to an additional term $-\varrho(2\pi)^3\,\delta(\boldsymbol{k})\,\delta(\omega)$ on the right-hand side of equation (3.2.1). This term excludes the infinity for forward scattering. For a solid, where $\boldsymbol{R}(l) = \boldsymbol{x}(l) + \boldsymbol{u}(l)$ (cf. (1.2.1)), in the harmonic approximation the scattering function for one-phonon processes reads (see (A3.95))

$$S(\boldsymbol{k}, \omega) = \frac{e^{-2W}}{2\pi N} \sum_{ll'} \int_{-\infty}^{\infty} dt\, e^{i\omega t} \langle e^{i\boldsymbol{k}(\boldsymbol{x}(l)-\boldsymbol{x}(l'))} (\boldsymbol{k}\boldsymbol{u}(l, t))\, (\boldsymbol{k}\boldsymbol{u}(l', 0))\rangle, \qquad (3.2.2)$$

where W is the Debye-Waller exponent (A3.83), which is assumed here to be configuration independent.

When the frequency dependence of $S(\boldsymbol{k}, \omega)$ exhibits sharp peaks, the frequency of these peaks, as a function of wave number $\boldsymbol{k}$, defines a dispersion curve for elementary excitations in the system (see, for example, (A3.97)). This curve is called dispersion relation for the density fluctuation. Zwanzig (1966) has shown that phonon-like excitations in liquids and glasses can be constructed without a reference lattice by changing variables from coordinates and momenta to momenta and forces in the dynamical description. It turns out that the longitudinal Fourier components of the current

$$J_{\parallel}(\boldsymbol{k}, t) = \frac{1}{M} \sum_{l} \boldsymbol{k}\boldsymbol{p}(l, t)\, e^{i\boldsymbol{k}\boldsymbol{R}(l,t)} \qquad (3.2.3)$$

become the collective coordinates governing the phonon-like excitations in such systems. Here $\boldsymbol{p}(l, t)$ is the momentum of the lth atom at time t, and M is the mass of an atom. M is assumed here to be independent of l. In view of the role of $J_{\parallel}(\boldsymbol{k}, t)$, one may expect that the longitudinal current correlation function

$$C_{\parallel}(\boldsymbol{k}, t) = \frac{1}{N} \langle J_{\parallel}^*(\boldsymbol{k}, t)\, J_{\parallel}(\boldsymbol{k}, 0)\rangle \qquad (3.2.4)$$

exhibits a (damped) oscillatory time behaviour. Thus, its Fourier transform

$$C_{\parallel}(\boldsymbol{k}, \omega) = \omega^2 S(\boldsymbol{k}, \omega) \qquad (3.2.5)$$

can be used to derive a dispersion curve by studying the peaks of $C_{\parallel}(\boldsymbol{k}, \omega)$ as a function of $\boldsymbol{k}$. This yields the so-called dispersion relation for the current fluctuation. Recently, arguments have been put forward in favour of determining dispersion relations from $S(\boldsymbol{k}, \omega)$ rather than from $C_{\parallel}(\boldsymbol{k}, \omega)$, because the latter quantity exhibits sometimes spurious peaks, which cannot be identified with collective modes (see, for example, YOSHIDA and TAKENO, 1978).

In the following we also need the static structure factor, which is defined by

$$S(\boldsymbol{k}) = \int_{-\infty}^{\infty} \mathrm{d}\omega S(\boldsymbol{k}, \omega) = \frac{1}{N} \sum_{ll'} \langle \mathrm{e}^{-i\boldsymbol{k}\boldsymbol{R}(l)} \, \mathrm{e}^{i\boldsymbol{k}\boldsymbol{R}(l')} \rangle. \tag{3.2.6}$$

In a crystal $S(\boldsymbol{k})$ exhibits δ-peaks at the sites of the reciprocal lattice, while in a system with structural disorder, in general, there is only one strong peak at $k_0 = 2\pi/a$ (a — mean nearest neighbour spacing). The latter peak reflects the existence of short-range order.

Figure 3.7 shows the static structure factor calculated for a model structure of 800 atoms, which describes a two-component metallic glass ($Mg_{70}Zn_{30}$). The

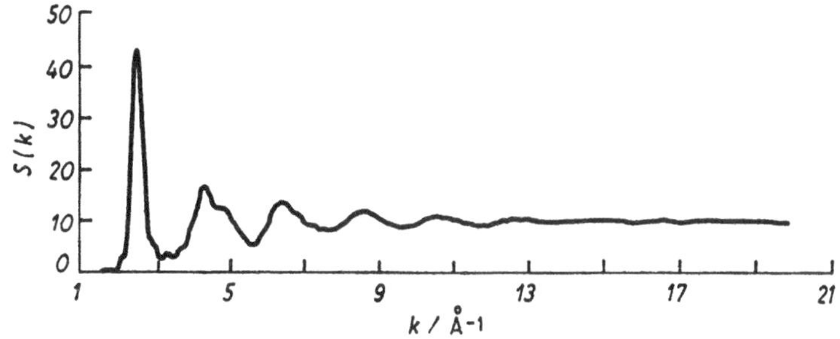

Fig. 3.7. Static structure factor $S(k)$ for a 800-atom model describing a two-component metallic glass ($Mg_{70}Zn_{30}$) (after VON HEIMENDAHL, 1979)

atoms are in force free positions (relaxed model), and periodic boundary conditions are imposed on the model. The vanishing oscillatory behaviour of $S(\boldsymbol{k})$ at large wave numbers reflects the absence of long-range order. Figure 3.8 shows the dynamic structure factor calculated numerically for longitudinal excitations in this model structure. The dispersion relation (ω versus k dependence) resulting from this figure exhibits a clear maximum, but a minimum following this maximum cannot be really identified, because the $S(k, \omega)$ spectra become too broad. The dispersion relation is shown in figure 3.9, where also that for transverse excitations is plotted. For the latter excitations no maximum can be clearly identified[1].

Figure 3.10 shows the dynamic structure factor calculated for a 500-atom amorphous model with periodic boundary conditions. As k increases there is a minimum in the frequency corresponding to the maximum scattering, which indicates some kind of "roton-like" behaviour. Such behaviour is characteristic for superfluid He (see below). Note, however, that also for the present

[1] Similar results, as described above for the metallic glass $Mg_{70}Zn_{30}$, were obtained by YAMAMOTO et al. (1980) for amorphous Fe by a 1450 atom random relaxed model.

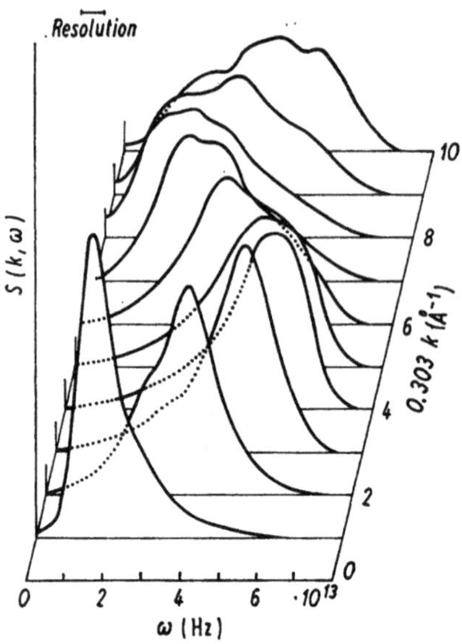

Fig. 3.8. Dynamic structure factor $S(k, \omega)$ for longitudinal excitations in a 800-atom model describing a two-component metallic glass ($Mg_{70}Zn_{30}$) (after von Heimendahl, 1979)

model a minimum is not well defined, as the $S(k, \omega)$ spectra become very broad beyond the maximum of $\omega(k)$ (see also Kobayashi and Takeuchi, 1980).

Up to now, no roton-like behaviour has been experimentally observed in non-crystalline materials. Transverse collective modes have recently been studied by neutron scattering in the metallic glasses $Mg_{70}Zn_{30}$ (Suck et al., 1980) and $Ca_{70}Mg_{30}$ (Suck et al., 1981). Figure 3.11 shows experimental results obtained from neutron scattering on vitreous GeO_2. The approximate existence of longitudinal and traverse acoustic phonon branches is indicated in the region about the first maximum of the static structure factor.

Now let us turn to collective motions in simple liquids. The dynamical behaviour of simple liquids may be qualitatively divided into three regions according to the time and distance scale considered. For times very short compared with the collision time, which is of the order of 10^{-11} to 10^{-12} s, and distances very small compared with the mean interatomic separation, the motion is nearly that of free particles. In the opposite limit of slow space and time variation, i.e. in the so-called hydrodynamic region, the motion is a collective one. In dense fluids the transient local structures give rise to an intermediate region, called zero-sound region, in which the behaviour may be approximately characterized in terms of high frequency phonon-like collective modes. The typical frequencies of these modes are of the order of 10^{13} s^{-1} and can be studied by neutron scattering.

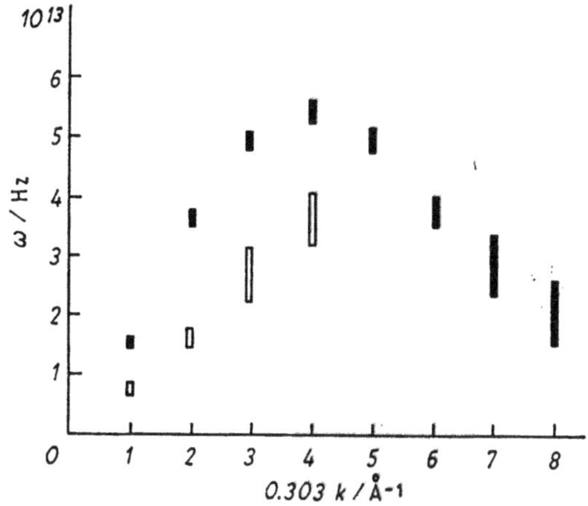

Fig. 3.9. Dispersion relations (full bars: longitudinal; open bars: transverse) for a 800-atom model describing a two-component metallic glass ($Mg_{70}Zn_{30}$) (after von Heimendahl, 1979)

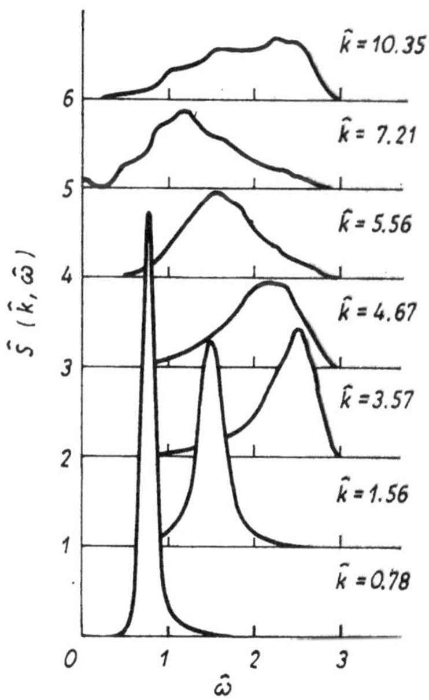

Fig. 3.10. Normalized dynamic structure factor $\hat{S}(k, \omega) = S(k, \omega)\, M\omega/k^2 \times [\bar{n}(\hbar\omega/kT) + 1]$ for a 500-atom amorphous model, plotted in dimensionless variables $\hat{k} = k\sigma$, $\hat{\omega} = \omega/(M\alpha)^{1/2}$ (M — atomic mass, α — force constant, 1.1σ — average nearest neighbour distance, $\bar{n}$ — population number (1.4.50)) (after Rehr and Alben, 1977)

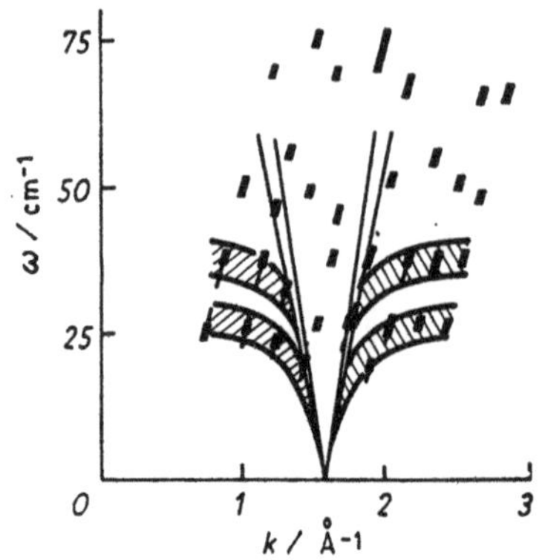

Fig. 3.11. Dispersion relations for vitreous GeO_2 from neutron scattering experiments. Squares represent peak positions of $S(k, \omega)$ and bars refer to peak edges at half-heights (after LEADBETTER and LITCHINSKY, 1970)

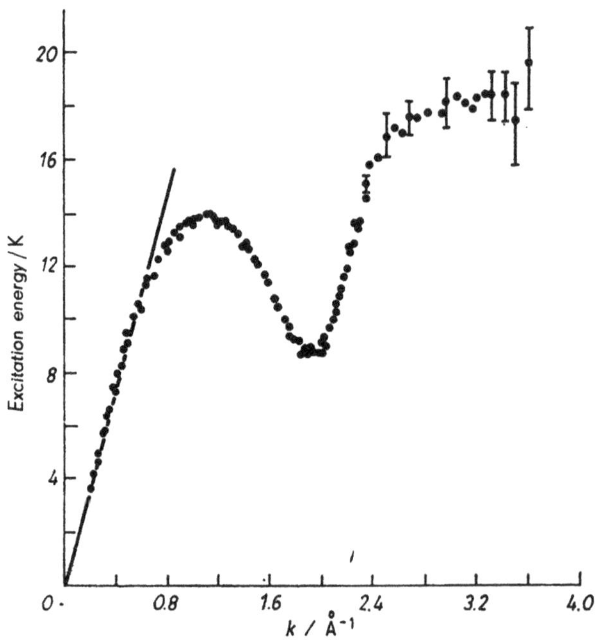

Fig. 3.12. The dispersion curve for He II for the elementary excitations at $T = 1.1$ K (after COWLEY and WOODS, 1971)

Loosely speaking, in the zero sound region of liquids the atoms perform nearly harmonic vibrations about a frozen liquid-like configuration. A strong damping of the collective modes arises from a diffusive decay of the equilibrium configuration. This decay leads to lifetimes of phonons in liquids, which are much shorter than those of phonons in non-crystalline solids. According to neutron scattering studies and computer simulation, in liquid metal systems (e.g. liquid Rb) phonon-like excitations exist up to $k \simeq 0.6k_0$, where k_0 is the position of the main peak in $S(k)$. In inert-gas liquids such excitations exist only up to much lower k-values. This fact is usually attributed to the relatively softer core of the pair potential for a liquid metal than for inert-gas liquids. A hard core produces large angle scattering of the atoms, which is assumed to destroy the coherence necessary for collective atomic motions (for more details on collective motions in simple liquids see the review by COPLEY and LOVESEY, 1975).

In contrast to simple classic liquids, in superfluid He the peaks observed in $S(\boldsymbol{k}, \omega)$ are sharp at low temperatures, so that the idea of phonon-like excitations is quite valid. The dispersion curve of superfluid He is shown in figure 3.12. It exhibits an initial linear rise with a slope corresponding to the velocity of sound followed by a maximum and a minimum, the so-called roton-

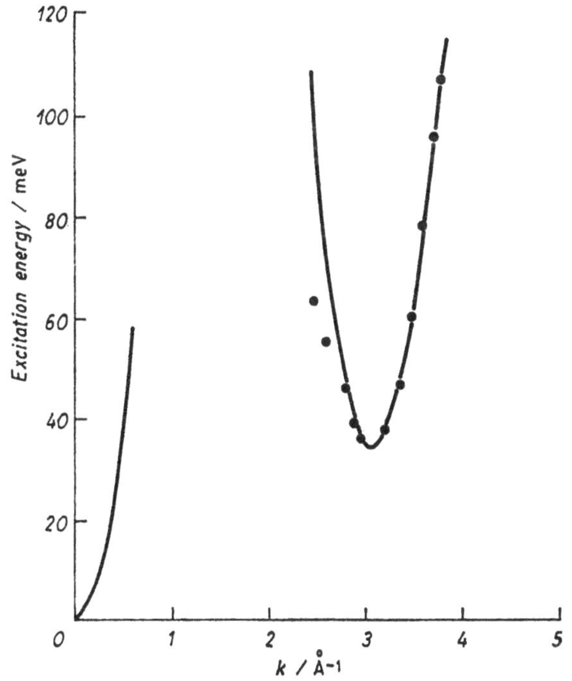

Fig. 3.13. The dispersion curve for spin waves in amorphous ferromagnetic Co_4P (after MOOK et al., 1975)

minimum. This minimum is a prerequisite for suprafluidity (for more details on excitations in liquid He see the review by WOODS and COWLEY, 1973).

The problem of spin-waves in structurally disordered ferromagnetic materials is mathematically quite similar to that of phonons in non-crystalline materials. In this context the roton-like behaviour of spin-waves, observed in amorphous ferromagnets by neutron scattering, is of great interest (fig. 3.13).

To conclude this section, note that in contrast to simple (classic) liquids the collective motions of a harmonic solid with arbitrary disorder are per definition long lived. Little is known, however, about these true eigenmodes. Therefore, it is reasonable to use the idea of plane-wave excitations also in the case of disordered solids, since such excitations can be studied by neutron scattering. The dispersion and damping of plane-wave excitations found in experiments can then be used to calculate quantities of interest, such as thermodynamic quantities.

3.2.2. Investigation of dispersion and damping by analytical methods

The purpose of this section is to investigate phonon-like excitations in a monatomic structurally disordered material. On assuming harmonic vibrations, the dynamic structure factor $S(\boldsymbol{k}, \omega)$ is given by (3.2.2). By means of (2.1.14) and (2.1.22) this expression may be rewritten as

$$S(\boldsymbol{k}, \omega) = -\mathrm{e}^{-2W} \frac{\hbar}{\pi N} (1 - \mathrm{e}^{-\hbar\omega\beta})^{-1}$$
$$\times \sum_{\alpha\alpha' ll'} k_\alpha k_{\alpha'} \langle \mathrm{e}^{-i\boldsymbol{k}(\boldsymbol{x}(l) - \boldsymbol{x}(l'))} \operatorname{Im} G_{\alpha\alpha'}(ll', \omega + i\varepsilon) \rangle, \quad \varepsilon \to +0, \tag{3.2.7}$$

where $\langle \dots \rangle$ denotes the configuration average. Using (2.1.38) we see that

$$S(\boldsymbol{k}, \omega) \propto \operatorname{Im} \left\{ \frac{1}{N} \sum_{ll'} \langle \mathrm{e}^{-i\boldsymbol{k}(\boldsymbol{x}(l) - \boldsymbol{x}(l'))}\, G(ll', \omega + i\varepsilon) \rangle \right\}, \quad \varepsilon \to +0. \tag{3.2.8}$$

Thus, an investigation of phonon-like excitations reduces to a study of the configuration averaged displacement-displacement Green function in $\boldsymbol{k}$-representation. Before studying this quantity, we define here some functions characterizing the structure of the disordered system in question.

To evaluate configuration averages we introduce the N-particle probability density $P(\boldsymbol{x}_1 \boldsymbol{x}_2 \dots \boldsymbol{x}_N)$ with

$$\int \mathrm{d}\boldsymbol{x}_1\, \mathrm{d}\boldsymbol{x}_2 \dots \mathrm{d}\boldsymbol{x}_N P(\boldsymbol{x}_1 \boldsymbol{x}_2 \dots \boldsymbol{x}_N) = 1 \tag{3.2.9}$$

(N — number of atoms in the system, $\boldsymbol{x}_l \equiv \boldsymbol{x}(l)$). $P(\boldsymbol{x}_1 \boldsymbol{x}_2 \dots \boldsymbol{x}_N)\, \mathrm{d}\boldsymbol{x}_1\, \mathrm{d}\boldsymbol{x}_2 \dots \mathrm{d}\boldsymbol{x}_N$ denotes the probability of finding the first atom in the system in the volume element $\mathrm{d}\boldsymbol{x}_1$ centred about $\boldsymbol{x}_1$, the second atom in $\mathrm{d}\boldsymbol{x}_2$ at $\boldsymbol{x}_2$, and so on. If we are interested only in atoms $1, 2, \dots, n$ $(n < N)$, then the corresponding probability density is given by

$$P(\boldsymbol{x}_1 \boldsymbol{x}_2 \dots \boldsymbol{x}_n) = \int \mathrm{d}\boldsymbol{x}_{n+1} \dots \mathrm{d}\boldsymbol{x}_N P(\boldsymbol{x}_1 \boldsymbol{x}_2 \dots \boldsymbol{x}_N). \tag{3.2.10}$$

This function is normalized to unity, too.

We introduce here conditional probability densities by

$$P(\boldsymbol{x}_1 \dots \boldsymbol{x}_n \mid \boldsymbol{x}_{n+1} \dots \boldsymbol{x}_N) = P(\boldsymbol{x}_1 \dots \boldsymbol{x}_N)/P(\boldsymbol{x}_1 \dots \boldsymbol{x}_n) \qquad (n < N). \tag{3.2.11}$$

$P(\boldsymbol{x}_1 \dots \boldsymbol{x}_n \mid \boldsymbol{x}_{n+1} \dots \boldsymbol{x}_N)\, \mathrm{d}\boldsymbol{x}_{n+1} \dots \mathrm{d}\boldsymbol{x}_N$ is the probability of finding the $(n + i)$th atom in $\mathrm{d}\boldsymbol{x}_{n+i}$ at $\boldsymbol{x}_{n+i}$ $\left(i = 1, 2, \dots, (N - n)\right)$ if with certainty we know that the atom k is in $\mathrm{d}\boldsymbol{x}_k$ at $\boldsymbol{x}_k$ $(k = 1, 2, \dots, n)$. In particular, we have

$$\begin{aligned} P(\boldsymbol{x}_1 \dots \boldsymbol{x}_N) &= P(\boldsymbol{x}_1)\, P(\boldsymbol{x}_1 \mid \boldsymbol{x}_2 \dots \boldsymbol{x}_N) \\ &= P(\boldsymbol{x}_1)\, P(\boldsymbol{x}_1 \mid \boldsymbol{x}_2)\, P(\boldsymbol{x}_1\boldsymbol{x}_2 \mid \boldsymbol{x}_3 \dots \boldsymbol{x}_N). \end{aligned} \tag{3.2.12}$$

Using (3.2.10), we see that the conditional probabilities are normalized as

$$\int \mathrm{d}\boldsymbol{x}_{n+1} \dots \mathrm{d}\boldsymbol{x}_N P(\boldsymbol{x}_1 \dots \boldsymbol{x}_n \mid \boldsymbol{x}_{n+1} \dots \boldsymbol{x}_N) = 1. \tag{3.2.13}$$

The configuration average of a quantity $A(\boldsymbol{x}_1 \dots \boldsymbol{x}_N)$ is defined by

$$\langle A \rangle = \int \mathrm{d}\boldsymbol{x}_1 \dots \mathrm{d}\boldsymbol{x}_N A(\boldsymbol{x}_1 \dots \boldsymbol{x}_N)\, P(\boldsymbol{x}_1 \dots \boldsymbol{x}_N). \tag{3.2.14}$$

Similarly, a conditional average $\langle A \rangle_{\boldsymbol{x}_1 \dots \boldsymbol{x}_n}$ of A with the atoms at $\boldsymbol{x}_1 \dots \boldsymbol{x}_n$ held fixed is defined by

$$\langle A \rangle_{\boldsymbol{x}_1 \dots \boldsymbol{x}_n} = \int \mathrm{d}\boldsymbol{x}_{n+1} \dots \mathrm{d}\boldsymbol{x}_N A(\boldsymbol{x}_1 \dots \boldsymbol{x}_N)\, P(\boldsymbol{x}_1 \dots \boldsymbol{x}_n \mid \boldsymbol{x}_{n+1} \dots \boldsymbol{x}_N). \tag{3.2.15}$$

Now we introduce the pair, triple ... correlation functions by

$$g_2(\boldsymbol{x}\boldsymbol{x}') = \frac{1}{\varrho^2} [\langle n(\boldsymbol{x}) \{n(\boldsymbol{x}') - \delta(\boldsymbol{x} - \boldsymbol{x}')\}\rangle], \tag{3.2.16}$$

$$\begin{aligned} g_3(\boldsymbol{x}\boldsymbol{x}'\boldsymbol{x}'') &= \frac{1}{\varrho^3} [\langle n(\boldsymbol{x}) \{n(\boldsymbol{x}') - \delta(\boldsymbol{x} - \boldsymbol{x}')\} \\ &\quad \times \{n(\boldsymbol{x}'') - \delta(\boldsymbol{x} - \boldsymbol{x}'') - \delta(\boldsymbol{x}' - \boldsymbol{x}'')\}\rangle], \end{aligned} \tag{3.2.17}$$

where

$$n(\boldsymbol{x}) = \sum_l \delta\left(\boldsymbol{x} - \boldsymbol{x}(l)\right) \tag{3.2.18}$$

is the particle density number (cf. (A3.76)) and $\varrho = N/V$ (V — volume of the system). The functions (3.2.16), (3.2.17), ... can be determined experimentally, at least in principle. In practice, however, g_2 alone can be found from (scattering) experiments, and, in general, higher correlation functions must be approximated by certain products of g_2. It can be readily seen that

$$g_2(\boldsymbol{x}\boldsymbol{x}') = \frac{N(N-1)}{\varrho^2} P(\boldsymbol{x}\boldsymbol{x}'), \tag{3.2.19}$$

$$g_3(\boldsymbol{x}\boldsymbol{x}'\boldsymbol{x}'') = \frac{N(N-1)(N-2)}{\varrho^3} P(\boldsymbol{x}\boldsymbol{x}'\boldsymbol{x}''). \tag{3.2.20}$$

On calculating configuration averages these relations may be used to express the probability densities P by the "experimental" correlation functions.

If the system is homogeneous and isotropic, then $P(\boldsymbol{x}) = 1/V$ and $P(\boldsymbol{x} \mid \boldsymbol{x}')$ as well as $g_2(\boldsymbol{x}\boldsymbol{x}')$ depend only on $|\boldsymbol{x} - \boldsymbol{x}'|$. From (3.2.16) we get $g_2(|\boldsymbol{x}|) \to 1$ for $|\boldsymbol{x}| \to \infty$, and $g_2(|\boldsymbol{x}|) \to 0$ for $|\boldsymbol{x}| \to 0$. For such a system, it holds that

$$\varrho g_2(\boldsymbol{k}) = S_0(\boldsymbol{k}) - 1, \tag{3.2.21}$$

where $S_0(\boldsymbol{k})$ is the Fourier transform (cf. (A2.10)) of $\langle n(\boldsymbol{x})\, n(\boldsymbol{x}')\rangle/\varrho$. $S_0(\boldsymbol{k})$ is the so-called elastic structure factor. It reads (cf. (3.2.6))

$$S_0(\boldsymbol{k}) = \frac{1}{N} \sum_{ll'} \langle \mathrm{e}^{-i\boldsymbol{k}(\boldsymbol{x}(l)-\boldsymbol{x}(l'))}\rangle. \tag{3.2.22}$$

Various approaches for describing phonon-like excitations in structurally disordered systems have been proposed. However, at present a complete theory of such excitations is still far from realisation. Here we outline the approach given by BÖTTGER (1973a). It yields explicit non-self-consistent expressions for dispersion and damping.

The starting point is the equation for the Green function (2.1.27)

$$\Omega^2 G(ll') = \delta_{ll'} + \sum_{l''} \Phi(ll'')\, G(l''l'), \tag{3.2.23}$$

where the following short-hand notation has been used:

$$G(ll') = \big(G_{\alpha\alpha'}(ll', \omega)\big),\ \Phi(ll') = \big(\Phi_{\alpha\alpha'}(ll')\big),\ \delta_{ll'} = (\delta_{ll'}\delta_{\alpha\alpha'}) \text{ and } \Omega^2 = M\omega^2.$$

For simplicity, the interaction potential between the atoms is assumed to be given by a sum of pair potentials (cf. (1.4.45)):

$$\Phi = \frac{1}{2} \sum_{l \neq l'} \varphi(|\boldsymbol{R}(l) - \boldsymbol{R}(l')|). \tag{3.2.24}$$

For $l \neq l'$, the force constant $\Phi_{\alpha\alpha'}(ll')$ is then a function of $\big(\boldsymbol{x}(l) - \boldsymbol{x}(l')\big)$ only and may be expanded in a Fourier series:

$$\Phi_{\alpha\alpha'}(ll') = \frac{1}{V} \sum_{\boldsymbol{k}} \Phi_{\alpha\alpha'}(\boldsymbol{k})\, \mathrm{e}^{i\boldsymbol{k}(\boldsymbol{x}(l)-\boldsymbol{x}(l'))}, \tag{3.2.25}$$

where, assuming periodic boundary conditions, the density of allowed values in the wave-number space is uniform and equal to $V(2\pi)^{-3}$. $\Phi_{\alpha\alpha'}(\boldsymbol{k})$ is related to $\varphi(\boldsymbol{k})$, the $\boldsymbol{k}$th Fourier coefficient of $\varphi(|\boldsymbol{R}|)$, by

$$\Phi_{\alpha\alpha'}(\boldsymbol{k}) = k_\alpha k_{\alpha'} \varphi(\boldsymbol{k}). \tag{3.2.26}$$

Introducing a Green function $G(\boldsymbol{x}\boldsymbol{x}')$ by multiplying $G(ll')$ with $\delta(\boldsymbol{x} - \boldsymbol{x}(l)) \times \delta(\boldsymbol{x}' - \boldsymbol{x}(l'))$ and summing up over l and l', then for its Fourier transform

$$G(\boldsymbol{k}\boldsymbol{k}') = \frac{1}{N} \sum_{ll'} \mathrm{e}^{-i\boldsymbol{k}\boldsymbol{x}(l)} G(ll')\, \mathrm{e}^{i\boldsymbol{k}'\boldsymbol{x}(l')} \tag{3.2.27}$$

by (3.2.23), (1.3.52) and (3.2.25) we find

$$\Omega^2 G(\boldsymbol{k}\boldsymbol{k}') = A(\boldsymbol{k}\boldsymbol{k}') + \sum_{\boldsymbol{k}_1} \psi(\boldsymbol{k}\boldsymbol{k}_1)\, G(\boldsymbol{k}_1\boldsymbol{k}'), \tag{3.2.28}$$

where

$$A(\boldsymbol{k}\boldsymbol{k}') = \frac{1}{N} \sum_l \mathrm{e}^{-i\boldsymbol{x}(l)(\boldsymbol{k}-\boldsymbol{k}')} \tag{3.2.29}$$

is a random function depending on the atomic positions, and the quantity $\psi(\boldsymbol{k}\boldsymbol{k}_1)$ is the short-hand notation of $\psi_{\alpha\alpha_1}(\boldsymbol{k}\boldsymbol{k}_1)$, which is defined by

$$\psi_{\alpha\alpha_1}(\boldsymbol{k}\boldsymbol{k}_1) = \varrho\big(\Phi_{\alpha\alpha_1}(\boldsymbol{k}_1) - \Phi_{\alpha\alpha_1}(\boldsymbol{k} - \boldsymbol{k}_1)\big)\, A(\boldsymbol{k}\boldsymbol{k}_1). \tag{3.2.30}$$

In order to evaluate the configuration averaged Green function we introduce the following auxillary quantity

$$\Lambda = \sum_{\boldsymbol{k}_1\boldsymbol{k}_2} \lambda(\boldsymbol{k}_1\boldsymbol{k}_2)\, \psi(\boldsymbol{k}_1\boldsymbol{k}_2), \tag{3.2.31}$$

where the λ's are assumed to be independent of the particular configuration of atoms. According to the above short-hand notation, the sum over $\boldsymbol{k}_1$ and $\boldsymbol{k}_2$ in (3.2.31) involves that over α_1 and α_2, too. By means of the quantity (3.2.31), we define the Green function

$$\mathscr{G}_\lambda(\boldsymbol{k}\boldsymbol{k}') = \langle G(\boldsymbol{k}\boldsymbol{k}')\rangle_\lambda, \tag{3.2.32}$$

where

$$\langle\ldots\rangle_\lambda = \left\langle \ldots \frac{\mathrm{e}^\Lambda}{\langle \mathrm{e}^\Lambda\rangle} \right\rangle. \tag{3.2.33}$$

Here $\langle\ldots\rangle$ is the familiar averaging symbol. In the limit $\lambda \to 0$ the Green function (3.2.32) reduces to the configuration averaged Green function

$$\mathscr{G}(\boldsymbol{k}\boldsymbol{k}') = \langle G(\boldsymbol{k}\boldsymbol{k}')\rangle, \tag{3.2.34}$$

whose diagonal element is the quantity governing phonon-like excitations (cf. (3.2.8)).

Averaging equation (3.2.28) according to (3.2.33), we obtain

$$\Omega^2 \mathscr{G}_\lambda(\boldsymbol{k}\boldsymbol{k}') = \mathscr{A}_\lambda(\boldsymbol{k}\boldsymbol{k}') + \sum_{\boldsymbol{k}_1} \Sigma_\lambda(\boldsymbol{k}\boldsymbol{k}_1)\, \mathscr{G}_\lambda(\boldsymbol{k}_1\boldsymbol{k}'). \tag{3.2.35}$$

where

$$\mathscr{A}_\lambda(\boldsymbol{k}\boldsymbol{k}') = \langle A(\boldsymbol{k}\boldsymbol{k}')\rangle_\lambda. \tag{3.2.36}$$

Defining a self-energy $\Sigma_\lambda(\boldsymbol{k}\boldsymbol{k}')$ by

$$\sum_{\boldsymbol{k}_1} \Sigma_\lambda(\boldsymbol{k}\boldsymbol{k}_1)\, \mathscr{G}_\lambda(\boldsymbol{k}_1\boldsymbol{k}') = \sum_{\boldsymbol{k}_1} \langle \psi(\boldsymbol{k}\boldsymbol{k}_1)\, G(\boldsymbol{k}_1\boldsymbol{k}')\rangle_\lambda, \tag{3.2.37}$$

then for the Green function $\mathscr{G}_\lambda(\boldsymbol{k}\boldsymbol{k}')$ the following Dyson equation may be written down:

$$\Omega^2 \mathscr{G}_\lambda(\boldsymbol{k}\boldsymbol{k}') = \mathscr{A}_\lambda(\boldsymbol{k}\boldsymbol{k}') + \sum_{\boldsymbol{k}_1} \Sigma_\lambda(\boldsymbol{k}\boldsymbol{k}_1)\, \mathscr{G}_\lambda(\boldsymbol{k}_1\boldsymbol{k}'). \tag{3.2.38}$$

For evaluating the self-energy $\Sigma_\lambda(\boldsymbol{kk}')$, we use the relation

$$\frac{\partial}{\partial\lambda(\boldsymbol{kk}_1)}\,\mathcal{G}_\lambda(\boldsymbol{k}_1\boldsymbol{k}') = \langle\psi(\boldsymbol{kk}_1)\,G(\boldsymbol{k}_1\boldsymbol{k}')\rangle_\lambda - \langle\psi(\boldsymbol{kk}_1)\rangle_\lambda\,\mathcal{G}_\lambda(\boldsymbol{k}_1\boldsymbol{k}') \tag{3.2.39}$$

and the equations

$$\sum_{\boldsymbol{k}_1}\mathcal{G}_\lambda^{-1}(\boldsymbol{kk}_1)\,\mathcal{G}_\lambda(\boldsymbol{k}_1\boldsymbol{k}') = \delta_{\boldsymbol{kk}'}\,, \tag{3.2.40}$$

$$\sum_{\boldsymbol{k}_1}\mathcal{A}_\lambda^{-1}(\boldsymbol{kk}_1)\,\mathcal{A}_\lambda(\boldsymbol{k}_1\boldsymbol{k}') = \delta_{\boldsymbol{kk}'}\,, \tag{3.2.41}$$

which define the inverse of $\mathcal{G}_\lambda(\boldsymbol{kk}')$ and $\mathcal{A}_\lambda(\boldsymbol{kk}')$, respectively. By means of these equations and (3.2.38), we find the partial differential equation

$$\Sigma_\lambda(\boldsymbol{kk}') = \langle\psi(\boldsymbol{kk}')\rangle_\lambda + \sum_{\boldsymbol{k}_1\boldsymbol{k}_2\boldsymbol{k}_3\boldsymbol{k}_4}\mathcal{G}_\lambda(\boldsymbol{k}_1\boldsymbol{k}_2)\,\mathcal{A}_\lambda^{-1}(\boldsymbol{k}_2\boldsymbol{k}_3)$$
$$\times\left\{\left[\frac{\partial}{\partial\lambda(\boldsymbol{kk}_1)}\,\mathcal{A}_\lambda(\boldsymbol{k}_3\boldsymbol{k}_4)\right]\mathcal{G}_\lambda^{-1}(\boldsymbol{k}_4\boldsymbol{k}') + \delta_{\boldsymbol{k}_3\boldsymbol{k}_4}\frac{\partial}{\partial\lambda(\boldsymbol{kk}_1)}\,\Sigma_\lambda(\boldsymbol{k}_4\boldsymbol{k}')\right\}, \tag{3.2.42}$$

from which, at least on principle, the self-energy Σ_λ may be determined by an iteration procedure. In the limit $\lambda \to 0$ the self-energy Σ_λ reduces to Σ, the self-energy of the Dyson equation for the averaged Green function $\mathcal{G}(\boldsymbol{kk}')$:

$$\Omega^2\mathcal{G}(\boldsymbol{kk}') = \mathcal{A}(\boldsymbol{kk}') + \sum_{\boldsymbol{k}_1}\Sigma(\boldsymbol{kk}_1)\,\mathcal{G}(\boldsymbol{k}_1\boldsymbol{k}')\,. \tag{3.2.43}$$

By (3.2.14), (3.2.12) and (3.2.13), for a homogeneous system we find

$$\mathcal{A}(\boldsymbol{kk}') = \delta_{\boldsymbol{kk}'}\,, \tag{3.2.44}$$

$$\mathcal{G}(\boldsymbol{kk}') = \delta_{\boldsymbol{kk}'}\mathcal{G}(\boldsymbol{k})\,, \tag{3.2.45}$$

$$\Sigma(\boldsymbol{kk}') = \delta_{\boldsymbol{kk}'}\Sigma(\boldsymbol{k})\,. \tag{3.2.46}$$

Hence, (3.2.43) becomes for such a system

$$\Omega^2\mathcal{G}(\boldsymbol{k}) = 1 + \Sigma(\boldsymbol{k})\,\mathcal{G}(\boldsymbol{k})\,. \tag{3.2.47}$$

A first-order expression for Σ_λ may be obtained from (3.2.42) by dropping the term on the right-hand side which involves the partial derivative of Σ_λ with respect to λ. The resulting expression reads

$$\Sigma_\lambda^{(1)}(\boldsymbol{kk}') = \sum_{\boldsymbol{k}_1\boldsymbol{k}_2\boldsymbol{k}_3\boldsymbol{k}_4}\mathcal{G}_\lambda(\boldsymbol{k}_1\boldsymbol{k}_2)\,\mathcal{A}_\lambda^{-1}(\boldsymbol{k}_2\boldsymbol{k}_3)\,\langle A(\boldsymbol{k}_3\boldsymbol{k}_4)\,\psi(\boldsymbol{kk}_1)\rangle_\lambda\,\mathcal{G}_\lambda^{-1}(\boldsymbol{k}_4\boldsymbol{k}')\,. \tag{3.2.48}$$

Taking the limit $\lambda \to 0$, from (3.2.48) the following first-order expression for the self-energy Σ is obtained:

$$\Sigma^{(1)}(\boldsymbol{k}) = \sum_{\boldsymbol{k}_1}\mathcal{G}(\boldsymbol{k}_1)\,\langle A(\boldsymbol{k}_1\boldsymbol{k})\,\psi(\boldsymbol{kk}_1)\rangle\,\mathcal{G}^{-1}(\boldsymbol{k})\,. \tag{3.2.49}$$

If the approximation

$$\mathscr{G}(\boldsymbol{k}) \simeq 1/\Omega^2 \tag{3.2.50}$$

is made, then (3.2.49) reduces to

$$\Sigma^{(1)}(\boldsymbol{k}) \approx \sum_{\boldsymbol{k}_1} \langle A(\boldsymbol{k}_1\boldsymbol{k})\, \psi(\boldsymbol{k}\boldsymbol{k}_1)\rangle. \tag{3.2.51}$$

On account of

$$\langle A(\boldsymbol{k}_1\boldsymbol{k})\, A(\boldsymbol{k}\boldsymbol{k}_1)\rangle = \frac{1}{N}\, S_0(\boldsymbol{k}_1 - \boldsymbol{k}), \tag{3.2.52}$$

which results from (3.2.29), (3.2.22), and (3.2.21), we see that (3.2.51) may be rewritten as

$$\Sigma^{(1)}_{\alpha\alpha'}(\boldsymbol{k}) = \varrho \int \mathrm{d}\boldsymbol{x}\; g_2(\boldsymbol{x})\, [\nabla_\alpha \nabla_{\alpha'} \varphi(\boldsymbol{x})]\, [1 - \mathrm{e}^{-i\boldsymbol{k}\boldsymbol{x}}]. \tag{3.2.53}$$

To obtain a second-order expression for Σ_λ we use in (3.2.48) the approximation $\mathscr{G}_\lambda(\boldsymbol{k}\boldsymbol{k}') \approx \mathscr{A}_\lambda(\boldsymbol{k}\boldsymbol{k}')/\Omega^2$ and insert the resulting expression into the right-hand side of equation (3.2.42). Taking the limit $\lambda \to 0$ and accounting for (3.2.52), we find (cf. Böttger, 1973a)

$$\Sigma^{(2)}_{\alpha\alpha'}(\boldsymbol{k}) = \Sigma^{(1)}_{\alpha\alpha'}(\boldsymbol{k}) + \varrho^2 \sum_{\alpha_1\alpha_2\boldsymbol{k}_1\boldsymbol{k}_2} [\Phi_{\alpha\alpha_1}(\boldsymbol{k}_1) - \Phi_{\alpha\alpha_1}(\boldsymbol{k} - \boldsymbol{k}_1)]\, \mathscr{G}_{\alpha_1\alpha_2}(\boldsymbol{k}_1)$$
$$\times [\Phi_{\alpha_2\alpha'}(\boldsymbol{k}_2) - \Phi_{\alpha_2\alpha'}(\boldsymbol{k}_1 - \boldsymbol{k}_2)]\, \frac{1}{N^3} \int \mathrm{d}\boldsymbol{x}\, \mathrm{d}\boldsymbol{x}'\, \mathrm{d}\boldsymbol{x}''\, \mathrm{e}^{-i[(\boldsymbol{k}_2-\boldsymbol{k})\boldsymbol{x} + (\boldsymbol{k}_1-\boldsymbol{k}_2)\boldsymbol{x}' + (\boldsymbol{k}-\boldsymbol{k}_1)\boldsymbol{x}'']}$$
$$\times \big(\varrho^3 g_3(\boldsymbol{x}\boldsymbol{x}'\boldsymbol{x}'') - \varrho^3 g_2(\boldsymbol{x}\boldsymbol{x}')\, g_2(\boldsymbol{x}\boldsymbol{x}'') + \varrho^2 g_2(\boldsymbol{x}\boldsymbol{x}')\, \delta(\boldsymbol{x}'' - \boldsymbol{x}')\big). \tag{3.2.54}$$

Dispersion and damping of phonon-like excitations may be defined either by the $\boldsymbol{k}$-dependence of the maximum frequency and the half-width of $-\mathrm{Im}\, \mathscr{G}(\boldsymbol{k}, \omega + i\varepsilon)$ (i.e. $S(\boldsymbol{k}, \omega)$, cf. section 3.2.1.) or by the real and imaginary parts of the poles of $\mathscr{G}(\boldsymbol{k}, \omega + i\varepsilon)$ in the lower complex frequency half-plane (cf. section 2.1.). The former definition is suggested by neutron scattering, and the latter is the familiar definition of elementary excitations. Obviously, a damping of phonons occurs only if the self-energy Σ has a finite imaginary part. $\Sigma^{(1)}(\boldsymbol{k})$ from (3.2.53) is a real quantity, i.e. it yields undamped phonons. In the literature this expression has been obtained by various methods (see, for example, Böttger, 1974, and references therein). It is the self-energy in the so-called quasi-crystalline approximation (QCA). The QCA is the analogue to the virtual crystal approximation (VCA) for compositionally disordered systems. The second-order expression $\Sigma^{(2)}$ from (3.2.54) is a complex quantity. Consequently, it accounts for damping of plane-wave excitations. We see that in this approximation the damping is governed by two- and three-body correlations.

Figure 3.14 shows the phonon dispersion curves for liquid argon calculated in the QCA by using a Lennard-Jones potential. We see that, in contrast to the transverse modes, the longitudinal modes exhibit a well-defined mininum in the $\omega(\boldsymbol{k})$-dependence near the first maximum of the elastic structure factor. Hence, the QCA yields for the longitudinal modes a roton minimum (cf. section 3.2.1.). It may be concluded that such behaviour arises from the local disorder inherent in spatially disordered systems, and that it occurs therefore — with some restrictions — in almost all non-crystalline solids and simple liquids, including liquid helium, too. However, this conclusion can be proved

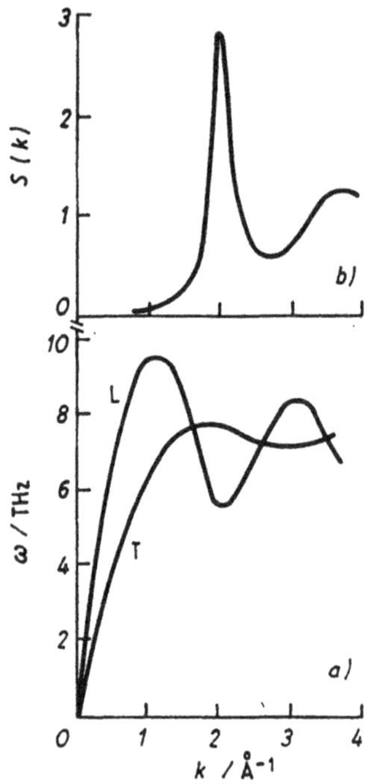

Fig. 3.14. a) Calculated dispersion curves for longitudinal (L) and transverse (T) phonons in liquid argon; b) the structure factor $S(k)$ of liquid argon (after Takeno and Goda, 1971)

only by going beyond the QCA, i.e. by calculating the damping and changes of the dispersion due to higher order terms. Note that it is much harder to go beyond the QCA than beyond the VCA. This is due to the lack of a periodic reference lattice in the case of a structurally disordered system. Although in recent years some promising attempts have been made to go beyond the QCA, particularly to develop a self-consistent theory of excitations in structurally disordered systems (see Djordjevic et al., 1979, Roth, 1979, Aloisio et al., 1981, Itoh et al., 1981), little is known about the validity of the QCA.

To study within the QCA the effect of local disorder on the phonon dispersion curve, let us consider a monatomic one-dimensional system described by the pair correlation function

$$\varrho g_2(x) = \sum_{l(\neq 0)} \frac{1}{\sqrt{\alpha^2 a\,|x(l)|}}\, e^{-(x-x(l))^2/\alpha^2 a |x(l)|}\,. \tag{3.2.55}$$

Here $x(l) = la$, a is the mean nearest neighbour spacing, and $\sqrt{\alpha^2 a}$ describes the fluctuation of the nearest neighbour spacing relative to its mean value. Equation (3.2.55) describes a type of distorted crystal having a more or less marked short-range order but no long-range order. Figure 3.15 shows the QCA

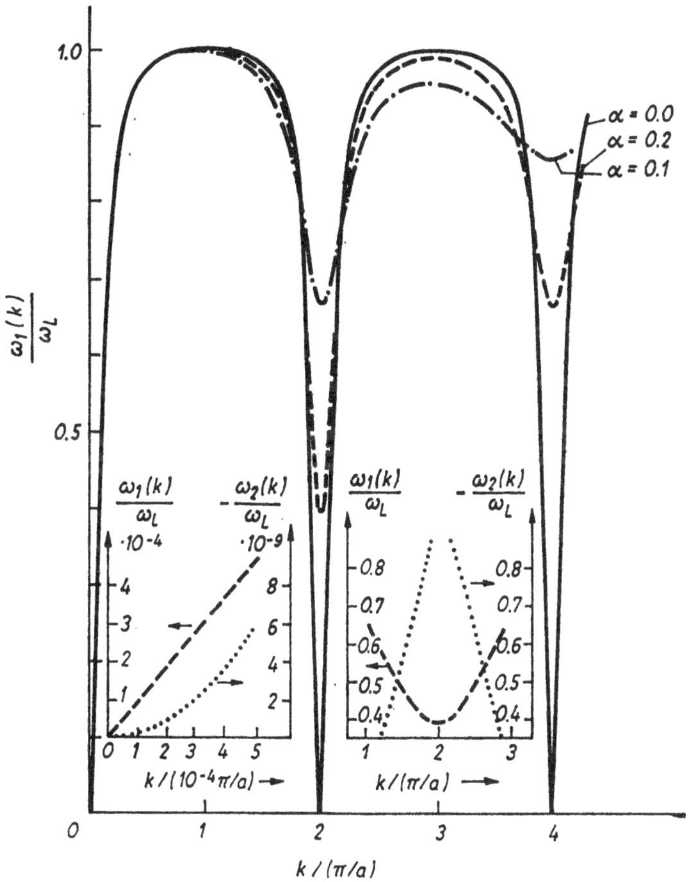

Fig. 3.15. Dispersion, $\omega_1(k)$, and damping, $\omega_2(k)$, of phonon-like excitations in a one-dimensional lattice with glass-like disorder (see text). The left inset ($\alpha = 0.1$) shows the behaviour at low k-values, the right inset ($\alpha = 0.1$) shows the behaviour around the roton minimum. α is the disorder parameter, ω_L is the maximum lattice frequency and a is the mean nearest neighbour distance (after LEIHKAUF and BÖTTGER, to be published)

dispersion curve calculated by means of (3.2.55) and by means of a Morse potential

$$\varphi(x) = D\{e^{-2\gamma(x-x_0)} - 2e^{-\gamma(x-x_0)}\}, \tag{3.2.56}$$

where D, γ and x_0 are parameters characterizing the potential. As can be seen in this figure, a decrease of α (i.e. a decrease of disorder) leads to a decrease of the frequency gap connected with the roton minimum. At $\alpha = 0$ (absent disorder) no gap exists. A calculation of the damping on the basis of (3.2.54) requires the triple correlation function. Using the approximation

$$\langle n(\boldsymbol{x})\, n(\boldsymbol{x}')\, n(\boldsymbol{x}'')\rangle \simeq \frac{1}{\varrho} \langle n(\boldsymbol{x})\, n(\boldsymbol{x}')\rangle \langle n(\boldsymbol{x}')\, n(\boldsymbol{x}'')\rangle$$

or equivalently

$$\varrho^3 g_3(\boldsymbol{x}\boldsymbol{x}'\boldsymbol{x}'') \simeq \varrho^2 g_2(\boldsymbol{x}\boldsymbol{x}')\, \{\varrho g_2(\boldsymbol{x}'\boldsymbol{x}'') - \delta(\boldsymbol{x}'' - \boldsymbol{x})\} \tag{3.2.57}$$

one obtains the damping plotted in figure 3.15. For small k (where $\omega = ck$, c — velocity of sound) the damping behaves as k^2, i.e. the long wavelength modes are only weakly damped (cf. section 3.1.1.). This behaviour is called Rayleigh-like (Brenig and Martin, 1975) or hydrodynamic-like (Kim and Nelkin, 1973). For a corresponding three-dimensional system the damping of the long wavelength phonons is proportional to k^4 (i.e. ω^{-4}). In this context it should be noted that the thermal conductivity of almost all glassy materials shows a ω^{-4} dependence of the phonon mean free path at temperatures above 1 K. The interpretation of this result is still a matter of controversy (see, for example, Dietsche and Kinder, 1979).

If the roton states are not too strongly damped, they give rise to an additional low frequency peak in the vibrational state density. This peak may be expected to affect appreciably the low temperature behaviour of the heat capacity (cf. (1.4.9)). Furthermore, one may expect an anomalous low temperature behaviour of the thermal conductivity to arise from resonance scattering of the Debye phonons on the roton states. Thus, a study of dispersion and damping of phonon-like excitations in structurally disordered systems is an important problem.

Figure 3.16 shows the temperature dependence of the specific heat calculated for a model of a structurally disordered system by assuming that structural disorder (disorder of the positions of the atoms) and dynamic disorder (disorder of the force constants) are not correlated with each other. This model predicts that at low temperatures the heat capacity strongly deviates from the familiar Debye T^3-dependence (cf. (1.4.35)), which is, at least to some extent, due to the lorentzian force constant distribution used. By a simple analytical approach to the vibrational state density of structurally disordered solids Montgomery (1980) found increased T^3 and T^5 specific heat contributions at low temperatures due to force constant variations.

Historically, Zeller and Pohl (1971) first reported anomalous low temperature thermal properties of amorphous solids, which are now a very active field of research (see, e.g., the reviews by Böttger, 1974, Phillips, 1978,

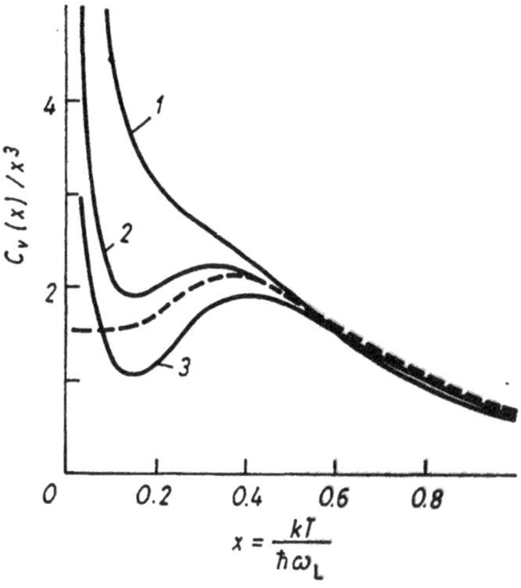

Fig. 3.16. Specific heat at low temperatures calculated for a three-dimensional model with structural and dynamic disorder, assuming the dispersion relation $\omega^2(k) = 2\,\frac{\gamma - i\lambda}{M}\left(3 - \sum_{\alpha=1}^{3} f(k_\alpha)\cos k_\alpha a\right)$, where $f(k_\alpha) = e^{-(\sigma k_\alpha)^2/2}$, a — lattice constant of the reference crystal, γ — nearest neighbour force constant of the reference crystal, M — atomic mass. λ accounts for phonon damping by force constant disorder, $f(k_\alpha)$ accounts for disorder of the atomic positions. Approximating C_v at $T \to 0$ by $C_v \propto \left(\frac{kT}{\hbar\omega_L}\right)^b$ (ω_L — maximum frequency of the lattice), then curve (1): $\sigma = 0.02$, $\lambda = 0.2$; $b = 2.23$; curve (2): $\sigma = 0.1$, $\lambda = 0.1$; $b = 2.27$; curve (3): $\sigma = 0.2$, $\lambda = 0.05$; $b = 2.33$; dashed curve: $\sigma = \lambda = 0$; $b = 3.0$ (Debye behaviour) (after KLEINERT, unpublished results, cf. KLEINERT and LEIHKAUF, 1980, 1981)

GOLDING, 1980, and BHATTACHARYYA, 1981). For example, at temperatures $T \leqq 1$ K the specific heat C_v of various amorphous materials is roughly linear in T rather than following Debye's T^3-law as crystals. Furthermore, at about 10 K C_v/T^3 exhibits a large "excess" peak, which may be ascribed both to the volume difference between glassy and crystalline modifications and to the effect of disorder on the vibrational density of states (cf. GODA, 1982). The roton states are expected to affect thermal behaviour above about 1 K (cf. KLEINERT and LEIHKAUF, 1981).

The excitations leading to $C_v \propto T$ (observed in amorphous dielectrics and metals and even in some superionic materials) are commonly interpreted (ANDERSON et al., 1972, PHILLIPS, 1972) as tunnelling of atoms or groups of atoms between two or more possible equilibrium positions (fig. 3.17). An asymmetric double-well potential (fig. 3.18) is in general used for describing local rearrangements of atoms (cf. GILROY and PHILLIPS, 1981). The motion of an atom in such a potential may be approximately described as an oscillation

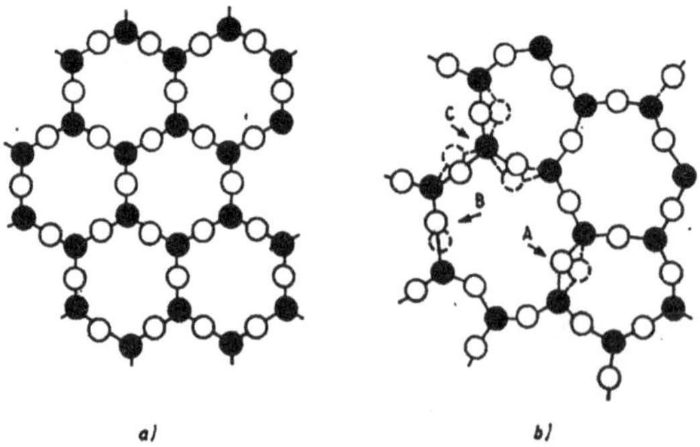

Fig. 3.17. Schematic representation of a) crystalline and b) glassy quartz structure with three possible two-state defects in the glass (denoted by A, B and C) (after HUNKLINGER, 1974)

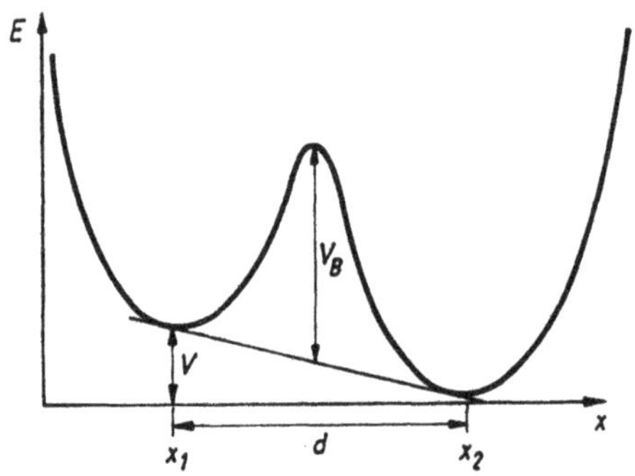

Fig. 3.18. Asymmetric double-well potential (with barrier height V_B, asymmetry V, and well separation d) for two-state defects in glassy materials

around either of the two potential minima. If the potential barrier between the two minima is sufficiently low, transitions between the two positions can occur by tunnelling, whereby a resonant phonon is absorbed or emitted. The tunnelling model has the benefit that not can it only explain the main features of the low temperature thermal anomalies mentioned above, but it can also describe other effects, such as the anomalous ultrasonic attenuation by resonant absorption observed in glasses at low temperatures (see ANTHONY and ANDERSON, 1979, and references therein) and the low-frequency light scattering "excess" observed in glasses (NEMANICH, 1977, and references therein). Specific heat measurements (JONES et al., 1978, LÖHNEYSEN and SCHINK, 1982, LOPONEN et al., 1982) support the idea that a "closed" structure (a-As, a-Ge) might inhibit local atomic rearrangement. Note that the theoretical development of the tunnelling model has been mostly phenomenological. A microscopic description should reveal the relation between structure and low-frequency excitations (cf. BRAWER, 1981, PIETRONERO et al., 1981, SETHNA, 1981).

Note that the problem of magnon-like excitations in an amorphous ferromagnet may be formulated quite similarly as the phonon problem described before (see KANEYOSHI, 1978, and HANDRICH and KOBE, 1980, chapter 6). Hopping conduction is also a mathematically related problem (see BÖTTGER and BRYKSIN, 1982).

3.3. Vibrational spectra of noncrystalline solids

3.3.1. Computer results

At present the two most efficient theoretical approaches to the problem of vibrational properties of noncrystalline solids are: (i) investigation of large clusters of atoms by numerical techniques, (ii) use of analytic methods for studying small clusters of atoms embedded in an effective medium. Here, we consider the former approach. The latter will be the subject of the following section.

The former approach requires the existence of some model for the structure of the material of interest. For noncrystalline materials two types of continuous random models have been developed. The first type is based on a dense packing of hard spheres. It has been used for describing simple monatomic liquids and densely packed amorphous solids, such as amorphous metals. The second type includes random networks, which are suitable for describing low coordination, covalently bonded materials, such as compound glasses (SiO_2) and elemental amorphous semiconductors (Ge). The models developed so far contain in general some hundreds of atoms. They are either hand-built or computer generated. The positions of the atoms are chosen such that the experimental radial distribution function (pair correlation function) and the density of the material are well described by the model. In the best models available at present the atomic positions are relaxed by computers by using a suitable interatomic potential and by minimizing the elastic energy. The relaxation procedure leads to equilibrium positions of the atoms within the topological constraints of the structure (see, e.g., VISSCHER and GUBERNATIS, 1980; for a review of the structure of noncrystalline materials see GASKELL, 1979).

First, let us consider results obtained for the hand-built SiO_2-glass models of Bell and Dean (see the reviews of BELL, 1976, and BELL et al., 1980). These models are based on SiO_4-tetrahedra joined at the corners, so that the Si-atoms are four-fold coordinated and the O-atoms two-fold coordinated. They contain up to 600 atoms each. The dynamical matrices are generated by measuring the coordinates from the models and by using suitable interatomic force constants. The density of the vibrational states (fig. 3.19a)) was computed by the negative eigenvalue theorem and some groups of eigenstates by the method of inverse iteration (cf. section 2.1.2.). The latter groups were used to calculate the participation ratio $p(\omega)$ (3.1.2), which gives information on the spatial

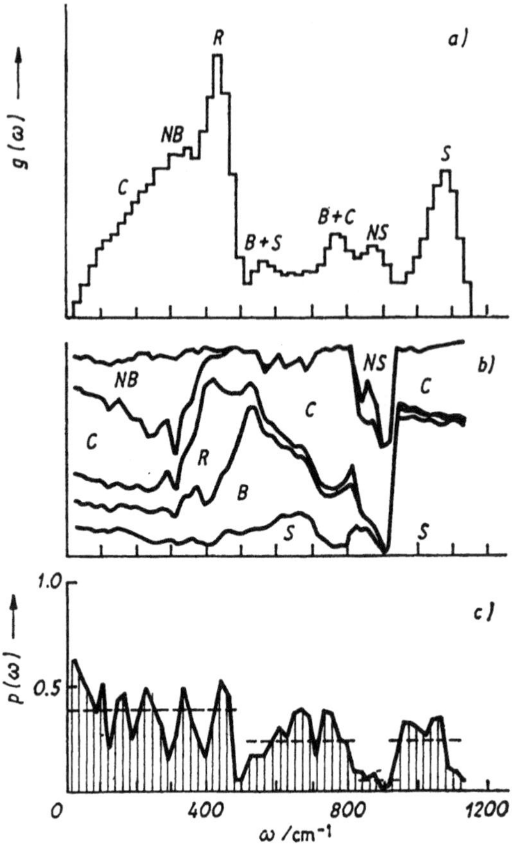

Fig. 3.19. Analysis of the vibrational modes for a (free-end) random network model of vitreous SiO_2. a) Density of vibrational states; b) analysis of energy contributions to the normal modes throughout the spectrum; c) variation of spatial localization of the modes with frequency (after DEAN, 1972)

localization of eigenmodes (fig. 3.19c)). They can also be used for calculating the kinetic energy associated with a normal mode of a given frequency. By dividing this kinetic energy into different classes, a normal mode assignment was obtained. In figure 3.19b) each class is represented by an area whose vertical height, at any particular frequency, is a measure of the amount of the total vibrational energy appropriate to that class.

The motion of a so-called bridging O-atom (fig. 3.20a)) is decomposed into components along local directions labelled B (bond-bending), S (bond-streching), and R (bond-rocking). The B-direction is parallel to the bisector of the Si—O—Si angle, the S-direction is perpendicular to the bisector but in the Si—O—Si plane, and the R-direction is perpendicular to the B- and S-directions. For so-called non-bridging O-atoms (fig. 3.20b)) *NS* refers to the direction along the Si—O bond and *NB* to all directions perpendicular to this bond. *C* represents the energy contribution from the Si-atoms. This contribution is not subdivided further, since the motion of a four-fold coordinated Si-atom cannot be easily decomposed into simple characteristic partial motions.

The mode assignment described here is a very useful tool for characterizing the normal modes in a noncrystalline material, since owing to the lack of long-range order in such a material the conception of longitudinal and transverse normal modes is not appropriate. This assignment has also been used for interpreting the experimental bands of infrared, Raman and inelastic neutron

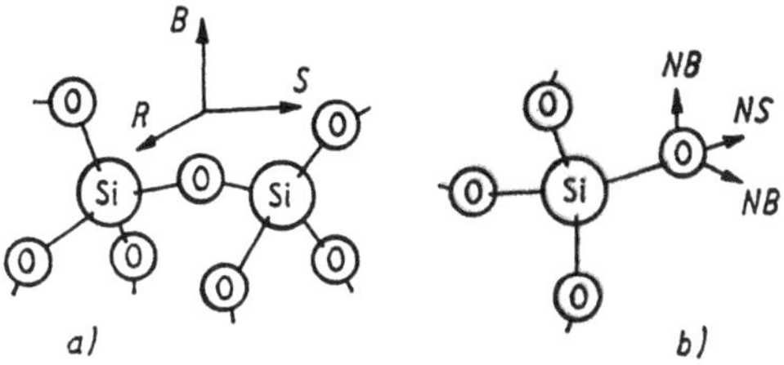

Fig. 3.20. Indication of the local coordinate system for a) a bridging and b) a non-bridging oxygen atom in the random network model of vitreous SiO_2 whose vibrational modes are analysed in figure 3.19 (after Dean, 1972)

scattering spectra, which, in general, coincide well with bands of the vibrational state density. Note that models with only some hundreds of atoms give no significant results in the low frequency region.

Let us yet draw attention to figure 3.19c). As can be seen from this figure, almost all models exhibit rather strong localization. In particular, the modes involving non-bridging atoms are very strongly localized. Such atoms occur at the surface of a model with "free end" boundary conditions, and in a real material also in the volume at structural defects. The localization behaviour shown in figure 3.19c) agrees qualitatively with that discussed in section 3.1.

Now let us turn to elemental amorphous semiconductors such as Si and Ge. Analysis of the radial distribution functions shows that such material has a distorted tetrahedral structure with each atom four-fold coordinated. The radial distribution function, however, does not unambiguously determine the structure. Hence, a number of topological differences are possible for models of tetrahedrally coordinated amorphous semiconductors. One of the most interesting of these differences is the sizes of the rings obtained if one leaves an atom along one bond and returns to it along the shortest possible path.

In recent years several competing forms of a continuous-random-network model with four-fold bonding have been constructed (see, for example, Steinhardt et al., 1974). They vary in the ring statistics and in the degree of bond-angle and bond-length deviation. For instance, the hand-built 440-atom model of Polk (1971) contains 0.38 five-fold, 0.91 six-fold and 1.04 seven-fold rings per atom, and the hand-built 238-atom model of Connell and Temkin (1974) contains 2.3 six-fold rings and no odd-numbered rings. For comparison, crystalline Si or Ge contains six-fold rings only, with two such rings per atom. Interesting information on the role of ring statistics and bond-angle and bond-length deviations can be obtained by calculating the vibrational state density and experimental spectra for different random network models, and by comparing after that the results obtained with each other and with experimental results.

Computer results for vibrational spectra of amorphous elemental tetrahedrally coordinated semiconductors have been obtained (i) by diagonalizing the dynamical matrices of relatively small clusters of atoms (containing about 60 atoms) with periodic boundary conditions (see ALBEN et al., 1975) and (ii) by using the equation-of-motion method, which enables spectra to be obtained for the largest structural models available at present (containing about 500 atoms) without having to diagonalize large matrices (see BEEMAN and ALBEN, 1977). Before discussing some results obtained for amorphous elemental semiconductors, let us describe the equation-of-motion method.

Starting point of this method is the equation of motion (1.3.2) with the initial conditions

$$\left.\begin{aligned} u_\alpha(l, t = 0) &= u_\alpha{}^0(l), \\ \dot{u}_\alpha(l, t = 0) &= 0. \end{aligned}\right\} \tag{3.3.1}$$

The equation of motion is numerically integrated forward in time to obtain the displacements for times up to a time τ. The resulting displacements are used to compute the function

$$F(\omega) = \frac{2}{\pi} \int_0^\tau \mathrm{d}t \sum_{l\alpha} A_\alpha(l)\, u_\alpha(l, t)\, (\cos \omega t)\, \mathrm{e}^{-\lambda t^2}, \tag{3.3.2}$$

where λ is a suitable convergence parameter, and the $A_\alpha(l)$'s are a set of numbers.

The $A_\alpha(l)$'s together with the $u_\alpha{}^0(l)$'s can be appropriately chosen so that $F(\omega)$ gives the vibrational state density or the infrared spectrum or the Raman spectrum. To see this, we account for (1.3.8) and (3.3.1) and write

$$u_\alpha(l, t) = \big(M(l)\big)^{-1/2} \sum_s q_s B^{(s)}(l) \cos \omega_s t, \tag{3.3.3}$$

where the $B_\alpha{}^{(s)}(l)$'s are the (real) orthonormalized eigenvectors from equation (1.3.11), q_s is the (real) amplitude, and ω_s is the (positive) frequency of the sth normal mode. Using the orthonormality condition (1.3.12), we obtain from (3.3.3) and (3.3.1)

$$q_s = \sum_{l'\alpha'} B_{\alpha'}^{(s)}(l') \big(M(l')\big)^{1/2} u_{\alpha'}^0(l'). \tag{3.3.4}$$

Hence, substituting (3.3.4) into (3.3.3), we can write the displacements at any time in terms of the initial displacements as

$$u_\alpha(l, t) = \big(M(l)\big)^{-1/2} \sum_{sl'\alpha'} B_\alpha^{(s)}(l)\, B_{\alpha'}^{(s)}(l') \big(M(l')\big)^{1/2} u_{\alpha'}^0(l') \cos \omega_s t. \tag{3.3.5}$$

Inserting (3.3.5) into (3.3.2), we find

$$F(\omega) = \sum_s \left[\sum_{ll'\alpha\alpha'} B_\alpha^{(s)}(l)\, B_{\alpha'}^{(s)}(l')\, \left(M(l')/M(l)\right)^{1/2} A_\alpha(l)\, u_{\alpha'}^0(l') \right] \times [\Delta(\omega - \omega_s) + \Delta(\omega + \omega_s)], \tag{3.3.6}$$

where

$$\Delta(\omega) = \frac{1}{\pi} \int_0^\tau \mathrm{d}t(\cos \omega t)\, \mathrm{e}^{-\lambda t^2}. \tag{3.3.7}$$

In the limits $\tau \to \infty$ and $\lambda \to +0$ the function $\Delta(\omega)$ becomes a δ-function. For finite τ and λ, $\Delta(\omega)$ becomes a gaussian-like function with a width inversely proportional to τ. In general, τ is chosen, such that this width is only some percent of the full spectral width.

If, for instance, we are interested in the vibrational state density, then we put in (3.3.6)

$$A_\alpha(l) = u_\alpha{}^0(l) = \sqrt{2} \cos \Theta_{l\alpha}, \tag{3.3.8}$$

where the $\Theta_{l\alpha}$'s are random angles uniformly distributed between 0 and 2π. Computing $F(\omega)$ for different sets of Θ's and accounting for

$$2\langle \cos \Theta_{l\alpha} \cos \Theta_{l'\alpha'} \rangle = \delta_{ll'} \delta_{\alpha\alpha'}, \tag{3.3.9}$$

where $\langle \ldots \rangle$ denotes the average over such different sets, we find from (3.3.6) and (3.3.8)

$$\langle F(\omega) \rangle = \sum_s [\Delta(\omega - \omega_s) + \Delta(\omega + \omega_s)]. \tag{3.3.10}$$

Hence, $\langle F(\omega) \rangle / 3N$ (N — number of atoms in the system) is just the broadened vibrational state density (cf. (1.4.11)). Similarly, by an appropriate choice of the parameters $A_\alpha(l)$ and $u_\alpha{}^0(l)$, the infrared and Raman spectra can be obtained from (3.3.2) (see Beeman and Alben, 1977).

In figure 3.21 we show the vibrational state density and the one-phonon infrared spectrum computed by the equation-of-motion method for three structural models of Ge, each with about 500 atoms: a 525-atom sphere cut out of a perfect (diamond) lattice, a 500-atom computer-built random network model, and a 519-atom hand-built and computer relaxed random network model. The obvious good overall agreement between the main features of the vibrational state density of the different models confirms the conception that the vibrational spectrum is decisively governed by the short-range order. The peaks at the bottom and the top of the state density correspond to crystalline TA and TO branches. The central peaks correspond to crystalline LA and LO branches. Unlike the TA and TO peaks, the LA and LO peaks depend sensitively on the topology of the structure. The dip between the latter peaks has been found to arise from six-fold rings. For large distortions of the bond angles and bond lengths the structure in the LA-LO region does get washed out.

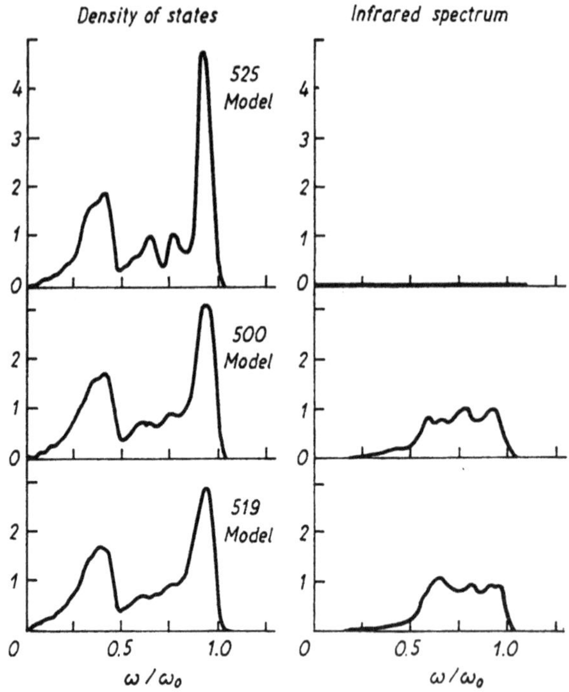

Fig. 3.21. Density of vibrational states and infrared absorption spectra for three Ge models. Distortions and number of five membered rings in the models increase from top to bottom. The 525-atom model is a sphere cut from a perfect diamond lattice; the 500-atom model is computer-built; the 519-atom model is hand-built and computer relaxed. The frequency scale is normalized to the frequency of the total symmetric mode of the Ge crystal ω_0 (after Beeman and Alben, 1977)

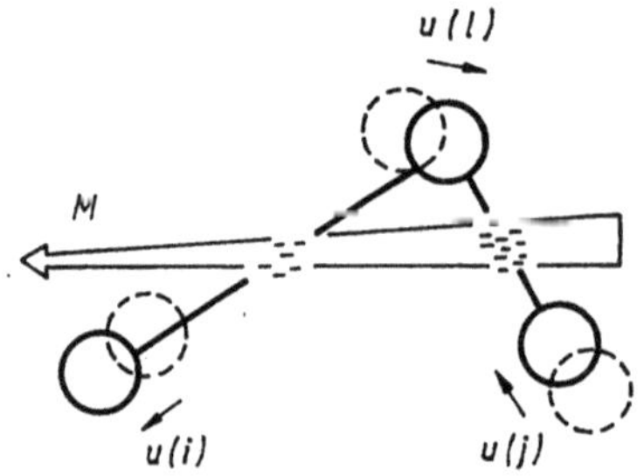

Fig. 3.22. Mechanism for infrared activity in homopolar materials. During a vibration bond charge moves from extended to compressed bonds, resulting in a local electric-dipole moment $\boldsymbol{M}$ (after Alben et al., 1975)

Note that in calculating the infrared spectrum of a homopolar solid, such as Ge or Si, we cannot simply use a point charge model as, for example, in the case of SiO_2. For the former solids in general one uses a model where "dynamic" charges arise from local electric dipole moments induced by the transfer of charge from extended to compressed bonds (fig. 3.22). In the case of a crystalline structure the superimposed contributions from the bonds connected with a given atom cancel each other. Hence, a homopolar crystal does not show a one-phonon infrared spectrum (fig. 3.21). However, in a distorted structure the mechanism described gives rise to a net dynamic charge (for more details see ALBEN et al., 1975).

As can be seen from fig. 3.21, the infrared spectrum of a distorted structure extends over the entire vibrational spectrum, which reflects the fact that the $\boldsymbol{k} = 0$ selection rule is invalid in such a system (cf. section 2.3.3.). Note that for the low frequency modes there is little net intensity due to the little associated bond lengths changes.

For more details on computer results obtained for amorphous elemental semiconductors the reader is referred to ALBEN et al. (1975), BEEMAN and ALBEN (1977) (systems with four-fold tetrahedral coordination), and POLLARD and JOANNOPOULOS (1978) (systems with three-fold pyramidal coordination), see also the review by WEAIRE and TAYLOR (1980).

The vibrational spectrum of a densely packed metallic glass model was investigated by HEIMENDAHL and THORPE (1975) and by YAMAMOTO et al. (1980).

3.3.2. The cluster-Bethe lattice method

The major difficulty in trying to understand properties of non-crystalline materials by examining small clusters of atoms (having less than about 100 atoms) is that the number of surface atoms exceeds the number of interior atoms, so that many properties become governed by the surface atoms. To simulate the part of the material outside a cluster, it is necessary to impose appropriate boundary conditions upon the surface atoms of the cluster. The most important boundary conditions used so far are: (i) free-end boundary conditions (conception of molecular units interacting weakly; LUCOVSKY and MARTIN, 1972, see also SEN and THORPE, 1977, CONNELL and LUCOVSKY, 1978, THORPE and GALEENER, 1980), (ii) periodic boundary conditions (ALBEN et al., 1975), (iii) embedding of the cluster in a structureless effective medium whose properties are determined self-consistently (THORPE 1973), and (iv) use of a Bethe lattice for simulating the effect of the atoms outside the cluster (YNDURAIN and SEN, 1976, YNDURAIN, 1978). Especially the latter approach, the cluster-Bethe lattice method, has been found to be very effective and is the subject of this section.

A Bethe lattice (Cayley tree) can be thought of as an infinite, non-periodic connected system of atoms, where every atom has the same nearest-neighbour configuration and where closed rings of bonds are absent. Such a lattice

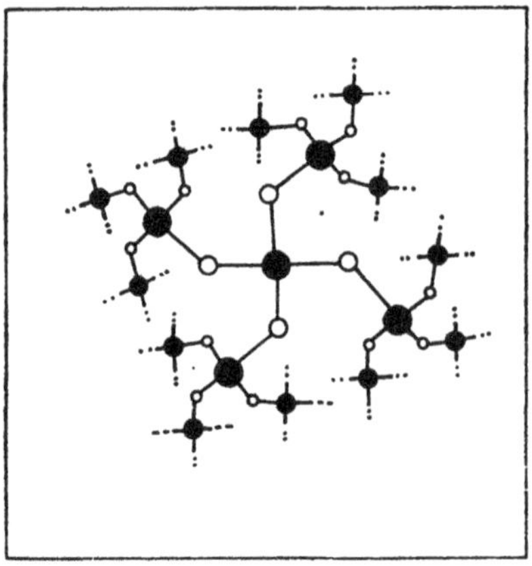

Fig. 3.23. Schematic representation of the topology of a Bethe lattice for SiO_2. The black dots and open circles represent the Si- and O-atoms, respectively (after JOANNOPOULOS, 1979)

branches out, like a tree, to infinity. This is shown schematically in figure 3.23. The Bethe lattice itself is a poor model for an actual material, except in cases where lattice topology (rings of bonds) is unimportant. The replacement of an actual material by a Bethe lattice is called Bethe-Peierls approximation.

The choice of a Bethe lattice as boundary condition for a cluster is schematically shown in figure 3.24. In figure 3.24a) a particular atom of a continuous random network is chosen as a reference atom. This atom is removed from the system together with a small surrounding cluster of atoms. Each of the dangling bonds on the surface of the cluster (fig. 3.24b)) is then connected with a Bethe lattice having the same coordination as the original system (fig. 3.24c)). The new system is infinite and non-periodic like the original one (fig. 3.24d)).

There are mainly three reasons for using a Bethe lattice as boundary condition for a cluster: (i) a Bethe lattice can be solved exactly, (ii) it preserves the coordination, and (iii) its density of states is smooth and featureless (because of the lack of rings of bonds). Owing to the latter fact any structure found in the state density of the cluster-Bethe lattice may be associated with the topology of the atoms in the cluster.

To illustrate the Bethe-Peierls approximation, let us study lattice vibrations of a tetrahedrally bonded crystal. To transform such a crystal into a Bethe lattice we proceed as follows (fig. 3.25): We pick out any atom as reference atom and label it 0. We label its first neighbour by 1 and ν_1 where ν_1 runs from 1 to 4 and indicates the direction from the reference atom to the atom in question. The second nearest neighbours of the atom 0 are labelled by 2, ν_1 and ν_2, where ν_1 and ν_2 indicate the sequence of directions used to get from the atom 0 to the atom addressed. The third nearest neighbours are labelled by 3, ν_1, ν_2, ν_3 and so on. Declaring every atom labelled in this manner

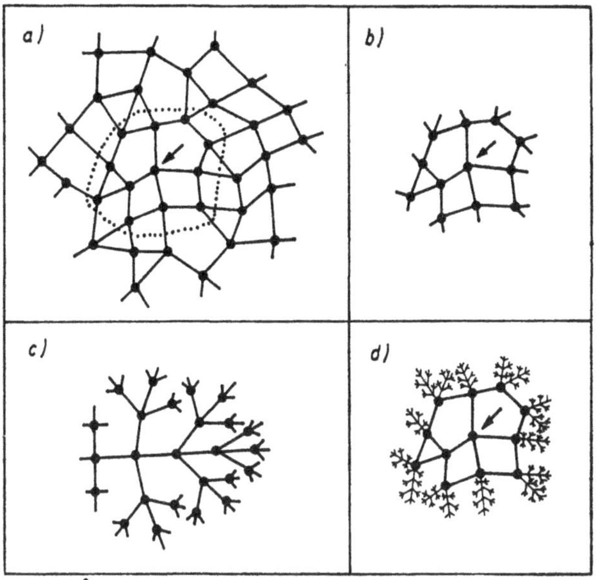

Fig. 3.24. Choice of a Bethe lattice as boundary condition for a cluster (see text) (after JOANNOPOULOS, 1979)

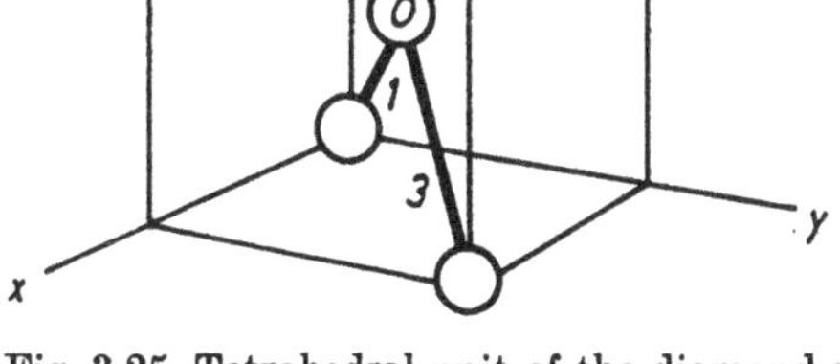

Fig. 3.25. Tetrahedral unit of the diamond structure

distinct, we obtain the Bethe lattice. For a study of the dynamics of this lattice we postulate that each nearest-neighbour bond may be characterized by a 3×3 matrix of force constants $\boldsymbol{\Phi}_\nu$, which connects the atoms at either end of the νth bond (ν runs from 1 to 4). $\boldsymbol{\Phi}_\nu$ is a symmetric matrix.

Denoting by $\boldsymbol{G}_{00}$ the part of the Green function connecting the atom 0 to itself and by $\boldsymbol{G}_{10}^{\nu_1}$ that part connecting the atom 0 to the first nearest neighbour

$(1, \nu_1)$ and so on, then we obtain from (2.1.28) the following infinite set of linear equations:

$$\left.\begin{aligned}
(M\omega^2\mathbf{1} - \boldsymbol{\Phi}_0)\, \boldsymbol{G}_{00} &= \mathbf{1} + \sum_{\nu_1} \boldsymbol{\Phi}_{\nu_1} \boldsymbol{G}_{10}^{\nu_1},\\
(M\omega^2\mathbf{1} - \boldsymbol{\Phi}_0)\, \boldsymbol{G}_{10}^{\nu_1} &= \boldsymbol{\Phi}_{\nu_1} \boldsymbol{G}_{00} + \sum_{\nu_2(\neq\nu_1)} \boldsymbol{\Phi}_{\nu_2} \boldsymbol{G}_{20}^{\nu_1\nu_2},\\
(M\omega^2\mathbf{1} - \boldsymbol{\Phi}_0)\, \boldsymbol{G}_{20}^{\nu_1\nu_2} &= \boldsymbol{\Phi}_{\nu_2} \boldsymbol{G}_{01}^{\nu_1} + \sum_{\nu_3(\neq\nu_2)} \boldsymbol{\Phi}_{\nu_3} \boldsymbol{G}_{30}^{\nu_1\nu_2\nu_3},\\
&\;\vdots
\end{aligned}\right\} \tag{3.3.11}$$

where M is the mass of an atom and $\boldsymbol{\Phi}_0$ is the matrix of force constants connecting an atom with itself, given by (cf. (1.2.7))

$$\boldsymbol{\Phi}_0 = -\sum_{\nu=1}^{4} \boldsymbol{\Phi}_\nu. \tag{3.3.12}$$

We see that the infinite set of equations (3.3.11) is such that the sequence of four consecutive equations ($\nu = 1, 2, 3, 4$) repeats itself up to infinity. This set of equations has a solution of the form

$$\left.\begin{aligned}
\boldsymbol{G}_{10}^{\nu_1} &= \boldsymbol{T}_{\nu_1} \boldsymbol{G}_{00},\\
\boldsymbol{G}_{20}^{\nu_1\nu_2} &= \boldsymbol{T}_{\nu_2} \boldsymbol{G}_{10}^{\nu_1},\\
&\;\vdots
\end{aligned}\right\} \tag{3.3.13}$$

provided that the four so-called transfer matrices $\boldsymbol{T}_\nu$ satisfy the equation

$$(M\omega^2\mathbf{1} - \boldsymbol{\Phi}_0)\, \boldsymbol{T}_\nu = \boldsymbol{\Phi}_\nu + \sum_{\nu'(\neq\nu)} \boldsymbol{\Phi}_{\nu'} \boldsymbol{T}_{\nu'} \boldsymbol{T}_\nu. \tag{3.3.14}$$

Once the transfer matrices are found, the Green function can be easily calculated. $\boldsymbol{G}_{00}$ needed for that is given by (from eqs. (3.3.11) and (3.3.13))

$$\boldsymbol{G}_{00} = [M\omega^2\mathbf{1} - \boldsymbol{\Phi}_0 - \sum_{\nu} \boldsymbol{\Phi}_\nu \boldsymbol{T}_\nu]^{-1}. \tag{3.3.15}$$

According to eq. (3.3.13), the remainder of the Green function is then constructed by repeatedly applying the transfer matrices.

To proceed we assume the Born model for the lattice potential (cf., e.g., Alben et al., 1975)

$$\Phi = \frac{3}{4}\,\beta \sum_{l\Delta} [(\boldsymbol{u}_l - \boldsymbol{u}_{l\Delta})\, \boldsymbol{r}_\Delta(l)]^2 + \frac{1}{4}\,(\alpha - \beta) \sum_{l\Delta} (\boldsymbol{u}_l - \boldsymbol{u}_{l\Delta})^2, \tag{3.3.16}$$

where the sums are on the atoms l and their nearest neighbours Δ; $\boldsymbol{r}_\Delta(l)$ is the unit vector from the equilibrium position of the lth atom to that of its neighbour Δ; $\boldsymbol{u}_l$ and $\boldsymbol{u}_{l\Delta}$ are the displacement vectors of these atoms; $(2\beta + \alpha)$ governs the central force and $(\alpha - \beta)$ the noncentral one. Using (3.3.16), the four force constant matrices become (cf. fig. 3.25)

$$\boldsymbol{\Phi}_1 = \begin{bmatrix} -\alpha & -\beta & -\beta \\ -\beta & -\alpha & -\beta \\ -\beta & -\beta & -\alpha \end{bmatrix}, \qquad \boldsymbol{\Phi}_2 = \begin{bmatrix} -\alpha & \beta & \beta \\ \beta & -\alpha & -\beta \\ \beta & -\beta & -\alpha \end{bmatrix},$$
$$\boldsymbol{\Phi}_3 = \begin{bmatrix} -\alpha & -\beta & \beta \\ -\beta & -\alpha & \beta \\ \beta & \beta & -\alpha \end{bmatrix}, \qquad \boldsymbol{\Phi}_4 = \begin{bmatrix} -\alpha & \beta & -\beta \\ \beta & -\alpha & \beta \\ -\beta & \beta & \alpha \end{bmatrix}. \tag{3.3.17}$$

To solve equation (3.3.14) we use the matrices

$$\boldsymbol{S}_1 = \begin{bmatrix} 1 & 0 & 0 \\ 0 & 1 & 0 \\ 0 & 0 & 1 \end{bmatrix}, \qquad \boldsymbol{S}_2 = \begin{bmatrix} -1 & 0 & 0 \\ 0 & 1 & 0 \\ 0 & 0 & 1 \end{bmatrix},$$
$$\boldsymbol{S}_3 = \begin{bmatrix} -1 & 0 & 0 \\ 0 & -1 & 0 \\ 0 & 0 & 1 \end{bmatrix}, \qquad \boldsymbol{S}_4 = \begin{bmatrix} 1 & 0 & 0 \\ 0 & -1 & 0 \\ 0 & 0 & 1 \end{bmatrix}, \tag{3.3.18}$$

which transform the T-matrices into each other as

$$\boldsymbol{T}_2 = \boldsymbol{S}_2\boldsymbol{T}_1\boldsymbol{S}_2, \qquad \boldsymbol{T}_3 = \boldsymbol{S}_3\boldsymbol{T}_1\boldsymbol{S}_3, \qquad \boldsymbol{T}_4 = \boldsymbol{S}_4\boldsymbol{T}_1\boldsymbol{S}_4. \tag{3.3.19}$$

Hence, on account of (3.3.19), (3.3.14) reduces to

$$(M\omega^2\boldsymbol{1} - \boldsymbol{\Phi}_0)\,\boldsymbol{T}_1 = \boldsymbol{\Phi}_1 + \sum_{\nu(\neq 1)} \boldsymbol{\Phi}_\nu\boldsymbol{S}_\nu\boldsymbol{T}_1\boldsymbol{S}_\nu\boldsymbol{T}_1. \tag{3.3.20}$$

Owing to the symmetry of the matrices $\boldsymbol{\Phi}_\nu$ and $\boldsymbol{S}_\nu$, involved in (3.3.20), $\boldsymbol{T}_1$ can be shown to have the following form:

$$\boldsymbol{T}_1 = \begin{bmatrix} t_0 & t_1 & t_1 \\ t_1 & t_0 & t_1 \\ t_1 & t_1 & t_0 \end{bmatrix}. \tag{3.3.21}$$

Substituting (3.3.21) into (3.3.20), a set of two coupled quadratic equations for t_0 and t_1 results, which must be solved to obtain $\boldsymbol{T}_1$.

Consider now the density of vibrational states, which is given by (cf. (2.1.42))

$$g(\omega) = -\frac{2\omega M}{3\pi N} \sum_j \operatorname{Im} G_{jj}(\omega), \tag{3.3.22}$$

where N is the number of atoms in the system and G_{jj} is the diagonal element of the Green function associated with the jth degree of freedom (according to (2.1.42) we may have $j = (l, \alpha)$, where $\alpha = x, y, z$; but instead of cartesian coordinates we may also use coordinates adapted to local symmetry (cf. fig. 3.18)). Each term in the sum (3.3.22) is referred to as a "local density of states". The latter quantity shows how much the particular degree of freedom participates in the states with frequency ω. It gives an excellent feel for the nature of the corresponding motions, and it can be relatively easily calculated. To obtain it we need G_{00}, the part of the Green function connecting the atom 0 (reference atom) to itself. In the case of using the Bethe-Peierls approximation G_{00} is computed as described above. In the case of the cluster-Bethe lattice approximation we obtain from (2.1.28) for the Green function an infinite set of coupled equations. This set contains Green function elements connecting the atom 0 with atoms of the cluster and connecting the atom 0 with atoms outside the cluster. By using the transfer matrix the latter elements may be traced back to the former. In this manner the information about the medium surrounding the cluster is carried into the cluster. The resulting set of equations for calculating G_{00} is finite and can be solved exactly (JOANNOPOULOS and YNDURAIN, 1974, YNDURAIN and SEN, 1976).

In the case of a disordered system the final expression for the local state density must be averaged over the various cluster configurations. In the presence of structural disorder, this can be done by using a continuous random network model, and by averaging over the various local configurations in it.

Hence, the procedure of attaching Bethe lattices to the boundary of a cluster is straightforward, and the most difficult task in the cluster-Bethe lattice method is solving the Bethe lattice itself.

In figure 3.26, for a diamond structure, the vibrational state density calculated in the Bethe-Peierls approximation is compared with that of a perfect infinite crystal. We see that the TA and TO peaks are well approximated by

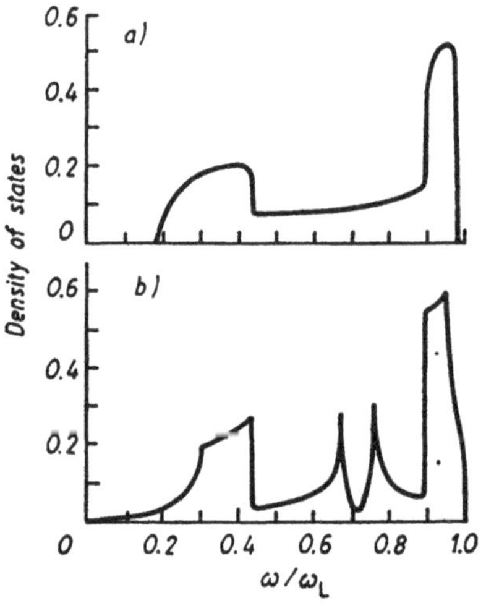

Fig. 3.26. Density of vibrational states for a crystal with diamond structure. a) Bethe-Peierls approximation; b) exact. ω_L is the maximum frequency of the crystal (after YNDURAIN and SEN, 1976, the exact results are from THORPE, 1973)

the Bethe-Peierls approximation. The longitudinal peaks, however, do not appear in this approximation because of the lack of closed rings of bonds (cf. section 3.3.1.). On the other hand, the cluster-Bethe lattice approximation also reproduces well the longitudinal peaks, although for a cluster these peaks are not so sharp as for a crystal (fig. 3.27).

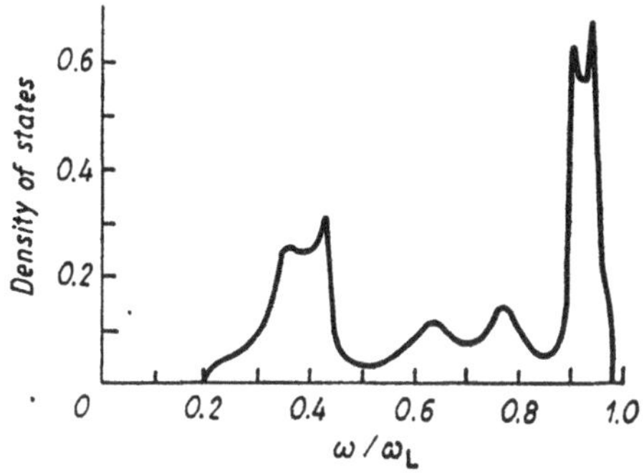

Fig. 3.27. Density of vibrational states for a crystal with diamond structure calculated by the cluster-Bethe lattice approximation for a 59-atom cluster. ω_L denotes the maximum frequency of the crystal (after YNDURAIN and SEN, 1976)

Figure 3.28 shows the local densities of vibrational states for a SiO_2-Bethe lattice. If one artificially broadens the vibrational state density of a SiO_2-Bethe lattice with a gaussian curve of width 75 cm^{-1}, the resulting state density agrees well with that of glassy SiO_2 obtained from neutron scattering and from the Bell and Dean model (cf. section 3.3.1.). This can be seen in figure 3.29. The gaussian curve used accounts for bond angle disorder. Note that Si—O—Si bond angle variations of about 10° are typical for glassy SiO_2. A change in this angle causes a small displacement of the bonds in the state density of a Bethe lattice. Thus, bond-angle fluctuations should cause a smearing of the state density in a glassy material. As mentioned above, the broadened Bethe lattice density of states is very similar to that of the random

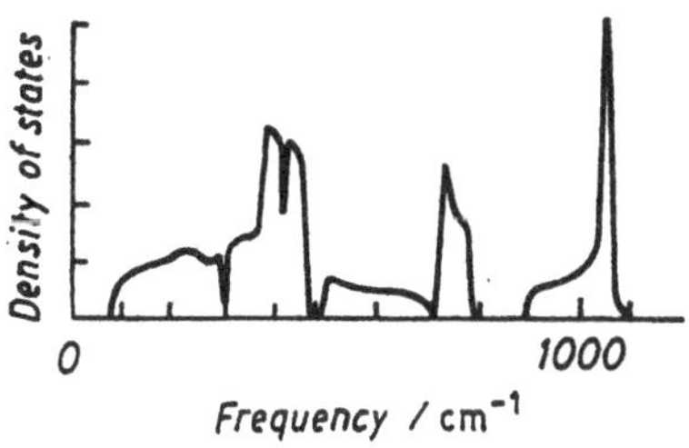

Fig. 3.28. Density of vibrational states calculated for a silicondioxide Bethe lattice (after LAUGHLIN and JOANNOPOULOS, 1977)

network of Bell and Dean, which shows that bond angle disorder destroys any explicit manifestation of the existence of rings of bonds in the density of states of glassy SiO_2. This is obviously due to the absence of small rings of bonds (12-fold rings of bonds exist in quartz).

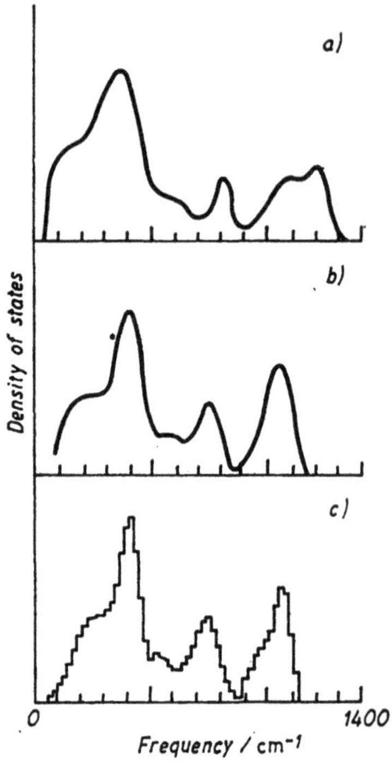

Fig. 3.29. Density of vibrational states for glassy silicondioxide. a) Neutron data, showing the density of states modulated by a weakly frequency-dependent matrix element. b) Bethe lattice state density artificially broadened (see text). c) Model of Bell and Dean (cf. section 3.3.1.) (after LAUGHLIN and JOANNOPOULOS, 1977, the experimental data are from LEADBETTER and STRINGFELLOW, 1972)

It is worth noticing that the cluster-Bethe lattice method is a valuable method also for investigating systems with compositional disorder (KLEINERT, 1979) and systems with compositional and structural disorder (YNDURAIN, 1978). In the case of compositional disorder the Bethe lattice connected with a cluster may be described in the VCA, ATA or CPA. — The use of the cluster-Bethe lattice method for studying infrared absorption, Raman scattering and inelastic neutron scattering spectra is problematic, because a calculation of these quantities requires us to take a sum containing diagonal as well as nondiagonal elements of the Green function (cf. (2.2.10)). In the case of localized states one may expect that reasonable results can be obtained by restricting this sum to cluster atoms only. For extended states, however, the sum must be extended over all atoms of the system consisting of a cluster embedded in an infinite Bethe lattice. Hence, in the latter case the computed spectra will be governed by the Bethe lattice contribution (KLEINERT, 1979). Note that a Bethe lattice can also be used for examining effects of chemical impurities on a surface (LAUGHLIN and JOANNOPOULOS, 1978), effects of interfaces (JOANNOPOULOS, 1979) and vibrational excitations of defect sites in amorphous materials (POLLARD and JOANNOPOULOS, 1980, 1981).

Note on localization added in proof:

Let us yet draw attention to a new direction in the theory of localization, which is based on some similarity between the behaviour of one-particle states in disordered systems near E_c (energy separating extended and localized states) and the critical behaviour of a magnetic system near T_c (temperature of magnetic phase transition): at E_c the localization length and at T_c the correlation length diverges. This similarity led to application of ideas and methods of the theory of phase transitions (scaling arguments, renormalization group methods) to the localization problem (WEGNER, 1976, ABRAHAMS et al., 1979, ANDERSON et al., 1980, SARKER and DOMANY, 1981, HARRIS and LUBENSKY, 1981). The frontier of this active research field is still fast advancing. Most of these localization studies indicate that (for on-diagonal disorder) one-particle states in a random potential are always localized for $d \leq 2$ dimensions irrespective of the strength of disorder (cf. VOLLHARDT and WÖLFLE, 1982). This is also supported by numerical localization studies (cf. KRAMER et al., 1981).

To the authors knowledge, scaling theories of localization have not been applied to vibrational systems so far, although the study of mode behaviour in mixed crystals by SCHMELTZER and BESERMAN (1981) by means of the real space renormalization group method is already a first step in this direction. In the latter study the transition between one-mode and two-mode behaviour is related to the transition between extended and localized mode behaviour. Note that rather good agreement with CPA and REI-model results have been achieved.

Note that renormalization group method has also been used for studying local density of vibrational states in a disordered chain (GONCALVES DA SILVA and KOILLER, 1981).

CHAPTER 4
ANHARMONIC CRYSTALS AND STRUCTURAL PHASE TRANSITIONS

4.1. Weakly anharmonic crystals

4.1.1. Perturbation theory of anharmonic crystals

When anharmonicity is present, vibrational excitations of a crystal can no longer be described by non-interacting quasi-particles or phonons. However, provided that the anharmonicity is not too strong, the anharmonicity may be accounted for by perturbation theory, taking the harmonic Hamiltonian for the unperturbed system. Anharmonicity gives rise to the phenomenon of thermal expansion and to interactions between phonons. The latter interactions open up channels for the decay of phonons and lead to an energy shift and a damping of phonons. The temperature dependence of each phonon energy arises from thermal expansion and from anharmonic interactions between phonons. The latter mechanism gives rise to a temperature dependence of phonon energies also under isochoric conditions, which can be used to distinguish this effect from effects of thermal strain.

The account of anharmonicity is a typical many-body problem requiring appropriate many-body techniques. Here, we outline the approach of Maradudin and Fein (1962) and Cowley (1963), which has been widely used in recent years for investigating many problems of weakly anharmonic crystals, such as thermal expansion, infrared absorption, Raman scattering, neutron scattering, and so on (see, for example, the review of Barron and Klein, 1974).

The total Hamiltonian of a anharmonic crystal, $\mathscr{H}$, reads

$$\mathscr{H} = H + H_{\mathrm{A}}, \tag{4.1.1}$$

where H denotes the Hamiltonian in the harmonic approximation (cf. (1.3.66)) and H_{A} denotes the anharmonic part of the Hamiltonian (cf. (1.3.70)). To shorten the notation we replace the pair of indices ($\boldsymbol{k}j$) by a single index λ, with the convention that $-\lambda$ denotes $-\boldsymbol{k}j$. Thus we have

$$H = \sum_{\lambda} \hbar\omega_{\lambda}[b_{\lambda}^{+}b_{\lambda} + 1/2], \tag{4.1.2}$$

$$H_{\mathrm{A}} = \sum_{\lambda} V(\lambda)\, A_{\lambda} + \sum_{n=3}^{\infty} \sum_{\lambda_1 \ldots \lambda_n} V(\lambda_1 \ldots \lambda_n)\, A_{\lambda_1} \ldots A_{\lambda_n}, \tag{4.1.3}$$

where (cf. (1.3.72))

$$A_{\lambda} = b_{\lambda} + b_{-\lambda}^{+} \tag{4.1.4}$$

and (cf. (1.3.73))

$$V(\lambda_1 \ldots \lambda_n) = V^{*}(-\lambda_1 \ldots -\lambda_n). \tag{4.1.5}$$

The linear response of a lattice to external electromagnetic radiation and the scattering properties of a lattice in lowest nonzero approximation are governed by functions of the type (cf. appendix 3)

$$J_{AB}(\omega) = \int_{-\infty}^{\infty} dt\, e^{i\omega t} \langle A(t)\, B(0)\rangle, \tag{4.1.6}$$

where A and B are operators and

$$A(t) = e^{\frac{i}{\hbar}\mathscr{H}t} A\, e^{-\frac{i}{\hbar}\mathscr{H}t}, \tag{4.1.7}$$

$$\langle\ldots\rangle = \mathrm{Tr}\,\{e^{-\beta\mathscr{H}}\ldots\}/\mathrm{Tr}\,\{e^{-\beta\mathscr{H}}\}, \qquad \beta = 1/kT. \tag{4.1.8}$$

To calculate such correlation functions we introduce, as in section 2.1.3., time-dependent Green functions by

$$G^{r,a}(AB, t) = \mp\frac{i}{\hbar}\, \Theta(\pm t)\, \langle[A(t), B(0)]\rangle \equiv \langle\langle A(t); B(0)\rangle\rangle^{r,a}, \tag{4.1.9}$$

where the upper and lower signs define the retarded (G^r) and advanced (G^a) Green function, respectively.

The Fourier transforms

$$G^{r,a}(AB, \omega) = \int_{-\infty}^{\infty} dt\, e^{i\omega t} G^{r,a}(AB, t) \tag{4.1.10}$$

have the property that $G^r(\omega)$ $\big(G^a(\omega)\big)$ is analytic in the upper (lower) complex frequency half plane. That is, from these two analytic parts we may compose a Green function $G(AB, z)$ analytic in the whole complex frequency plane except on the real axis (z — complex frequency).

Analogous to (2.1.19) and (2.1.22) we have

$$G(AB, z) = \frac{1}{2\pi\hbar} \int_{-\infty}^{\infty} d\omega\, \frac{(1 - e^{-\beta\hbar\omega})}{z - \omega}\, J_{AB}(\omega) \tag{4.1.11}$$

and

$$J_{AB}(\omega) = i\hbar\, \frac{[G(AB, \omega + i\varepsilon) - G(AB, \omega - i\varepsilon)]}{(1 - e^{-\beta\hbar\omega})}, \qquad \varepsilon \to +0. \tag{4.1.12}$$

Now we introduce a so-called temperature Green function by

$$G(AB, \tau) = \langle T_\tau A(\tau)\, B(0)\rangle, \qquad -\beta < \tau < \beta, \tag{4.1.13}$$

where A and B are again arbitrary operators,

$$A(\tau) = e^{\mathscr{H}\tau} A\, e^{-\mathscr{H}\tau}, \tag{4.1.14}$$

T_τ is the time-ordering operator with the property

$$T_\tau A(\tau)\, B(\tau') = \Theta(\tau - \tau')\, A(\tau)\, B(\tau') + \Theta(\tau' - \tau)\, B(\tau')\, A(\tau), \tag{4.1.15}$$

and $\langle\ldots\rangle$ is given by (4.1.8).

$G(AB, \tau)$ has the following periodic property:

$$G(AB, \tau + \beta) = G(AB, \tau), \quad -\beta < \tau < 0. \tag{4.1.16}$$

This follows from the cyclic property of the trace, since for $\tau < 0$

$$\begin{aligned}\mathrm{Tr}\,\{B(0)\,A(\tau)\} &= \mathrm{Tr}\,\{\mathrm{e}^{-\beta\mathscr{H}} B\,\mathrm{e}^{\mathscr{H}\tau} A\,\mathrm{e}^{-\mathscr{H}\tau}\} = \mathrm{Tr}\,\{\mathrm{e}^{\mathscr{H}\tau} A\,\mathrm{e}^{-(\tau+\beta)\mathscr{H}} B\} \\ &= \mathrm{Tr}\,\{\mathrm{e}^{-\beta\mathscr{H}}\,\mathrm{e}^{(\tau+\beta)\mathscr{H}} A\,\mathrm{e}^{-(\tau+\beta)\mathscr{H}} B\} \\ &= \mathrm{Tr}\,\{A(\tau+\beta)\,B(0)\}.\end{aligned} \tag{4.1.17}$$

Since the temperature Green function is only required in the region $-\beta < \tau < \beta$, it can be continued periodically and expanded as a Fourier series:

$$G(AB, \tau) = \sum_{n=-\infty}^{\infty} G(AB, i\omega_n) \exp(-i\hbar\omega_n\tau), \tag{4.1.18}$$

where, on account of (4.1.16),

$$\omega_n = 2\pi n/\beta\hbar. \tag{4.1.19}$$

The Fourier coefficients are given by

$$\left.\begin{aligned}G(AB, i\omega_n) &= \frac{1}{2\beta}\int_{-\beta}^{\beta} \mathrm{d}\tau\, G(AB, \tau) \exp(i\hbar\omega_n\tau) \\ &= \frac{1}{\beta}\int_{0}^{\beta} \mathrm{d}\tau \langle A(\tau)\,B(0)\rangle \exp(i\hbar\omega_n\tau),\end{aligned}\right\} \tag{4.1.20}$$

where we have accounted for (4.1.16) and $\exp(i\beta\hbar\omega_n) = 1$ for all n.

From (4.1.6) we obtain

$$\langle A(t)\,B(0)\rangle = \frac{1}{2\pi}\int_{-\infty}^{\infty} \mathrm{d}\omega\, \mathrm{e}^{-i\omega t} J_{AB}(\omega). \tag{4.1.21}$$

By putting $z = -i\hbar\tau$ in this equation we get

$$\langle A(\tau)\,B(0)\rangle = \frac{1}{2\pi}\int_{-\infty}^{\infty} \mathrm{d}\omega\, \mathrm{e}^{-\hbar\omega\tau} J_{AB}(\omega). \tag{4.1.22}$$

Substituting (4.1.22) into (4.1.20), we find

$$G(AB, i\omega_n) = \frac{1}{2\pi\beta\hbar}\int_{-\infty}^{\infty} \mathrm{d}\omega\, \frac{(1 - \mathrm{e}^{-\beta\hbar\omega})}{(\omega - i\omega_n)} J_{AB}(\omega). \tag{4.1.23}$$

Hence, comparing (4.1.11) and (4.1.23), we obtain

$$G^{r,a}(AB, \omega) = -\beta G(AB, \omega \pm i\varepsilon), \qquad \varepsilon \to +0, \tag{4.1.24}$$

where the upper (lower) sign holds for $G^r(G^a)$. Here we have used the fact that the integral (4.1.23), considered as a function of $i\omega_n$, can be continued analytically.

To use anharmonic perturbation theory for calculating the temperature Green function (4.1.13), we introduce the operator (analogue to the S-matrix)

$$S(\tau, \tau_0) = e^{H\tau} e^{-\mathscr{H}(\tau-\tau_0)} e^{-H\tau_0}. \tag{4.1.25}$$

Note that S is not unitary, but it satisfies the group property

$$S(\tau_1, \tau_2) S(\tau_2, \tau_3) = S(\tau_1, \tau_3) \tag{4.1.26}$$

and the boundary condition

$$S(\tau_0, \tau_0) = 1. \tag{4.1.27}$$

Hence, on account of (4.1.26) and (4.1.27), the inverse operator $S^{-1}(\tau_1, \tau_2)$ is given by

$$S^{-1}(\tau_1, \tau_2) = S(\tau_2, \tau_1). \tag{4.1.28}$$

Differentiating (4.1.25) with respect to τ, we obtain

$$\frac{\partial}{\partial\tau} S(\tau, \tau_0) = -\tilde{H}_A(\tau) S(\tau, \tau_0), \tag{4.1.29}$$

where the tilde denotes

$$\tilde{H}_A(\tau) = e^{H\tau} H_A e^{-H\tau}. \tag{4.1.30}$$

Solving (4.1.26) by iteration, we find for $\tau > \tau_0$

$$\begin{aligned} S(\tau, \tau_0) &= 1 + \sum_{n=1}^{\infty} (-1)^n \int_{\tau_0}^{\tau} d\tau_1 \int_{\tau_0}^{\tau_1} d\tau_2 \dots \int_{\tau_0}^{\tau_{n-1}} d\tau_n \tilde{H}_A(\tau_1) \tilde{H}_A(\tau_2) \dots \tilde{H}_A(\tau_n) \\ &= 1 + \sum_{n=1}^{\infty} \frac{(-1)^n}{n!} \int_{\tau_0}^{\tau} d\tau_1 \dots \int_{\tau_0}^{\tau} d\tau_n T_\tau[\tilde{H}_A(\tau_1) \dots \tilde{H}_A(\tau_n)] \\ &= T_\tau \exp\left(-\int_{\tau_0}^{\tau} d\tau' \tilde{H}_A(\tau')\right). \end{aligned} \tag{4.1.31}$$

In particular, (4.1.25) may be rewritten as

$$e^{-\mathscr{H}\tau} = e^{-H\tau} S(\tau, 0) \equiv e^{-H\tau} S(\tau). \tag{4.1.32}$$

If τ is made equal to β, (4.1.32) and (4.1.31) provide a perturbation expansion for the partition function:

$$Z = \operatorname{Tr}\{e^{-\beta\mathscr{H}}\} = \operatorname{Tr}\{e^{-\beta H} S(\beta)\} = Z_0 \langle S(\beta) \rangle_0, \tag{4.1.33}$$

where

$$\langle \ldots \rangle_0 = \mathrm{Tr}\,\{\mathrm{e}^{-\beta H} \ldots\} \tag{4.1.34}$$

and

$$Z_0 = \mathrm{Tr}\,\{\mathrm{e}^{-\beta H}\}. \tag{4.1.35}$$

For $\tau > 0$, by (4.1.13), (4.1.32), (4.1.28), (4.1.33), the temperature Green function becomes

$$\begin{aligned} G(AB, \tau) &= Z^{-1}\,\mathrm{Tr}\,\{\mathrm{e}^{-\beta\mathscr{H}}\,\mathrm{e}^{\mathscr{H}\tau}A\,\mathrm{e}^{-\mathscr{H}\tau}B\} \\ &= Z_0^{-1}\langle S(\beta)\rangle_0^{-1}\,\mathrm{Tr}\,\{\mathrm{e}^{-\beta H}S(\beta, 0)\,S(0, \tau)\,\mathrm{e}^{H\tau}A\,\mathrm{e}^{-H\tau}S(\tau, 0)\,B\} \\ &= \langle S(\beta, \tau)\,\tilde{A}(\tau)\,S(\tau, 0)\,\tilde{B}(0)\rangle_0/\langle S(\beta)\rangle_0. \end{aligned} \tag{4.1.36}$$

A similar expression results for $\tau < 0$. Hence, using (4.1.26) and the time ordering operator T_τ, the temperature Green function may be rewritten as

$$G(AB, \tau) = \langle T_\tau \tilde{A}(\tau)\,\tilde{B}(0)\,S(\beta)\rangle_0/\langle S(\beta)\rangle_0, \tag{4.1.37}$$

where T_τ orders the operators in the expansion of $S(\beta)$ simultaneously with $\tilde{A}(\tau)$ and $\tilde{B}(0)$.

On assuming that the operators A and B can be expressed as sums of products of operators A_λ, the average in the numerator of equation (4.1.37) has to be taken over a linear combination of ordered products of $\tilde{A}_\lambda(\tau_i)$'s. Similar expressions occur in evaluating $\langle S(\beta)\rangle_0$. Such an expression can be calculated by a diagram technique based on (generalized) Wick's theorem (see, for example, Fetter and Walecka, 1971, section 24).

The Wick's theorem states that for a system with a Hamiltonian of the type (4.1.2) the thermal average of a product of operators B_λ (B_λ represents either b_λ or b_λ^+) can be expressed as a sum of contributions from all possible pairing schemes of this product:

$$\begin{aligned} \langle B_{\lambda_1}B_{\lambda_2} \ldots B_{\lambda_{2p}}\rangle_0 &= \langle B_{\lambda_1}B_{\lambda_2}\rangle_0\,\langle B_{\lambda_3}B_{\lambda_4}\rangle_0 \ldots \langle B_{\lambda_{2p-1}}B_{\lambda_{2p}}\rangle_0 \\ &+ \text{similar contributions from all other pairing schemes.} \end{aligned} \tag{4.1.38}$$

Before we prove Wick's theorem, let us give some formulae that will be needed. For the Hamiltonian (4.1.2), the statistical operator for calculating $\langle \ldots \rangle_0$ can be rewritten as

$$\hat{\varrho} = \mathrm{e}^{-\beta H}/\mathrm{Tr}\,\{\mathrm{e}^{-\beta H}\} = \prod_\lambda \hat{\varrho}(\lambda), \tag{4.1.39}$$

where

$$\hat{\varrho}(\lambda) = \mathrm{e}^{-\beta\hbar\omega_\lambda b_\lambda^+ b_\lambda} \Big/ \sum_{n_\lambda=0}^{\infty} \mathrm{e}^{-\beta\hbar\omega_\lambda n_\lambda}. \tag{4.1.40}$$

Since the statistical operator (4.1.39) has only diagonal elements, we find for the thermal average $\langle B_{\lambda_1}B_{\lambda_2}\rangle_0$ (B_λ represents again either b_λ or b_λ^+)

$$\langle B_{\lambda_1}B_{\lambda_2}\rangle_0 = [B_{\lambda_1}, B_{\lambda_2}]/(1 - \mathrm{e}^{\pm\beta\hbar\omega_{\lambda_2}}). \tag{4.1.41}$$

Furthermore, we shall need the relations

$$\left.\begin{aligned} b_\lambda^+\hat{\varrho} &= \hat{\varrho}b_\lambda^+ \,\mathrm{e}^{\beta\hbar\omega_\lambda}, \\ b_\lambda\hat{\varrho} &= \hat{\varrho}b_\lambda \,\mathrm{e}^{-\beta\hbar\omega_\lambda}. \end{aligned}\right\} \tag{4.1.42}$$

To prove these relations let us consider the operator

$$b_\lambda(\alpha) = \mathrm{e}^{\alpha b_\lambda^+ b_\lambda} b_\lambda \,\mathrm{e}^{-\alpha b_\lambda^+ b_\lambda}, \qquad b_\lambda(0) \equiv b_\lambda. \tag{4.1.43}$$

Differentiating (4.1.43) with respect to α, we obtain

$$\frac{\mathrm{d}}{\mathrm{d}\alpha} b_\lambda(\alpha) = -b_\lambda(\alpha). \tag{4.1.44}$$

Integrating (4.1.44), we find

$$b_\lambda(\alpha) = b_\lambda \,\mathrm{e}^{-\alpha} \tag{4.1.45}$$

With the aid of this equation and of (4.1.43), (4.1.39), and (4.1.40), (4.1.42) can be readily verified.

Now let us turn to the proof of Wick's theorem. We take the identity

$$\begin{aligned} [B_{\lambda_1}, B_{\lambda_2} \dots B_{\lambda_{2p}}] &= [B_{\lambda_1}, B_{\lambda_2}]\, B_{\lambda_3} \dots B_{\lambda_{2p}} \\ &\quad + B_{\lambda_2}[B_{\lambda_1}, B_{\lambda_3}]\, B_{\lambda_4} \dots B_{\lambda_{2p}} + \cdots \end{aligned} \tag{4.1.46}$$

and evaluate the thermal average of both sides. On account of

$$\langle B_{\lambda_2} \dots B_{\lambda_{2p}} B_{\lambda_1}\rangle_0 = \mathrm{e}^{\pm\beta\hbar\omega_{\lambda_1}} \langle B_{\lambda_1} \dots B_{\lambda_{2p}}\rangle_0, \tag{4.1.47}$$

which holds owing to (4.1.42), we obtain from the thermal average of (4.1.46)

$$\begin{aligned} (1 - \mathrm{e}^{\pm\beta\hbar\omega_{\lambda_1}})\, \langle B_{\lambda_1} \dots B_{\lambda_{2p}}\rangle_0 &= [B_{\lambda_1}, B_{\lambda_2}]\, \langle B_{\lambda_3} \dots B_{\lambda_{2p}}\rangle_0 \\ &\quad + [B_{\lambda_1}, B_{\lambda_3}]\, \langle B_{\lambda_2} \dots B_{\lambda_{2p}}\rangle_0 + \cdots. \end{aligned} \tag{4.1.48}$$

Accounting for (4.1.41), we find from (4.1.48)

$$\begin{aligned} \langle B_{\lambda_1} \dots B_{\lambda_{2p}}\rangle_0 &= \langle B_{\lambda_1} B_{\lambda_2}\rangle_0\, \langle B_{\lambda_3} \dots B_{\lambda_{2p}}\rangle_0 \\ &\quad + \langle B_{\lambda_1} B_{\lambda_3}\rangle_0\, \langle B_{\lambda_2} \dots B_{\lambda_{2p}}\rangle_0 + \cdots. \end{aligned} \tag{4.1.49}$$

Continuing this procedure, we obtain Wick's theorem.

Wick's theorem also holds for a T_τ-product of operators:

$$\begin{aligned} &\langle T_\tau \tilde{B}_{\lambda_1}(\tau_1)\, \tilde{B}_{\lambda_2}(\tau_2) \dots \tilde{B}_{\lambda_{2p}}(\tau_{2p})\rangle_0 \\ &= \langle T_\tau \tilde{B}_{\lambda_1}(\tau_1)\, \tilde{B}_{\lambda_2}(\tau_2)\rangle_0\, \langle T_\tau \tilde{B}_{\lambda_3}(\tau_3)\, \tilde{B}_{\lambda_4}(\tau_4)\rangle_0 \dots \langle T_\tau \tilde{B}_{\lambda_{2p-1}}(\tau_{2p-1})\, \tilde{B}_{\lambda_{2p}}(\tau_{2p})\rangle_0 \\ &\quad + \text{similar contributions from all other pairing schemes.} \end{aligned} \tag{4.1.50}$$

This can be seen by using $\langle T_\tau \dots\rangle_0 = T_\tau\langle\dots\rangle_0$ and

$$\tilde{B}_\tau(\tau) = B_\lambda \,\mathrm{e}^{\pm\hbar\omega_\lambda\tau}, \tag{4.1.51}$$

where the upper (lower) sign holds for $B_\lambda = b_\lambda^+(b_\lambda)$, and with the aid of (4.1.38). Equation (4.1.51) can be proved analogously as (4.1.42).

When we use Wick's theorem for evaluating the temperature Green function (4.1.37) or $\langle S(\beta)\rangle_0$, we have to deal with products of quantities

$$\langle T_\tau \tilde{A}_\lambda(\tau_1)\, \tilde{A}_{\lambda'}(\tau_2)\rangle_0 = \delta_{\lambda,-\lambda'} g(\lambda, \tau_1 - \tau_2), \tag{4.1.52}$$

where $g(\lambda, \tau_1 - \tau_2)$ is the harmonic one-phonon Green function (phonon propagator) defined in the range $-\beta < \tau < \beta$ by

$$\begin{aligned} g(\lambda, \tau) &= \langle T_\tau \tilde{A}_\lambda(\tau)\, \tilde{A}_{-\lambda}(0)\rangle_0 \\ &= \bar{n}_\lambda \exp\left(|\tau|\, \hbar\omega_\lambda\right) + (\bar{n}_\lambda + 1) \exp\left(-|\tau|\, \hbar\omega_\lambda\right). \end{aligned} \tag{4.1.53}$$

Here $\bar{n}_\lambda$ is the phonon population number of the mode λ, given by (1.4.50). The Fourier transform of $g(\lambda, \tau)$ reads

$$g(\lambda, i\omega_n) = \frac{1}{\beta\hbar}\left\{\frac{1}{\omega_\lambda - i\omega_n} + \frac{1}{\omega_\lambda + i\omega_n}\right\} = \frac{2\omega_\lambda}{\beta\hbar(\omega_\lambda^2 + \omega_n^2)}. \tag{4.1.54}$$

The harmonic Green function (4.1.54) considered as a function of the variable $i\omega_n$ is seen to exhibit poles when $i\omega_n = \pm\omega_\lambda$, i.e. the poles occur at the eigenfrequencies of the harmonic system. On account of anharmonicity we expect the vibrations to be altered in frequency by an amount Δ and to have a finite lifetime $1/\Gamma$, so that the poles of the analogous Green function for an anharmonic crystal may be expected to be at $i\omega_n = \pm(\omega_\lambda + \Delta + i\Gamma)$.

Now let us study the one phonon Green function

$$G(\lambda\lambda', \tau) = \langle T_\tau A_\lambda(\tau)\, A_{-\lambda'}(0)\rangle \tag{4.1.55}$$

for an anharmonic crystal. Using (4.1.37), (4.1.31) and (4.1.50) for $G(\lambda\lambda', i\omega_n)$, a perturbation expansion can be obtained whose terms of any order can be represented by diagrams and evaluated as follows (Maradudin and Fein, 1962, Cowley, 1963):

1. Represent a phonon entering a region of interaction by a line labelled λ and a phonon leaving this region by a line labelled λ' ($\boldsymbol{k} = \boldsymbol{k}'$, because the momentum is conserved in all interactions).
2. Within the region of interaction mark n interaction vertices (for an nth order term). Draw all topologically distinct diagrams of phonon lines joining these vertices.
3. Associate indices λ and frequencies $i\omega_n$ with the phonon lines in such a way that the internal lines at each vertex satisfy conservation of crystal momentum (positive momentum entering, negative leaving).
4. With each phonon line associate a factor $g(\lambda, i\omega_n)$ (from (4.1.54)).
5. With each vertex connected with s lines associate a factor $V(\lambda_1 \ldots \lambda_s)$.
6. Multiply by $(-1)^n\, \beta^n/n!$ (β^n arises from the n integrations).
7. Multiply by a weighting factor based on the number of possible pairings of phonon operators. (Equivalent diagrams may result from (i) pairings of different operators associated with fixed vertices and (ii) pairings of

operators belonging to different vertices. The latter possibility corresponds to an interchange of integration variables).

8. Sum all intermediate indices λ and ω_n.

Diagrams containing parts not connected with at least one external phonon line are termed disconnected diagrams. The contribution of these diagrams can be shown to be canceled by the denominator $\langle S(\beta)\rangle_0$, giving finally

$$G(\lambda\lambda', \tau) = \langle T_\tau \tilde{A}_\lambda(\tau)\, A_{-\lambda'}(0)\, S(\beta)\rangle_{0,\mathrm{conn.to}A_\lambda A_{-\lambda'}}, \tag{4.1.56}$$

where "conn. to $A_\lambda A_{-\lambda'}$" denotes diagrams such that all parts are connected to at least one external phonon line. To see this, consider a given nth order diagram having $(n - m)$ vertices not connected with any external phonon line. The contribution of this diagram to $G(\lambda\lambda', \tau)$ can be written symbolically as

$$\frac{(-1)^n}{n!} \int\limits_0^\beta \mathrm{d}\tau_1 \dots \int\limits_0^\beta \mathrm{d}\tau_m \langle T_\tau \tilde{A}_\lambda(\tau)\, \tilde{A}_{-\lambda'}(0)\, \tilde{H}_A(\tau_1) \dots \tilde{H}_A(\tau_m)\rangle_{0,\mathrm{conn.to}A_\lambda A_{-\lambda'}}$$

$$\times \int\limits_0^\beta \mathrm{d}\tau_{m+1} \dots \int\limits_0^\beta \mathrm{d}\tau_n \langle T_\tau \tilde{H}_A(\tau_{m+1}) \dots \tilde{H}_A(\tau_n)\rangle_0, \tag{4.1.57}$$

where the weighting factor, arising from an interchange of integration variables between the connected and the disconnected parts of the diagram, is not yet included. There are $n!/m!(n - m)!$ such interchanges leaving the diagram invariant. Multiplying (4.1.57) by this factor and summing all disconnected diagrams having the same part of m vertices connected with external phonon lines, then we obtain

$$\frac{(-1)^m}{m!} \int\limits_0^\beta \mathrm{d}\tau_1 \dots \int\limits_0^\beta \mathrm{d}\tau_m \langle T_\tau \tilde{A}_\lambda(\tau)\, \tilde{A}_{-\lambda'}(0)\, \tilde{H}_A(\tau_1) \dots \tilde{H}_A(\tau_m)\rangle_{0,\mathrm{conn.to}A_\lambda A_{-\lambda'}}$$

$$\times \left\{1 + \sum_{n-m=1}^{\infty} \frac{(-1)^{n-m}}{(n - m)!} \int\limits_0^\beta \mathrm{d}\tau_{m+1} \dots \int\limits_0^\beta \mathrm{d}\tau_n \langle T_\tau \tilde{H}_A(\tau_{m+1}) \dots \tilde{H}_A(\tau_n)\rangle_0\right\}. \tag{4.1.58}$$

Comparing the expression in the braces of (4.1.58) with (4.1.31), we see that this expression equals $\langle S(\beta)\rangle_0$. Hence, the sum of all possible diagrams equals the product of $\langle S(\beta)\rangle_0$ and the sum of all possible connected diagrams. Consequently, $G(\lambda\lambda', \tau)$ is given by the sum of all connected diagrams.

Some typical connected diagrams are shown in figure 4.1. These diagrams may be divided into two classes. Proper diagrams a) and b) cannot be separated into two parts by cutting a single internal phonon line, while improper diagrams c) and d) can be separated into two disconnected parts in this way. The

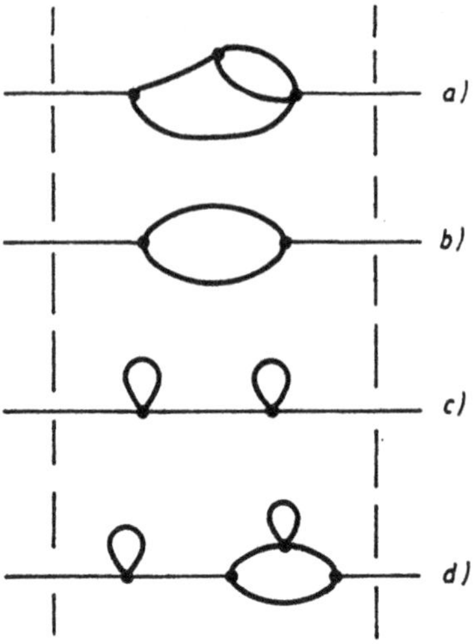

Fig. 4.1. Some typical connected diagrams. a), b) Proper diagrams; c), d) improper diagrams

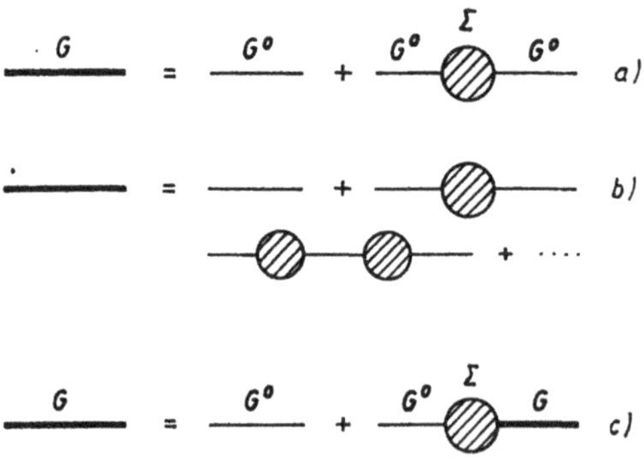

Fig. 4.2. The Dyson equation for the one phonon Green function. a) The sum of all proper diagrams; b), c) the sum of all proper and improper diagrams

total contribution of all proper diagrams to the Fourier transform of the Green function is (cf. fig. 4.2a))

$$G(\boldsymbol{k}jj', i\omega_n) \simeq g(\boldsymbol{k}j, i\omega_n)\, \delta_{jj'} + g(\boldsymbol{k}j, i\omega_n)\, \Sigma(\boldsymbol{k}jj', i\omega_n)\, g(\boldsymbol{k}j', i\omega_n), \tag{4.1.59}$$

or in an obviously condensed notation

$$\boldsymbol{G}(\boldsymbol{k}) \simeq \boldsymbol{G}^0(\boldsymbol{k}) + \boldsymbol{G}^0(\boldsymbol{k})\, \boldsymbol{\Sigma}(\boldsymbol{k})\, \boldsymbol{G}^0(\boldsymbol{k}), \tag{4.1.60}$$

where $\boldsymbol{\Sigma}$, the so-called self-energy matrix, is the sum of all proper diagrams (without external phonon lines). On account of proper and improper diagrams we have (cf. fig. 4.2b))

$$G = G^0 + G^0 \Sigma G^0 + G^0 \Sigma G^0 \Sigma G^0 + \cdots, \tag{4.1.61}$$

where, because of the crystal momentum conservation at each vertex, the wave vector of any single internal phonon line in an improper diagram is the same as that of the external lines. The infinite series implied in (4.1.61) may be written concisely as (cf. fig. 4.2c))

$$G(\boldsymbol{k}) = G^0(\boldsymbol{k}) + G^0(\boldsymbol{k})\, \Sigma(\boldsymbol{k})\, G(\boldsymbol{k}), \tag{4.1.62}$$

which is the so-called Dyson equation.

Using (4.1.5) and (4.1.54) Σ can be shown to be a hermitian matrix. Owing to this property it can be diagonalized for any ω_n^2 by choosing a new set of base coordinates. However, in general, it cannot be diagonalized simultaneously for all ω_n, so that non-diagonal elements remain, which mix modes belonging to different branches j. Assuming the non-diagonal elements of Σ to be small, then the diagonal elements of G become approximately

$$G(\boldsymbol{k}jj, i\omega_n) \equiv G(\lambda, i\omega_n) \simeq \{G_0^{-1}(\lambda, i\omega_n) - \Sigma(\lambda, i\omega_n)\}^{-1}$$
$$= \frac{2\omega_\lambda/\beta\hbar}{-(i\omega_n)^2 + \omega_\lambda^2 - (2\omega_\lambda/\beta\hbar)\,\Sigma(\lambda, i\omega_n)}. \tag{4.1.63}$$

As mentioned above, physical properties depend upon the analytic continuation of the temperature Green function to the real frequency axis. Near the real axis we have

$$\Sigma(\lambda\lambda', \omega \pm i\varepsilon) \xrightarrow{\varepsilon\to+0} -\beta\hbar\{\Delta(\lambda\lambda', \omega) \pm i\Gamma(\lambda\lambda', \omega)\}, \tag{4.1.64}$$

where Δ and Γ are hermitian matrices. By (4.1.24), (4.1.63) and (4.1.64), the diagonal elements of the retarded Green function become

$$G^r(\lambda, \omega) \simeq -\frac{2\omega_\lambda/\hbar}{-\omega^2 + \omega_\lambda^2 + 2\omega_\lambda\Delta_\lambda(\omega) - 2i\omega_\lambda\Gamma_\lambda(\omega)}, \tag{4.1.65}$$

where

$$\Delta_\lambda(\omega) \equiv \Delta(\lambda\lambda, \omega), \qquad \Gamma_\lambda(\omega) \equiv \Gamma(\lambda\lambda, \omega).$$

This Green function is similar to that of a harmonic crystal, if ω_λ^2 is replaced by $\tilde{\omega}_\lambda^2$, such that

$$\tilde{\omega}_\lambda^2 = \omega_\lambda^2 + 2\omega_\lambda\big(\Delta_\lambda(\omega) - i\Gamma_\lambda(\omega)\big). \tag{4.1.66}$$

According to the usual quasi-particle conception (see, for example, Kirzhnits, 1967, section 24), the real and imaginary parts of the poles of the one-particle Green function (4.1.65) respectively determine dispersion and damping of plane-wave-like excitations (phonons). Provided that Δ_λ and Γ_λ vary slowly with ω, dispersion and damping of these excitations are given by $\omega_\lambda + \Delta_\lambda(\omega_\lambda)$ (quasi-harmonic frequency) and $\Gamma_\lambda(\omega_\lambda)$, respectively.

In general, the self-energy Σ contains two contributions: (i) from thermal strain and (ii) from anharmonic interaction between the phonons. The proce-

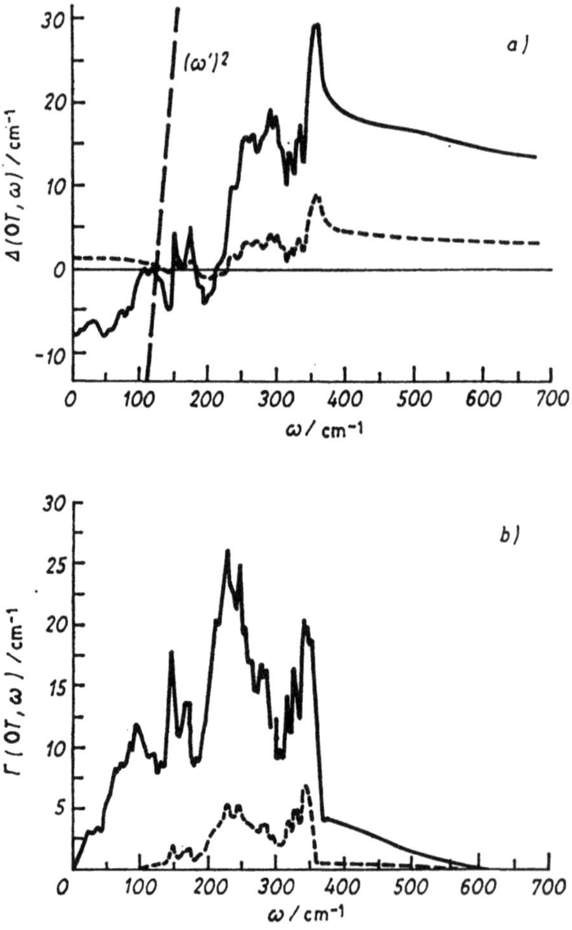

Fig. 4.3. Calculated frequency dependence of the anharmonic contribution of the self-energy for the long wavelength transverse optic modes of KBr at 5 K (dashed line) and 300 K (solid line): a) $\Delta(\mathbf{0}T, \omega)$, b) $\Gamma(\mathbf{0}T, \omega)$. In a) the intersection of the line

$$\omega'^2 = \left(\frac{\omega^2 - [\omega_T^2 - 2\omega_T \Delta^E(\mathbf{0}T)]}{2\omega_T}\right)_{290\,\mathrm{K}}$$

(ω_T — frequency of the long wavelength transverse optic mode in harmonic approximation, $\Delta^E(\mathbf{0}T)$ — thermal strain contribution to the real part of the self-energy) gives the renormalized long wavelength transverse optic mode frequency for 290 K (after LOWNDES and RASTOGI, 1976) dure for taking into account both these contributions is to expand the lattice potential in terms of displacements due to strain and those due to vibrations. Detailed expressions for the self-energy Σ have been evaluated by a number of authors. Quantitative estimates of these expressions require complex calculations (LOWNDES and RASTOGI, 1976, MONGA et al., 1979, and references therein). High-pressure spectroscopic studies on specific phonons can be used for measuring both thermal-strain and anharmonic self-energies (LOWNDES and RASTOGI, 1976, and references therein).

Unlike the thermal-strain contribution to the self-energy the anharmonic contribution often strongly depends on ω. The ω-dependence of the latter contribution is illustrated by fig. 4.3, where for the $\boldsymbol{k} \approx 0$ transverse-optic modes in KBr the real and imaginary parts of the anharmonic contribution to the self-energy are plotted. These results were obtained by accounting for cubic and quartic anharmonicity to the second order (for absent first term on the right-hand side in (4.1.3)). Of course, the anharmonicity also affects the dielectric constant. For instance, for an ionic crystal with rocksalt structure an account of anharmonicity requires that we replace in (1.5.82) the strictly harmonic long wavelength squared transverse frequency ω_T^2 by (cf. (4.1.66))

$$\tilde{\omega}_T^2 = \omega_T^2 + 2\omega_T\big(\Delta(0T, \omega) - i\Gamma(0T, \omega)\big), \tag{4.1.67}$$

where the index T labels the transverse-optic branch. Thus, the imaginary part of the dielectric constant, $\varepsilon_2(\omega)$, exhibits a broad band peaked at $\omega_T + \Delta(0T, \omega_T)$ (see fig. 4.4), instead of a δ-function peak at ω_T as in the harmonic case.

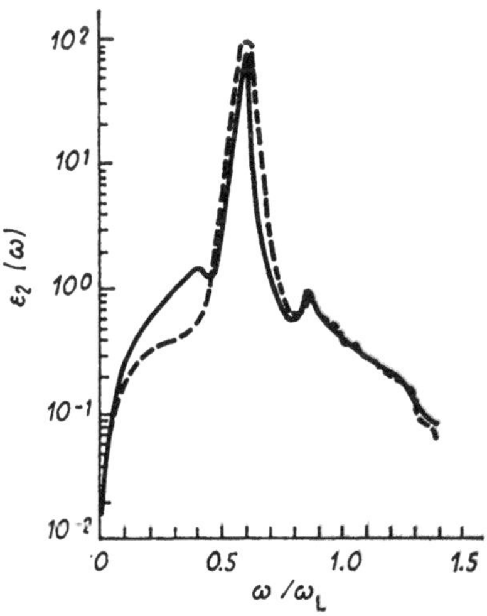

Fig. 4.4. Plots of theoretical (solid line) and experimental (dashed line) results of the imaginary part of the dielectric constant $\varepsilon_2(\omega)$ for NaCl at $T = 300$ K. ω_L is the maximum frequency of the crystal (after MONGA et al., 1979, the experimental results are from GENZEL et al., 1959, and GEICK, 1962)

The technique described above has been used for studying various physical phenomena such as infrared, Raman and neutron scattering spectra (see BARRON and KLEIN, 1974, and references therein). In particular, it has also been used for studying structural phase transitions[1] in some ferroelectric and anti-ferroelectric crystals which arise from instabilities of these crystals against particular vibrational modes phonon softening (COWLEY, 1965, COWLEY and COOMBS, 1973, BRUCE and COWLEY, 1973). For a ferroelectric crystal, for example, the dielectric constant follows a Curie temperature dependence above the transition:

$$\varepsilon = c/(T - T_c). \tag{4.1.68}$$

Here T_c is the transition temperature and c is a constant. The mode in question is a long wavelength transverse optic mode having a temperature dependence above T_c of the form

$$\tilde{\omega}_\lambda^2 = A(T - T_c), \tag{4.1.69}$$

where A is a constant. Such a dependence can be understood on the basis of equation (4.1.67). At high temperatures both the thermal-strain and the anharmonic contributions to the self-energy become proportional to T, so that $2\omega_\lambda \Delta_\lambda(\omega)$ is of the form AT. If the square of the strictly harmonic frequency, ω_λ^2, is negative and written as $-AT$, (4.1.67) becomes identical to (4.1.69). Note that this approach to the transitions described has been criticized, since in the presence of imaginary frequencies the harmonic basis represents an unstable equilibrium state for the crystal, and the perturbation expansion does not in principle exist (cf. GILLIS and KOEHLER, 1971).

We close this section by drawing attention to equation of motion and functional derivative techniques for studying effects of anharmonicity (cf. BARRON and KLEIN, 1974); to a diagram technique for anharmonic crystals in a strong radiation field (BÖTTGER, 1968); to the long-standing problem of two-phonon bound states (cf. KIMBALL et al., 1981); to anharmonic motion of isolated defects in crystals (cf. MARADUDIN, 1966b); to the problem of anharmonic disordered crystals (cf. CARON, 1976, DIEHL, 1977), and to a new type of exact periodic solutions found for lattices exhibiting structural phase transitions (BILZ et al., 1982).

4.1.2. Second sound and diffusive heat conduction

A weakly anharmonic crystal may be considered as a weakly interacting phonon gas, where scattering processes occur and phonons can decay. For a phonon with frequency ω_λ these processes lead to a frequency shift Δ_λ and to

[1] For the reader desiring a general picture of structural phase transitions (theoretical as well as experimental studies) the reviews of COWLEY (1980), BRUCE (1980a) and BRUCE and COWLEY (1980) would be of particular interest.

a damping Γ_λ (cf. section 4.1.1.). The average phonon lifetime τ, which we need in the following, is given by the reciprocal of the average damping constant Γ_λ.

On applying an external perturbation with frequency ω to the vibrating crystal, the following two frequency regimes may be distinguished:

$$\omega\tau \gg 1 \quad - \quad \text{the collisionless regime,} \tag{4.1.70}$$

which is sometimes also called zero sound regime, and

$$\omega\tau \ll 1 \quad - \quad \text{the collision-dominated regime,} \tag{4.1.71}$$

which is also called hydrodynamic regime. The condition (4.1.70) is usually fulfilled in neutron scattering and high-frequency ultrasonic experiments, the condition (4.1.71) in heat conduction experiments.

In the collision-dominated or hydrodynamic regime there are two different types of relaxation processes for establishing thermal equilibrium after having applied an external perturbation: (i) relaxation to a state of local thermal equilibrium by normal processes (cf. the discussion after equation (1.3.76)), followed by (ii) relaxation to total thermal equilibrium by umklapp processes. The existence of a state of local thermal equilibrium allows us to describe the system by hydrodynamic equations. This approach implies the assumption that all modes besides hydrodynamic ones (i.e. besides collective modes slowly varying in space and time) average out within the time scale of interest. Examples of hydrodynamic modes are propagating hydrodynamic waves (slightly renormalized phonons) and heat diffusion, which represents the evolution of the phonon number per unit volume (phonon density).

Heat diffusion is the usual heat transport mechanism in insulating crystals. However, there exists an exeptional type of heat transport in a narrow temperature window: the second sound. This phenomenon is a wave-like disturbance of the phonon density and can be interpreted as ordinary sound (first sound) in the phonon gas. The conditions for its existence are: (i) sufficiently strong normal processes for establishing local thermal equilibrium and (ii) absence of processes that destroy conservation of the quasi-momentum. Note that in the absence of umklapp processes the energy and the quasi-momentum are conserved quantities. In the presence of such processes the energy is still a conserved quantity, while the quasi-momentum is damped. In addition to umklapp processes, crystal imperfections, such as impurities and dislocations, also suppress second sound. Therefore, for studying second sound one needs perfect crystals and low temperatures (see below).

The hydrodynamic treatment of heat transport (for reviews see Beck et al., 1974, Beck, 1975, and Gurevich, 1980) bases on the assumption that there exists a function $n(\lambda, \boldsymbol{r}t)$, where $\lambda \equiv (\boldsymbol{k}, j)$, describing the distribution of wave packets of phonons with polarization j and wave vector $\boldsymbol{k}$, which can be localized around position $\boldsymbol{r}$ in a region smaller than a characteristic wavelength of the external perturbation. The time evolution of $n(\lambda, \boldsymbol{r}t)$ is governed by the

Boltzmann-Peierls equation

$$\frac{\partial n}{\partial t} + \boldsymbol{v}_\lambda \frac{\partial n}{\partial \boldsymbol{r}} = C[n], \tag{4.1.72}$$

where

$$\boldsymbol{v}_\lambda \equiv v_j(\boldsymbol{k}) = \frac{\partial \omega_j(\boldsymbol{k})}{\partial \boldsymbol{k}} \tag{4.1.73}$$

denotes the group velocity. The collision operator $C = C_N + C_R$, where $C_R = C_I + C_B + C_U$, describes normal processes (C_N), umklapp processes (C_U), and scattering from imperfections (C_I) and boundaries (C_B) of the crystal. The resistive part, C_R, contains quasi-momentum non-conserving processes.

We define the following quantities:

$$E(\boldsymbol{r}t) = V^{-1} \sum_\lambda \hbar\omega_\lambda n(\lambda, \boldsymbol{r}t), \tag{4.1.74}$$

$$S_\alpha(\boldsymbol{r}t) = V^{-1} \sum_\lambda \hbar\omega_\lambda (\boldsymbol{v}_\lambda)_\alpha\, n(\lambda, \boldsymbol{r}t), \tag{4.1.75}$$

$$P_\alpha(\boldsymbol{r}t) = V^{-1} \sum_\lambda \hbar\lambda_\alpha n(\lambda, \boldsymbol{r}t) \qquad (\lambda_\alpha \equiv k_\alpha), \tag{4.1.76}$$

$$T_{\alpha\alpha'}(\boldsymbol{r}t) = V^{-1} \sum_\lambda \hbar\lambda_\alpha (\boldsymbol{v}_\lambda)_{\alpha'}\, n(\lambda, \boldsymbol{r}t), \tag{4.1.77}$$

which can be interpreted as densities of energy, energy current, momentum and momentum flux (V — volume of the crystal). Multiplication of equation (4.1.72) by $\hbar\omega_\lambda$ and $\hbar\lambda_\alpha$, respectively, and summation over λ yields

$$\frac{\partial}{\partial t} E + \frac{\partial}{\partial \boldsymbol{r}} \boldsymbol{S} = 0, \tag{4.1.78}$$

$$\frac{\partial P_\alpha}{\partial t} + \sum_{\alpha'} \frac{\partial T_{\alpha\alpha'}}{\partial r_{\alpha'}} = \sum_\lambda \hbar\lambda_\alpha C_R[n]. \tag{4.1.79}$$

In writing down these equations we have taken into account that in (4.1.78) a term $\sum_\lambda \hbar\omega_\lambda C[n]$ vanishes due to energy conservation and that to the right-hand side of (4.1.79) only quasi-momentum non-conserving processes contribute.

In order to obtain hydrodynamic equations from (4.1.78) and (4.1.79), we have to solve (4.1.72). This is a difficult task. For simplicity, we use the so-called relaxation time approximation for C (Thellung and Weiss, 1969). To make this approximation plausible, let us note that for $C_R = 0$ energy and quasi-momentum are conserved quantities, i.e. in total thermal equlibrium for $C_R = 0$ the distribution function reads

$$n(\lambda) = \{\exp\left[\beta\hbar(\omega_\lambda - \boldsymbol{u}\boldsymbol{k})\right] - 1\}^{-1}, \tag{4.1.80}$$

where $\boldsymbol{u}$ has the meaning of a drift velocity of the phonon gas. It is reasonable to assume that for $C_{\mathrm{R}} = 0$ the solution of (4.1.72) can be approximated by the "drifting" local equilibrium expression

$$n_{\mathrm{LE}}^{\mathrm{d}}(\lambda, \boldsymbol{rt}) = \left\{\exp\left[\beta(\boldsymbol{rt})\, \hbar\left(\omega_\lambda - \boldsymbol{u}(\boldsymbol{rt})\, \boldsymbol{k}\right)\right] - 1\right\}^{-1}, \tag{4.1.81}$$

where $\beta(\boldsymbol{rt})$ and $\boldsymbol{u}(\boldsymbol{rt})$ denote the reciprocal temperature and the drift velocity, respectively, which are assumed to vary slowly in space and time.

A term $C_{\mathrm{R}} \neq 0$ destroys a phonon drift. Hence, for $C_{\mathrm{R}} \neq 0$ the distribution function for total thermal equilibrium is given by (4.1.80) with $\boldsymbol{u} = \boldsymbol{0}$ (cf. (1.4.50)) and for local thermal equilibrium by (4.1.81) with $\boldsymbol{u}(\boldsymbol{rt}) = \boldsymbol{0}$:

$$n_{\mathrm{LE}}(\lambda, \boldsymbol{rt}) = \{\exp\left[\beta(\boldsymbol{rt})\, \hbar\omega_\lambda\right] - 1\}^{-1}. \tag{4.1.82}$$

This suggests the following expression for C:

$$C[n] = -\frac{n - n_{\mathrm{LE}}^{\mathrm{d}}}{\tau_{\mathrm{N}}(\lambda)} - \frac{n - n_{\mathrm{LE}}}{\tau_{\mathrm{R}}(\lambda)}. \tag{4.1.83}$$

According to this expression, normal processes drive a nonequilibrium state back to drifting local equilibrium within time τ_{N}, and resistive processes drive it back to local equilibrium within time τ_{R}. Note that for $C_{\mathrm{R}} \geqq C_{\mathrm{U}}$ the first term on the right-hand side of (4.1.83) can be omitted. In this case we have to put $\boldsymbol{u} = \boldsymbol{0}$ in all the following expressions.

Now we assume $\boldsymbol{u}(\boldsymbol{rt})$ and $\delta\beta(\boldsymbol{rt})$, which denotes the deviation of $\beta(\boldsymbol{rt})$ from its equilibrium value β_0, to be small and linearize n_{LE} and $n_{\mathrm{LE}}^{\mathrm{d}}$ as

$$n_{\mathrm{LE}}(\lambda, \boldsymbol{rt}) \approx n^0(\lambda) - m(\lambda)\, \omega_\lambda \delta\beta(\boldsymbol{rt})/\beta_0, \tag{4.1.84}$$

$$n_{\mathrm{LE}}^{d}(\lambda, \boldsymbol{rt}) \approx n^0(\lambda) - m(\lambda)\left[\omega_\lambda \delta\beta(\boldsymbol{rt})/\beta_0 - \boldsymbol{u}(\boldsymbol{rt})\, \boldsymbol{k}\right]. \tag{4.1.85}$$

Here $n^0(\lambda)$ denotes the equilibrium distribution function given by (4.1.80) with $\boldsymbol{u} = \boldsymbol{0}$ and $\beta = \beta_0$, and

$$m(\lambda) = -\mathrm{d}n^0(\lambda)/\mathrm{d}\omega_\lambda = -\beta_0 n^0(\lambda)\left(n^0(\lambda) + 1\right). \tag{4.1.86}$$

Furthermore, we write

$$n(\lambda, \boldsymbol{rt}) \mathrel{\hat=} n^0(\lambda) + m(\lambda)\, g(\lambda, \boldsymbol{rt}). \tag{4.1.87}$$

Inserting (4.1.84) to (4.1.87) into (4.1.72) with C from (4.1.83), the equation governing the unknown function g becomes

$$\left[\frac{\partial}{\partial t} + \boldsymbol{v}_\lambda \frac{\partial}{\partial \boldsymbol{r}} + \tau^{-1}\right] g = -\frac{\omega_\lambda}{\tau} \frac{\delta\beta}{\beta_0} + \frac{1}{\tau_{\mathrm{N}}} \boldsymbol{uk}, \tag{4.1.88}$$

where

$$\tau^{-1}(\lambda) = \tau_{\mathrm{N}}^{-1}(\lambda) + \tau_{\mathrm{R}}^{-1}(\lambda). \tag{4.1.89}$$

Upon Fourier transforming all quantities according to (A2.9), we obtain from (4.1.88)

$$\begin{aligned} g(\lambda, \boldsymbol{q}\omega) &= [-i\omega + i\boldsymbol{q}\boldsymbol{v}_\lambda + \tau^{-1}(\lambda)]^{-1} \\ &\quad\times \left[\frac{-\omega_\lambda}{\tau(\lambda)} \frac{\delta\beta(\boldsymbol{q}\omega)}{\beta_0} + \frac{1}{\tau_{\mathrm{N}}(\lambda)} \boldsymbol{u}(\boldsymbol{q}\omega)\, \boldsymbol{k}\right] \\ &\approx [1 + i\omega\tau - i\boldsymbol{v}_\lambda \boldsymbol{q}\tau + \cdots] \left[-\omega_\lambda \frac{\delta\beta}{\beta_0} + \frac{\tau}{\tau_{\mathrm{N}}} \boldsymbol{u}\boldsymbol{k}\right]. \end{aligned} \tag{4.1.90}$$

The last line of equation (4.1.90) implies the assumption $\omega\tau \ll 1$ and $\boldsymbol{v}_\lambda \boldsymbol{q}\tau \ll 1$, which means that the distribution function is assumed to vary slowly in space and time.

The distribution function (4.1.90) contains the two unknown functions $\delta\beta$ and $\boldsymbol{u}$, which must be determined such that they satisfy (4.1.78) and (4.1.79). For $C_{\mathrm{R}} \geqq C_{\mathrm{N}}$ only (4.1.78) is needed, since in this case $\boldsymbol{u} = 0$. To obtain the hydrodynamic equations for $\delta\beta$ and $\boldsymbol{u}$, we substitute (4.1.84), (4.1.86) and (4.1.87) into (4.1.74) to (4.1.77) and (4.1.83), and use (4.1.78), (4.1.79) and (4.1.90).

Let us first ignore all terms proportional to τ and assume that $\tau = \tau_{\mathrm{N}}$, i.e. that resistive scattering processes are absent. Then, for a crystal with inversion symmetry, we obtain the following hydrodynamic equations:

$$i\omega\langle\omega_\lambda^2\rangle\, \delta\beta/\beta_0 + i \sum_{\alpha\alpha'} q_\alpha u_{\alpha'} \langle\omega_\lambda (\boldsymbol{v}_\lambda)_\alpha\, k_{\alpha'}\rangle = 0, \tag{4.1.91}$$

$$i \sum_{\alpha'} q_{\alpha'} \langle\omega_\lambda (\boldsymbol{v}_\lambda)_{\alpha'}\, k_\alpha\rangle\, \delta\beta/\beta_0 + i\omega \sum_{\alpha'} u_{\alpha'} \langle k_\alpha k_{\alpha'}\rangle = 0, \tag{4.1.92}$$

where we have used the short-hand notation

$$\langle\ldots\rangle = \frac{1}{V} \sum_\lambda g(\lambda) \ldots \tag{4.1.93}$$

The equations (4.1.91) and (4.1.92) have non-trivial solutions for $\delta\beta$ and u_α, provided that the determinant vanishes. This requirement yields for a cubic crystal

$$\omega^2 = q^2 \frac{\langle\omega_\lambda \boldsymbol{v}_\lambda \boldsymbol{k}\rangle^2}{3\langle\boldsymbol{k}^2\rangle\,\langle\omega_\lambda^2\rangle} \equiv c_{\mathrm{II}}^2 q^2, \tag{4.1.94}$$

which represents the dispersion relation for an undamped second sound wave. Approximating the phonon spectrum by a Debye model, $\omega_\lambda = c_j\, |\boldsymbol{k}|$ (cf. (1.4.34)), and replacing the sum over the Brillonin zone by an integral (with an infinite upper limit, since we only consider temperatures much lower than the Debye temperature) we get (SUSSMANN and THELLUNG, 1963)

$$c_{\mathrm{II}}^2 = \frac{1}{3} \Big(\sum_j c_j^{-3}\Big) \Big/ \Big(\sum_j c_j^{-5}\Big). \tag{4.1.95}$$

Next we account for contributions of order τ and also for those arising from resistive scattering. Thus, for a cubic crystal with inversion symmetry, we obtain the following hydrodynamic equations:

$$(\delta\beta/\beta_0)\left[i\omega\langle\omega_\lambda^2\rangle - \frac{1}{3}\langle\omega_\lambda^2 v_\lambda^2 \tau\rangle\right]$$
$$+ \frac{1}{3}(\boldsymbol{qu})\,[i\langle\omega_\lambda \boldsymbol{kv}_\lambda(1-\gamma)\rangle - \omega\langle\omega_\lambda \boldsymbol{kv}_\lambda \tau(1-\gamma)\rangle] = 0, \tag{4.1.96}$$

$$(\delta\beta/\beta_0)\,[-iq_\alpha\langle\omega_\lambda \boldsymbol{kv}_\lambda(1-\gamma)\rangle + \omega q_\alpha\langle\omega_\lambda \boldsymbol{kv}_\lambda \tau\rangle]$$
$$+ u_\alpha[-i\omega\langle k^2\rangle + q^2 B + \langle k^2(1-\gamma)\,\tau_\mathrm{R}^{-1}\rangle$$
$$+ i\omega\langle k^2(1-\gamma)\,\gamma\rangle + q_\alpha \boldsymbol{qu} A = 0, \tag{4.1.97}$$

where

$$A = \frac{1}{5}\,[3\langle(\boldsymbol{kv}_\lambda)^2\,(1-\gamma)\,\tau\rangle - \langle k^2 v_\lambda^2(1-\gamma)\,\tau\rangle], \tag{4.1.98}$$

$$B = \frac{1}{5}\,[2\langle k^2 v^2(1-\gamma)\,\tau\rangle - \langle(\boldsymbol{kv})^2\,(1-\gamma)\,\tau\rangle], \tag{4.1.99}$$

$$\gamma = \tau/\tau_\mathrm{R} = 1 - \tau/\tau_\mathrm{N}. \tag{4.1.100}$$

In deriving (4.1.96) and (4.1.97) contributions of order τ to E and P_α have been neglected, in order to obtain a result in agreement with a more sophisticated theoretical analysis (cf. BECK, 1975).

We assume that normal processes are still much more frequent than resistive ones, i.e. $\tau_\mathrm{N} \ll \tau_\mathrm{R}$ and $\tau \approx \tau_\mathrm{N}$. The condition of vanishing determinant of equations (4.1.96) and (4.1.97) would then read

$$\omega^2 + i\omega\left[\overline{\tau_\mathrm{R}^{-1}} + q^2 c_\mathrm{II}^2 \bar{\tau}_\mathrm{N}\right] - q^2 c_\mathrm{II}^2 \approx 0, \tag{4.1.101}$$

where

$$\overline{\tau_\mathrm{R}^{-1}} = \langle k^2 \tau_\mathrm{R}^{-1}\rangle/\langle k^2\rangle, \tag{4.1.102}$$

$$\bar{\tau}_\mathrm{N} = \langle\omega_\lambda^2 v_\lambda^2 \tau_\mathrm{N}\rangle\,\langle k^2\rangle/\langle\omega_\lambda \boldsymbol{kv}_\lambda\rangle^2$$
$$+ \frac{1}{5}\,[\langle k^2 v_\lambda^2 \tau_\mathrm{N}\rangle + 2\langle(\boldsymbol{kv}_\lambda)^2\,\tau_\mathrm{N}\rangle]\,\langle\omega_\lambda^2\rangle/\langle\omega_\lambda \boldsymbol{kv}_\lambda\rangle^2$$
$$- 2\langle\omega_\lambda \boldsymbol{kv}_\lambda \tau_\mathrm{N}\rangle/\langle\omega_\lambda \boldsymbol{kv}_\lambda\rangle. \tag{4.1.103}$$

From (4.1.101) we obtain the following dispersion relation:

$$\omega \approx \pm c_\mathrm{II} q\left[1 - \frac{1}{4}\left(\frac{\overline{\tau_\mathrm{R}^{-1}}}{q c_\mathrm{II}} + \bar{\tau}_\mathrm{N} q c_\mathrm{II}\right)^2\right]^{1/2} - \frac{i}{2}\left[\overline{\tau_\mathrm{R}^{-1}} + q^2 c_\mathrm{II}^2 \bar{\tau}_\mathrm{N}\right]. \tag{4.1.104}$$

We see that the damping of second sound has two sources: normal and resistive scattering. The damping will be weak, if (i) there are many normal processes,

$$\omega\bar{\tau}_{\mathrm{N}} \ll 1, \tag{4.1.105}$$

and (ii) resistive processes are rare,

$$\omega\bar{\tau}_{\mathrm{R}} \equiv \omega\left(\overline{\tau_{\mathrm{R}}^{-1}}\right)^{-1} \gg 1. \tag{4.1.106}$$

From (4.1.105) and (4.1.106) we obtain the "window" condition of second sound:

$$\omega\bar{\tau}_{\mathrm{N}} \ll 1 \ll \omega\bar{\tau}_{\mathrm{R}}. \tag{4.1.107}$$

Note that at low temperatures the average relaxation times behave as (see, for example, BECK, 1975)

$$\bar{\tau}_{\mathrm{N}} \approx A_{\mathrm{N}}T^{-5}, \tag{4.1.108}$$

$$\overline{\tau_{\mathrm{R}}^{-1}} \approx A_{\mathrm{B}}/L + A_{\mathrm{I}}T^4 + A_{\mathrm{U}}T^5 \exp\left(-\alpha\hbar\omega_{\mathrm{D}}/kT\right) \qquad (\alpha \approx 1/2). \tag{4.1.109}$$

Here L is a characteristic sample length, and the subscripts have the same meaning as in the text following equation (4.1.73), α is a number of order unity. The condition (4.1.107) requires large L, small A_{I}, and A_N small enough for the normal processes to be still sufficiently frequent at the temperature T, where resistive processes are negligible. We see from (4.1.109) that umklapp processes die out exponentially at low temperatures (which can be obtained by (1.3.76), (1.3.54) and (1.4.50)).

If, however, $\overline{\tau_{\mathrm{R}}^{-1}}/qc_{\mathrm{II}} \gg 1$, i.e. if the inequality (4.1.106) is reversed, then we find from (4.1.101) that

$$\omega \approx -iq^2c_{\mathrm{II}}\bar{\tau}_{\mathrm{R}}. \tag{4.1.110}$$

This corresponds to diffusive heat transport instead of a wave-like heat propagation (for more details on diffusive heat transport see BECK, 1975, and references therein; see also A3.4).

By means of heat-pulse experiments second sound was first detected in ^{4}He (ACKERMAN et al., 1966) and subsequently also observed in ordinary crystals, such as NaF (MC NELLY et al., 1970, and JACKSON et al., 1970) and Bi (NARAYANAMURTI and DYNES, 1972). Recently, in NaF second sound was studied more directly by forced thermal light scattering (POHL and IRNIGER, 1976).

Second sound is supposed to be a normal low-temperature phenomenon, provided that the crystal is sufficiently perfect. It is not restricted only to the gas of acoustic phonons, but it may also occur in the gas of optical phonons. The latter situation is expected in crystals (perovskite crystals) where a soft optical mode leads to a structural phase transition. When the frequency of the soft mode becomes very small, the corresponding displacement must be viewed as a hydrodynamic mode (cf. SCHNEIDER and STOLL, 1978a).

Heat transport in dielectric crystals can be described by means of two "interacting fluids": the elastic deformation (elastic displacement field) and the gas of thermal phonons (phonon density) (for a review see ENZ, 1974). Via their interaction the phonon gas attenuates a sound wave, and a deformation affects the phonon gas (by changing force constants and thus vibrational frequencies). At low temperatures this coupling is weak. Hence, the second sound can be described as collective excitations of the phonon gas alone. In structural phase transitions, near the critical temperature, the coupling between the order parameter (displacement connected with the soft mode) and the phonon gas becomes important (cf. SZABO, 1976).

Note that a very effective method for studying heat transport in crystals is the Green function method. It turns out that the spectral density exhibits an additional low-frequency resonance (accordingly, the Green function exhibits an additional pole) corresponding to second sound or heat diffusion (see BECK, et al., 1974, and BECK, 1975, and references therein). Furthermore, molecular-dynamics techniques has also been used for studying second sound (SCHNEIDER and STOLL, 1978a).

4.2. Strongly anharmonic crystals in the self-consistent harmonic approximation

The traditional theory of lattice dynamics is based on the assumption of small vibrational amplitudes of the atoms about their mean positions. The ratio between these amplitudes and the interatomic spacing is used as a small expansion parameter in a perturbation treatment. Computations on rare gas crystals (DE WETTE and NIJBOER, 1965) show, however, that such an expansion does not exist for He at any temperature and that it is only of slow convergence for the other rare gas crystals near their melting temperature. For instance, for Kr the harmonic approximation is quite good at $T = 0$ K, but above one-third of the melting temperature, where the vibrational amplitudes become about 10% of the interatomic spacing, a perturbation treatment breaks down if only a finite number of anharmonic terms is accounted for.

In He crystals, owing to the light atomic masses, the zero point vibrational amplitudes are about 30% of the interatomic spacing. The pressure of the zero point kinetic energy is so large that the lattice constant of He crystals is expanded beyond the inflection point of the interatomic pair potential, as illustrated in figure 4.5. Thus, for He the harmonic theory yields imaginary frequencies, so that a perturbation series does not exist. The situation is similar, but less dramatic, for solid molecular hydrogen. Materials whose basic behaviour is dominated by quantum-mechanical effects, such as zero point vibrations, are commonly called quantum solids or quantum crystals (solid ^3He, ^4He, H_2).

Strong anharmonic effects also occur in systems with structural phase transitions including ferroelectric materials. Here, near the phase transitions, a group of modes will in general be imaginary in the harmonic approximation, so that anharmonic effects may also not be calculated by using the perturbation theory.

Furthermore, some kinds of impurities in crystals, such as Li in KCl, represent extremely anharmonic systems, too.

For all the systems considered above, harmonic theory or low order perturbation expansion are not appropriate. Therefore, a more powerful approximation is needed for strongly anharmonic systems. Among such

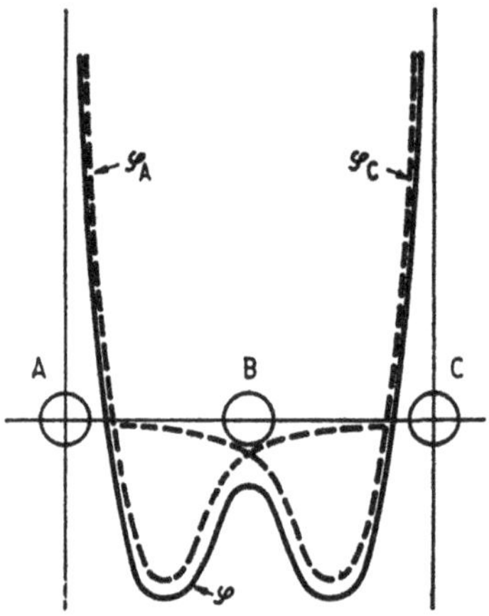

Fig. 4.5. Schematic representation of the potential, for a one-dimensional model of a strongly anharmonic system. φ_A (φ_C) is the potential due to atom A (C). $\varphi = \varphi_A + \varphi_C$ is the potential in which atom B moves

approximations the self-consistent harmonic approximation plays a basic role. This approximation and extensions of it have been derived by making use of different techniques (see Werthamer, 1970, Klein and Horton, 1972, Horner, 1974, and references therein). We follow here the approach of Plakida and Siklos (1969) and Takeno (1970) (for a review see Plakida, 1973), which is based on the method of the time-dependent Green functions.

In the case of a strongly anharmonic solid we decompose the instantaneous position $\boldsymbol{R}(l)$ of the lth atom as (cf. (1.2.1))

$$\boldsymbol{R}(l) = \langle \boldsymbol{R}(l) \rangle + \boldsymbol{u}(l) \equiv \boldsymbol{r}(l) + \boldsymbol{u}(l). \tag{4.2.1}$$

Here $\langle \ldots \rangle$ denotes the thermal average defined by (4.1.18). $\boldsymbol{r}(l)$ is the mean or equilibrium position of the lth atom and $\boldsymbol{u}(l)$ is the displacement of the lth atom from its mean position. Note that the $\boldsymbol{r}(l)$'s do not coincide with the equilibrium positions determined by the minima of the potential energy (cf. section 1.2.). For the set of atomic positions $\{\boldsymbol{r}(l)\}$ the free energy has a minimum (see below).

On expanding the potential energy Φ in powers of the displacements (cf. (1.2.2)), the Hamiltonian of the anharmonic vibrational systems becomes

$$\left.\begin{aligned}\mathcal{H} = \frac{1}{2} \sum_x \frac{p^2(x)}{M(x)} + \sum_{n=1}^{\infty} \frac{1}{n!} \sum_{x_1 x_2 \dots x_n} \Phi(x_1 x_2 \dots x_n)\, u(x_1)\, u(x_2) \dots u(x_n) \\ + \text{constant terms.}\end{aligned}\right\} \tag{4.2.2}$$

Here $u(x)$, $p(x)$ and $M(x)$ denote respectively the displacement $u_\alpha(l)$, the momentum $p_\alpha(l)$ and the mass $M(l)$ of the lth atom, and

$$\Phi(x_1 x_2 \dots x_n) = [\nabla(x_1)\, \nabla(x_2) \dots \nabla(x_n)\, \Phi]_0 \tag{4.2.3}$$

is the nth order force constant, where the subscript 0 means that the expression in the brackets is evaluated with all atomic coordinates at the equilibrium positions $\{\boldsymbol{r}(l)\}$.

To obtain the equation of motion of the time-dependent displacement–displacement Green function (cf. (4.1.9))

$$G^{\mathrm{r,a}}(xx', t) = \langle\langle u(xt); u(x'0)\rangle\rangle^{\mathrm{r,a}} \tag{4.2.4}$$

we use (2.1.23) (replacing H by $\mathcal{H}$) and the equations

$$[u(x), \mathcal{H}] = i\hbar p(x)/M(x) \tag{4.2.5}$$

and

$$[p(x), \mathcal{H}] = -i\hbar \sum_{n=1}^{\infty} \frac{1}{n!} \sum_{x_1 x_2 \dots x_n} \Phi(x x_1 x_2 \dots x_n)\, u(x_1)\, u(x_2) \dots u(x_n), \tag{4.2.6}$$

which were derived by (4.2.2) and (2.1.26), and by making use of the fact that the force constants (4.2.3) are symmetric functions of $x_1, x_2, \dots, x_n$. Thus, we find

$$\begin{aligned}-M(x) \frac{\mathrm{d}^2}{\mathrm{d}t^2} G(xx', t) = \delta(t)\, \delta_{xx'} + \sum_{n=1}^{\infty} \frac{1}{n!} \\ \times \sum_{x_1 x_2 \dots x_n} \Phi(x x_1 x_2 \dots x_n)\, \langle\langle u(x_1 t)\, u(x_2 t) \dots u(x_n t); u(x'0)\rangle\rangle,\end{aligned} \tag{4.2.7}$$

where $\delta_{xx'} \equiv \delta_{ll'} \delta_{\alpha\alpha'}$ and where we have suppressed the indices r, a at the Green functions. Fourier transforming (cf. (2.1.17)) this equation, we obtain

$$\begin{aligned}M(x)\, \omega^2 G(xx', \omega) = \delta_{xx'} + \sum_{n=1}^{\infty} \frac{1}{n!} \\ \times \sum_{x_1 x_2 \dots x_n} \Phi(x x_1 x_2 \dots x_n)\, \langle\langle u(x_1)\, u(x_2) \dots u(x_n); u(x')\rangle\rangle_\omega .\end{aligned} \tag{4.2.8}$$

When we retain only the term with $n = 1$, then (4.2.8) agrees with (2.1.28), which describes the lattice vibrations in the harmonic approximation.

Inspired by Wick's theorem (cf. section 4.1.1.) here we use the following decoupling approximation for the Green functions on the right-hand side of equation (4.2.8) (Takeno, 1970):

$$\left.\begin{aligned}&\langle\langle u(x_1)\, u(x_2) \ldots u(x_n);\, u(x')\rangle\rangle_\omega\\ &= \sum_{i=1}^{n} \Big\langle \prod_{j(\neq i)} u(x_j)\Big\rangle \langle\langle u(x_i);\, u(x')\rangle\rangle_\omega\\ &\quad + \sum_{i<j} \Big\langle \prod_{k(\neq ij)} u(x_k)\Big\rangle \langle\langle u(x_i)\, u(x_j);\, u(x')\rangle\rangle_\omega\\ &\quad + \sum_{i<j<k} \Big\langle \prod_{l(\neq ijk)} u(x_l)\Big\rangle \langle\langle u(x_i)\,(u(x_j)\, u(x_k);\, u(x')\rangle\rangle_\omega + \cdots\end{aligned}\right\} \tag{4.2.9}$$

Inserting this approximation into (4.2.8), we obtain

$$\begin{aligned}M(x)\, \omega^2 G(xx', \omega) = \delta_{xx'} &+ \sum_{n=1}^{\infty} \frac{1}{n!}\\ &\times \sum_{x_1 x_2 \ldots x_n} \tilde{\Phi}(x x_1 x_2 \ldots x_n)\, \langle\langle u(x_1)\, u(x_2) \ldots u(x_n);\, u(x')\rangle\rangle_\omega,\end{aligned} \tag{4.2.10}$$

where

$$\tilde{\Phi}(x_1 x_2 \ldots x_n) = \nabla(x_1)\, \nabla(x_2) \ldots \nabla(x_n)\, \langle \Phi(\boldsymbol{R})\rangle. \tag{4.2.11}$$

Here

$$\langle \Phi(\boldsymbol{R})\rangle = \langle \exp \sum_x u(x)\, \nabla(x)\rangle\, \Phi(\boldsymbol{r}) \tag{4.2.12}$$

$\big(\boldsymbol{R} \equiv \{\boldsymbol{R}(l)\},\ \boldsymbol{r} \equiv \boldsymbol{r}(l)\}\big)$ is the "effective potential" associated with the "bare potential" $\Phi(\boldsymbol{r})$, and $\tilde{\Phi}(x_1 x_2 \ldots x_n)$ are "effective force constants". Equation (4.2.10) has been obtained by means of the following relations:

$$\begin{aligned}&\sum_{n=1}^{\infty} \frac{1}{n!} \sum_{x_1 x_2 \ldots x_n} \Phi(x x_1 x_2 \ldots x_n) \sum_{\substack{i_1 < i_2 < \cdots < i_m\\ =1}}^{n} \Big\langle \prod_{j(\neq i_1 i_1 \ldots i_m)} u(x_j)\Big\rangle\\ &\times \langle\langle u(x_{i_1})\, u(x_{i_2}) \ldots u(x_{i_m});\, u(x')\rangle\rangle_\omega\\ &= (1/m!) \sum_{x_1' x_1' \ldots x_m'} \tilde{\Phi}(x x_1' x_2' \ldots x_m')\, \langle\langle u(x_1')\, u(x_2') \ldots u(x_m');\, u(x')\rangle\rangle_\omega.\end{aligned} \tag{4.2.13}$$

The physical meaning of the effective potential (4.2.17) and the effective force constants (4.2.10) may be understood as follows. In an anharmonic solid all atoms will be engaged in vibrations of sizable amplitude about their mean positions, such that an atom can experience decidedly anharmonic forces due to the surrounding atoms. In calculating these forces the surrounding atoms cannot be regarded as rigidly fixed, because the forces depend on how the latter atoms are moving. Thus, the vibration of some single atom is governed by a potential formed by all the other atoms, whose motions cause a random thermal modulation of the potential of the single atom. — We must demand that the motions of all atoms are equivalent and self-consistent.

Accordingly, in an anharmonic solid the atomic motion is governed by effective potentials or effective force constants, rather than by bare potentials or bare force constants.

If the series on the right-hand side of equation (4.2.10) is truncated at $n = 1$, then noninteracting renormalized harmonic phonons are obtained. To account for scattering processes of the renormalized phonons, truncation at $n > 1$ must be performed. Here we restrict ourselves to truncation at $n = 1$. According to (4.2.10), in this approximation we have

$$M(x)\,\omega^2 G(xx', \omega) = \delta_{xx'} + \sum_{x_1} \tilde{\Phi}(xx_1)\, G(x_1x', \omega), \tag{4.2.14}$$

where, from (4.2.13),

$$\tilde{\Phi}(xx') = \Phi(xx') + \sum_{n=1}^{\infty} \frac{1}{n!} \sum_{x_1x_2\ldots x_n} \Phi(xx'x_1x_2 \ldots x_n)\, \langle u(x_1)\, u(x_2) \ldots u(x_n)\rangle. \tag{4.2.15}$$

Hence, in contrast to the usual harmonic approximation, in the renormalized harmonic approximation (4.2.14) the force constants are the thermal average of the second derivatives of the potential rather than the equilibrium value of the second derivatives (given by the first term on the right-hand side of equation (4.2.15)).

In the following the correlation functions occurring in (4.2.15) are approximated in the spirit of Wick's theorem (cf. section 4.1.1.) as

$$\begin{aligned}\langle u(x_1)\, u(x_2) \ldots u(x_n)\rangle &\approx (n-1)\, \langle u(x_1)\, u(x_2)\rangle\, \langle u(x_3) \ldots u(x_n)\rangle \\ &\approx (n-1)\,(n-3) \ldots 1 \langle u(x_1)\, u(x_2)\rangle \ldots \\ &\quad\times \langle u(x_{n-1})\, u(x_n)\rangle,\end{aligned} \tag{4.2.16}$$

where we have also accounted for the symmetry of the force constants with respect to interchanges of indices.

Inserting (4.2.16) into (4.3.15), we obtain

$$\tilde{\Phi}(xx') = \nabla(x)\, \nabla(x') \exp\left\{\frac{1}{2} \sum_{x_1x_2} \langle u(x_1)\, u(x_2)\rangle\, \nabla(x_1)\, \nabla(x_2)\right\} \Phi(\boldsymbol{r}). \tag{4.2.17}$$

At this stage, the term "self-consistent harmonic approximation" (SHA) becomes clear: The Green function $\boldsymbol{G}$ is governed by the force constants $\tilde{\Phi}$ from (4.2.17), which in turn depend on this Green function via the correlation functions $\langle u(x)\, u(x')\rangle$ (cf. (2.1.22)). In practice, one begins the self-consistent procedure with a guess for the correlation functions occurring in (4.2.17). This yields the force constants, which, when inserted in (4.2.14), permit the calculation of $\boldsymbol{G}$, which in turn allows us then to calculate the correlation functions by means of (2.1.22). The procedure is then repeated, until convergence is achieved.

In the following we consider in more detail the case where $\Phi(\boldsymbol{r})$ is composed of pair potentials (cf. (1.4.45)) as

$$\Phi(\boldsymbol{r}) = \frac{1}{2} \sum_{ll'} \varphi\big(\boldsymbol{r}(l) - \boldsymbol{r}(l')\big). \tag{4.2.18}$$

Inserting this expression into (4.2.17) we obtain

$$\tilde{\Phi}(xx') = \nabla(x)\, \nabla(x')\, \tilde{\Phi}(\boldsymbol{r}), \tag{4.2.19}$$

where

$$\tilde{\Phi}(\boldsymbol{r}) = \frac{1}{2} \sum_{ll'} \tilde{\varphi}\big(\boldsymbol{r}(l) - \boldsymbol{r}(l')\big) \tag{4.2.20}$$

with

$$\tilde{\varphi}\big(\boldsymbol{r}(l) - \boldsymbol{r}(l')\big) = \exp\left\{\frac{1}{2} \sum_{\alpha\alpha'} C^{ll'}_{\alpha\alpha'} \nabla_\alpha(l)\, \nabla_{\alpha'}(l')\right\} \varphi\big(\boldsymbol{r}(l) - \boldsymbol{r}(l')\big). \tag{4.2.21}$$

Here $C^{ll'}_{\alpha\alpha'}$ is the 3×3 displacement-displacement correlation matrix

$$C^{ll'}_{\alpha\alpha'} = \Big\langle \big(u_\alpha(l) - u_\alpha(l')\big) \big(u_{\alpha'}(l) - u_{\alpha'}(l')\big) \Big\rangle. \tag{4.2.22}$$

To make the physical meaning of equation (4.2.21) more clear we Fourier transform the pair potential

$$\varphi(\boldsymbol{r}) = \frac{1}{V} \sum_{\boldsymbol{k}} \varphi(\boldsymbol{k})\, \mathrm{e}^{i\boldsymbol{k}\boldsymbol{r}}, \tag{4.2.23}$$

$$\varphi(\boldsymbol{k}) = \int \mathrm{d}\boldsymbol{r} \varphi(\boldsymbol{r})\, \mathrm{e}^{-i\boldsymbol{k}\boldsymbol{r}}. \tag{4.2.24}$$

Inserting now (4.2.23) into (4.2.21) and carrying out the integration over $\boldsymbol{k}$, we obtain by (4.2.24)

$$\begin{aligned} \tilde{\varphi}\big(\boldsymbol{r}(l) - \boldsymbol{r}(l')\big) &= \frac{1}{V} \sum_{\boldsymbol{k}} \varphi(\boldsymbol{k}) \exp\left\{-\frac{1}{2} \Big\langle \big(\boldsymbol{k}(\boldsymbol{u}(l) - \boldsymbol{u}(l'))\big)^2 \Big\rangle\right\} \\ &\quad \times \exp\big[i\boldsymbol{k}\big(\boldsymbol{r}(l) - \boldsymbol{r}(l')\big)\big] \\ &= \{(2\pi)^3 \det |C^{ll'}_{\alpha\alpha'}|\}^{-1/2} \int \mathrm{d}\boldsymbol{r} \varphi\big(\boldsymbol{r}(l) - \boldsymbol{r}(l') + \boldsymbol{r}\big) \\ &\quad \times \exp\left[\frac{1}{2} \sum_{\alpha\alpha'} r_\alpha (C^{-1\,ll'}_{\alpha\alpha'})\, r_{\alpha'}\right]. \end{aligned} \tag{4.2.25}$$

According to this equation, in the SHA the atomic motion is governed by an effective pair interaction potential $\tilde{\varphi}$, which is given by the original bare potential φ averaged with a gaussian distribution for the interatomic spacing. The effective potential, as function of the interatomic spacing, is smoother than the bare potential: its depth is smaller and its minimum is at larger spacing than at the bare potential. Since the width of the gaussian distribution is governed by the displacement-displacement correlation matrix, the difference between the two potentials $\tilde{\varphi}$ and φ is especially strong at high temperatures and in the case of large zero point energy.

To illustrate the self-consistent potential (4.2.25) let us consider nearest neighbour interaction in a one-dimensional chain with a Morse potential as bare interaction potential. We then have

$$C_{\alpha\alpha'}^{ll+1} = \delta_{\alpha\alpha'}\langle(u(l) - u(l+1))^2\rangle \equiv \delta_{\alpha\alpha'}\overline{u^2(l)}, \tag{4.2.26}$$

where $\overline{u^2(l)}$ is the mean square relative displacement of neighboured atoms. Expanding the exponent in (4.2.21) in a series and using the expression (3.2.56) for the Morse potential, we obtain by (4.2.26) the following effective pair potential:

$$\tilde{\varphi}(z) = D[\mathrm{e}^{-2\gamma x_0 z}\,\mathrm{e}^{2y} - 2\mathrm{e}^{-\gamma x_0 z}\,\mathrm{e}^{y/2}], \tag{4.2.27}$$

where $y = \gamma^2\overline{u^2(l)}$, $z = x/x_0 - 1$, and D, γ and x_0 are the parameters characterizing the Morse potential. In figure 4.6 $\tilde{\varphi}(z)/D$ is shown for $\gamma x_0 = 6$ and various values y. The tendency described above is well illustrated by this figure.

In treating lattice dynamics of solid He, the problem is not only the strong anharmonicity but also the need to account for short-range correlations arising from the hard character of the interatomic potential. In general, any

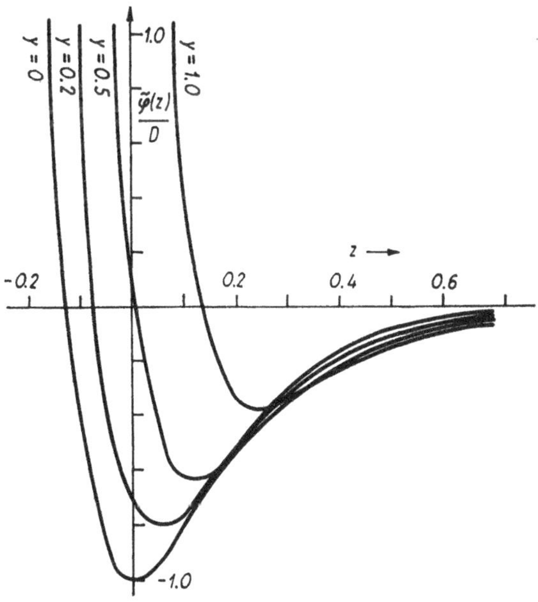

Fig. 4.6. Pair potential in self-consistent harmonic approximation for a one-dimensional chain with a Morse potential as bare interaction potential (see text) (after PLAKIDA, 1973)

realistic potential has, at least, a quite hard core. In particular, the Lennard-Jones potential (1.4.46) has an infinitely hard core. According to (4.2.25), the effective pair potential is given by an integral over the bare pair potential convoluted with a gaussian function. The occurrence of the gaussian function implies the assumption of harmonic wave functions to describe the atomic vibrations. However, such an assumption is not adequate for treating solids with hard core potentials, since in this case the wave functions will be shaped so that the atomic motions will avoid the hard cores. The gaussian does not cut off the integrand in (4.2.25) quickly enough so that the contribution of small distances is unreasonably large for hard core potentials, or, in other words, the gaussian allows, unphysically, a penetration of the atoms into the hard core region of their neighbours. Various methods have been proposed for introducing short-range correlations into the phonon-problem (for reviews see GUYER, 1969, and KOEHLER, 1975). For instance, in the approach of MEISSNER (1968) the effective harmonic force constants are given by an expression like (4.2.25), but where also a short-range correlation function is included (see also ANDERSON et al., 1977, and references therein).

In treating lattice dynamics of ferroelectrics, complications arise from long-range electrostatic interaction and from the fact that the interaction potential between the atoms is not known. For a study of phase transitions in ferroelectrics by the SHA the reader is referred to GILLIS and KOEHLER (1971, 1972).

Before leaving the self-consistent phonon theory, we derive some relations determining the equilibrium positions $r(x)$ used in this section. We assume the crystal to be subjected to some external forces giving rise to the following additional contribution to the Hamiltonian:

$$H' = -\sum_x f(x)\, R(x)\,, \tag{4.2.28}$$

where $f(x) \equiv f_\alpha(l)$ is the αth cartesian component of the external force acting on the lth atom. The equilibrium positions $r(x)$ are determined by the requirement that no net average force acts on a given atom in its equilibrium position, that is

$$\frac{\mathrm{d}}{\mathrm{d}t}\langle p(x)\rangle = \frac{i}{\hbar}\langle[\mathcal{H} + H', p(x)]\rangle = -\nabla(x)\,\langle\Phi(\boldsymbol{R})\rangle + f(x) = 0\,. \tag{4.2.29}$$

In particular, for absent external forces (4.2.29) yields the equilibrium condition

$$\nabla(x)\,\langle\Phi(\boldsymbol{R})\rangle = 0\,. \tag{4.2.30}$$

Now the crystal is subjected to a small homogeneous deformation $u_{\alpha\alpha'}$, which is equivalent to a small atomic displacement $\delta r_\alpha(l) = \sum_{\alpha'} u_{\alpha\alpha'} r_\alpha(l)$. The work performed thereby by the forces $f(x)$ is

$$-\delta\langle H'\rangle = V\sum_{\alpha\alpha'} u_{\alpha\alpha'}\sigma_{\alpha\alpha'} = \sum_{l\alpha} f_\alpha(l)\,\delta r_\alpha(l) = \sum_{l\alpha\alpha'} f_\alpha(l)\, u_{\alpha\alpha'} r_{\alpha'}(l)\,, \tag{4.2.31}$$

where $\sigma_{\alpha\alpha'}$ is the stress tensor. Because of the free choice of the $u_{\alpha\alpha'}$'s, we find from (4.2.31) by (4.2.29)

$$\sigma_{\alpha\alpha'} = \frac{1}{V} \sum_l f_\alpha(l)\, r_{\alpha'}(l) = \frac{1}{V} \sum_l \nabla_\alpha(l) \langle \Phi(\boldsymbol{R}) \rangle\, r_{\alpha'}(l) \tag{4.2.32}$$

and for external hydrostatic pressure p

$$p = -\frac{1}{3} \sum_x \sigma_{\alpha\alpha} = -\frac{1}{3V} \sum_x \nabla(x) \langle \Phi(\boldsymbol{R}) \rangle\, r(x). \tag{4.2.33}$$

This equation can be used either to calculate the lattice constant for fixed pressure or to calculate the pressure for a given choice of the lattice constant. Note that an external pressure must be taken into account in the case of solid helium.

Now let us point out that the free energy has a minimum for the atomic positions $r(x)$. The partition function of the crystal in the presence of the external forces $f(x)$ reads

$$Z = \operatorname{Tr} \exp \left\{-\beta[\mathcal{H} - \sum_x f(x)\, R(x)]\right\} \qquad (\beta = 1/kT), \tag{4.2.34}$$

and the free enthalpy is

$$G = -kT \ln Z. \tag{4.2.35}$$

Differentiating (4.2.35) with respect to $f(x)$, we find

$$\frac{\partial G}{\partial f(x)} = -\langle R(x) \rangle \equiv -r(x). \tag{4.2.36}$$

This equation governs the structure of the crystal for given forces and represents therefore the equilibrium condition. Knowing G, we may define the free energy F as function of the $r(x)$'s as

$$F\big(r(x), T\big) = G + \sum_{x'} f(x')\, r(x') \tag{4.2.37}$$

with $f(x) = f\big(r(x)\big)$. Differentiating (4.2.37) with respect to $r(x)$ we obtain

$$\frac{\partial F}{\partial r(x)} = \sum_{x'} \frac{\partial G}{\partial f(x')} \frac{\partial f(x')}{\partial r(x)} + \sum_{x'} \frac{\partial f(x')}{\partial r(x)}\, r(x') + f(x) \tag{4.2.38}$$

or, on account of (4.2.36),

$$\frac{\partial F}{\partial r(x)} = f(x), \tag{4.2.39}$$

which represents another form of the equilibrium condition. According to (4.2.39), for absent external forces, F becomes minimum for positions $r(x)$ (for more details see Leibfried and Ludwig, 1961).

Concluding this section let us draw attention to some recent papers dealing with the SHA and its application. SHUKLA et al. (1981) proposed to approximate (within the SHA) functions of phonon frequencies, such as the correlation function $\langle u(x)\, u(x')\rangle$, by functions of an average phonon frequency (being in its turn a function of $\langle u(x)\, u(x')\rangle$). This procedure turns out to be useful for obtaining thermodynamic properties (cf. PASKIN et al., 1982). It is much easier to handle than the actual SHA, since unlike the SHA it does not involve extensive calculations of the phonon frequencies.

SUGIYAMA and OKAMOTO (1980) studied the effect of external force on surface lattice dynamics and NISHIOKA and LEE (1981) studied fracture strength under stress within the framework of the SHA by using the self-consistent Einstein model (independent oscillator model) proposed by Matsubara and co-workers (MATSUBARA and KAMIYA, 1977, MATSUSHITA and MATSUBARA, 1978). An important feature of the latter model is that it can also be used for investigating anharmonic crystals with lattice defects.

WOJTCZAK et al. (1982) studied lattice constant fluctuations in anharmonic crystals (arising from random distribution of lattice sites at a given moment) by using a self-consistent phonon theory (PLAKIDA and SIKLOŚ, 1978), which also accounts for damping of the phonons obtained in the SHA.

4.3. Lattice solitons

4.3.1. Some general features of solitons

For an anharmonic lattice potential the atomic displacements are governed by a nonlinear equation of motion. By a perturbation treatment of nonlinearity only one type of solution can be obtained: spatially extended travelling waves. However, such a treatment fails to provide another important type of solution: spatially localized (particle-like) large-amplitude travelling waves (solitary waves or solitons) that retain their shape during propagation and that are functionally independent to a high degree (if not exactly so). The existence of these two types of solutions is characteristic for a class of nonlinear equations describing discrete as well as continuous nonlinear dispersive media.

The study of nonlinear lattices began with the famous numerical investigation of one-dimensional lattices with cubic and quartic anharmonicities by FERMI, PASTA and ULAM (1955). These authors intended to test the ergodic behaviour of such lattices. They expected to observe a sharing of the energy, initially stored in a single mode, with all the other modes due to lattice anharmonicity.

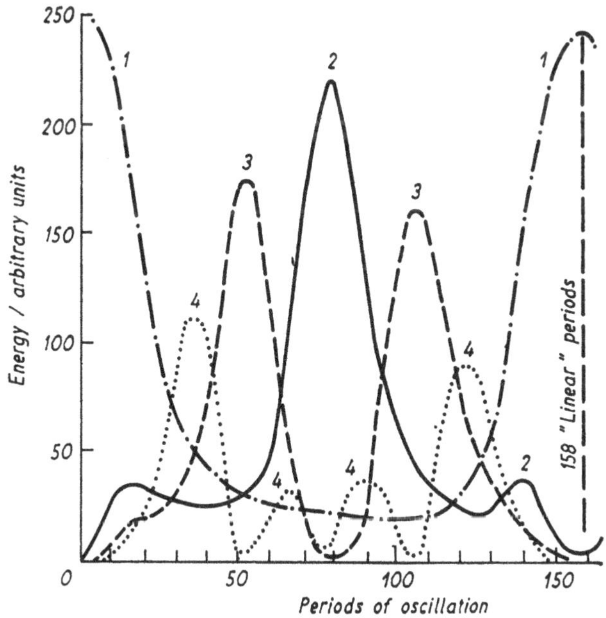

Fig. 4.7. The energy in the low frequency modes of a 32-atom non-linear one-dimensional lattice with cubic anharmonicity. The energy is initially put into the mode 1 (after FERMI et al., 1955)

However, they obtained the surprising result that no tendency towards thermalization occurs. The energy, initially put into the lowest frequency mode, was shared only amongst a small group of other low frequency modes and after 158 "linear periods" (for a 32-atom one-dimensional lattice) the energy is almost entirely back in the first mode (fig. 4.7).

Stimulated by this unexpected result, ZABUSKY and KRUSKAL (1965) studied numerically the Korteweg-de Vries (KdV) equation (see below), which approximately describes anharmonic vibrations of a one-dimensional lattice in

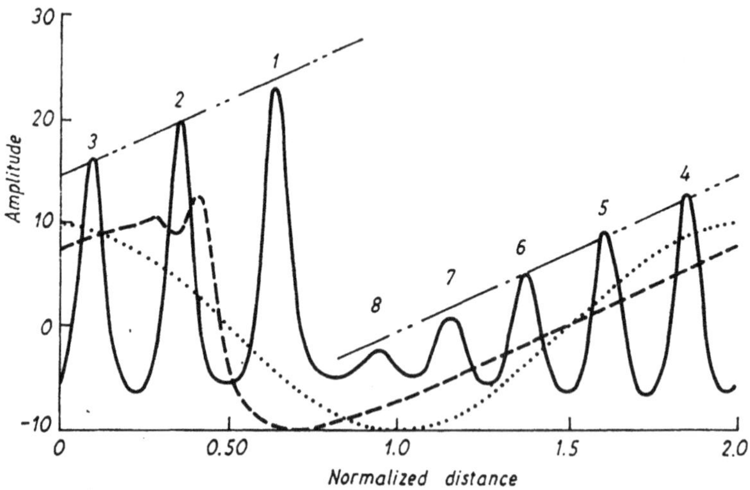

Fig. 4.8. Solutions of the Korteweg-de Vries equation $\dot{u} + uu' + \delta u''' = 0$ ($\delta = 0.022$) with periodic boundary conditions (the periodic unit is of length 2) at three different times: ····· $t = 0$, - - - - - $t = t_B$, ——— $t = 3.6 t_B$ (after ZABUSKY and KRUSKAL, 1965)

the continuum limit. Using periodic boundary conditions and an initial cosine atomic displacement pattern (and $\dot{u}(x, 0) = 0$) they observed the remarkable behaviour shown in figure 4.8. The cosine breaks to spikes of roughly sech^2 form, for which Zabusky and Kruskal coined the term "soliton". The solitons collide and pass through each other without change of shape or velocity, and after some time the initial cosine is recreated. This recurrence phenomenon is similar to that found by FERMI, PASTA and ULAM (1955). Recurrence and, accordingly, nonergodicity are consequences of the solitary wave solutions of the nonlinear systems at hand.

The nonlinear equations with soliton solutions show the following features in common:

(i) Pulse-like solutions break to trains of single bell-shaped spikes (solitons).
(ii) The speed of an individual soliton often depends on the amplitude of the wave; in general, it increases with increasing amplitude.

(iii) Solitons do not scatter upon collision. They emerge from the collision having the shape and velocity with which they entered, but they exhibit a small shift of the phase.

In contrast to the term "soliton", the term "solitary wave" denotes also localized traveling waves possessing only approximately (if at all) the property (iii).

The standard nonlinear wave equations with soliton solutions are (in dimensionless units):

(i) the "Korteweg-de Vries" equation

$$\dot{u} + 6uu' + u''' = 0, \tag{4.3.1}$$

(ii) the "nonlinear Schrödinger" equation

$$i\dot{u} + 2u\,|u|^2 + u'' = 0, \tag{4.3.2}$$

(iii) the "sine-Gordon" equation

$$u'' - \ddot{u} = \sin u\,. \tag{4.3.3}$$

Here x is the space variable, t is the time variable and u is the "amplitude". Furthermore, it is $\dot{u} \equiv \partial u/\partial t$ and $u' \equiv \partial u/\partial x$.

As an example, let us consider the KdV equation in more detail. If we neglect in this equation the nonlinear term, then the term u''' gives rise to the dispersion relation $\omega(k) = -k^3$. In the absence of the nonlinear term, the dispersive term (like a dissipative term) spreads out an initial pulse, as illustrated in figure 4.9a).

On the other hand, if we neglect the dispersive term in eq. (4.4.1), but retain the nonlinear one, then the resulting equation has the following im-

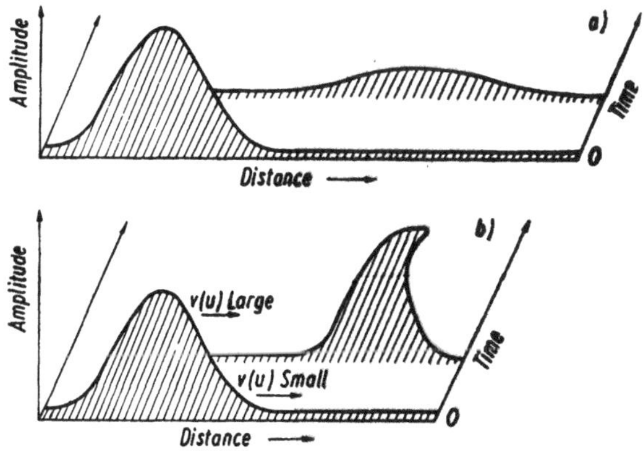

Fig. 4.9. Schematic plot of the time dependence of an initial pulse governed by (4.3.1), if a) the nonlinear term and b) the dispersive term is neglected in this equation. $v(u)$ is the amplitude dependent velocity

plicit solution

$$u = u(v(u)\,t - x), \tag{4.3.4}$$

where $v(u) = 6u$ is a velocity depending on u. Since this velocity increases with increasing u, the front of an initially bell-shaped pulse steepens and tends to break, as illustrated in figure 4.9b).

The KdV equation (4.3.1) has a bell-shaped solitary wave solution of the following form:

$$u = 2\varkappa^2 \operatorname{sech}^2 \left(\varkappa(x - 4\varkappa^2 t)\right), \tag{4.3.5}$$

which moves with a velocity $v = 4\varkappa^2$ being proportional to the amplitude. Hence, the existence of localized travelling solutions with permanent profile is obviously a result of a balance between dispersion and nonlinearity.

The solution (4.3.5) can be shown to be actually a soliton. For illustration of the soliton behaviour upon collision figure 4.10 shows the nondestructive collision of ion-acoustic plasma pulses, which are also described by the KdV equation.

Note that solitary solutions of the sine-Gordon equation (4.3.3), which correspond to a rotation in u by 2π as x goes from $-\infty$ to $+\infty$ ("kink" solutions, cf. fig. 4.11), are given by

$$u = 4 \tan^{-1} \left[\exp \pm \left(\frac{x - vt}{\sqrt{1 - v^2}}\right)\right]. \tag{4.3.6}$$

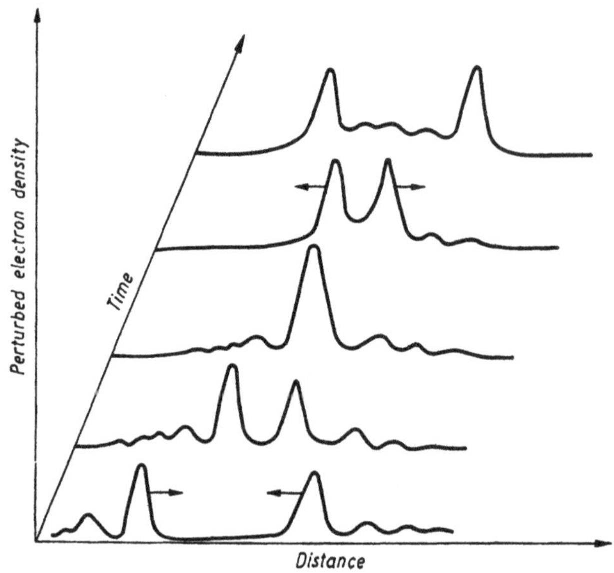

Fig. 4.10. Nondestructive collision of ion-acoustic plasma pulses (after IKEZI et al., 1970)

Here the + sign corresponds to the positive sense of rotation and the − sign to the negative sense of rotation. The former solution can be considered as a soliton and the latter as an "anti-soliton". Since the total rotation must be conserved, in any collision solitons and anti-solitons are created and destroyed in pairs. Note that the sine-Gordon equation is invariant to a Lorentz transformation.

It is worthwhile noticing that recently some progress has been made in extending results obtained for solitons in one-dimensional systems to mul-

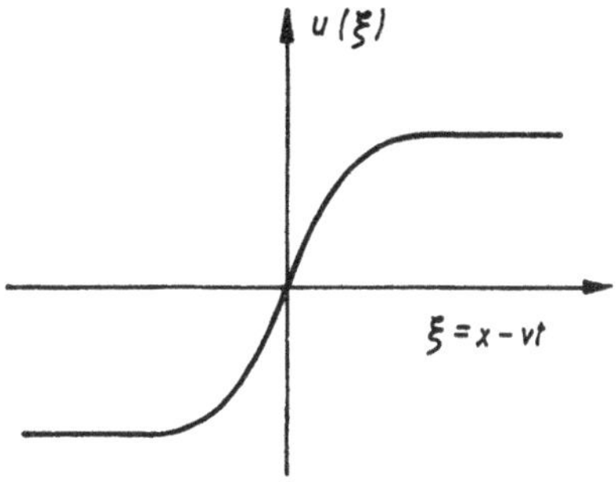

Fig. 4.11. Schematic plot of a kink shaped solitary wave

tiple dimensional systems. For instance, computer experiments performed on (idealized) three-dimensional lattices suggest that fairly stable solitary waves may also exist in actual solids (SCHNEIDER and STOLL, 1978b, BATTEH and POWELL, 1979). This means that energy transport and the manner of return to thermal equilibrium after a disturbance may be governed by solitary waves in actual solids, too.

For the mathematical background of the soliton phenomenon the reader is referred to WITTHAM (1974), ZACHAROV et al. (1980), EILENBERGER (1981), and MC LAUGHLIN (1981), and for the general soliton concept and its application to various fields of physics (elementary particle physics, solid state physics, plasma physics, nonlinear optics etc.) to SCOTT et al. (1973), BULLOUGH (1977), REBBI (1979), and, in particular, with respect to solitons in condensed matter to BISHOP et al. (1980).

4.3.2. The Toda lattice

The so-called Toda lattice (or exponential lattice), introduced by TODA (1967), is a one-dimensional lattice with exponential nearest neighbour interaction. In recent years the Toda lattice has been extensively studied, because it is an attractive model of a nonergodic nonlinear dynamical system. It is a completely integrable system. It has so-called N-soliton solutions and admits anharmonicity of arbitrary strength. In the continuum limit it is described by the Korteweg-de Vries equation (for reviews of the Toda lattice see TODA, 1975, 1976, 1981).

In the Toda lattice nearest neighbour "atoms" are connected by springs with the potential

$$\varphi(r) = \frac{a}{b} e^{-br} + ar \qquad (ab > 0), \tag{4.3.7}$$

where r denotes the elongation of a spring beyond its natural length. For $a, b > 0$ the first term on the right-hand side of equation (4.3.7) represents a repulsive force and the second term an attractive one (fig. 4.12), while for $a, b < 0$ the first term represents a strong attractive force and the second a weak repulsive one. In the following, we restrict ourselves to $a, b > 0$.

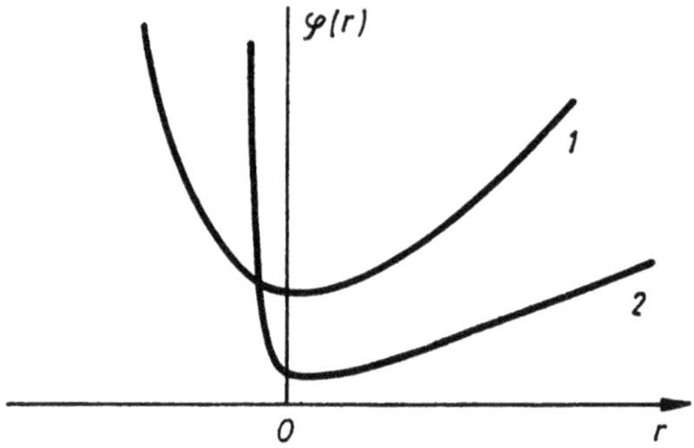

Fig. 4.12. Anharmonic potential of the exponential type (4.3.7) for $a, b > 0$. 1 — small b, 2 — large b (hard sphere)

For small b, $\varphi(r)$ becomes

$$\varphi(r) = \frac{ab}{2}\left(r^2 - \frac{b}{3} r^3 + \cdots\right). \tag{4.3.8}$$

Thus, in the limit $b \to 0$ (ab finite) the Toda lattice reduces to a harmonic lattice with force constants $\gamma = ab$.

The equation of motion of the Toda lattice reads

$$M\ddot{u}_l = a \{\exp[-b(u_l - u_{l-1})] - \exp[-b(u_{l+1} - u_l)]\}, \tag{4.3.9}$$

where M is the mass of an atom, and u_l denotes the displacement of the lth atom from its equilibrium position.

Introducing the relative displacement of neighboured atoms by

$$r_l = u_l - u_{l-1} \tag{4.3.10}$$

and substracting the equation for u_{l-1} of (4.3.9), we obtain

$$M\ddot{r}_l = a[2e^{-br_l} - e^{-br_{l-1}} - e^{-br_{l+1}}]. \tag{4.3.11}$$

It is convenient to write

$$e^{-br_l} - 1 = \ddot{S}_l/a. \tag{4.3.12}$$

Then (4.3.11) yields

$$\ln(1 + \ddot{S}_l/a) = \frac{b}{M}(S_{l-1} + S_{l+1} - 2S_l) \tag{4.3.13}$$

and

$$r_l = \frac{1}{M}(2S_l - S_{l-1} - S_{l+1}). \tag{4.3.14}$$

Here we have chosen the integration constants to be zero, so as to obtain the simplest relation between r_l and S_l.

Introducing further a function ψ_l by

$$S_l = \frac{M}{b}\ln\psi_l, \tag{4.3.15}$$

then we find from (4.3.13) and (4.3.12)

$$1 + \frac{M}{ab}(\ddot{\psi}_l\dot{\psi}_l - \dot{\psi}_l^2)/\psi_l^2 = \psi_{l-1}\psi_{l+1}/\psi_l^2 \tag{4.3.16}$$

and

$$e^{-br_l} - 1 = \frac{M}{ab}(\ddot{\psi}_l\psi_l - \dot{\psi}_l^2)/\psi_l^2. \tag{4.3.17}$$

To obtain a one-soliton solution we write

$$\psi_l = 1 + A\,e^{\alpha l - \beta t}, \qquad A, \alpha > 0. \tag{4.3.18}$$

Substituting this expression into (4.3.16), we see that (4.3.18) is a solution of (4.3.16), provided that

$$\beta^2 = \frac{4ab}{M}\sinh^2\frac{\alpha}{2}. \tag{4.3.19}$$

Inserting (4.3.18) into (4.3.17), we obtain the one-soliton solution

$$e^{-br_l} - 1 = \frac{M\beta^2}{4ab}\operatorname{sech}^2\frac{1}{2}(\alpha l - \beta t + \delta), \tag{4.3.20}$$

where

$$\delta = \ln A.$$

The speed of the soliton is

$$c = \frac{\beta}{\alpha}L = c_0\sinh(\alpha/2)/(\alpha/2), \tag{4.3.21}$$

where

$$c_0 = \pm\sqrt{\frac{ab}{M}}\,L. \tag{4.3.22}$$

Here L denotes the natural length of a spring.

A two-soliton solution can be derived from

$$\psi_l = 1 + A_1 \, e^{\alpha_1 l - \beta_1 t} + A_2 \, e^{\alpha_2 l - \beta_2 t} + e^{(\alpha_1 + \alpha_2) l - (\beta_1 + \beta_2) t}, \tag{4.3.23}$$

where α_i, $A_i > 0$ ($i = 1, 2$) and $\alpha_1 > \alpha_2$, without loss of generality. Inserting (4.3.23) into (4.3.16), we find that (4.3.23) is a solution if

$$\beta_i^2 = \frac{4ab}{M} \sinh^2 \frac{\alpha_i}{2}, \qquad i = 1, 2, \tag{4.3.24}$$

and

$$A_1 A_2 = \frac{4 \sinh^2 \frac{1}{2} (\alpha_1 + \alpha_2) - \frac{M}{ab} (\beta_1 + \beta_2)^2}{\frac{M}{ab} (\beta_1 - \beta_2)^2 - 4 \sinh^2 \frac{1}{2} (\alpha_1 - \alpha_2)}. \tag{4.3.25}$$

There are four possible combinations of the signs of the roots of (4.3.24), but only two of them yield independent solutions ($\beta_1\beta_2 > 0$ and $\beta_1\beta_2 < 0$). Here, we restrict ourselves to the case $\beta_1\beta_2 > 0$. From (4.3.24) and (4.3.25) we find

$$\left.\begin{aligned} &\beta_1 = 2 \sqrt{\frac{ab}{M}} \sinh \frac{\alpha_1}{2}, \qquad \beta_2 = 2 \sqrt{\frac{ab}{M}} \sinh \frac{\alpha_2}{2}, \\ &A_1 A_2 = \left(\frac{\sinh \frac{1}{4} (\alpha_1 + \alpha_2)}{\sinh \frac{1}{4} (\alpha_1 - \alpha_2)} \right)^2. \end{aligned}\right\} \tag{4.3.26}$$

Let us first consider ψ_l in the region, where $\varphi_1 = \alpha_1 l - \beta_1 t \approx 0$. Then

$$\varphi_2 = \alpha_2 l - \beta_2 t = \frac{\alpha_2}{\alpha_1} \varphi_1 + \epsilon t, \tag{4.3.27}$$

where

$$\epsilon = \frac{\alpha_2}{\alpha_1} \beta_1 - \beta_2 > 0, \tag{4.3.28}$$

since $\beta_1/\beta_2 = \sinh (\alpha_1/2)/\sinh (\alpha_2/2) > \alpha_1/\alpha_2$ for $\alpha_1 > \alpha_2 > 0$. Thus, eq. (4.3.23) becomes

$$\left.\begin{aligned} &\psi_l \simeq 1 + A_1 \exp (\alpha_1 l - \beta_1 t) \qquad \text{for } t \to -\infty, \\ &\psi_l \simeq A_2 + \exp (\alpha_1 l - \beta_1 t) \qquad \text{for } t \to +\infty. \end{aligned}\right\} \tag{4.3.29}$$

Similarly, in the region where $\varphi_2 = \alpha_2 l - \beta_2 t \approx 0$, we have

$$\left.\begin{aligned} &\psi_l \simeq A_1 + \exp (\alpha_2 l - \beta_2 t) \qquad \text{for } t \to -\infty, \\ &\psi_l \simeq 1 + A_2 \exp (\alpha_2 l - \beta_2 t) \qquad \text{for } t \to +\infty. \end{aligned}\right\} \tag{4.3.30}$$

According to (4.3.17), (4.3.29) and (4.3.30) (cf. (4.3.18) and (4.3.20)) it holds asymptotically that

$$
\begin{aligned}
e^{-br_l} - 1 &= \frac{M}{4ab} \beta_i^2 \operatorname{sech}^2 \frac{1}{2} (\alpha_i l - \beta_i t \pm \delta^-), \quad i = \begin{pmatrix}1\\2\end{pmatrix} \quad \text{for} \quad t \to -\infty, \\
e^{-br_l} - 1 &= \frac{M}{4ab} \beta_i^2 \operatorname{sech}^2 \frac{1}{2} (\alpha_i l - \beta_i t \mp \delta^+), \quad i = \begin{pmatrix}1\\2\end{pmatrix} \quad \text{for} \quad t \to +\infty,
\end{aligned}
\tag{4.3.31}
$$

where

$$\delta^- = \ln A_1 \qquad \text{and} \qquad \delta^+ = \ln A_2. \tag{4.3.32}$$

Thus, at $t \to -\infty$ this solution describes two solitons arranged in such a manner that the slower soliton (2) is in front of the faster soliton (1), and at $t \to +\infty$ the arrangement is reversed (fig. 4.13). This implies that the two solitons are stable in the collision process[1].

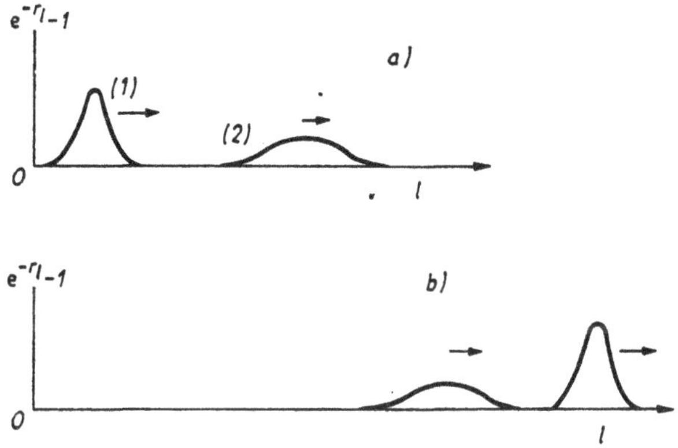

Fig. 4.13. Two-soliton solution of the Toda lattice for $\beta_1\beta_2 > 0$ (see text) a) at $t \to -\infty$, b) at $t \to +\infty$. 1 — fast soliton, 2 — slow soliton

Note that in the case where $\beta_1\beta_2 < 0$, two solitons are running in opposite directions. For more details on this case and for the N-soliton problem the reader is referred to Toda (1975, 1976) and to references therein.

Now let us derive a periodic solution of the Toda lattice. To do this we use the following addition theorem for the Jacobian elliptic functions:

$$\operatorname{sn}^2 (u + v) - \operatorname{sn}^2 (u - v) = 2 \frac{d}{dv} \frac{\operatorname{sn} u \operatorname{cn} u \operatorname{dn} u \operatorname{sn}^2 v}{1 - k^2 \operatorname{sn}^2 u \operatorname{sn}^2 v}, \tag{4.3.33}$$

---

[1] It is worth noticing that by numerical studies on a one-dimensional chain with a Lennard-Jones nearest neighbour interaction potential (cf. (1.4.46)) for solitary waves collision behaviour similar to the exact solution behaviour of the Toda lattice was found (Valkering and de Lange, 1980).

where the Jacobian elliptic functions sn, cn and dn are defined by (for the properties of these functions used in the following see, for example, TODA, 1970, appendix 1)

$$u = \int_0^{\varphi} \frac{d\Theta}{\sqrt{1 - k^2 \sin^2 \Theta}} = \int_0^{\sin\varphi} \frac{dx}{\sqrt{(1 - x^2)(1 - k^2x^2)}}, \tag{4.3.34}$$

$$\operatorname{sn} u = \sin \varphi = \operatorname{sn}(u, k),$$

$$\operatorname{cn} u = \cos \varphi = \operatorname{cn}(u, k),$$

$$\operatorname{dn} u = \sqrt{1 - k^2 \operatorname{sn}^2 u} = \operatorname{dn}(u, k).$$

Integrating (4.3.33) and accounting for $\operatorname{dn}^2 u = 1 - k^2 \operatorname{sn}^2 u$ and $d(\operatorname{dn} u)/du = -k^2 \operatorname{sn} u \operatorname{cn} u$, we obtain

$$\epsilon(u + v) + \epsilon(u - v) - 2\epsilon(u) = \frac{\epsilon''(u)}{(1/\operatorname{sn}^2 v) - 1 + \epsilon'(u)}, \tag{4.3.35}$$

where we have introduced the function

$$\epsilon(u) = \int_0^u du' \operatorname{dn}^2 u'. \tag{4.3.36}$$

The Jacobian zeta function $Z(u)$, or zn u, given by

$$Z(u) = \epsilon(u) - Eu/K \tag{4.3.37}$$

is a periodic function with period $2K$. Here K and E are the complete elliptic integrals

$$K = K(k) = \int_0^{\pi/2} \frac{d\Theta}{\sqrt{1 - k^2 \sin^2 \Theta}} = \int_0^1 \frac{dx}{\sqrt{(1 - x^2)(1 - k^2x^2)}}, \tag{4.3.38}$$

$$E = E(k) = \int_0^{\pi/2} d\Theta \sqrt{1 - k^2 \sin^2 \Theta} = \int_0^1 dx \sqrt{\frac{1 - k^2x^2}{1 - x^2}}. \tag{4.3.39}$$

By (4.3.35) and (4.3.37), we find

$$\frac{Z''(u)}{(1/\operatorname{sn}^2 v) - 1 + E/K - Z'(u)} = Z(u + v) + Z(u - v) - 2Z(u). \tag{4.3.40}$$

To compare (4.3.40) with the equation of motion of the Toda lattice, we write

$$S_l(t) = \int^t dt' s_l(t') \tag{4.3.41}$$

and differentiate (4.3.13) with respect to time. Thus we obtain

$$\frac{\mathrm{d}}{\mathrm{d}t} \ln(1 + \dot{s}_l/a) = \frac{\ddot{s}_l}{a + \dot{s}_l} = \frac{b}{M} (s_{l-1} + s_{l+1} - 2s_l). \tag{4.3.42}$$

Furthermore, with the aid of (4.3.41), (4.3.12) becomes

$$\mathrm{e}^{-br_l} - 1 = \dot{s}_l/a. \tag{4.3.43}$$

Comparing (4.3.40) and (4.3.42), we obtain

$$s_l = 2 \frac{M}{b} K\nu Z(u), \tag{4.3.44}$$

$$u = 2(l/\lambda \pm \nu t) K, \tag{4.3.45}$$

$$v = 2K/\lambda, \tag{4.3.46}$$

$$\left(1/(\mathrm{sn}^2 v) - 1 + E/K\right) (2K\nu)^2 \frac{M}{ab} = 1. \tag{4.3.47}$$

By (4.3.43), (4.3.44) to (4.3.47) and (4.3.37), we finally find the (so-called *cn*oidal wave) solution

$$\mathrm{e}^{-br_l} - 1 = \frac{M}{ab} (2K\nu)^2 \{\mathrm{dn}^2 [2(l/\lambda \mp \nu t) K] - E/K\}, \tag{4.3.48}$$

where the frequency ν and the wavelength λ satisfy the dispersion relation (4.3.47)

$$2K\nu = \sqrt{\frac{ab}{M}} \left(\frac{1}{\mathrm{sn}^2 (2K/\lambda)} - 1 + \frac{E}{K}\right)^{-1/2}. \tag{4.3.49}$$

The function $\mathrm{dn}^2 (2xK) - E/K$ is a periodic function of x with the period 1. Its Fourier expansion reads

$$\mathrm{dn}^2 (2xK) - \frac{E}{K} = \frac{\pi^2}{K^2} \sum_{n=1}^{\infty} \frac{n \cos 2\pi n x}{\sinh (\pi n K'/K)}, \tag{4.3.50}$$

where

$$K' = K\left(\sqrt{1 - k^2}\right). \tag{4.3.51}$$

For a very small modulus k, K' is very large and holds approximately

$$\left.\begin{aligned} &\mathrm{e}^{-\pi K'/K} \simeq k^2/16, \\ &K \simeq \frac{\pi}{2}, \\ &\mathrm{sn}\, x \simeq \sin x. \end{aligned}\right\} \tag{4.3.52}$$

In this case the cnoidal wave (4.3.48) reduces to a cosinus-wave in a harmonic lattice:

$$r_l \simeq -\frac{M\omega^2 k^2}{8ab^2} \cos\left(\frac{2\pi l}{\lambda} - \omega t\right) \tag{4.3.53}$$

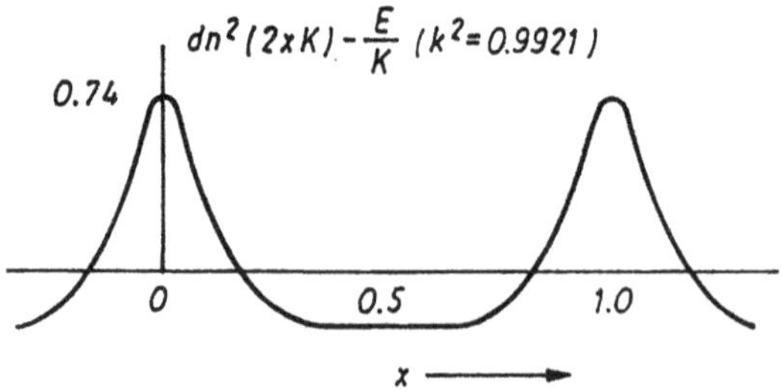

Fig. 4.14. Shape of a cnoidal wave

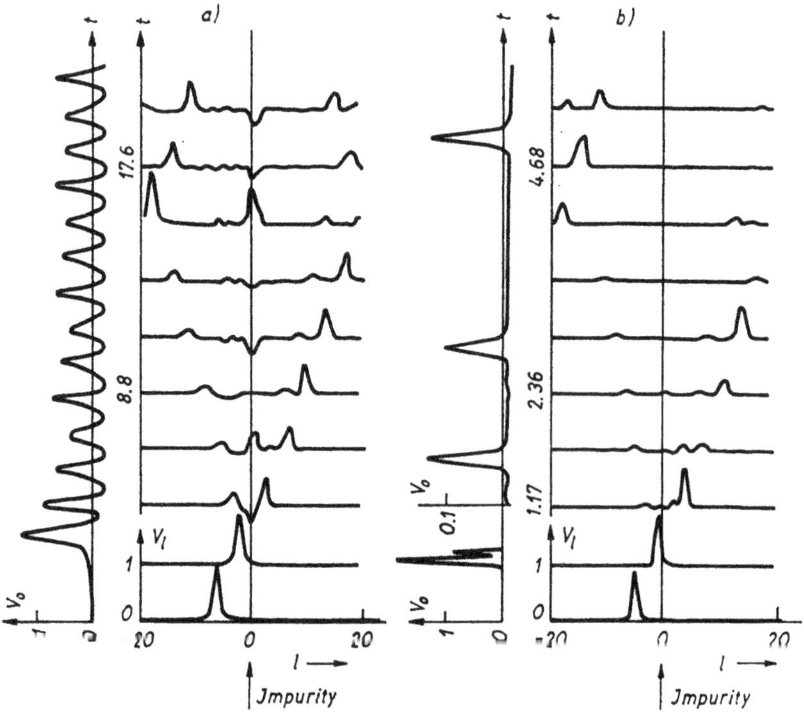

Fig. 4.15. Impurity scattering of a soliton in the Toda lattice for a strong coupling of the impurity. $V_l = a_l(e^{-b_l r_l} - 1)$, $a_l = a\{1 + (a' - 1)(\delta_{l0} + \delta_{l1})\}$, $b_l = b\{1 + (b' - 1)(\delta_{l0} + \delta_{l1})\}$, $a' = b' = 5$ (cf. (4.3.7)); a) $\alpha = 4$, b) $\alpha = 8$ (cf. (4.3.7)). The amplitudes are normalized to that of the incident soliton. The time is in units $(M/ab)^{1/2}$. For a very steep incident soliton (large α) the localized mode tends to be excited less effectively (after Nakamura, 1978)

with

$$\omega = 2\sqrt{ab/M}\sin(\pi/\lambda). \tag{4.3.54}$$

Thus, the modulus k governs the amplitude of the wave.

On the other hand, if we use the identity

$$\mathrm{dn}^2(2xK) - \frac{E}{K} = \left(\frac{\pi}{2K'}\right)^2 \sum_{n=-\infty}^{\infty} \mathrm{sech}^2\left[\frac{\pi K}{K'}(x-n)\right] - \frac{\pi}{2KK'}, \tag{4.3.55}$$

the cnoidal wave can be expressed as

$$\mathrm{e}^{-br_l} - 1 = \frac{M}{ab}\left[\sum_{n=-\infty}^{\infty} \beta^2 \,\mathrm{sech}^2\left[\alpha(l-\lambda n) - \beta t\right] - 2\beta\nu\right], \tag{4.3.56}$$

where

$$\alpha = \pi K/\lambda K', \quad \beta = \pi K\nu/K'. \tag{4.3.57}$$

Hence, the cnoidal wave represents an infinite sequence of equidistant sech^2-pulses (solitons) (fig. 4.14). These solitons are not independent of each other. Their speed is not given by (4.3.21) but by the dispersion relation (4.3.49). For $\lambda \to \infty$, $k \to 1$ and α finite the wave train reduces to a single soliton in an infinite lattice.

By numerical experiments Nakamura (1978) pointed out that an impurity atom in a Toda lattice may produce a fairly well-defined (undamped) localized mode at the impurity site. The localization is very strong for a strong coupling of the impurity, while it is weaker for a light impurity mass (cf. Yoshida and Sakuma, 1982). Furthermore, under certain circumstances, a soliton incident on an impurity can produce reflected and/or transmitted solitary waves, which are accompanied by ripple-like waves (fig. 4.15). Further problems of current interest are the diatomic Toda lattice (cf. Mokross and Büttner, 1981) and the statistical mechanics of the Toda lattice (Mertens and Büttner, 1981, Diederich. 1981).

4.3.3. Solitons and phonons in a double-well potential model for structural phase transitions

A widely used one-component d-dimensional model for structural (ferrodistortive) phase transitions is defined by the following classic Hamiltonian:

$$H = \sum_l \frac{p_l^2}{2M} + \sum_l \left[\frac{A}{2}u_l^2 + \frac{B}{4}u_l^4\right] + \frac{C}{2}\sum_{\substack{ll'\\(\mathrm{n.n.})}} (u_l - u_{l'})^2 \tag{4.3.58}$$

with $A < 0$; $B, C > 0$ (see, for example, Bruce, 1978, Schneider and Stoll, 1978b, Bruce, 1980a and Beale et al., 1981). Here u_l and p_l respectively are the displacement and momentum of the atom being placed at the lth site of a regular (simple cubic, for $d = 3$) lattice. M is the atomic mass, and the third summation in (4.3.58) is over nearest neighbours only. This Hamiltonian represents a set of harmonically coupled oscillators having identical double-well single-particle potentials (fig. 4.16). It simulates a crystal with two sub-lattices, where the atoms of the one sublattice, which are assumed to be held fixed, produce double-well potentials for the (mobile) atoms of the other sublattice.

The model (4.3.58) can be shown (SARBACH, 1977) to exhibit (for $d > 1$) a phase transition at a non-zero temperature T_c to a homogeneously ordered

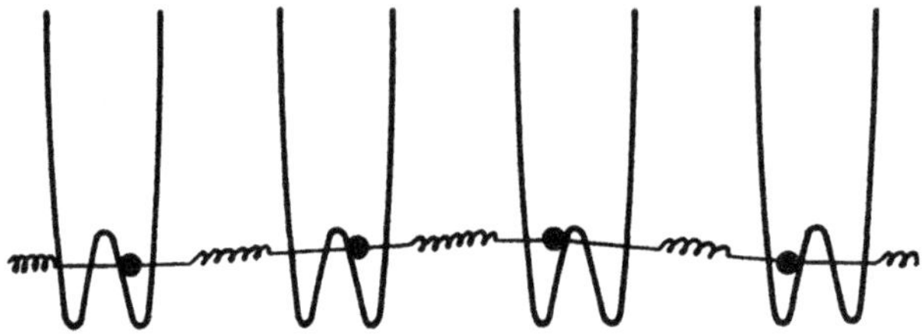

Fig. 4.16. Schematic representation of the model (4.3.58) for $d = 1$ (after BRUCE, 1980)

phase characterized by a non-vanishing average displacement ($\langle u_l \rangle \neq 0$). It describes also the two extreme cases of such phase transitions: the "order-disorder" and the "displacive" transitions. The difference between these two limiting cases can be well understood by means of the model shown in figure 4.16. At low temperatures, all mobile atoms are located at the bottom of, say, the left-hand well. At higher temperatures, however, two different situations may be found. Either, the most probable atomic positions are at the top of the energy barrier (displacive transitions) or in the vicinity of the bottom of a well, whereby the wells are now equally occupied (order-disorder transitions).

This classification of the distortive transitions may be made according to the ratio $\zeta = E_0/E_c$, where $E_0 = A^2/4B$ and $E_c = 4C\,|A|/B$ respectively denote the depth of the wells (height of the barrier) and the interaction energy between nearest neighbours displaced to opposite wells (cf. figs. 4.17 and 4.18) (the minima of a double-well potential are at $\pm u_0 = \pm\sqrt{|A|/B}$). Note that for $d = 3$ kT_c/E_c is of order unity (SARBACH, 1977). Hence, in the order-disorder limit, defined by $\zeta \gg 1$, except at temperatures $T \gg T_c$ each (mobile) atom will be localized near the bottom of a well, so that within the harmonic approximation all vibrations about the high-temperature equilibrium positions (i.e. about the top of the barrier) are unstable. In the order-disorder limit the fundamental dynamic processes are hopping processes between adjacent wells. This situation can be described by a pseudo-spin model and, at sufficiently high temperatures, by a model consisting of noninteracting double-well oscillators. — In the displacive limit, where $\zeta \ll 1$, only a small group of

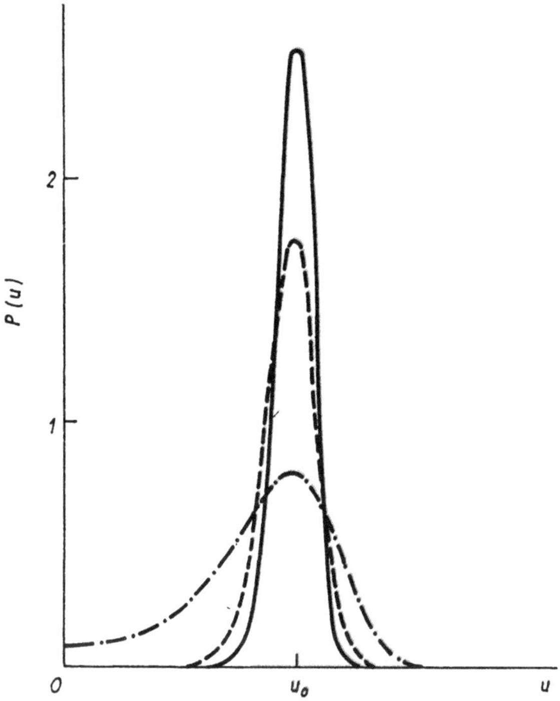

Fig. 4.17. Position probability density $P(u)$ of a particle for an order-disorder case ($E_0/E_c = 6.25$) of the model (4.3.58) for $d = 1$ at different temperatures (after AUBRY, 1975)

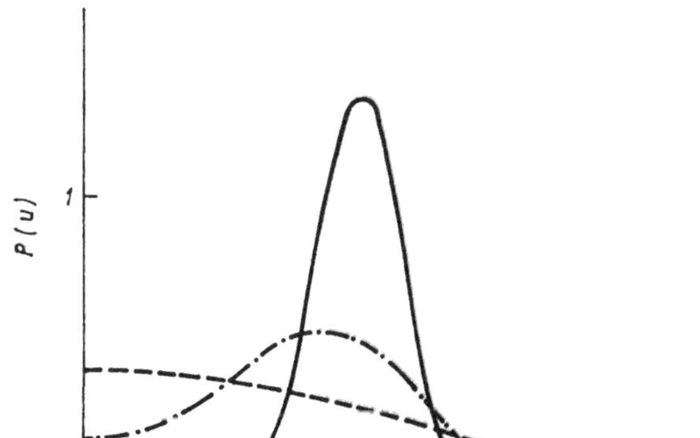

Fig. 4.18. Position probability density $P(u)$ of a particle for a displacive case ($E_0/E_c = 0.031$) of the model (4.3.58) for $d = 1$ at different temperatures; —— $kT/E_0 = 0.5$; -·-·- $kT/E_0 = 2$, --- $kT/E_0 = 10$ (after AUBRY, 1975)

long wavelength vibrations about the high-temperature equilibrium positions is unstable. Thus, the soft phonon mode concept is appropriate for describing this limiting case.

Here we restrict ourselves to a study of the displacive limit ($\zeta \ll 1$) of a one-dimensional system (KRUMHANSL and SCHRIEFFER, 1975, AUBRY, 1975, VARMA, 1976). Because for $\zeta \ll 1$ the difference $|u_l - u_{l+1}|$ is relatively small, we assume that the Hamiltonian (4.3.58) may be replaced by a continuum representation

$$H \simeq \int \frac{dx}{L} \left[\frac{p^2(x)}{2M} + \frac{A}{2} u^2(x) + \frac{B}{4} u^4(x) + \frac{Mc_0^2}{2} \left(\frac{\partial u(x)}{\partial x}\right)^2\right] \qquad (4.3.59)$$
$$\equiv \int dx \mathcal{H}(p(x), u(x), \partial u(x)/\partial x),$$

where L is the lattice spacing, $x_l = lL \equiv x$ is the continuous space variable and $c_0^2 \equiv 2L^2C/M$.

By the canonical equations

$$\dot{p} = -\frac{\partial \mathcal{H}}{\partial u} + \frac{\partial}{\partial x} \frac{\partial \mathcal{H}}{\partial(\partial u/\partial x)},$$
$$\dot{u} = \frac{\partial \mathcal{H}}{\partial p}, \qquad (4.3.60)$$

and the Hamilton density $\mathcal{H}$ from (4.3.59) we obtain the following equation of motion for the displacement field $u(x)$:

$$M\ddot{u} + Au + Bu^3 - Mc_0^2u'' = 0. \qquad (4.3.61)$$

If we look for solutions of equation (4.4.61) of permanent profile, i.e. the velocity v is constant and the solutions depend only on a single variable $x - vt$, then (4.3.61) reduces to

$$\frac{d^2\eta}{ds^2} + \eta - \eta^3 = 0, \qquad (4.3.62)$$

where we have used the dimensionless variables

$$u/u_0 = \eta,$$
$$(x - vt)/\xi = s \qquad (4.3.63)$$

with

$$\xi^2 = M(c_0^2 - v^2)/|A| \qquad \text{(length squared)}. \qquad (4.3.64)$$

First, let us consider solutions of equation (4.3.62), which describe small amplitude oscillations. For $\eta^3 \ll \eta \ll 1$ (4.3.62) reduces to the equation

$$\eta'' + \eta = 0, \qquad (4.3.65)$$

which has solutions of the form

$$\eta = \alpha \sin(s + \Theta), \qquad (4.3.66)$$

where α is the amplitude and Θ is the phase. Using (4.3.66) and (4.3.63), we obtain

$$u = \alpha u_0 \sin [(x - vt)\, \xi^{-1} + \Theta], \tag{4.3.67}$$

which is a phonon with wave number $k = \xi^{-1}$, frequency v/ξ and phase velocity v. From (4.3.67) and (4.3.64) we get the dispersion relation

$$\omega^2(k) = v^2k^2 = c_0{}^2k^2 + A/M. \tag{4.3.68}$$

Since $A < 0$, the frequency $\omega(k)$ will only be real for $k \geqq (|A|/Mc_0{}^2)^{1/2}$. The solutions (4.4.67) describe small amplitude oscillations about $u = 0$.

Small amplitude oscillations about $\pm u_0$ can be obtained by inserting

$$\eta = \pm 1 + \gamma \tag{4.3.69}$$

into (4.3.62) and retaining the terms linear in γ only. This yields

$$\gamma'' - 2\gamma = 0, \tag{4.3.70}$$

which has solutions of the form

$$\gamma = \alpha \sin \left(i \sqrt{2}\, s + \Theta\right). \tag{4.3.71}$$

By (4.3.63), (4.3.69), (4.3.71) and (4.3.64) we find

$$u = \pm u_0 + \alpha u_0 \sin \left(kx - \omega(k)\, t + \Theta\right) \tag{4.3.72}$$

and the dispersion relation

$$\omega^2(k) = v^2k^2 = c_0{}^2k^2 + 2\,|A|/M. \tag{4.3.73}$$

Unlike (4.3.68), (4.3.73) yields real frequencies for all $k \geqq 0$. Note that the model at hand has no acoustic modes. This is due to the rigidity of the sublattice setting up the local double-well potentials.

Now let us turn to large amplitude solutions of equation (4.3.62), i.e. to the regime $\eta^3 \simeq \eta \simeq \pm 1$. As can be readily verified, a particular solution of this equation is

$$\eta = \tanh \left(s/\sqrt{2}\right), \tag{4.3.74}$$

which corresponds to

$$u = u_0 \tanh \left[(x - vt)/\sqrt{2}\xi\right], \tag{4.3.75}$$

where ξ is defined in (4.3.64). This is a "kink" solution (or domain wall) of the kind shown in figure 4.11. For $(x - vt) \lessapprox 0$ and $\xi > 0$ the displacement is constant at $-u_0$, and for $(x - vt) \gtrapprox 0$ and $\xi > 0$ it is at $+u_0$. It moves with velocity v, where $|v| < c_0$ (cf. (4.3.64)). For $\xi > 0$ the harmonic "springs" between nearest neighbour (mobile) atoms are delated in the kink region, whereas for $\xi < 0$ they are compressed. Note that the continuum approximation (4.3.59) is valid if the kink thickness is greater than about three lattice spacings (Schmidt, 1979).

In terms of elliptic functions the two types of solutions (4.3.66) and (4.3.74) can be connected with each other. To see this, note that (4.3.62) is equivalent to

$$\left(\frac{\mathrm{d}\eta}{\mathrm{d}\sigma}\right)^2 = (a^2 - \eta^2)(b^2 - \eta^2) \qquad \left(\sigma \equiv s/\sqrt{2}\right) \tag{4.3.76}$$

with

$$a^2 = 1 - \left[1 - \left(\frac{\mathrm{d}\eta}{\mathrm{d}\sigma}\right)^2_{\eta=0}\right]^{1/2},$$
$$b^2 = 1 + \left[1 + \left(\frac{\mathrm{d}\eta}{\mathrm{d}\sigma}\right)^2_{\eta=0}\right]^{1/2}, \tag{4.3.77}$$

which can be verified by differentiating (4.3.76) with respect to σ and using (4.3.62). Integrating (4.3.76), we obtain the elliptic integral

$$\sigma = \frac{1}{b}\int_0^{\eta/a} \frac{\mathrm{d}x}{\sqrt{(1 - x^2)(1 - k^2x^2)}}, \tag{4.3.78}$$

where the modulus k is given by

$$k = \frac{a}{b} = \left[\frac{1 - \left(1 - 2\left(\frac{d\eta}{\mathrm{d}s}\right)^2_{\eta=0}\right)^{1/2}}{1 + \left(1 - 2\left(\frac{\mathrm{d}\eta}{\mathrm{d}s}\right)^2_{\eta=0}\right)^{1/2}}\right]^{1/2}. \tag{4.3.79}$$

Hence, according to (4.4.34), we have

$$\eta = a\,\mathrm{sn}\,(b\sigma). \tag{4.3.80}$$

The elliptic sine function sn, is periodic in $4K$, where K is the complete elliptic integral (4.4.38) of modulus k.

In the small amplitude limit $(\mathrm{d}\eta/\mathrm{d}s)_{\eta=0} \ll 1$ we have $k \simeq 0$ and thus $4K = 2\pi$. In this case (4.3.80) becomes (cf. (4.3.34))

$$\eta \simeq \alpha \sin s, \tag{4.3.81}$$

which agrees with thes olution (4.4.66). On the other hand, for large amplitudes, $k \to 1$ and $K \to \infty$. In particular, if $(\mathrm{d}\eta/\mathrm{d}s)_{\eta=0} = \sqrt{2}$, then the solution (4.3.80) approaches

$$\eta = \tanh\left(s/\sqrt{2}\right), \tag{4.3.82}$$

which is the domain wall solution (4.3.74). Shapes of small and large amplitude solutions are plotted in figure 4.19.

Now let us study the stability of the kink (solitary wave) solution (4.3.75). First, we note that the equation of motion (4.3.61) is invariant under the

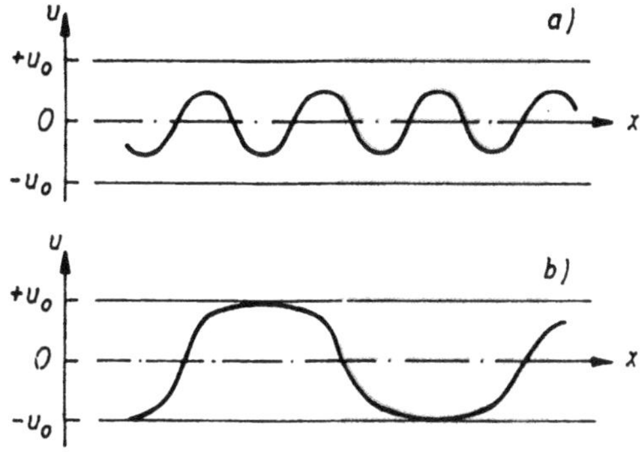

Fig. 4.19. a) Small and b) large amplitude solutions (4.3.80)

Lorentz transformation

$$\begin{pmatrix} x \\ t \end{pmatrix} \to \frac{1}{\sqrt{1 - v^2/c_0^2}} \begin{pmatrix} x - vt \\ t - vx/c_0^2 \end{pmatrix}, \tag{4.3.83}$$

where v is smaller than c_0 in magnitude. Introducing the dimensionless variables

$$\tau = \sqrt{\frac{|A|}{M}}\, t, \qquad y = \frac{1}{\sqrt{2}\, c_0} \sqrt{\frac{|A|}{M}}\, x, \qquad w = \sqrt{\frac{B}{|A|}}\, u, \tag{4.3.84}$$

then (4.4.61) takes the form

$$\ddot{w} - \frac{1}{2} w'' - w + w^3 = 0, \tag{4.3.85}$$

where $\dot{w}$ and w' are the derivatives with respect to τ and y, respectively. The double-well potential now has minima at $w = \pm 1$ and a barrier height of 1/4.

The stability of the solution (4.3.75) is assured when a small perturbation remains small at large times. Such a perturbation may be treated by assuming a solution of the form (Aubry, 1976)

$$w(y, t) = w_0(y) + v(x, t) \tag{4.3.86}$$

with $w_0(y) = \tanh y$ (cf. (4.3.75)) and $|w(y, t)| \ll 1$. Here we study the stability in a system, where the kink is at rest. Owing to (4.3.83), the stability of $w_0(y)$ implies the stability of any other solution (4.3.75).

Inserting (4.3.86) into (4.3.85) we find

$$\ddot{w} - \frac{1}{2} w'' + \left(2 - \frac{3}{\cosh^2 y}\right) w + 3(\tanh y)\, w^2 + w^3 = 0. \tag{4.3.87}$$

To first order in w, with $w(y, t) = e^{i\omega t} f(y)$, (4.3.87) reduces to the Schrödinger-like eigenvalue equation

$$-\frac{1}{2} f'' + \left(2 - \frac{3}{\cosh^2 y}\right) f = \omega^2 f, \tag{4.3.88}$$

which has the following solutions:

$$f_{\mathrm{T}}(y) = \frac{\sqrt{3}}{2}\,\frac{1}{\cosh^2 y}, \qquad \omega_{\mathrm{T}}^2 = 0, \tag{4.3.89}$$

$$f_{\mathrm{L}}(y) = \sqrt{\frac{3}{2}}\,\frac{\sinh y}{\cosh^2 y}, \qquad \omega_{\mathrm{L}}^2 = \frac{3}{2}, \tag{4.3.90}$$

$$\begin{aligned} f_k(y) &= \frac{\mathrm{e}^{iky}(3\tanh^2 y - 1 - k^2 - 3ik\tanh y)}{[2\pi(4 + 5k^2 + k^4)]^{1/2}}, \\ \omega_k^2 &= \frac{1}{2}\,k^2 + 2 \qquad (-\infty < k < +\infty). \end{aligned} \tag{4.3.91}$$

The system (4.3.89) to (4.3.91) can be shown to form a complete orthogonal set of eigenfunctions:

$$\begin{aligned} &\int \mathrm{d}y f_k(y)\, f_{k'}^*(y) = \delta(k - k'), \qquad \int \mathrm{d}y f_{\mathrm{L}}(y) f_{\mathrm{T}}(y) = 0, \\ &\int \mathrm{d}y f_{\mathrm{L}}^2(y) = \int \mathrm{d}y f_{\mathrm{T}}^2(y) = 1, \\ &\int \mathrm{d}k f_k(y)\, f_k^*(y') + f_{\mathrm{T}}(y)\, f_{\mathrm{T}}^*(y') + f_{\mathrm{L}}(y)\, f_{\mathrm{L}}^*(y') = \delta(y - y'). \end{aligned} \tag{4.3.92}$$

Thus, we have obtained two bound states: f_{T} and f_{L}. Since (to first order in a) $\tanh(y + a) = \tanh y + a(\cosh y)^{-2}$, f_{T} describes a small translation of the kink $w_0(y)$. f_{T} is therefore referred to as "translation mode"[1]. The fact that f_{T} allows kink motion to be described, is of great importance in a treatment of kink motion in the presence of external perturbations (cf. Fogel et al., 1977) and kink-phonon interaction (Hasenfratz and Klein, 1977, Theodorakopoulos, 1979, Currie et al., 1980), which is based on an expansion of the solution in the complete set (4.4.89) to (4.4.91) and which uses perturbation theory for determining the expansion coefficients. f_{L} describes a localized mode, which corresponds to an internal vibration of the kink. — The continuum solutions f_k (phonons) resemble the linearized solutions (4.3.72), except for a localized perturbation in the vicinity of the kink. By examining the asymptotic form of these solutions for $y \to \pm\infty$, one finds that they only suffer a phase shift $\delta_k = 2 \operatorname{arc\,tan} [3k/(2 - k^2)]$ when passing through the kink (owing to the reflectionless "potential" in (4.4.88)). The dispersion relations (4.3.91) and (4.3.73) are identical.

Hence, the solutions (4.3.89) to (4.3.91) establish the stability of the kink, since all eigenvalues ω_i^2 ($i = \mathrm{T}, \mathrm{L}, k$) are non-negative. Note that in (4.3.87) the term in w^2 describes the kink-phonon interaction and the term in w^3 contributes to the phonon self-energy. In the presence of fluctuations in the phonon amplitude (thermal phonons) kink-phonon collisions cause the kink to perform a stochastic motion (Koehler et al., 1975, Theodorakopoulos, 1979) (cf. fig. 4.20).

[1] The zero frequency translation mode is a so-called Goldstone mode. In the system at hand it restores the translation invariance broken by the presence of the kink.

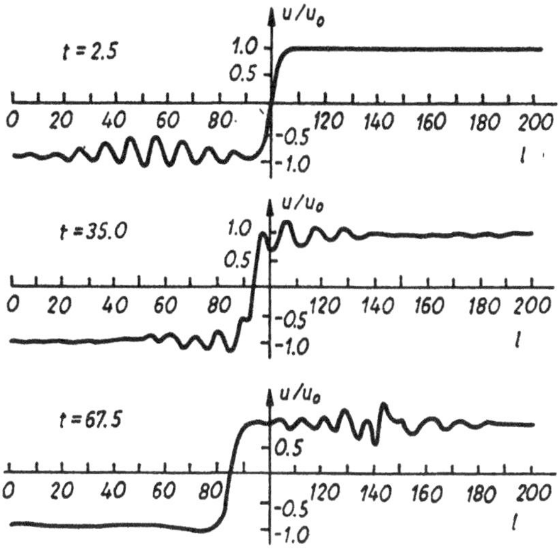

Fig. 4.20. Results of a computer experiment simulating the interaction of a phonon wave packet with a kink. The wave packet approaches the kink from the left. It shifts the kink when passing through it. The time is in units $(M/|A|)^{1/2}$ (cf. (4.3.84)) (after HASENFRATZ and KLEIN, 1977)

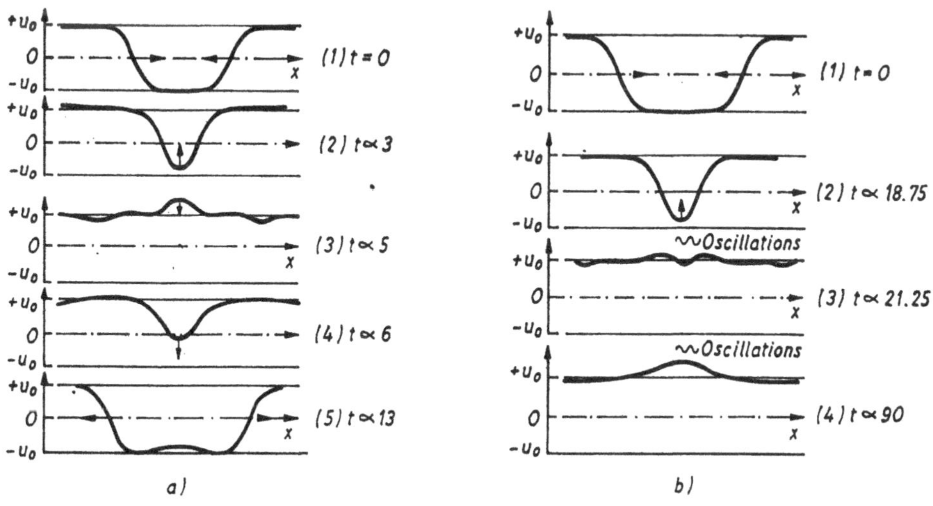

Fig. 4.21. Results of molecular dynamics computer simulation of collisions between anti-kinks and kinks: a) without annihilation ($v = \pm 0.5c_0$), b) with annihilation ($v = \pm 0.1c_0$) (after AUBRY, 1976)

Now let us turn to the multikink problem. First, note that in the 1d system being considered, in general kinks have not the soliton property of crossing each other without annihilation. Figure 4.21 shows results of molecular dynamics computer simulation of collisions between anti-kinks (taking u from $+u_0$ to $-u_0$) and kinks (taking u from $-u_0$ to $+u_0$). We see that for some values v the usual soliton properties were found (fig. 4.21a)), but for other values v kink and anti-kink destroy each other (fig. 4.21b)). The latter process is reversible: phonons may create kink-anti-kink pairs. Furthermore, studies of molecular dynamics (KOEHLER et al., 1975) reveal for low temperatures a displacement pattern as shown in figure 4.22. This pattern corresponds to a set

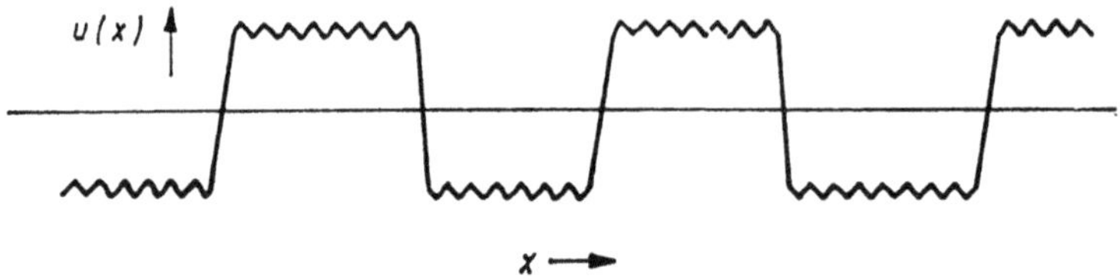

Fig. 4.22. Schematic plot of the low temperature displacement pattern for a one-dimensional system (4.3.59)

of a large number of clusters (domains) separated from each other by kinks and anti-kinks (domain walls). Superimposed on the clusters are fluctuations (phonons). The density of domain walls increases with temperature. At high temperatures the domain walls are completely destroyed by thermal fluctuations.

It should be noted that in one-dimensional systems without long-range forces true long-range order is not possible at finite temperatures (LANDAU and LIFSHITZ, 1959). Consequently, for a system undergoing a phase transition at T_c, it is not possible to explore behaviour through a 1d model. In particular, the 1d system (4.3.59) has the drawback that the temperature, where the density of kinks goes to zero, and T_c are the same, which will obviously not hold in higher dimensions. Therefore, it is important to investigate also higher dimensional systems (4.3.58).

A three-dimensional one-component model (4.3.58) was studied by SCHNEIDER and STOLL (1978b) using molecular dynamics technique. This model exhibits at (finite) T_c a distortive continuous phase transition (i.e., for $T \to T_c$ the order parameter, $\langle u_l \rangle$, continuously goes to zero). It shows clusters and associated cluster walls (fig. 4.23). In a cluster the atoms have a local displacement with a sign opposite to the zero-temperature order parameter. At the cluster walls the local displacement changes sign, analogous to the 1d case. Below T_c with increasing T the cluster size increases by approaching T_c, since the number of atoms that can overcome the potential barrier increases. Above T_c the number of clusters with positive and negative u_l is equal on average, because here $\langle u_l \rangle = 0$. Furthermore, at $T > T_c$ the cluster size decreases with increasing temperature and the system starts to behave like independent

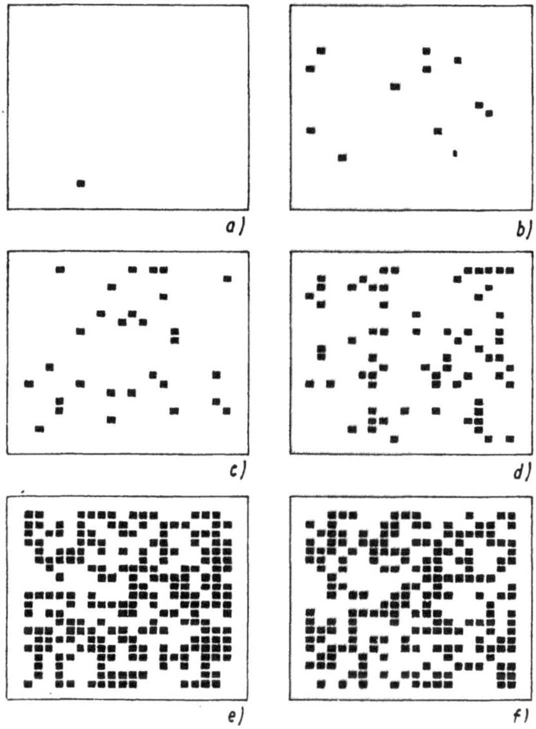

Fig. 4.23. Instantaneous cluster configuration in a cubic crystal described by the Hamiltonian (4.3.58) in a plane perpendicular to [100]. The squares designate atoms having a local displacement, u_l, with a sign opposite to the zero temperature value $\langle u_l \rangle_{T=0}$. a) $kT = 2$, b) $= 4$, c) $= 5$, d) $= 6$, e) $= 8$, f) $= 15$; $kT_c \approx 7.1$ (from Schneider and Stoll, 1978b)

quartic anharmonic oscillators. — Evidently, kink-like solutions (domain walls) play an important role also in three-dimensional systems, especially near T_c. Note that domain effects are lost in any mean field models (cf. Bishop and Krumhansl, 1975).

Figures 4.22 and 4.23 show only a snapshot of spatial displacement patterns. These patterns are time-dependent: the domain walls move through the system, which is caused by some collective "flipping" of the atoms from one well to the other; and the clusters have a finite lifetime only. Moreover, the atoms perform oscillations about some instantaneous mean displaced positions. These oscillations are fast compared with the flipping motion. Thus, the local order parameter is non-zero for a time scale long in comparison with typical inverse phonon frequencies. The finite cluster lifetime means that, in contrast to the Fermi-Pasta-Ulam-problem (cf. section 4.4.1.), the system at hand may approach an ergodic state.

Krumhansl and Schrieffer (1975) elaborated the statistical mechanics of the 1d model (4.4.59), using a classic functional-integral method (the transfer-operator method described in section 4.4.2.) and found that at low temperatures the domain walls (kinks) together with the oscillations (phonons) can be interpreted as elementary excitations of the system. In recent years quantum and classic statistical mechanics have been applied to kinks of a variety of 1d systems. These investigations have firmly established solitons as a new class of elementary excitations in 1d systems (Bishop, 1978, 1981, Takayama and Maki, 1979, 1980, Currie et al., 1980, and references therein). The extent to which this idea applies in higher dimensions is as yet uncertain.[1]

Concluding this section, let us draw attention to the observation of Riste et al. (1971) that the dynamical structure factor (cf. appendix 3) of the perovskite $SrTiO_3$, which is a paradigm for a system exhibiting a displacive phase transition, shows unusual features as $T \to T_c^+$: in addition to the two peaks expected for a softening of quasi-harmonic phonons (cf. section 4.1.1.) a central peak (at $\omega = 0$) evolves. This central peak is extremely narrow, and it grows critically in intensity on approaching T_c. The phonon frequencies continue to soften, but attain at T_c a finite frequency (for a review see Bruce and Cowley, 1980).

Although the origin of this phenomenon is not yet settled completely, it seems likely that the central peak is mainly due to defects (Halperin and Varma, 1976, Petschek, 1980, Bruce and Cowley, 1980, and references therein), but there is little doubt that cluster dynamics (domain wall motion) also contribute to the central peak (Krumhansl and Schrieffer, 1975, Schneider and Stoll, 1976; Sahni and Mazenko, 1979, Hammer et al., 1981). According to the latter mechanism the central peak represents a cluster (short range order) induced dynamic precursor of the Bragg peak to appear below T_c. The short-range order is assumed to stabilize the frequency of the softening phonon so that a finite frequency at T_c results (Bruce et al., 1979, and Bruce et al., 1980, Aksenov et al., 1981). An analytical theory of the central peak phenomenon is still lacking.

Note that in systems with a small classic value of the zero-temperature order parameter, quantum fluctuations (zero point vibrations) can suppress the occurrence of a distortive phase transition (cf. Schneider et al., 1976, Müller and Burkard, 1979). Furthermore, note that statistical mechanics of dynamical properties of nonlinear system is much less understood than statistical mechanics of static properties of (simple) such systems. On the model (4.3.58) the former problem was studied by molecular-dynamics technique by Schneider and Stoll (1981).

---

[1] For $d > 1$ the energy of solitons (being proportional to their macroscopically large area) appears to be too large for them to be included among the elementary excitations (cf. Leung et al. 1980, Mattis, 1981). However, Ventura (1981) pointed out that above some critical temperature spontaneous soliton generation should be allowed by soliton induced change of entropy.

4.4. Electron-phonon driven lattice instabilities, charge density wave excitations and incommensurate structures

4.4.1. Mean field theory of the Peierls instability in one-dimensional metals and charge density wave excitations

As pointed out in section 1.6.3., for one-dimensional metallic systems the microscopic phonon theory predicts a giant Kohn anomaly (phonon softening) for modes with wave number $2k_F$ ($\hbar k_F$ — Fermi momentum). This anomaly arises from the logarithmic singularity occurring at $2k_F$ in the dielectric function of such a system (cf. (1.6.48) and fig. 1.16).

For one-dimensional metals Peierls (1955) pointed out that it is energetically favourable to distort the lattice periodically, with a wavelength λ given by

$$2\pi/\lambda = 2k_F. \tag{4.4.1}$$

In the electronic spectrum this distortion opens a gap, 2Δ, whose size is proportional to the amplitude of the $2k_F$ distortion (fig. 4.24). This gap separates

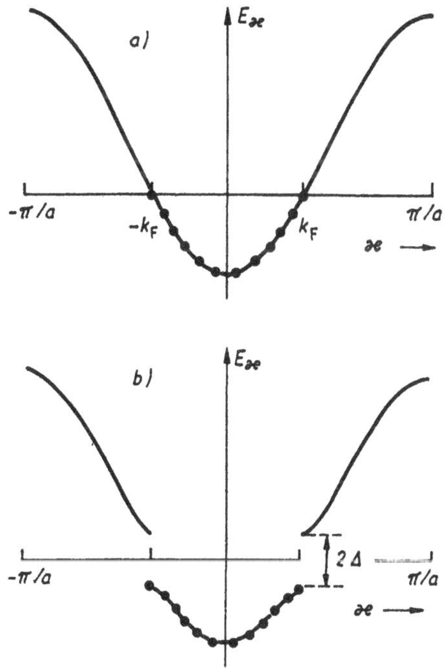

Fig. 4.24. a) Electronic energy, $E_\varkappa$, as function of wave number $\varkappa$ for a one-dimensional metal with states filled for $|\varkappa| \leq k_F$. b) The creation of a gap in $E_\varkappa$ of width 2Δ at $\pm k_F$ due to a $2k_F$ Peierls distortion. a — lattice constant

(at $T = 0$) filled and empty electronic states, so that at low temperatures the one-dimensional "metal" behaves like an insulator.

In a system with such a Peierls distortion a charge density wave (CDW) occurs, since the electronic charge density undergoes a $2k_F$ distortion in response to the $2k_F$ lattice distortion. If the wavelength of the CDW is "incommensurate"[1] with the lattice and if other "pinning" mechanisms (see below) are absent, the CDW is free to move through the system, so that a supercurrent may result (Fröhlich, 1954).

The phenomena mentioned before have attracted great interest in recent years, and a great deal of progress has been made in the theoretical understanding of them in the last decade. Typical quasi one-dimensional metals are the tetrathiafulvalene tetracyanoquinodimethane (TTF-TCNQ) and $K_2Pt(CN)_4Br_{0.3} \cdot 3H_2O$ (KPC). Both these materials exhibit a giant Kohn anomaly and a Peierls instability (for reviews of one-dimensional metals see, for example, Toombs, 1978, Berlinsky, 1979, and Friend and Jerome, 1979, and for a concise situation report see Heeger, 1982).

To treat the Kohn anomaly and the Peierls instability in a one-dimensional metal we start from the following (Fröhlich-)Hamiltonian:

$$H = \sum_{\varkappa} E_{\varkappa} c_{\varkappa}^{+} c_{\varkappa} + \sum_{k>0} \hbar\omega_k (b_k^{+} b_k + b_{-k}^{+} b_{-k}) + \left(g/\sqrt{N}\right) \sum_{k>0} (\varrho_k A_k + \varrho_{-k} A_{-k}), \tag{4.4.2}$$

where

$$\varrho_k = \sum_{\varkappa} c_{\varkappa+k}^{+} c_{\varkappa}, \tag{4.4.3}$$

$$A_k = b_k + b_{-k}^{+}. \tag{4.4.4}$$

Here $c_\varkappa^+$ ($c_\varkappa$) and b_k^+ (b_k) are, respectively, the electron and phonon creation (destruction) operators ($k, \varkappa$ — wave numbers); $E_\varkappa$ is the electron energy measured from the Fermi energy, ω_k is the unperturbed phonon frequency, g is the electron-phonon coupling constant, N is the number of atoms, and ϱ_k is the electron density wave operator.

We make use of retarded double time Green functions, defined by (cf. (2.1.18))

$$\langle\langle A(t); B(t')\rangle\rangle^{\pm} = -\frac{i}{\hbar}\,\Theta(t-t')\,\langle[A(t), B(t')]_{\pm}\rangle, \tag{4.4.5}$$

whose Fourier transforms obeys the equation of motion (cf. (3.1.23))

$$\hbar z \langle\langle A; B\rangle\rangle_z^{\pm} = \langle[A, B]_{\pm}\rangle + \langle\langle[A, H]_-; B\rangle\rangle_z^{\pm}, \tag{4.4.6}$$

where z is the complex frequency ($\operatorname{Im} z > 0$). Here A and B are any two operators, and $[A, B]_+$ and $[A, B]_-$ respectively denote anticommutators and commutators.

---

[1] Any two periodic structures with periodicities a and b, respectively, are "incommensurate" if there are no integers M and N such that $Ma = Nb$ holds.

The temperature dependence of the soft mode frequency can be obtained from an inspection of the poles of the phonon Green function

$$G(k, z) = \langle\langle A_k; A_{-k}\rangle\rangle_z^-, \tag{4.4.7}$$

for which holds

$$G(k, z) = G^0(k, z) + D^0(k, z)\,(g^2/N) \sum_\varkappa K(k\varkappa, z)\, G^0(k, z). \tag{4.4.8}$$

Here G^0 is the unperturbed ($g = 0$) phonon Green function (cf. (4.1.54), (4.1.24))

$$G^0(k, z) = \frac{2\omega_k/\hbar}{z^2 - \omega_k^2} \tag{4.4.9}$$

and $K(k\varkappa, z)$ is the electron-hole Green function

$$K(k\varkappa, z) = \langle\langle c_\varkappa^+ c_{\varkappa+k}; c_{\varkappa+k}^+ c_\varkappa\rangle\rangle_z^-. \tag{4.4.10}$$

On deriving equation (4.4.8) we have used (4.4.6) and the fact that the phonon operators obey the commutation relations (1.3.64) and the electron operators analogous anticommutation relations.

Here we approximate $K(k\varkappa, z)$ as follows:

$$K = K^0 + K^0G^0K^0 + K^0G^0K^0G^0K^0 + \cdots, \tag{4.4.11}$$

where K^0 is the unperturbed electron-hole Green function (4.4.10), which (by Wick's theorem) can be shown to read

$$K^0(k\varkappa, z) = \frac{f_0(E_\varkappa) - f_0(E_{\varkappa+k})}{E_\varkappa - E_{\varkappa+k} + \hbar z}. \tag{4.4.12}$$

Here $f_0(E_\varkappa) = \langle c_\varkappa^+ c_\varkappa\rangle_0$ ($\langle\ldots\rangle_0$ — thermal average for $g = 0$) is the Fermi distribution function (A2.75).

For the corresponding electron-hole temperature Green function (cf. section 4.2.1.) (4.4.11) represents a partial summation of an infinite series of diagrams describing electron-hole pair decays into a phonon, which then decays back into an electron-hole pair one and more times. Equation (4.4.11) is a mean field approximation (see below).

Inserting (4.4.11) into (4.4.8) we find the Dyson equation

$$G(k, z) = G^0(k, z) + G^0(k, z)\, \Sigma(k, z)\, G(k, z) \tag{4.4.13}$$

with the self-energy

$$\Sigma(k, z) = (g^2/N) \sum_\varkappa K^0(k\varkappa, z). \tag{4.4.14}$$

(4.4.13) can be rewritten as

$$G(k, z) = \frac{2\omega_k/\hbar}{z^2 - \omega_k^2 - 2\omega_k\Sigma(k, z)/\hbar}. \tag{4.4.15}$$

The denominator of equation (4.4.15) vanishes at $z = \Omega_k$, where

$$\Omega_k^2 = \omega_k^2 + 2\omega_k \Sigma(k, \Omega_k)/\hbar . \tag{4.4.16}$$

The roots of this equation determine the renormalized phonon frequencies.

When we compare (4.4.12) and (A2.77) (in the latter equation we account for $|\langle \boldsymbol{k}\nu| \, e^{i\boldsymbol{q}\boldsymbol{r}} \, |\boldsymbol{k} - \boldsymbol{q}\nu'\rangle|^2 = 1 + O(q^2)$ for $\nu = \nu'$), we see that the self-energy (4.4.14) is proportional to the electronic susceptibility calculated in self-consistent field (mean field) approximation, i.e. we recover here the behaviour (soft mode at $2k_F$) described in section 1.6.3.

Since Ω_k from (4.4.16) is generally much smaller than any structure in the electron-hole spectrum, we may put $\Omega_k = 0$ in $\Sigma(2k_F, \Omega_{2k_F})$ in (4.4.16). In calculating $\Sigma(2k_F, 0)$, $E_\varkappa$ (for $|\varkappa|/k_F \ll 1$) can be expanded up to a term linear in $\varkappa$:

$$E_\varkappa = \hbar(|\varkappa| - k_F)\, v_F , \tag{4.4.17}$$

where $v_F = k_F/m$ (m — mass of a free electron) is the Fermi velocity. Using 4.4.12), (4.4.14), and (4.4.17) we find

$$\Sigma(2k_F, 0) = -\frac{g^2 a}{\pi\hbar v_F} \int\limits_0^{E_B/2kT} dx \, \frac{\tan x}{x} . \tag{4.4.18}$$

Here $-E_B < E < E_B$ is the range over which the approximation (4.4.17) holds, and a is the lattice constant. Neglecting the contribution to $\Sigma(2k_F, 0)$ from outside this range, the evaluation of the integral (4.4.18) yields

$$\Sigma(2k_F, 0) = \frac{g^2 a}{\pi\hbar v_F} \ln{(kT/1.14 E_B)} . \tag{4.4.19}$$

Accordingly, $\Sigma(2k_F, 0)$ diverges logarithmically with temperature.

By (4.4.16) and (4.4.19) we find phonon softening of the form (Rice and Strässler, 1973; cf. fig. 4.25)

$$\Omega_{2k_F}^2 = \lambda\omega_{2k_F}^2 \ln{(T/T_c)}, \tag{4.4.20}$$

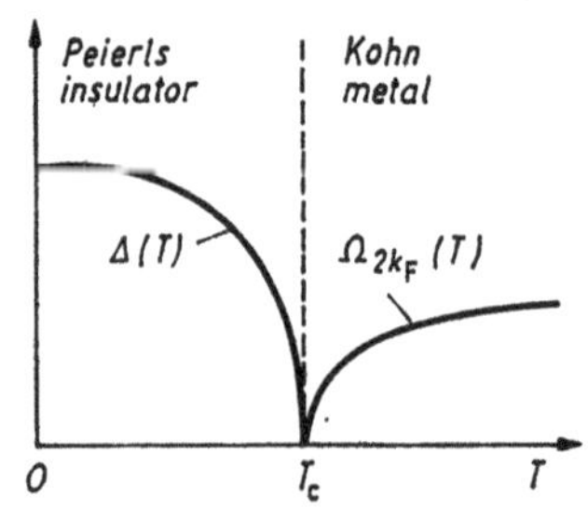

Fig. 4.25. Schematic representation of the temperature dependence of the soft mode frequency Ω_{2k_F} and the Peierls gap Δ (after Rice and Strässler, 1973)

where

$$kT_c = 1.14E_B\,e^{-1/\lambda}, \tag{4.4.21}$$

$$\lambda = g^2(2a/\pi v_F)/\hbar\omega_{2k_F} = g^2N(0)/\hbar\omega_{2k_F}. \tag{4.4.22}$$

Here $N(0)$ is the density of states at the Fermi energy, defined per atom and per both spin states.

The vanishing of Ω_{2k_F} at $T \to T_c^+$ indicates a lattice instability (Peierls instability), which results in a continuous phase transition to a distorted lattice at $T < T_c$. Below the critical temperature T_c the $2k_F$-lattice distortion can be thought of as phonon condensation into a coherent state, such that $\langle b_{2k_F}\rangle = \langle b_{-2k_F}\rangle \neq 0$.

To describe the behaviour of the system at $T < T_c$ we introduce the matrix Green functions (LEE et al., 1974, cf. BERLINSKY, 1979)

$$K_{mn}(\varkappa, z) = \langle\langle c_{\varkappa+nQ/2}; c^+_{\varkappa+mQ/2}\rangle\rangle_z^+, \tag{4.4.23}$$

$$G_{mn}(k, z) = \langle\langle A_{k+nQ}; A_{-k-mQ}\rangle\rangle_z^-, \tag{4.4.24}$$

where $m, n = \pm$ and $Q = 2k_F$. For electrons, the momentum is measured from $\pm Q/2$ and for phonons from $\pm Q$.

First, let us consider the electron problem. Using (4.4.6), we obtain the equations of motion for $K_{mn}(\varkappa, z)$. These equations involve quantities such as

$$\langle\langle A_{-k}c_{\varkappa+k+nQ/2}; c^+_{\varkappa+mQ/2}\rangle\rangle_z^+, \tag{4.4.25}$$

which we decouple by putting

$$\langle A_k\rangle = \langle A_{-k}\rangle = \delta_{kQ}\Delta\sqrt{N}/g. \tag{4.4.26}$$

We find then

$$\begin{aligned}\hbar z K_{++}(\varkappa, z) &= 1 + E_{\varkappa+Q/2}K_{++}(\varkappa, z) + \Delta K_{+-}(\varkappa, z)\\ &\quad + \Delta\langle\langle c_{\varkappa+3Q/2}; c^+_{\varkappa+Q/2}\rangle\rangle_z^+,\\ \hbar z K_{+-}(\varkappa, z) &= E_{\varkappa-Q/2}K_{+-}(\varkappa, z) + \Delta K_{++}(\varkappa, z)\\ &\quad + \Delta\langle\langle c_{\varkappa-3Q/2}; c^+_{\varkappa-Q/2}\rangle\rangle_z^+\end{aligned} \tag{4.4.27}$$

and similar equations for $K_{-,n}(\varkappa, z)$.

Let us assume a free electron or jellium (positively charged elastic medium) model. For such a model states with wave numbers $\varkappa \pm Q/2$ and $\varkappa \pm 3Q/2$ ($|\varkappa| \ll k_F$) have very different energies. Hence, in determining the poles of $K_{mn}(\varkappa, z)$, we may ignore the last terms in (4.4.27), because these terms never become resonant. The solution of equation (4.4.27) then reads

$$\boldsymbol{K}(\varkappa, z) = D^{-1}(\varkappa, z)\begin{bmatrix}\hbar z - E_{\varkappa-Q/2} & \Delta\\ \Delta & \hbar z - E_{\varkappa+Q/2}\end{bmatrix}, \tag{4.4.28}$$

where

$$D(\varkappa, z) = (\hbar z - E_{\varkappa+Q/2})(\hbar z - E_{\varkappa-Q/2}) - \Delta^2. \tag{4.4.29}$$

By (4.4.17), we find from $D(\varkappa, z) = 0$ the following electron energies in the distorted lattice (cf. fig. 24):

$$E_\pm(\varkappa) = \pm[(\hbar v_F \varkappa)^2 + \Delta^2]^{1/2} \equiv \pm E(\varkappa). \tag{4.4.30}$$

Hence, the energy gap introduced at the Fermi energy is 2Δ and is proportional to the periodic lattice distortion (cf. (4.4.26)). The temperature dependence of Δ can be obtained by calculating the free energy change due to the Peierls distortion and minimizing it with respect to Δ (Rice and Strässler, 1973; cf. fig. 4.25). For zero temperature one finds

$$\Delta(0) = 2E_B\, \mathrm{e}^{-1/\lambda} = 1.76kT_c \tag{4.4.31}$$

for $\Delta(0) \ll E_B$.

Now let us turn to the phonon problem in the distorted phase. For this phase the phonon Green function $G_{mn}(k, z)$ can be calculated in the same way as for the undistorted phase. We find the following Dyson equation for $G_{m,+}(k, z)$:

$$\left.\begin{aligned} G_{++} &= G^0_{++}[1 + \Sigma_{+-,+-}G_{++} + \Sigma_{+-,-+}G_{-+}], \\ G_{-+} &= G^0_{--}[\Sigma_{-+,-+}G_{-+} + \Sigma_{-+,+-}G_{++}], \end{aligned}\right\} \tag{4.4.32}$$

and similar expressions for $G_{m,-}$. Here we have

$$G_{mn} \equiv G_{mn}(k, z), \tag{4.4.33}$$

$$G^0_{mn} \equiv G^0_{mn}(k, z) = \delta_{mn} G^0(k + mQ, z), \tag{4.4.34}$$

$$\Sigma_{mm',nn'} \equiv \Sigma_{mm',nn'}(k, z) = (g^2/N) \sum_\varkappa K^0_{mm',nn'}(k\varkappa, z), \tag{4.4.35}$$

where $K^0_{mm',nn'}(k\varkappa, z)$ denotes

$$K_{mm',nn'}(k\varkappa, z) = \langle\langle c^+_{\varkappa+n'Q/2} c_{\varkappa+k+nQ/2}; c^+_{\varkappa+k+mQ/2} c_{\varkappa+m'Q/2} \rangle\rangle_z^- \tag{4.4.36}$$

for absent interaction with phonons, but for present Peierls distortion. By Wick's theorem and a relation similar to (2.1.22) $K^0_{mm',nn'}$ can be expressed by the Green functions (4.4.28) as

$$K^0_{mm',nn'}(k\varkappa, z) = \int \frac{\mathrm{d}\omega}{2\pi} \int \frac{\mathrm{d}\omega'}{2\pi} \frac{A_{n'm'}(\varkappa, \omega)\, A_{mn}(\varkappa + k, \omega')}{\hbar(z + \omega - \omega')} \times [f_0(\hbar\omega) - f_0(\hbar\omega')], \tag{4.4.37}$$

where

$$A_{mn}(\varkappa, \omega) = -2\hbar \operatorname{Im} K_{mn}(\varkappa, \omega + i\varepsilon), \quad \varepsilon \to +0. \tag{4.4.38}$$

By (4.4.28) we obtain for the latter quantities

$$A(\varkappa, \omega) = \frac{2\pi}{2E(\varkappa)} \left\{ \begin{bmatrix} E(\varkappa) + \hbar v_F \varkappa & \Delta \\ \Delta & E(\varkappa) - \hbar v_F \varkappa \end{bmatrix} \delta\big(\omega - E(\varkappa)/\hbar\big) + \begin{bmatrix} E(\varkappa) - \hbar v_F \varkappa & -\Delta \\ -\Delta & E(\varkappa) + \hbar v_F \varkappa \end{bmatrix} \delta\big(\omega + E(\varkappa)/\hbar\big) \right\}. \quad (4.4.39)$$

To proceed we assume the electron bands to be symmetric about the Fermi energy. Then it holds

$$\Sigma_{+-,+-} = \Sigma_{-+,-+}, \quad \Sigma_{+-,-+} = \Sigma_{-+,+-}. \quad (4.4.40)$$

Furthermore, we neglect the dispersion of the unperturbed phonon frequency, i.e. we put $\omega_{Q\pm k} \approx \omega_Q$ for small k. Defining

$$G^{\pm}(k, z) = G_{++}(k, z) \pm G_{-+}(k, z), \quad (4.4.41)$$

then we find from (4.4.32), (4.4.40), (4.4.34) and (4.4.9)

$$G^{\pm}(k, z) = \frac{2\omega_Q/\hbar}{z^2 - \omega_Q^2 - 2\omega_Q \Sigma^{\pm}(k, z)/\hbar} \quad (4.4.42)$$

with the self-energies

$$\Sigma^{\pm}(k, z) = \Sigma_{+-,+-}(k, z) \pm \Sigma_{+-,-+}(k, z). \quad (4.4.43)$$

An explicit calculation of $\Sigma^{\pm}$, for $T = z = 0$ to first order in k^2, yields for the dispersion relations of the "$\pm$ modes", resulting from the zeros of the denominator of equation (4.4.42) (Lee et al., 1974, Berlinsky, 1979):

$$\Omega_+^2(k) = \lambda\omega_Q \left[1 + \frac{1}{12} (\hbar v_F k/\Delta)^2\right], \quad (4.4.44)$$

$$\Omega_-^2(k) = (\lambda\omega_Q^2/4)(\hbar v_F k/\Delta)^2. \quad (4.4.45)$$

We see that the frequency of the "$-$ mode" goes to zero for $k = 0$, while the frequency of the "$+$ mode" remains finite for $k = 0$.

Physically, these modes represent collective excitations of the CDW. In particular, the "$-$ mode" describes phase motions of the CDW and the "$+$ mode" amplitude motions. Therefore, they are called phase mode or phason and amplitude mode or amplitudon, respectively. To clarify the physical meaning of the "$\pm$ modes" we introduce for the $\pm 2k_F$ phonons new operators (Schuster, 1975)

$$(b_Q + b_{-Q})/\sqrt{2} = R_Q + R_Q^0/\sqrt{2}, \quad (4.4.46)$$

$$(b_Q - b_{-Q})/i\sqrt{2} = \Phi_Q + \Phi_Q^0/\sqrt{2}. \quad (4.4.47)$$

Here R_Q^0 and Φ_Q^0 are real constants describing the mean field part of the amplitude and the phase of the CDW. R_Q and Φ_Q are Bose operators

$$[R_Q, R_{Q'}^+] = \delta_{QQ'}, \qquad [\Phi_Q, \Phi_{Q'}^+] = \delta_{QQ'}. \quad (4.4.48)$$

For a jellium model or an incommensurate structure we may put, without loss of generality, $\Phi_Q^0 = 0$. Then the lattice displacement δR with wave number Q at point R can be written as

$$\left.\begin{aligned}\delta R &= (\hbar/2NM\omega_Q)^{1/2}\,(b_Q + b_{-Q}^+)\,e^{iQR} + \text{h.c.}\\ &= (\hbar/2NM\omega_Q)^{1/2}\left[R_Q^0 + (R_Q + R_Q^+)/\sqrt{2}\right.\\ &\quad \left.+\, i(\Phi_Q + \Phi_Q^+)/\sqrt{2}\right] e^{iQR} + \text{h.c.},\end{aligned}\right\} \tag{4.4.49}$$

where M is the atomic mass. For $R_Q^0 \neq 0$ we have to first order in the quantities R_Q and Φ_Q

$$\begin{aligned}\delta R &= (\hbar/2NM\omega_Q)^{1/2}\left[R_Q^0 + (R_Q + R_Q^+)/\sqrt{2}\right]\\ &\quad \times \exp\left[iQR + i(\Phi_Q + \Phi_Q^+)/\sqrt{2}\,R_Q^0\right].\end{aligned} \tag{4.4.50}$$

Consequently, R_Q and Φ_Q respectively describe amplitude and phase motions. From (4.4.7), (4.4.4), (4.4.40), and (4.4.46) G^+ and G^- can be readily written in terms of the amplitude and phase operators, respectively. Hence, it is true that the "+ mode" describes amplitude and the "− mode" phase motions. Note that the amplitude and phase modes are proper eigenmodes of the system, which result from the removal of the degeneracy of the $\pm 2k_F$ phonons due to the Peierls distortion. Above T_c', where $R_Q^0 = 0$ and $\Phi_Q^0 = 0$, (4.4.46) and (4.4.47) merely transform the degenerate $\pm 2k_F$ phonons into new degenerate bosons.

The phase determines the position of the CDW in the laboratory frame. In the jellium model, considered before, the phase of the $2k_F$ distortion is arbitrary due to the translational invariance of the model. Therefore, the phase mode has zero frequency, and the CDW and the periodic lattice distortion can move freely together along the chain. At low temperatures this should render possible a supercurrent (cf. Patton and Sham, 1979, Kaveh, 1981).

In an actual system, however, discreteness of the lattice (commensurability), impurities or interchain interaction can pin the CDW relative to the lattice, so that the phase of the CDW is no longer arbitrary and the infinite dc conductivity becomes finite.

In the following we consider the effect of commensurability (of the CDW with the lattice) on the phase mode. — Let M be the smallest integer, for which

$$E_{\varkappa+MQ} = E_\varkappa \tag{4.4.51}$$

for all $\varkappa$. An account of the lattice requires M equations, like those of equation (4.4.27), for the Green functions $K_{mn}(\varkappa, z)$, where $1 \leq m,\ n \leq M$. Δ must now be taken to be complex, $\Delta = |\Delta|\,e^{i\Phi}$, where the phase Φ determines the relative position of the CDW in the lattice. To obtain the electron energies we put the determinant of the coefficient matrix $K_{mn}(\varkappa, z)$ equal to zero. Lee et al. (1974) pointed out that the only Φ-dependence in the resulting secular equation is of the form (cf. section 4.4.3.)

$$2\,|\Delta|^M\,(\cos M\Phi - 1), \tag{4.4.52}$$

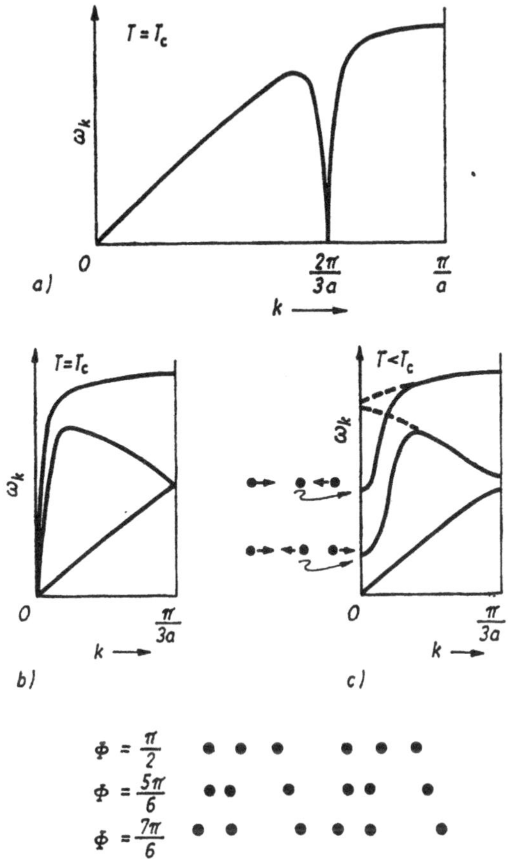

Fig. 4.26. Schematic plot of the phonon dispersion curves of a (commensurate) one-dimensional metal with a 1/3-filled band ($M = 3$): a) at the Peierls transition temperature, T_c, in the Brillouin zone of the undistorted system, b) the same as a), however, in the Brillouin zone of the distorted system, c) below the Peierls transition. As indicated by the displacement pattern, the lowest optical mode is the phase mode, the second lowest mode is the amplitude mode; the dashed line represents the usual dispersion of a linear chain with three atoms in the unit cell (a — lattice constant). In d) the lattice configurations for some values of the phase are shown; $\Phi = \frac{\pi}{2}$ and $\Phi = \frac{7\pi}{6}$ (which differ only by a primitive translation of the lattice) correspond to the minimum of the energy, $\Phi = \frac{5\pi}{6}$ to the maximum of the energy (after Schulz, 1978)

for $M > 2$. For small Φ the expression (4.4.52) gives a correction to the total energy per electron of order $(|\Delta|^2/E_B)\,(\mathrm{e}\,|\Delta|/W)^{M-2}\,(M\Phi)^2/2$, where W is the electronic band width and $\mathrm{e} = 2.718\ldots$ Using this value as spring energy, the phase mode obtains the finite pinning frequency

$$\omega_\mathrm{T} \approx \lambda^{1/2}\omega_Q M(\mathrm{e}\,|\Delta|/W)^{M/2-1}, \tag{4.4.53}$$

which, by (4.4.31), turns out to be proportional to $\mathrm{e}^{-M/2\lambda}$. Consequently, pinning by commensurability becomes negligible for large M.

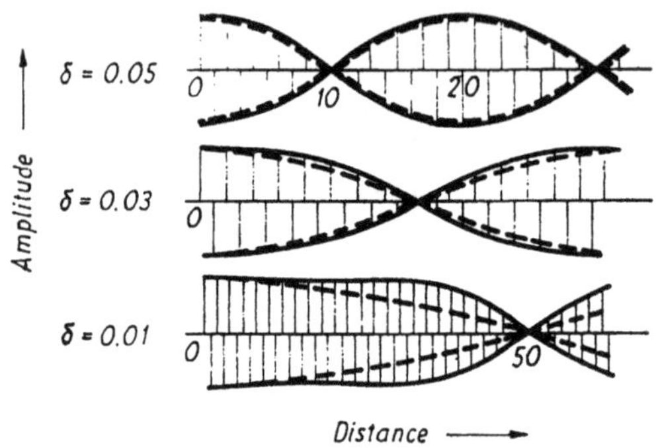

Fig. 4.27. Effects of harmonics on the displacement pattern of an incommensurate one-dimensional metal in the Peierls phase for various values of $\delta = \dfrac{\pi}{a} - 2k_\mathrm{F}$ ($\hbar k_\mathrm{F}$ — Fermi momentum, a — lattice constant). The amplitude of the lth atom at R_l, $\delta R_l \propto \cos 2k_\mathrm{F} R_l$, is normalized to that of the 0th atom. The thin solid and the dashed curve, respectively, represent the envelope function obtained with and without the harmonics. Owing to the harmonics, the envelope function shows a tendency to a change from a sinusoidal to a rectangular curve (i.e. to an antiphase domain-like structure) with an infinite periode for $\delta \to 0$ (commensurate phase) (after Kotani, 1977)

An explicit calculation of the pinning potential for $M = 3$ can be found in the paper by Buzdin and Bulajewskij (1977). For such a system the phonon dispersion curves at $T = T_\mathrm{c}$ and $T < T_\mathrm{c}$ are shown in figure 4.26. Note that a term (4.4.52) can also give rise to phase solitary waves (Horovitz and Krumhansl, 1978). For effects of interchain interaction and of impurities on the phase mode the reader is referred, for example, to Bak and Brazovsky (1978).

Concluding this section we show that a lattice distortion with wave number Q should, in general, coexist with its harmonics, i.e. with distortions of wave numbers nQ (n integers). To achieve this we consider the relations

$$\langle A_{-k}\rangle = -\frac{2g}{\hbar\omega_k\sqrt{N}}\langle\varrho_k\rangle \tag{4.4.54}$$

and

$$(\hbar z - E_\varkappa) \langle\langle c_\varkappa; c_{\varkappa'}^+ \rangle\rangle_z^+$$

$$= \delta_{\varkappa\varkappa'} + \frac{g}{\sqrt{N}} \sum_{k>0} \{\langle A_k \rangle \langle\langle c_{\varkappa-k}; c_{\varkappa'}^+ \rangle\rangle_z^+ + \langle A_{-k} \rangle \langle\langle c_{\varkappa+k}; c_{\varkappa'}^+ \rangle\rangle_z^+\}, \quad (4.4.55)$$

which we have obtained by the equation of motion of a Heisenberg operator and by (4.4.6), respectively. From (4.4.54) it follows that, if $\{A_{\pm Q}\} \neq 0$, then $\langle c_{\varkappa\mp Q}^+ c_\varkappa \rangle \neq 0$. But, according to (4.4.55), the existence of the "anomalous" averages $\langle c_{\varkappa\pm Q}^+ c_\varkappa \rangle$ requires $\langle c_{\varkappa\pm nQ}^+ c_\varkappa \rangle \neq 0$, too. Hence, by (4.4.54) we obtain also $\langle A_{\pm nQ} \rangle \neq 0$.

Consequently, a system of pairs of equations (4.4.54) and (4.4.55) must be solved, which is finite (M pairs) for a commensurate system, but infinite for an incommensurate one. A system of the latter type is defined by the property that no integers M and N exist, such that

$$Ma = N\lambda, \quad (4.4.56)$$

where a is the lattice constant and λ is defined by (4.4.1). The effect of harmonics on the displacement pattern of an incommensurate system is shown in figure 4.27. For more details on the problem of harmonics see LUKIN (1977) and KOTANI (1977). For a recent paper on phonons in CDW state see NAKANISHI et al. (1982).

4.4.2. Phenomenological theory of order parameter fluctuations in a Peierls system

As pointed out in the preceding section, mean field theory predicts for a one-dimensional metal a structural phase transition (Peierls transition) at a finite temperature. It is well known, however, that in one-dimensional systems with short-range forces fluctuations prevent phase transitions (i.e., a long-range order) at non-zero temperature (LANDAU and LIFSHITZ, 1959). Following SCALPINO et al. (1972) (see also DIETERICH, 1976), here we account for fluctuations by making use of the functional generalization of the Ginzburg-Landau theory of phase transitions. Starting point of this approach is the free energy $F[\Psi(x)]$ as a functional of the space-dependent order parameter $\Psi(x)$. Fluctuations are then treated by using $F[\Psi(x)]$ as energy functional for a canonical average over all possible $\Psi(x)$.

In a one-dimensional Peierls system the average displacement of the atom at x_l may be written as (cf. (1.3.60))

$$\langle u(x_l) \rangle = \frac{1}{2} [\Psi(x_l) e^{i2k_F x_l} + \text{c.c.}], \quad (4.4.57)$$

where

$$\Psi(x) = \sum_k \left(\frac{\hbar}{2MN\omega_{2k_F+k}} \right)^{1/2} e^{ikx} \langle b_{2k_F+k} + b_{-(2k_F+k)}^+ \rangle \quad (4.4.58)$$

is the appropriate order parameter for the Peierls transition. If we ignore fluctuations, then (cf. (4.4.26)) $\Psi\Big/\left(\frac{\hbar}{2Mg^2\omega_{2k_F}}\right)^{1/2}$ agrees with the mean field value Δ. For a half-filled band $\Psi(x)$ is real[1], since the bare phonons $2k_F + k$ and $-2k_F + k$ ($k_F = \pi/a$) are equivalent. In all other cases $\Psi(x)$ is complex. The charge density of the CDW, $\varrho(x)$, can be expressed by the order parameter as

$$\varrho = \varrho_0 + \alpha \operatorname{Re}(\Psi e^{i2k_Fx}), \tag{4.4.59}$$

where ϱ_0 is the average background density and α is a proportionality factor.

For the free energy we make the following Landau ansatz:

$$F[\Psi(x)] = \int \frac{dx}{L}\left[a\,|\Psi(x)|^2 + b\,|\Psi(x)|^4 + c\left|\frac{d\Psi(x)}{dx}\right|^2\right], \tag{4.4.60}$$

where L is the lattice spacing, $a = a'[(T/T_c) - 1]$ (T_c — mean field Peierls transition temperature) and $a', b, c > 0$. The latter coefficients can be taken from a mean field calculation of the free energy by starting from the Hamiltonian (4.4.2) (see ALLENDER et al., 1974). Note that the free energy functional (4.4.60) is appropriate for systems lacking pinning mechanisms. In the presence of such mechanisms a corresponding pinning potential must be added.

Before we deal with static aspects of order-parameter fluctuations, let us briefly consider some aspects of the dynamics of such fluctuations. To this end the free energy functional (4.4.60) is taken as an effective anharmonic potential, which is completed to a (classic) Hamiltonian by adding the following kinetic energy term:

$$T = \frac{M}{2}\int \frac{dx}{L}\left|\frac{\partial\Psi(x,t)}{\partial t}\right|^2, \tag{4.4.61}$$

where M is some mass. It is worth noting that the resulting Hamiltonian corresponds to the continuum limit $\left(L\sum_l \to \int dx\right)$ of the discrete model:

$$H = \sum_l \left[\frac{M}{2}\,|\dot{\Psi}_l|^2 + a\,|\Psi_l|^2 + b\,|\Psi_l|^4 + c\left|\frac{\Psi_l - \Psi_{l+1}}{L}\right|^2\right], \tag{4.4.62}$$

where $\Psi_l \equiv \Psi(x_l)$. For real Ψ and $a < 0$ the model (4.4.62) agrees with the coupled double-well potential model (4.2.58) (for $d = 1$). For complex Ψ, it describes a one-dimensional chain of particles sitting in two-dimensional anharmonic potentials.

Let us consider small amplitude oscillations of the order parameter about its equilibrium value $\bar{\Psi}$, which is determined from $\delta F/\delta\Psi^*(x) = 0$. $\bar{\Psi}$ is

---

[1] Owing to interaction between electrons, either by virtual phonon exchange or by Coulomb interaction, the order parameter of a Peierls condensate for a half-filled band becomes also complex (HOROVITZ, 1980).

given by

$$\overline{\Psi} = \begin{cases} 0, \\ (|a|/2b)^{1/2}, \end{cases} \quad \text{if} \quad T \gtrless T_c. \tag{4.4.63}$$

For a complex order parameter we write

$$\Psi(x) = \left(\overline{\Psi} + \delta\Psi(x)\right) e^{i\Phi(x)} \tag{4.4.64}$$

and linearize the equation of motion

$$\frac{M}{2} \ddot{\Psi}(x) = -\frac{1}{L} \frac{\delta F}{\delta \Psi^*(x)} = -(a + 2b\,|\Psi(x)|^2)\,\Psi(x) + c\Psi''(x) \tag{4.4.65}$$

with respect to $\delta\Psi$ and the derivatives of Φ. In this approximation we obtain decoupled equations for the amplitude and phase oscillations. We find the following dispersion relations for the amplitude and phase modes:

$$\Omega_A^2(k) = \frac{2}{M} \begin{cases} a + ck^2, \\ 2\,|a| + ck^2, \end{cases} \quad \text{if} \quad T \gtrless T_c, \tag{4.4.66}$$

$$\Omega_P^2(k) = \frac{2}{M}\, ck^2, \quad \text{if} \quad T < T_c. \tag{4.4.67}$$

These relations agree qualitatively with the dispersion relations (4.4.44) and (4.4.45) found for an incommensurate CDW. — For a detailed study of dynamical aspects of the model (4.4.62) by the molecular dynamics technique the reader is referred to Koehler and Lee (1977).

Now we turn to the correlation function of the order-parameter

$$\langle \Psi(x)\,\Psi^*(0)\rangle = \frac{\int \delta\Psi\, e^{-\beta F[\Psi]}\, \Psi(x)\, \Psi^*(0)}{\int \delta\Psi\, e^{-\beta F[\Psi]}}, \tag{4.4.68}$$

which for large x has the form

$$\langle \Psi(x)\,\Psi^*(0)\rangle \propto \exp\left(-|x|/\xi(T)\right). \tag{4.4.69}$$

Here $\xi(T)$ is the correlation length characterizing the spatial extent of the fluctuations at a given temperature T.

By the transfer-operator technique, the functional integrations occurring in (4.4.68) can be replaced by an eigenvalue problem as follows (Scalapino et al., 1972). Let us consider a chain of length Λ with N atoms. Dividing Λ into N segments of length L and using cyclic boundary conditions, then the partition function $Z = \int \delta\Psi\, e^{-\beta F[\Psi]}$ can be written as

$$Z = \int \prod_{l=1}^{N} \mathrm{d}\Psi_l\, e^{-\beta f(\Psi_{l+1}, \Psi_l)} \tag{4.4.70}$$

with

$$f(\Psi_{l+1}, \Psi_l) = a\,|\Psi_{l+1}|^2 + b\,|\Psi_{l+1}|^4 + c\left|\frac{\Psi_{l+1} - \Psi_l}{L}\right|^2, \tag{4.4.71}$$

where the boundary conditions imply $\Psi_{N+1} = \Psi_1$. For a complex field $\mathrm{d}\Psi_l = \mathrm{d}(\mathrm{Re}\,\Psi_l)\,\mathrm{d}(\mathrm{Im}\,\Psi_l)$.

To proceed we introduce an additional variable Ψ_1' and write

$$Z = \int \mathrm{d}\Psi_1'\,\mathrm{d}\Psi_1\,\mathrm{d}\Psi_2 \dots \mathrm{d}\Psi_N \delta(\Psi_1 - \Psi_1')\,\mathrm{e}^{-\beta f(\Psi_1',\Psi_N)} \times \mathrm{e}^{-\beta f(\Psi_N,\Psi_{N-1})} \dots \mathrm{e}^{-\beta f(\Psi_2,\Psi_1)}. \tag{4.4.72}$$

Expanding the δ-function in a complete set of normalized eigenstates

$$\delta(\Psi_1 - \Psi_1') = \sum_n F_n^*(\Psi_1')\,F_n(\Psi_1), \tag{4.4.73}$$

we obtain

$$Z = \sum_n \int \mathrm{d}\Psi_1' \dots \mathrm{d}\Psi_n F_n^*(\Psi_1')\,\mathrm{e}^{-\beta f(\Psi_1',\Psi_N)} \dots \mathrm{e}^{-\beta f(\Psi_2,\Psi_1)} F_n(\Psi_1). \tag{4.4.74}$$

This expression may be directly evaluated if the F_n's are eigenfunctions of the transfer operator

$$\int \mathrm{d}\Psi_l\,\mathrm{e}^{-\beta f(\Psi_{l+1},\Psi_l)}\,F_n(\Psi_l) = \mathrm{e}^{-\beta\epsilon_n} F_n(\Psi_{l+1}). \tag{4.4.75}$$

We find then

$$Z = \sum_n \mathrm{e}^{-(\Lambda/L)\beta\epsilon_n}. \tag{4.4.76}$$

In the thermodynamic limit, where $\Lambda \to \infty$, Z is dominated by the lowest eigenvalue ϵ_0.

Similarly, the correlation function (4.4.68) can be expressed in terms of the eigenfunctions F_n as

$$\langle \Psi(x)\,\Psi^*(0)\rangle = \sum_n |\langle F_n|\,\Psi\,|F_0\rangle|^2\,\mathrm{e}^{-\beta\frac{|x|}{L}(\epsilon_n - \epsilon_0)}. \tag{4.4.77}$$

For large distance, i.e. $|x| \gg L$, from (4.4.69) and (4.4.77) we find for the correlation length

$$\xi = L\,\frac{kT}{\epsilon_1 - \epsilon_0}, \tag{4.4.78}$$

where ϵ_1 is the lowest excited state coupled to F_0 by $\langle F_1|\,\Psi\,|F_0\rangle$.

The transfer-operator eigenvalue equation (4.4.75) can be reduced to a Schrödinger-like equation as follows. On the left-hand side of (4.4.75) $F_n(\Psi_l)$ is expanded as

$$F_n(\Psi_l) = F_n(\Psi_{l+1}) + (\Psi_l - \Psi_{l+1})\,F_n'(\Psi_{l+1}) + \frac{1}{2}(\Psi_l - \Psi_{l+1})^2\,F_n''(\Psi_{l+1}) + \cdots, \tag{4.4.79}$$

and the integration over Ψ_l is then carried out. If only the non-zero lowest order terms are accounted for, we find

$$\int \mathrm{d}\Psi_l \, \mathrm{e}^{-\beta f(\Psi_{l+1},\Psi_l)} \, F_n(\Psi_l)$$
$$= \mathrm{e}^{-\beta(a|\Psi_{l+1}|^2 + b|\Psi_{l+1}|^4)} \sqrt{\frac{\pi L^2}{\beta c}} \left(1 + \frac{1}{4} \frac{L^2}{\beta c} \frac{\partial^2}{\partial \Psi_{l+1}^2}\right) F_n(\Psi_{l+1}). \tag{4.4.80}$$

When we exponentiate the derivative term and combine it with the exponential term, then the transfer operator eigenvalue equation (4.4.75) becomes

$$\mathrm{e}^{-\beta H} F_n = \mathrm{e}^{-\beta \epsilon_n} F_n \tag{4.4.81}$$

with the effective one-particle Hamiltonian

$$H = -\frac{1}{4} \frac{L^2}{\beta^2 c} \frac{\partial^2}{\partial \Psi^2} + a \, |\Psi|^2 + b \, |\Psi|^4 - \frac{1}{\beta} \ln \sqrt{\frac{\pi L^2}{\beta c}} \tag{4.4.82}$$

where for a complex order parameter $\partial^2/\partial\Psi^2$ means

$$\frac{\partial^2}{\partial \Psi^2} \equiv \frac{\partial^2}{\partial (\mathrm{Re}\,\Psi)^2} + \frac{\partial^2}{\partial (\mathrm{Im}\,\Psi)^2} = \frac{\partial^2}{\partial \, |\Psi|^2} + \frac{1}{|\Psi|} \frac{\partial}{\partial \, |\Psi|} + \frac{1}{|\Psi|^2} \frac{\partial^2}{\partial \Phi^2} \tag{4.4.83}$$

with $\Psi = |\Psi| \, \mathrm{e}^{i\Phi}$.

For real Ψ, the anharmonic potential $a\Psi^2 + b\Psi^4$ is shown in figure 4.28 for two different temperatures. For $T > T_c$ the potential has a single minimum at $\Psi = 0$; but for $T < T_c$ it has two minima at $\Psi_0 = \pm\sqrt{|a|/2b}$. The effective mass involved in the Hamiltonian (4.4.82) is proportional to T^{-2}. Therefore, at low temperatures, where the effective mass is large, the eigenvalues begin near the bottoms of the potential wells and a small tunnelling splitting between the two lowest states of H occurs (fig. 4.28). One finds for a real order parameter (half-filled band; the Peierls distortion doubles the lattice spacing) that $\xi(T)$ increases exponentially with decreasing temperature; and for a com-

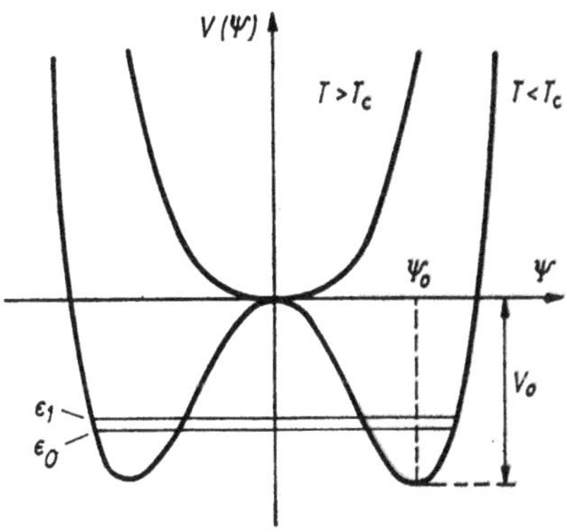

Fig. 4.28. Anharmonic potential $V(\Psi) = a\Psi^2 + b\Psi^4$ with $a \gtrless 0$ for $T \gtrless T_c$ and tunnelling splitting of the two lowest levels. $\Psi_0 = [|a|/2b]^{1/2}$ and $V_0 = a^2/4b$

plex order parameter that $\xi(T)$ diverges as T^{-1}, when $T \to 0$ (Lee et al., 1974, Dieterich, 1976). That is, in both these cases long-range order occurs only at $T = 0K$.

For real Ψ, $\xi(T)$ increases strongly at $T_c/4$. Different mechanisms have been proposed for obtaining a similar behaviour also for incommensurate systems (complex Ψ) (cf. Blunck, 1978, and references therein). Such behaviour could well explain the three-dimensional ordering observed in quasi one-dimensional materials, which arises from interchain interaction (Dieterich, 1976, and references therein). — The validity of replacing (4.4.80) by a "Schrödinger" equation with Hamiltonian (4.4.82) was studied by Schneider and Stoll, 1980).

4.4.3. Statics and dynamics of incommensurate systems near the "lock-in" transition

In recent years one has seen a growing interest in incommensurate crystals, i.e. in crystals with a periodic distortion associated with a wave vector, which cannot be expressed as a simple rational fraction of a reciprocal lattice vector. Beside one-dimensional Peierls systems, studied in sections 4.4.1. and 4.4.2., there are many other systems exhibiting incommensurate phases, such as two-dimensional charge density wave systems, spin density wave systems, helical magnetic structures, some ferroelectrics, some superionic conductors and some thin films being incommensurate with the substrate.

Figure 4.27 shows a behaviour, which is, in general, typical for an incommensurate structure: If there is a large discrepancy between the two periodicities present in the incommensurate structure (large δ), then the envelope of the displacement field is dominated by a single Fourier component ("plane-wave" limit). If, however, the two periodicities are nearly commensurate (small δ), then higher harmonics of the basic distortion must be taken into account, which leads to a domain-like displacement pattern ("domain" or "soliton" limit). In the latter situation the incommensurate structure consists of wide "locked-in" commensurate regions separated by narrow incommensurate regions, so-called discommensurations, which appear to be defects.

As will be shown below, the discommensurations are solutions of the sine-Gordon equation, i.e. they are "solitons". At the incommensurate-commensurate phase transition ("lock-in" transition) the soliton density goes smoothly to zero, indicating a second order transition (the nature of this transition is still a matter of controversy, cf. Bruce et al., 1978, Aubry, 1980, Novaco, 1980). The soliton density may serve as the order parameter of this transition.[1]

[1] In the simplest cases, the incommensurate phase is stable at temperatures $T_2 < T < T_1$ between a more symmetrical (disordered) phase ($T > T_1$) and the commensurate (fully transformed) phase ($T < T_2$). Note that the conception of a "lock-in" transition presented here is appropriate to transitions to commensurate phases whose wave vector is not a so-called Lifshitz wave vector, which is defined by the property that the (disordered phase) soft-mode dispersion has inversion symmetry (see, e.g., Bruce and Cowley, 1980).

Following POKROVSKIJ and TALAPOV (1978), THEODOROU and RICE (1978) and HOROVITZ et al. (1978), here we study the behaviour of an incommensurate system near the lock-in transition by the Frenkel-Kontorova model, which consists of a chain of atoms connected by springs situated in a sinusoidal potential and which has the potential energy

$$V = \frac{\gamma}{2} \sum_l (x_{l+1} - x_l - a)^2 + V_0 \sum_l \cos \frac{2\pi}{b} x_l, \tag{4.4.84}$$

where x_l is the position of the lth atom. This model was originally introduced as dislocation model (FRENKEL and KONTOROVA, 1939) and has also been used in the study of CDW systems (SOKOLOFF, 1977), thin films on surfaces (SNYMAN and VAN DER MERWE, 1974), superionic materials (PIETRONERO and STRÄSSLER, 1978, BEYELER et al., 1980), disordered solids (PIETRONERO and STRÄSSLER, 1979) etc. Here, we use it as a model for two interpenetrating incommensurate sublattices with chain structure, where one sublattice (with period a) experiences an external periodic potential due to the other sublattice (with period b).

Let us study the potential energy (4.4.84) for nearly commensurate situations. First we consider the case $a/b = 1 + \delta$ ($\delta \ll 1$). Making the transformation

$$x_l = lb + \frac{b}{2\pi} \Phi_l \tag{4.4.85}$$

and restricting ourselves to the continuum limit $\left(2\pi\delta l \equiv x, \Phi_l \to \Phi(x)\right)$, then the potential (4.4.84) becomes

$$V = \frac{\delta\gamma b^2}{4\pi} \int \mathrm{d}x \left[\left(\frac{\mathrm{d}\Phi}{\mathrm{d}x} - 1\right)^2 + \frac{2V_0}{\gamma b^2\delta^2} (\cos \Phi - 1)\right], \tag{4.4.86}$$

where we have added a constant term.

Now we consider the case $a/b = 1/2 + \delta$ ($\delta \ll 1$). To do so we write the potential (4.4.84) as

$$\begin{aligned} V = {} & \frac{\gamma b^2}{2(2\pi)^2} \sum_l [(\varphi_l - \psi_l + \varepsilon)^2 + (\varphi_l - \psi_{l-1} - \varepsilon)^2] \\ & + V_0 \sum_l (\cos \varphi_l - \cos \psi_l), \end{aligned} \tag{4.4.87}$$

where

$$\left.\begin{aligned} \varphi_l &= \frac{2\pi}{b} x_{2l} - \pi 2l, \\ \psi_l &= \frac{2\pi}{b} x_{2l+1} - \pi(2l + 1), \\ \varepsilon &= 2\pi\delta. \end{aligned}\right\} \tag{4.4.88}$$

Introducing $\alpha_l = \varphi_l - \psi_l$ and $\Phi_l = (\varphi_l + \psi_l)/2$, then in the continuum limit the potential (4.4.87) becomes

$$V = \frac{\gamma b^2}{(2\pi)^3 \delta} \int dx \left[\frac{1}{2} (\alpha + \varepsilon)^2 + \frac{1}{2} \left(\alpha - \varepsilon + 2\varepsilon \frac{d\Phi}{dx}\right)^2 - \frac{4\pi^2 V_0 \alpha}{\gamma b^2} \sin \Phi\right], \tag{4.4.89}$$

where we have retained the leading terms only (note that α is a small quantity). We determine α by minimizing V, which yields

$$\alpha = \frac{2\pi^2 V_0}{\gamma b^2} \sin \Phi - \varepsilon \frac{d\Phi}{dx}. \tag{4.4.90}$$

Inserting this expression into (4.4.89), then we obtain

$$V = \frac{\delta \gamma b^2}{2\pi} \int dx \left[\left(\frac{d\Phi}{dx} - 1\right)^2 + \left(\frac{\pi V_0}{\sqrt{2}\, \gamma b^2 \delta}\right)^2 (\cos 2\Phi - 1)\right]. \tag{4.4.91}$$

Similarly, for the general case $a/b = N/M + \delta$ (N, M — integers) one finds

$$V = C \int dx \left[\left(\frac{d\Phi}{dx} - 1\right)^2 + Y(\cos M\Phi - 1)\right], \tag{4.4.92}$$

where $Y \propto V_0{}^M/\delta^2$ and C is a constant. Φ is the phase (describing the relative displacements of the atoms of the two sublattices) averaged over M consecutive sites. — On comparing the "pinning" potential given by the second term on the right-hand side of equation (4.4.92) with (4.4.52), one should note that in treating a CDW system with the Frenkel-Kontorova model, the sinusoidal potential is identified with the CDW (SOKOLOFF, 1977). Note further that (4.4.92) agrees with MCMILLAN's (1976) "phase-modulation-only" ansatz (assumption of constant amplitude, as the energy associated with amplitude fluctuations is much larger than that of phase fluctuations) for the Landau free energy functional for describing the lock-in transition in a CDW system.

Replacing $M\Phi = \Theta$, then (4.4.92) becomes

$$V = M^{-2} C \int dx \left[\left(\frac{d\Theta}{dx} - M\right)^2 - 2\zeta(\cos \Theta - 1)\right], \tag{4.4.93}$$

where $2\zeta = YM^2$. To find the ground state configuration of the system, we minimize the potential V. As extreme value condition we obtain the (time-independent) sine-Gordon equation (pendulum equation)

$$\frac{d^2\Theta}{dx^2} - \zeta \sin \Theta = 0, \tag{4.4.94}$$

or equivalently

$$\frac{d}{dx} \left[\frac{1}{2} \left(\frac{d\Theta}{dx}\right)^2 + \zeta \cos \Theta\right] = 0, \tag{4.4.95}$$

which has the well-known solution

$$x - x_0 = \int\limits_0^{\Theta} \frac{\mathrm{d}\Theta'}{(A - 2\zeta \cos \Theta')^{1/2}}, \tag{4.4.96}$$

where x_0 and A are integration constants with $A \leqq 2\zeta$ and x_0 arbitrary.

The solution of equation (4.4.96) is schematically shown in figure 4.29. From this figure we see that λ (the period of cos $M\Phi$ as a function of x) increases with decreasing A ($A > 2\zeta$). For $A \to 2\zeta + 0$ we have $\lambda \to \infty$ and $\Theta \to 0$, which corresponds to the commensurate phase.

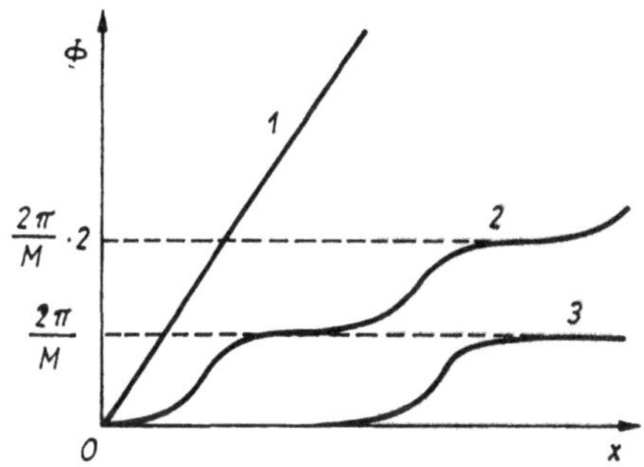

Fig. 4.29. Schematic representation of the dependence of the phase variable on the space coordinate x in an incommensurate system near the "lock-in" transition. The curves 1, 2, 3 correspond to increasing values of the constant ζ (see text) (after Bulajewskij and Khomskij, 1978)

We determine A by minimizing the potential energy of the system. To do this we insert (4.4.96) into (4.4.93) and account for $\mathrm{d}\Theta/\mathrm{d}x = (A - 2\zeta \cos \Theta)^{1/2}$. Thus we find (cf. Bulajewski and Khomskij, 1978)

$$\begin{aligned} \frac{VM^2}{C} &= \int \mathrm{d}x \left[A - 2M \frac{\mathrm{d}\Theta}{\mathrm{d}x} - 4\zeta \cos \Theta\right] \\ &= \frac{\Lambda}{\lambda} \int\limits_0^{\lambda} \mathrm{d}x \left[A - 2M \frac{\mathrm{d}\Theta}{\mathrm{d}x} - 4\zeta \cos \Theta\right] \\ &= \frac{\Lambda}{\lambda} \left[A\lambda - 4\pi M - 4\zeta \int\limits_0^{2\pi} \frac{\mathrm{d}\Theta \cos \Theta}{(A - 2\zeta \cos \Theta)^{1/2}}\right], \end{aligned} \tag{4.4.97}$$

where we have neglected a constant term and where Λ denotes the length of the chain.

By means of the complete elliptic integrals K and E, defined by (4.3.38) and (4.3.39), the period λ and the potential V may be expressed as

$$\lambda = \int_0^{2\pi} \frac{d\Theta}{(A - 2\zeta \cos \Theta)^{1/2}} = \frac{2\varkappa}{\zeta^{1/2}} K(\varkappa), \tag{4.4.98}$$

$$\frac{M^2 V}{\Lambda C} = A - \frac{2\pi M \zeta^{1/2}}{\varkappa K(\varkappa)} + 4\zeta \left[1 - \frac{2}{\varkappa^2} + \frac{2}{\varkappa^2} \frac{E(\varkappa)}{K(\varkappa)}\right], \tag{4.4.99}$$

where $\varkappa = 4\zeta/(A + 2\zeta)$. For $\varkappa \to 1$ $(A \to 2\zeta)$ we have $E(\varkappa) \to 1$ and $K(\varkappa) \to \ln \left(4/\sqrt{1 - \varkappa^2}\right)$. Using this asymptotic behaviour, we find that the potential (4.4.99) has a minimum for $A > 2\zeta$ (domain-like structure), if

$$\zeta < \zeta_c = \pi^2 M^2/16. \tag{4.4.100}$$

Otherwise $(\zeta \geqq \zeta_c)$ the minimum occurs at $A = 2\zeta$ (commensurate structure). The transition between these two phases (driven by a change of V_0 with temperature or of δ with pressure) is continuous and connected with $\lambda \to \infty$.

Consequently, when the two periodicities in the incommensurate structure at hand are nearly commensurate, the chain distorts in such a way that it becomes commensurate with the sinusoidal potential. The range of the parameter (discrepancy in the lattice constants), over which the lock-in transition occurs, increases as the force constant of the chain, γ, decreases and the strength of the sinusoidal potential, V_0, increases. When δ is larger than some critical value, then the ground state is an array of static solitons (domain walls), which are also called "stacking solitons". Note that, in contrast to the domain walls described in section 4.3.3., the stacking solitons are not elementary excitations, but a property of the ground state.

Now let us turn to the dynamical behaviour of the phase (i.e. to the phase mode) near the lock-in transition. To obtain the Hamiltonian needed, we add in the brackets of equation (4.4.93) a term $\frac{1}{c^2} \frac{\partial^2 \Theta}{\partial t^2}$ (c — speed of the phason in absence of commensurability effects). The classic equation of motion then becomes

$$\frac{\partial^2 \Theta}{\partial x^2} - \frac{1}{c^2} \frac{\partial^2 \Theta}{\partial t^2} - \zeta \sin \Theta = 0. \tag{4.4.101}$$

We look for a solution of this equation of the form (cf. (4.3.86))

$$\Theta(x, t) = \Theta_0(x) + \psi(x) e^{i\omega t}, \tag{4.4.102}$$

where $\Theta_0(x)$ is the solution of the static equation (4.4.94) and $\psi(x) e^{i\omega t}$ is assumed to be small as compared to $\Theta_0(x)$. Inserting (4.4.102) into (4.4.101) and retaining the linear terms in ψ only, we obtain for ψ the following Schrödinger-like equation:

$$-\frac{\partial^2 \psi}{\partial x^2} - \zeta \cos \Theta_0(x) = \omega^2 \psi. \tag{4.4.103}$$

This equation can be reduced to Lamé's equation, which is exactly soluble (see, e.g., POKROVSKIJ and TALAPOV, 1978). Note that the potential occurring in (4.4.103), $\zeta \cos \Theta_0(x)$, is periodic (cf. fig. 4.29). Therefore, we may take advantage of Bloch's theorem to write

$$\psi(x) = e^{ikx}u(x), \tag{4.4.104}$$

where $u(x)$ is periodic with period λ, and k is the wave number confined to the first "Brillouin" zone $-\frac{\pi}{\lambda} < k < \frac{\pi}{\lambda}$.

The phason dispersion relation, resulting from (4.4.103) and (4.4.104), is schematically shown in figure 4.30. It splits into two branches, which are separated by a frequency gap. The lower branch may be interpreted as collective motions of the soliton lattice. To see this, let us consider the limiting case $\zeta \to \zeta_c$, where only a single soliton is present in the system. Here we have (cf. (4.3.6))

$$\Theta_0(x) = 4 \tan^{-1} \exp\left(-\sqrt{\zeta}\, x\right), \tag{4.4.105}$$

which gives rise to the "potential" $\zeta\left[1 - 2/\cosh^2\left(x\sqrt{\zeta}\right)\right]$, having a bound state with eigenfrequency $\omega^2 = 0$ (being the "translational mode"; see, e.g.,

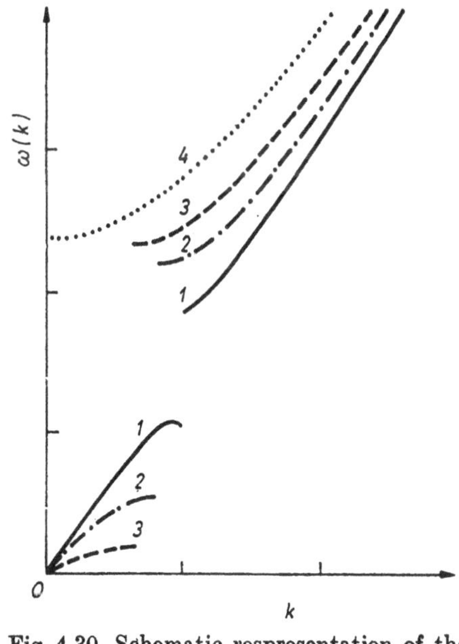

Fig. 4.30. Schematic respresentation of the dispersion relation of the phase modes in an incommensurate system near the lock-in transition. The curves 1 to 4 correspond to decreasing values δ (disrepancy of the lattice periods). For curve 4 $\delta = 0$; commensurate phase. The frequency gap occurs at wave number $k = \frac{\pi}{\lambda}$ (λ — domain size) (after MC MILLAN, 1977)

Fogel et al., 1977, and cf. (4.3.88) and (4.3.89)). Note that this state may be viewed as Goldstone mode connected with the breaking of the continuous translational symmetry of the system by the soliton (cf. the foot-note on page 252).

The upper branch of the dispersion curve may be interpreted as phase modes inside the commensurate domains. If ζ approaches ζ_c ($\zeta > \zeta_c$), then the "Brillouin zone" boundary moves toward the origin. In the limit $\zeta \to \zeta_c$ $\left(\lambda \to \infty \text{ and } \frac{\pi}{\lambda} \to 0\right)$ only the upper branch remains. The frequency gap is continuous at the lock-in transition.

Note that the theory presented here is a mean field theory, i.e. fluctuation effects have not been accounted for.

Effects of the kind described in this section have been studied on several systems, such as the quasi two-dimensional CDW system $2H{-}TaSe_2$ (Mc Millan, 1977, Fleming et al., 1980, Suits et al., 1980, 1981), quasi one-dimensional TTF halides and the linear chain mercury compound $Hg_{3-\delta}AsF_6$ (Theodorou and Rice, 1978, Emery and Axe, 1978), "ω-phase" systems such as Zr—Nb alloys (Horovitz et al., 1978), ferroelectrics such as K_2SeO_4 (Iizumi et al., 1977, Axe et al., 1980) graphite intercalation compounds (Clark et al., 1981), the quasi one-dimensional superionic conductor β-eucrypite (Press et al., 1980) and rare gases adsorbed on graphite (Stephens et al., 1979, Fain et al., 1980).

Incommensurate systems are still a very active field of research, which continues to generate new materials, new phenomena and new problems. We close this section with a concise overview of some theoretical problems studied in this field in recent years.

Studies of the incommensurate—commensurate (IC) transition in discrete lattices (Aubry, 1980a, Bruce, 1980b) show that for weak sinusoidal (lattice or "substrate") potential the continuous approximation presented above is a reasonable approximation. Solely close to the IC transition the discrete lattice causes pinning of the incommensurate (I) phase (i.e. a pinning of the solitons) before the transition to the commensurate (C) phase occurs. However, above some critical value $V_0{}^c$ of the strength of the sinusoidal potential no mobile (flowing) incommensurate phase exists, the solitons are pinned by the discrete lattice. On plotting the period of the "adsorbate" at $V_0 \neq 0$ over that at $V_0 = 0$ (i.e. a in (4.4.84)), one finds a stair-like curve (the so-called "devil's stair case") since each C-phase exists in some finite interval of the latter period. Contrary to $V_0 > V_0{}^c$, at $V_0 \ll V_0{}^c$ the devil's stair case is incomplete, since between each two C-phases there is an I-phase giving rise to a continuous change in the above plot, which smears out the steps.

The IC transition is in general rather complicated, when neither the substrate nor the adsorbate is rigid (Fenton, 1981, Coutinho et al., 1981). In this case the IC transition exhibits behaviour reminiscent of the devil's staircase.

Quantum and thermal fluctuations turn out to be important in the vicinity of the IC transition (Pokrovskij and Talapov, 1979, Nattermann, 1980, Sahni, 1981). In the one-dimensional case, at finite temperatures thermal fluctuations destroy the C-phase, i.e. the system is always in the I-phase. In the two-dimensional case, the C-phase exists so far as the misfit δ between the two periodicities of the system is below some critical value, which decreases linearly with increasing temperature. This implies that growing fluctuations decrease the degree of commensurability. Thermal fluctuations can cause contact between neighbouring domain walls, which restricts the variety of possible domain wall configurations. They can also reduce the effect of the substrate potential so much that in the I-phase the modulation of the adsorbate is significantly changed (Chui and Weeks, 1980).

It is interesting to note that in an incommensurate system electron states are all localized if the strength of the sinusoidal potential V_0 exceeds some critical value, whereas for smaller values of V_0 extended and localized states occur (Aubry 1980b, Sokoloff, 1981, Soukoulis and Economou, 1982) (unlike the situation in disordered one-dimensional systems where all states are localized).

Lattice dynamics of one-dimensional incommensurate systems was studied by de Lange and Janssen (1981). Properties of solitons in the discrete Frenkel-Kontorova model (4.4.84) were numerically investigated by Joos (1982). For a review of the lock-in transition see Villain (1980).

APPENDIX 1

CRYSTAL SYMMETRY AND THE DYNAMICAL MATRIX

Using group theoretical methods in this appendix, symmetry and degeneracy of normal mode frequencies $\omega_j(\boldsymbol{k})$ are investigated. Furthermore, the dynamical matrix is brought into a block diagonalized form, which greatly simplifies any normal mode calculation. The developments in this appendix supposes a basic knowledge of group theory, which can be obtained from many sources, e.g. from WIGNER (1959).

A1.1. Crystal symmetry and groups

Consider an infinite crystal or a crystal with the cyclic boundary conditions (1.3.32). The symmetry operations (cf.. (1.2.19)) $\{\boldsymbol{S} \mid \boldsymbol{t}(\boldsymbol{S})\}$ with the property

$$\{\boldsymbol{S} \mid \boldsymbol{t}(\boldsymbol{S})\}\, \boldsymbol{x}(l\varkappa) = \boldsymbol{S}\boldsymbol{x}(l\varkappa) + \boldsymbol{t}(\boldsymbol{S}) \equiv \boldsymbol{x}(LK), \tag{A1.1}$$

where $\boldsymbol{x}(l\varkappa)$ is the position vector of the $\varkappa$th atom in the lth primitive unit cell and $\boldsymbol{t}(\boldsymbol{S}) = \boldsymbol{v}(\boldsymbol{S}) + \boldsymbol{x}(m)$, bring the crystal into coincidence with itself. Here $\boldsymbol{S}$ is an orthogonal matrix representation of a proper $(\det|\boldsymbol{S}| = 1)$ or improper $(\det|\boldsymbol{S}| = -1)$ rotation, $\boldsymbol{v}(\boldsymbol{S})$ denotes a fractional translation associated with $\boldsymbol{S}$, and $\boldsymbol{x}(m)$ denotes a translation by a lattice vector. These operations constitute a group, the "space group" G_{S}, which can be readily verified by taking into account that the multiplication rule of two elements reads

$$\{\boldsymbol{S}_1 \mid \boldsymbol{t}_1\}\, \{\boldsymbol{S}_2 \mid \boldsymbol{t}_2\} = \{\boldsymbol{S}_1\boldsymbol{S}_2 \mid \boldsymbol{S}_1\boldsymbol{t}_2 + \boldsymbol{t}_1\} \tag{A1.2}$$

and that the inverse element is given by

$$\{\boldsymbol{S} \mid \boldsymbol{t}\}^{-1} = \{\boldsymbol{S}^{-1} \mid -\boldsymbol{S}^{-1}\boldsymbol{t}\}. \tag{A1.3}$$

The operations $\{\boldsymbol{E} \mid \boldsymbol{x}(m)\}$ ($\boldsymbol{E}$ — identity operation) constitute the translation group G_{T}, which is a subgroup of G_{S}. Consider the conjugate of $\{\boldsymbol{E} \mid \boldsymbol{x}(m)\}$ by any operation in G_{S}

$$\{\boldsymbol{S} \mid \boldsymbol{v}(\boldsymbol{S}) + \boldsymbol{x}(n)\}^{-1}\, \{\boldsymbol{E} \mid \boldsymbol{x}(m)\}\, \{\boldsymbol{S} \mid \boldsymbol{v}(\boldsymbol{S}) + \boldsymbol{x}(n)\} = \{\boldsymbol{E} \mid \boldsymbol{x}(m')\}, \tag{A1.4}$$

where $\boldsymbol{x}(m') \equiv \boldsymbol{S}^{-1}\boldsymbol{x}(m)$ is a lattice vector, we see that G_{T} is closed under conjugation, i.e. it is normal. Hence, decomposing G_{S} with respect to G_{T} as

$$G_{\mathrm{S}} = G_{\mathrm{T}} + \{\boldsymbol{S}_2 \mid \boldsymbol{v}(\boldsymbol{S}_2)\}\, G_{\mathrm{T}} + \cdots + \{\boldsymbol{S}_{\mathrm{p}} \mid \boldsymbol{v}(\boldsymbol{S}_{\mathrm{p}})\}\, G_{\mathrm{T}}, \tag{A1.5}$$

the factor group $G_{\mathrm{S}}/G_{\mathrm{T}} = G_{\mathrm{F}}$ is obtained, whose elements are the cosets $\{\boldsymbol{S} \mid \boldsymbol{v}(\boldsymbol{S})\}\, G_{\mathrm{T}}$. G_{F} is isomorphic to the "point group" G_{P} defined by setting all translations equal to zero in the set of coset representatives $\{\boldsymbol{S} \mid \boldsymbol{v}(\boldsymbol{S})\}$, i.e. the elements of G_{P} are the rotations $\{\boldsymbol{S} \mid 0\}$.

In general, no point in the crystal will have symmetry G_P, except that all $\boldsymbol{v}(\boldsymbol{S}) = \boldsymbol{0}$. The fractional translations $\boldsymbol{v}(\boldsymbol{S})$ depend on the choice of the origin of the "basis" describing the positions ($\boldsymbol{x}(\varkappa)$ in (1.2.17)) of the atoms in a Wigner-Seitz cell. In the case of a "non-symmorphic" space group some or all $\boldsymbol{v}(\boldsymbol{S}) \neq \boldsymbol{0}$ irrespective of the choice of the origin, whereas in the case of a "symmorphic" space group the origin can be chosen, such that all $\boldsymbol{v}(\boldsymbol{S}) = \boldsymbol{0}$. Of the existing 230 space groups only 73 are symmorphic. A crystal is symmorphic if the basis in the Wigner Seitz primitive unit cell has the symmetry G_P.

All rotations leaving the Bravais lattice invariant constitute the "holoedric group". If the basis has the symmetry of the holoedric group, the point group coincides with the holoedric group.

The symmetry operations, which replicate the entire crystal and leave a point $\boldsymbol{x}$ fixed, constitute the "site group" $G_{\text{site}}(\boldsymbol{x})$ of this point.

Consider the symmetry operations in G_S, whose rotational part $\boldsymbol{S}$ has the property (cf. (1.3.49))

$$\boldsymbol{S}\boldsymbol{k}_1 = \boldsymbol{k}_1 + \boldsymbol{K}, \tag{A1.6}$$

where $\boldsymbol{k}_1$ is a vector in the first Brillouin zone and $\boldsymbol{K}$ is a vector of the reciprocal lattice (cf. (1.3.50)). The collection of these symmetry operations is said to form the "group of $\boldsymbol{k}_1$", $G_{\boldsymbol{k}_1} \cdot G_{\boldsymbol{k}_1}$ is a subgroup of G_S. In fact, on using (A1.2), (A1.6) we find for any two elements $\{\boldsymbol{R}_i \mid \boldsymbol{r}_i\}$, $\{\boldsymbol{R}_j \mid \boldsymbol{r}_j\}$ of $G_{\boldsymbol{k}_1}$

$$\{\boldsymbol{R}_i \mid \boldsymbol{r}_i\} \{\boldsymbol{R}_j \mid \boldsymbol{r}_j\} = \{\boldsymbol{R}_i\boldsymbol{R}_j \mid \boldsymbol{R}_i\boldsymbol{r}_j + \boldsymbol{r}_i\} = \{\boldsymbol{R}_p \mid \boldsymbol{r}_p\}, \tag{A1.7}$$

where

$$\left.\begin{aligned} &\boldsymbol{R}_i\boldsymbol{k}_1 = \boldsymbol{k}_1 + \boldsymbol{K}_i, \qquad \boldsymbol{R}_j\boldsymbol{k}_1 = \boldsymbol{k}_1 + \boldsymbol{K}_j, \\ &\boldsymbol{R}_p\boldsymbol{k}_1 = \boldsymbol{R}_i\boldsymbol{R}_j\boldsymbol{k}_1 = \boldsymbol{R}_i(\boldsymbol{k}_1 + \boldsymbol{K}_j) = \boldsymbol{k}_1 + \boldsymbol{K}_j + \boldsymbol{R}_i\boldsymbol{K}_j = \boldsymbol{k}_1 + \boldsymbol{K}_p, \end{aligned}\right\} \tag{A1.8}$$

i.e. $\{\boldsymbol{R}_p \mid \boldsymbol{r}_p\}$ is an element of $G_{\boldsymbol{k}_1}$. Thus G_S can be decomposed, for example, into left cosets of $G_{\boldsymbol{k}_1}$ as

$$G_S = G_{\boldsymbol{k}_1} + \{\boldsymbol{P}_2 \mid \boldsymbol{p}_2\}\, G_{\boldsymbol{k}_1} + \cdots + \{\boldsymbol{P}_s \mid \boldsymbol{p}_s\}\, G_{\boldsymbol{k}_1}, \tag{A1.9}$$

where $\{\boldsymbol{P}_i \mid \boldsymbol{p}_i\}$ denotes those operations in G_S whose rotational elements rotate $\boldsymbol{k}_1$ into a vector essentially different from $\boldsymbol{k}_1$, i.e. (A1.6) does not hold for these elements. The operations

$$\{\boldsymbol{P}_i \mid \boldsymbol{p}_i\}, \qquad i = 1, \ldots, s \qquad (\{\boldsymbol{P}_1 \mid \boldsymbol{p}_1\} \equiv \{\boldsymbol{E} \mid \boldsymbol{0}\}),$$

are characterized by

$$\boldsymbol{P}_i\boldsymbol{k}_1 = \boldsymbol{k}_i, \qquad i = 1, \ldots, s. \tag{A1.10}$$

The set vectors $\boldsymbol{k}_1, \boldsymbol{k}_2, \ldots, \boldsymbol{k}_s$ generated by (A1.10) is the so-called star of $\boldsymbol{k}_1$.

The rotational elements of $G_{\boldsymbol{k}_1}$ form a point group, $G^{P}_{\boldsymbol{k}_1}$, which is a subgroup of G_P.

A1.2. Symmetry of the dynamical matrix and the multiplier representation

In examining the transformation properties of the dynamical matrix we follow MARADUDIN and VOSKO (1968). The starting point is the following expression for the dynamical matrix (cf. (1.3.38)):

$$D_{\alpha\alpha'}(KK' \mid \boldsymbol{S}\boldsymbol{k}) = (M_K M_{K'})^{-1/2}\, \Phi_{\alpha\alpha'}(LK, L'K')\, \mathrm{e}^{-i\boldsymbol{S}\boldsymbol{k}(\boldsymbol{x}(L)-\boldsymbol{x}(L'))}. \tag{A1.11}$$

We relate in (A1.11) the sites (LK), $(L'K')$ to the sites $(l\varkappa)$, $(l'\varkappa')$ with the aid of (A1.1), and we account for $M_K = M_\varkappa$, (1.2.22), and the relations

$$(\boldsymbol{Sk})\,\boldsymbol{x}(L) = \boldsymbol{k}(\boldsymbol{S}^{-1}\boldsymbol{x}(L)), \tag{A1.12}$$

$$\boldsymbol{S}^{-1}\boldsymbol{x}(L) = \boldsymbol{x}(l) - \{\boldsymbol{S} \mid \boldsymbol{v}(\boldsymbol{S}) + \boldsymbol{x}(m)\}^{-1}\,\boldsymbol{x}(K) - \boldsymbol{x}(\varkappa) \tag{A1.13}$$

(cf. (A1.3), (A1.1)). Thus, we obtain

$$D_{\alpha\alpha'}(KK' \mid \boldsymbol{Sk}) = \sum_{\alpha_1\alpha_1'\varkappa_1\varkappa_1'} \Gamma_{\alpha\alpha_1}(K\varkappa_1 \mid \boldsymbol{k}, \{\boldsymbol{S} \mid \boldsymbol{v}(\boldsymbol{S}) + \boldsymbol{x}(m)\})$$
$$\times\, D_{\alpha_1\alpha_1'}(\varkappa_1\varkappa_1' \mid \boldsymbol{k})\, \Gamma_{\alpha'\alpha_1'}(K'\varkappa_1 \mid \boldsymbol{k}, \{\boldsymbol{S} \mid \boldsymbol{v}(\boldsymbol{S}) + \boldsymbol{x}(m)\}), \tag{A1.14}$$

where

$$\Gamma_{\alpha\alpha_1}(K\varkappa_1 \mid \boldsymbol{k}, \{\boldsymbol{S} \mid \boldsymbol{v}(\boldsymbol{S}) + \boldsymbol{x}(m)\})$$
$$= S_{\alpha\alpha_1}\delta_{\varkappa\varkappa_1} \exp\left(i\boldsymbol{k}[\{\boldsymbol{S} \mid \boldsymbol{v}(\boldsymbol{S}) + \boldsymbol{x}(m)\}^{-1}\,\boldsymbol{x}(K) - \boldsymbol{x}(\varkappa_1)]\right). \tag{A1.15}$$

In (A1.15) $\varkappa$ labels the kind of atom being brought into the position K by the symmetry operation $\{\boldsymbol{S} \mid \boldsymbol{v}(\boldsymbol{S}) + \boldsymbol{x}(m)\}$.

By (A1.15) $\boldsymbol{\Gamma}$ can be readily shown to be unitary:

$$\Gamma^{+}_{\alpha_1\alpha}(\varkappa_1 K \mid \boldsymbol{k}, \{\boldsymbol{S} \mid \boldsymbol{v}(\boldsymbol{S}) + \boldsymbol{x}(m)\}) = \Gamma^{*}_{\alpha\alpha_1}(K\varkappa_1 \mid \boldsymbol{k}, \{\boldsymbol{S} \mid \boldsymbol{v}(\boldsymbol{S}) + \boldsymbol{x}(m)\})$$
$$= S_{\alpha\alpha_1}\delta_{\varkappa\varkappa_1} \exp\left(-i\boldsymbol{k}[\{\boldsymbol{S} \mid \boldsymbol{v}(\boldsymbol{S}) + \boldsymbol{x}(m)\}\,\boldsymbol{x}(K) - \boldsymbol{x}(\varkappa_1)]\right)$$
$$= \Gamma^{-1}_{\alpha_1\alpha}(\varkappa_1 K \mid \boldsymbol{k}, \{\boldsymbol{S} \mid \boldsymbol{v}(\boldsymbol{S}) + \boldsymbol{x}(m)\}). \tag{A1.16}$$

Hence, $\boldsymbol{D}(\boldsymbol{Sk})$ and $\boldsymbol{D}(\boldsymbol{k})$ are related by a unitary transformation

$$\boldsymbol{D}(\boldsymbol{Sk}) = \boldsymbol{\Gamma}(\boldsymbol{k}, \{\boldsymbol{S} \mid \boldsymbol{v}(\boldsymbol{S}) + \boldsymbol{x}(m)\})\, \boldsymbol{D}(\boldsymbol{k})\, \boldsymbol{\Gamma}^{+}(\boldsymbol{k}, \{\boldsymbol{S} \mid \boldsymbol{v}(\boldsymbol{S}) + \boldsymbol{x}(m)\}). \tag{A1.17}$$

Since a unitary transformation does not affect the eigenvalues of a matrix, it holds for the normal mode frequencies

$$\omega_j^2(\boldsymbol{Sk}) = \omega_j^2(\boldsymbol{k}), \tag{A1.18}$$

i.e. $\omega_j^2(\boldsymbol{k})$ has the symmetry of the point group of the space group (cf. (1.3.48)).

In the following we denote by $\boldsymbol{R}$ the rotational elements of the symmetry operations of the space group $G_{\boldsymbol{k}}$. On account of the periodicity of the dynamical matrix

$$\boldsymbol{D}(\boldsymbol{k} + \boldsymbol{K}) = \boldsymbol{D}(\boldsymbol{k}), \tag{A1.19}$$

which results from the definition (1.3.38), we obtain for symmetry operations $\{\boldsymbol{R} \mid \boldsymbol{v}(\boldsymbol{R}) + \boldsymbol{x}(m)\}$ from (A1.17)

$$\boldsymbol{D}(\boldsymbol{k}) = \boldsymbol{\Gamma}(\boldsymbol{k}, \{\boldsymbol{R} \mid \boldsymbol{v}(\boldsymbol{R}) + \boldsymbol{x}(m)\})\, \boldsymbol{D}(\boldsymbol{k})\, \boldsymbol{\Gamma}^{+}(\boldsymbol{k}, \{\boldsymbol{R} \mid \boldsymbol{v}(\boldsymbol{R}) + \boldsymbol{x}(m)\}), \tag{A1.20}$$

i.e. the unitary matrices $\boldsymbol{\Gamma}(\boldsymbol{k}, \{\boldsymbol{R} \mid \boldsymbol{v}(\boldsymbol{R}) + \boldsymbol{x}(m)\})$ commute with the dynamical matrix. The $\boldsymbol{\Gamma}$-matrices (A1.15) can be shown (Maradudin and Vosko, 1968) to obey the multiplication rule

$$\boldsymbol{\Gamma}(\boldsymbol{k}, \{\boldsymbol{R}_1 \mid \boldsymbol{v}(\boldsymbol{R}_1) + \boldsymbol{x}(m_1)\}\,\{\boldsymbol{R}_2 \mid \boldsymbol{v}(\boldsymbol{R}_2) + \boldsymbol{x}(m_2)\})$$
$$= \boldsymbol{\Gamma}(\boldsymbol{k}, \{\boldsymbol{R}_1 \mid \boldsymbol{v}(\boldsymbol{R}_1) + \boldsymbol{x}(m_1)\})\, \boldsymbol{\Gamma}(\boldsymbol{k}, \{\boldsymbol{R}_2 \mid \boldsymbol{v}(\boldsymbol{R}_2) + \boldsymbol{x}(m_2)\}), \tag{A1.21}$$

i.e. they provide a $3r$-dimensional (r — number of atoms in a primitive unit cell) representation of $G_{\boldsymbol{k}}$, with respect to which $\boldsymbol{D}(\boldsymbol{k})$ is invariant. These matrices play an analogous role for $\boldsymbol{D}(\boldsymbol{k})$ as do, in the case of a Hamiltonian, the unitary transformations which correspond to symmetry operations leaving the Hamiltonian invariant (WIGNER, 1959). Similarly as in the latter case, the property (A1.20) greatly simplifies a solution of the eigenvalue problem for the dynamical matrix.

To proceed further we note that any element of $G_{\boldsymbol{k}}$, $\{\boldsymbol{R} \mid \boldsymbol{v}(\boldsymbol{R}) + \boldsymbol{x}(m)\}$, may be written as

$$\{\boldsymbol{R} \mid \boldsymbol{v}(\boldsymbol{R}) + \boldsymbol{x}(m)\} = \{\boldsymbol{E} \mid \boldsymbol{x}(m)\} \{\boldsymbol{R} \mid \boldsymbol{v}(\boldsymbol{R})\}. \tag{A1.22}$$

Here the first term on the right-hand side describes a pure translation by a lattice vector. Since $\exp\left(-i\boldsymbol{k}\boldsymbol{x}(m)\right)$ is a corresponding representation of such a translation, we may write

$$\boldsymbol{\Gamma}(\boldsymbol{k}, \{\boldsymbol{R} \mid \boldsymbol{v}(\boldsymbol{R}) + \boldsymbol{x}(m)\}) = \mathrm{e}^{-i\boldsymbol{k}\boldsymbol{x}(m)}\, \mathrm{e}^{-i\boldsymbol{k}\boldsymbol{v}(\boldsymbol{R})}\boldsymbol{T}(\boldsymbol{k}, \boldsymbol{R}). \tag{A1.23}$$

The matrices $\boldsymbol{T}(\boldsymbol{k}, \boldsymbol{R})$, defined in (A1.23), provide a unitary $3r$-dimensional reducible multiplier representation of the point group $G_{\boldsymbol{k}}^{\mathrm{P}}$. To see this we use (A1.23) to obtain

$$\left.\begin{aligned} &\boldsymbol{\Gamma}(\boldsymbol{k}, \{\boldsymbol{R}_i \mid \boldsymbol{v}(\boldsymbol{R}_i)\})\, \boldsymbol{\Gamma}(\boldsymbol{k}, \{\boldsymbol{R}_j \mid \boldsymbol{v}(\boldsymbol{R}_j)\}) \\ &= \mathrm{e}^{-i\boldsymbol{k}(\boldsymbol{v}(\boldsymbol{R}_i)+\boldsymbol{v}(\boldsymbol{R}_j))}\boldsymbol{T}(\boldsymbol{k}, \boldsymbol{R}_i)\, \boldsymbol{T}(\boldsymbol{k}, \boldsymbol{R}_j) \end{aligned}\right\} \tag{A1.24}$$

and

$$\left.\begin{aligned} &\boldsymbol{\Gamma}(\boldsymbol{k}, \{\boldsymbol{R}_i \mid \boldsymbol{v}(\boldsymbol{R}_i)\}\, \{\boldsymbol{R}_j \mid \boldsymbol{v}(\boldsymbol{R}_j)\}) \\ &= \mathrm{e}^{-i\boldsymbol{K}\boldsymbol{v}(\boldsymbol{R}_j)}\, \mathrm{e}^{-i\boldsymbol{k}(\boldsymbol{v}(\boldsymbol{R}_j)+\boldsymbol{v}(\boldsymbol{R}_i))}\boldsymbol{T}(\boldsymbol{k}, \boldsymbol{R}_i\boldsymbol{R}_j), \end{aligned}\right\} \tag{A1.25}$$

where

$$\boldsymbol{K} = \boldsymbol{R}_i^{-1}\boldsymbol{k} - \boldsymbol{k}. \tag{A1.26}$$

From (A1.21), (A1.24) and (A1.25) we find

$$\boldsymbol{T}(\boldsymbol{k}, \boldsymbol{R}_i)\, \boldsymbol{T}(\boldsymbol{k}, \boldsymbol{R}_j) = \mathrm{e}^{-i(\boldsymbol{R}_i^{-1}\boldsymbol{k} - \boldsymbol{k})\boldsymbol{v}(\boldsymbol{R}_j)}\boldsymbol{T}(\boldsymbol{k}, \boldsymbol{R}_i\boldsymbol{R}_j). \tag{A1.27}$$

According to this equation, the matrices $\boldsymbol{T}(\boldsymbol{k}, \boldsymbol{R})$ provide a representation of $G_{\boldsymbol{k}}^{\mathrm{P}}$. If we consider nonsymmorphic groups (at least one $\boldsymbol{v}(\boldsymbol{R}) \neq 0$!) or $\boldsymbol{k}$ on the surface of the Brillouin zone ($\boldsymbol{k} \neq \boldsymbol{0}$ in (A1.6)!), this representation is not a representation in the ordinary sense, due to the exponential factor occurring in (A1.27). Because of this factor the matrices $\boldsymbol{T}(\boldsymbol{k}, \boldsymbol{R})$ are said to form a multiplier representation of $G_{\boldsymbol{k}}^{\mathrm{P}}$.

By (A1.20) and (A1.25) the invariance of $\boldsymbol{D}(\boldsymbol{k})$ against space group operations can now be expressed as

$$\boldsymbol{D}(\boldsymbol{k}) = \boldsymbol{T}(\boldsymbol{k}, \boldsymbol{R})\, \boldsymbol{D}(\boldsymbol{k})\, \boldsymbol{T}^{-1}(\boldsymbol{k}, \boldsymbol{R}). \tag{A1.28}$$

This equation, or equivalently (A1.20), can be used to determine the independent, nonzero elements of the dynamical matrix $\boldsymbol{D}(\boldsymbol{k})$.

A1.3. Classification of the normal modes and block diagonalization of the dynamical matrix

For a Hamiltonian, in the absence of accidental degeneracy, the eigenfunctions connected with a given eigenvalue transform irreducibly under symmetry operations leaving the Hamiltonian invariant (see, e.g. WIGNER (1959)). Quite analogously, the eigenvectors of the dynamical matrix $\boldsymbol{D}(\boldsymbol{k})$ connected with a given eigenfrequency transform irreducibly under symmetry operations $\boldsymbol{T}(\boldsymbol{k}, \boldsymbol{R})$ defined in (A1.24).

On applying the operation $\boldsymbol{T}(\boldsymbol{k},\boldsymbol{R})$ to the eigenvalue equation (1.3.41) we obtain for every operation $\boldsymbol{R}$ of $G_{\boldsymbol{k}}^{\mathrm{P}}$

$$\omega_j^2(\boldsymbol{k})\, \boldsymbol{T}(\boldsymbol{k},\boldsymbol{R})\, \boldsymbol{e}(\boldsymbol{k}j) = \boldsymbol{D}(\boldsymbol{k})\, \boldsymbol{T}(\boldsymbol{k},\boldsymbol{R})\, \boldsymbol{e}(\boldsymbol{k}j), \tag{A1.29}$$

where we have used (A1.28). According to (A1.29), $\boldsymbol{T}(\boldsymbol{k},\boldsymbol{R})\,\boldsymbol{e}(\boldsymbol{k}j)$ must be a linear com. bination of the linear independent eigenvectors connected with the eigenvalue $\omega_j^2(\boldsymbol{k})$ Note that $\boldsymbol{D}(\boldsymbol{k})$ has $3r$ eigenvalues $\omega_j^2(\boldsymbol{k})$ for fixed $\boldsymbol{k}$, but they do not all differ. For convenience we replace in the following the single index j by a double index $\varrho\lambda$, where ϱ labels the distinct eigenvalues of $\boldsymbol{D}(\boldsymbol{k})$ for a given $\boldsymbol{k}$ and $\lambda = 1, 2, \ldots, d^{(\varrho)}$ the different eigenvectors connected with the eigenvalue $\omega^2(\boldsymbol{k})$. $d^{(\varrho)}$ is the degree of degeneracy of $\omega_\varrho^2(\boldsymbol{k})$. Thus, we can write

$$\boldsymbol{T}(\boldsymbol{k},\boldsymbol{R})\, \boldsymbol{e}(\boldsymbol{k}\varrho\lambda) = \sum_{\lambda'=1}^{d^{(\varrho)}} \tau_{\lambda'\lambda}^{(\varrho)}(\boldsymbol{k},\boldsymbol{R})\, \boldsymbol{e}(\boldsymbol{k}\varrho\lambda'), \tag{A1.30}$$

where $\tau_{\lambda'\lambda}^{(\varrho)}(\boldsymbol{k},\boldsymbol{R})$ can be shown to be a matrix element of a $d^{(\varrho)}$-dimensional unitary irreducible multiplier representation with the same multipliers as those in the representation provided by $\boldsymbol{T}(\boldsymbol{k},\boldsymbol{R})$ (Maradudin and Vosko, 1968). Hence, the eigenvalues and eigenvectors of $\boldsymbol{D}(\boldsymbol{k})$ can be classified by means of the irreducible multiplier representations of $G_{\boldsymbol{k}}^{\mathrm{P}}$. The degeneracy $d^{(\varrho)}$ is given by the dimension of the ϱth irreducible multiplier representation.

The reduction of the reducible representation of $G_{\boldsymbol{k}}^{\mathrm{P}}$ (provided by $\boldsymbol{T}(\boldsymbol{k},\boldsymbol{R})$) into its rreducible representations $\tau^{(\varrho)}(\boldsymbol{k},\boldsymbol{R})$ is performed by means of the familiar formula see e.g. Wigner, 1959)

$$m^{(\varrho)} = \frac{1}{h} \sum_{\boldsymbol{R}} \chi(\boldsymbol{k},\boldsymbol{R})\, \chi^{(\varrho)}(\boldsymbol{k},\boldsymbol{R})^*, \tag{A1.31}$$

which gives the number of times the ϱth irreducible representation $\tau^{(\varrho)}(\boldsymbol{k},\boldsymbol{R})$ is contained in $\boldsymbol{T}(\boldsymbol{k},\boldsymbol{R})$. Here $\chi(\boldsymbol{k},\boldsymbol{R})$ is the character of $\boldsymbol{T}(\boldsymbol{k},\boldsymbol{R})$ and $\chi^{(\varrho)}(\boldsymbol{k},\boldsymbol{R})$ is the character of $\tau^{(\varrho)}(\boldsymbol{k},\boldsymbol{R})$. h denotes the order of $G_{\boldsymbol{k}}^{\mathrm{P}}$. Since we may have $m^{(\varrho)} > 1$, it is convenient to introduce an additional index $\mu = 1, 2, \ldots, m^{(\varrho)}$ labeling the different eigenvalues corresponding to the ϱth irreducible representation. Thus, the eigenvalue equation (1.3.41) now reads

$$\left.\begin{aligned} &\omega_{\varrho\mu}^2(\boldsymbol{k})\, \boldsymbol{e}(\boldsymbol{k}\varrho\lambda\mu) = \boldsymbol{D}(\boldsymbol{k})\, \boldsymbol{e}(\boldsymbol{k}\varrho\lambda\mu), \\ &\lambda = 1, 2, \ldots, d^{(\varrho)}; \qquad \mu = 1, 2, \ldots, m^{(\varrho)}. \end{aligned}\right\} \tag{A1.32}$$

So-called symmetry coordinates play an important role in lattice dynamical calculations. They are displacement patterns of atoms in a unit cell of certain symmetry. If symmetry coordinates are used to express the potential energy instead of simple cartesian displacements of the atoms, the dynamical matrix will be of a simple block form, which often greatly simplifies the solution of the eigenvalue problem.

To find displacement patterns (symmetry coordinates) that transform under $\boldsymbol{T}(\boldsymbol{k},\boldsymbol{R})$ according to irreducible representations $\tau^{(\varrho)}(\boldsymbol{k},\boldsymbol{R})$, we define, in analogy to the familiar projection operator (Wigner, 1959), the operator

$$P_{\lambda\lambda'}^{(\varrho)}(\boldsymbol{k}) = \frac{d^{(\varrho)}}{h} \sum_{\boldsymbol{R}} \tau_{\lambda\lambda'}^{(\varrho)}(\boldsymbol{k},\boldsymbol{R})\, \boldsymbol{T}(\boldsymbol{k},\boldsymbol{R}). \tag{A1.33}$$

On applying this operator to any $3r$-dimensional column vector $\boldsymbol{\Psi}$, we obtain the quantity

$$\boldsymbol{\Psi}(\boldsymbol{k}\varrho\lambda) = P_{\lambda\lambda'}^{(\varrho)}(\boldsymbol{k})\, \boldsymbol{\Psi}, \tag{A1.34}$$

which can be readily shown (MARADUDIN and VOSKO, 1968) to transform as

$$T(\boldsymbol{k}, \boldsymbol{R})\, \Psi(\boldsymbol{k}\varrho\lambda) = \sum_{\lambda'=1}^{d^{(\varrho)}} \tau^{(\varrho)}_{\lambda'\lambda}(\boldsymbol{k}, \boldsymbol{R})\, \Psi(\boldsymbol{k}\varrho\lambda'). \tag{A1.35}$$

On comparing (A1.35) and (A1.30) we see that $\Psi(\boldsymbol{k}\varrho\lambda)$ transforms as $\boldsymbol{e}(\boldsymbol{k}\varrho\lambda)$, i.e. $\Psi(\boldsymbol{k}\varrho\lambda)$ must be a linear combination of the eigenvectors $\boldsymbol{e}(\boldsymbol{k}\varrho\lambda\mu)$:

$$\Psi(\boldsymbol{k}\varrho\lambda) = \sum_{\mu=1}^{m^{(\varrho)}} c_\mu \boldsymbol{e}(\boldsymbol{k}\varrho\lambda\mu). \tag{A1.36}$$

The coefficients c_μ cannot be determined by symmetry arguments. They depend on the special dynamical matrix $\boldsymbol{D}(\boldsymbol{k})$. Only if $m^{(\varrho)} = 1$, do we have $c_\mu = 1$, and Ψ is then an exact eigenvector of $\boldsymbol{D}(\boldsymbol{k})$.

To simplify the dynamical matrix we have to find the $m^{(\varrho)}$ linear independent vectors contained in $\Psi(\boldsymbol{k}\varrho\lambda)$. This can be managed, for example, as follows (WORLTON and WARREN, 1972). We put, without loss of generality, in (A1.33) $\lambda' = \lambda$ and obtain from (A1.34), by an appropriate choice of the vectors Ψ, the following $3r$ vectors:

$$\boldsymbol{P}_a^{(\varrho)}(\boldsymbol{k}\lambda) = \frac{d^{(\varrho)}}{h} \sum_{\boldsymbol{R}} \tau^{(\varrho)}_{\lambda\lambda}(\boldsymbol{k}\boldsymbol{R})^*\, \boldsymbol{t}_a(\boldsymbol{k}, \boldsymbol{R}), \qquad a = 1, 2, \ldots, 3r, \tag{A1.37}$$

where $\boldsymbol{t}_a(\boldsymbol{k}, \boldsymbol{R})$ is the ath column vector of the matrix $\boldsymbol{T}(\boldsymbol{k}, \boldsymbol{R})$. From these $3r$ vectors $m^{(\varrho)}$ linear independent orthogonal vectors (symmetry coordinates), $\Psi(\boldsymbol{k}\varrho\lambda\mu)$, can be obtained by Schmidt's orthonormalization procedure.

The vectors $\Psi(\boldsymbol{k}\varrho\lambda\mu)$ have the property that any matrix element of the dynamical matrix $\boldsymbol{D}(\boldsymbol{k})$, or equivalently of the matrix $\boldsymbol{T}^{-1}(\boldsymbol{k}, \boldsymbol{R})\, \boldsymbol{D}(\boldsymbol{k})\, \boldsymbol{T}(\boldsymbol{k}, \boldsymbol{R})$, taken between two vectors $\Psi(\boldsymbol{k}\varrho\lambda\mu)$ and $\Psi(\boldsymbol{k}\varrho'\lambda'\mu')$, vanishes, unless $\varrho' = \varrho$ and $\lambda' = \lambda$:

$$\begin{aligned}\bar{D}(\boldsymbol{k} \mid \varrho\lambda\mu; \varrho'\lambda'\mu') &\equiv \sum_{\substack{\varkappa\alpha\\ \varkappa'\alpha'}} \Psi_\alpha(\varkappa \mid \boldsymbol{k}\varrho\lambda\mu)^*\, D_{\alpha\alpha'}(\varkappa\varkappa' \mid \boldsymbol{k})\, \Psi_{\alpha'}(\varkappa' \mid \boldsymbol{k}\varrho'\lambda'\mu') \\ &= \frac{1}{d^{(\varrho)}}\, \delta_{\varrho\varrho'} \delta_{\lambda\lambda'} \sum_{\sigma=1}^{d^{(\varrho)}} \bar{D}(\boldsymbol{k} \mid \varrho\sigma\mu; \varrho'\sigma\mu'),\end{aligned} \tag{A1.38}$$

which can be proved by using (A1.28), (A1.35) and the orthonormality of the irreducible multiplier representations

$$\sum_{\boldsymbol{R}} \tau^{(\varrho)}_{\lambda\lambda_1}(\boldsymbol{k}, \boldsymbol{R})^*\, \tau^{(\varrho')}_{\lambda'\lambda_1'}(\boldsymbol{k}, \boldsymbol{R}) = \frac{h}{d^{(\varrho)}}\, \delta_{\varrho\varrho'} \delta_{\lambda\lambda'} \delta_{\lambda_1\lambda_1'}. \tag{A1.39}$$

According to (A1.38), $\bar{D}(\boldsymbol{k} \mid \varrho\lambda\mu; \varrho\lambda\mu')$ is independent of λ. If we now construct a unitary matrix $\boldsymbol{U}(\boldsymbol{k})$ whose columns are the vectors $\Psi(\boldsymbol{k}\varrho\lambda\mu)$, we obtain

$$\boldsymbol{U}^{-1}(\boldsymbol{k})\, \boldsymbol{D}(\boldsymbol{k})\, \boldsymbol{U}(\boldsymbol{k}) = \bar{\boldsymbol{D}}(\boldsymbol{k}), \tag{A1.40}$$

where

$$\bar{\boldsymbol{D}}(\boldsymbol{k}) = \begin{bmatrix} \bar{\boldsymbol{D}}^{(1)}(\boldsymbol{k}) & 0 & 0 & \cdot \\ 0 & \bar{\boldsymbol{D}}^{(2)}(\boldsymbol{k}) & 0 & \cdot \\ 0 & \cdot & \cdot & \cdot \\ \cdot & \cdot & \cdot & \cdot \end{bmatrix} \tag{A1.41}$$

with

$$\bar{D}^{(\varrho)}(\boldsymbol{k} \mid \mu\mu') = \sum_{\varkappa\alpha\varkappa'\alpha'} \psi_\alpha(\varkappa \mid \boldsymbol{k}\varrho\lambda\mu)^*\, D_{\alpha\alpha'}(\varkappa\varkappa' \mid \boldsymbol{k})\, \psi_{\alpha'}(\varkappa' \mid \boldsymbol{k}\varrho\lambda\mu'). \tag{A1.42}$$

$\bar{\boldsymbol{D}}^{(\varrho)}(\boldsymbol{k})$ is a $m^{(\varrho)} \times m^{(\varrho)}$ matrix. It occurs $d^{(\varrho)}$ times in the diagonal of $\bar{\boldsymbol{D}}(\boldsymbol{k})$. Since a unitary transformation does not affect the eigenvalues of a matrix, the block diagonal matrix $\bar{\boldsymbol{D}}(\boldsymbol{k})$ has the same eigenvalues as $\boldsymbol{D}(\boldsymbol{k})$. The eigenvalues of $\bar{\boldsymbol{D}}^{(\varrho)}(\boldsymbol{k})$ are just the $m^{(\varrho)}$ squares of the frequencies of the normal modes of the ϱth irreducible representation of $G_{\boldsymbol{k}}{}^{\mathrm{P}}$, which are $d^{(\varrho)}$ times degenerate.

Note that in addition to the degeneracy of the normal modes associated with the multi-dimensionality of the irreducible representations of $G_{\boldsymbol{k}}{}^{\mathrm{P}}$ time reversal symmetry can give rise to extra degeneracies (see Maradudin and Vosko, 1968).

For further application of group theory to lattice dynamics the reader is referred to the monograph of Birman (1974), where the problem of selection rules for infra-red absorption and Raman scattering is also investigated. The problem of generation of symmetry coordinates was investigated in detail by Hansen (1978).

Group theory can also be used for studying lattice dynamics of incommensurate crystals (i.e. of crystals modulated by a periodic deformation with a wave vector, which cannot be formed by simple rational fractions of reciprocal lattice vectors; cf. section 4.4.). To this end the incommensurate crystal is considered as a two-component system, with the basic structure as one component and the deformation as the other component. The relative motions between the two components give rise to additional dimensions. The proper symmetry group of such a system is a so-called superspace group, i.e. a crystallographic space group in more than three (maximal six) dimensions (Janner and Janssen, 1977, Janssen, 1979, cf. Rasing et al., 1982).

For application of group theory to vibrational properties of solids with isolated defects, see, e.g., Maradudin et al. (1971).

APPENDIX 2

LINEAR RESPONSE OF ELECTRONS IN A CRYSTAL TO AN ELECTROMAGNETIC FIELD

Here we derive some basic formulae, describing the response of a crystal to an electromagnetic field. This formula will be needed in the microscopic theory of phonons (section 1.6.). Consider an external electromagnetic field (described by the electric field $\boldsymbol{E}^{\text{ext}}(\boldsymbol{r}, t)$ and the magnetic induction $\boldsymbol{B}^{\text{ext}}(\boldsymbol{r}, t)$) originating from external charge and current densities $\varrho^{\text{ext}}(\boldsymbol{r}, t)$ and $\boldsymbol{j}^{\text{ext}}(\boldsymbol{r}, t)$, respectively. This field is assumed to induce in a medium (crystal) charge and current densities $\varrho^{\text{ind}}(\boldsymbol{r}, t)$ and $\boldsymbol{j}^{\text{ind}}(\boldsymbol{r}, t)$, which on their part are sources of an electromagnetic field $\boldsymbol{E}^{\text{ind}}(\boldsymbol{r}, t)$, $\boldsymbol{B}^{\text{ind}}(\boldsymbol{r}, t)$.

The internal field in the crystal

$$\boldsymbol{E}^{\text{int}}(\boldsymbol{r}, t) = \boldsymbol{E}^{\text{ext}}(\boldsymbol{r}, t) + \boldsymbol{E}^{\text{ind}}(\boldsymbol{r}, t), \tag{A2.1}$$

$$\boldsymbol{B}^{\text{int}}(\boldsymbol{r}, t) = \boldsymbol{B}^{\text{ext}}(\boldsymbol{r}, t) + \boldsymbol{B}^{\text{ind}}(\boldsymbol{r}, t) \tag{A2.2}$$

satisfies Maxwell's equations

$$\nabla \boldsymbol{E}^{\text{int}} = 4\pi\{\varrho^{\text{ext}} + \varrho^{\text{ind}}\}, \qquad \nabla \boldsymbol{B}^{\text{int}} = 0, \qquad \text{(A2.3), (A2.4)}$$

$$\nabla \times \boldsymbol{E}^{\text{int}} = -\frac{1}{c}\dot{\boldsymbol{B}}^{\text{int}}, \qquad \nabla \times \boldsymbol{B}^{\text{int}} = \frac{1}{c}\dot{\boldsymbol{E}}^{\text{int}} + \frac{4\pi}{c}\{\boldsymbol{j}^{\text{ext}} + \boldsymbol{j}^{\text{ind}}\}. \qquad \text{(A2.5), (A2.6)}$$

The total microscopic field in the medium $\boldsymbol{E}^{\text{micr}}(\boldsymbol{r}, t)$, $\boldsymbol{B}^{\text{micr}}(\boldsymbol{r}, t)$ is additively composed of the internal field $\boldsymbol{E}^{\text{int}}(\boldsymbol{r}, t)$, $\boldsymbol{B}^{\text{int}}(\boldsymbol{r}, t)$ and of the equilibrium field $\boldsymbol{E}^{\text{eq}}(\boldsymbol{r})$, $\boldsymbol{B}^{\text{eq}}(\boldsymbol{r})$, which exists in the crystal in the absence of an external field. Assuming permanent magnetic moments to be absent, then $\boldsymbol{B}^{\text{eq}}(\boldsymbol{r})$ vanishes. $\boldsymbol{E}^{\text{eq}}(\boldsymbol{r})$ originates from the equilibrium charge density $\varrho^{\text{eq}}(\boldsymbol{r})$. For reasons of neutrality, the spatial average over the volume of a primitive unit cell of $\varrho^{\text{eq}}(\boldsymbol{r})$ and $\boldsymbol{E}^{\text{eq}}(\boldsymbol{r})$ vanishes.

To get a closed set of equations for the internal electromagnetic field, we must complete (A2.3) to (A2.6) by equations relating $\boldsymbol{j}^{\text{ind}}$ and ϱ^{ind} to the internal electromagnetic field. For weak fields we may assume

$$\boldsymbol{j}^{\text{ind}}(\boldsymbol{r}, t) = \int_{-\infty}^{\infty} \mathrm{d}t' \int \mathrm{d}\boldsymbol{r}' \sigma(\boldsymbol{r}, \boldsymbol{r}', t - t')\, \boldsymbol{E}^{\text{int}}(\boldsymbol{r}', t'), \tag{A2.7}$$

where σ is the field independent so-called microscopic conductivity tensor. A corresponding relation for ϱ^{ind} can be obtained from (A2.7) by means of the continuity equation. Fourier transforming (A2.7), we obtain

$$\boldsymbol{j}^{\text{ind}}(\boldsymbol{q}, \omega) = \sum_{\boldsymbol{q}'} \sigma(\boldsymbol{q}, \boldsymbol{q}', \omega)\, \boldsymbol{E}^{\text{int}}(\boldsymbol{q}', \omega). \tag{A2.8}$$

On deriving (A2.8) we have accounted for the fact that a quantity $A(\boldsymbol{r}, t)$ is Fourier transformed as

$$A(\boldsymbol{r}, t) = \frac{1}{V} \sum_{\boldsymbol{q}} \int_{-\infty}^{\infty} \frac{d\omega}{2\pi} \, e^{i(\boldsymbol{qr} - \omega t)} A(\boldsymbol{q}, \omega), \tag{A2.9}$$

$$A(\boldsymbol{q}, \omega) = \int_V d\boldsymbol{r} \int_{-\infty}^{\infty} dt \; e^{-i(\boldsymbol{qr} - \omega t)} A(\boldsymbol{r}, t), \tag{A2.10}$$

where V is the periodicity volume. For a crystal, V is a parallelepiped whose edges are defined by the lattice vectors $L\boldsymbol{a}_1$, $L\boldsymbol{a}_2$, $L\boldsymbol{a}_3$ (cf. section 1.3.), and the wave vector $\boldsymbol{q}$ is given by

$$\boldsymbol{q} = \frac{m_1}{L} \boldsymbol{b}_1 + \frac{m_2}{L} \boldsymbol{b}_2 + \frac{m_3}{L} \boldsymbol{b}_3, \quad m_i \text{ — integers} \quad (i = 1, 2, 3), \tag{A2.11}$$

where $\boldsymbol{b}_i$ $(i = 1, 2, 3)$ are the unit vectors of the reciprocal lattice (cf. (1.3.34), (1.3.36)). The consistency of (A2.9), (A2.10) can be readily verified by using the orthogonality and closure conditions of plane waves

$$\frac{1}{V} \int_V d\boldsymbol{r} \; e^{i(\boldsymbol{q}' - \boldsymbol{q})\boldsymbol{r}} = \delta_{\boldsymbol{q}'\boldsymbol{q}}, \tag{A2.12}$$

$$\frac{1}{V} \sum_{\boldsymbol{q}} e^{i\boldsymbol{q}(\boldsymbol{r} - \boldsymbol{r}')} = \delta(\boldsymbol{r} - \boldsymbol{r}'). \tag{A2.13}$$

On deriving (A2.8) we have further used

$$\sigma(\boldsymbol{q}, \boldsymbol{q}', \omega) = \frac{1}{V} \iint_V d\boldsymbol{r} \, d\boldsymbol{r}' \int d(t - t') \, \sigma(\boldsymbol{r}, \boldsymbol{r}', t - t') \, e^{-i(\boldsymbol{qr} - \boldsymbol{q}'\boldsymbol{r}' - \omega(t - t'))}. \tag{A2.14}$$

In a crystal the translational symmetry ensures that

$$\sigma(\boldsymbol{r}, \boldsymbol{r}', t - t') = \sigma(\boldsymbol{r} + \boldsymbol{x}(l), \boldsymbol{r}' + \boldsymbol{x}(l), t - t') \tag{A2.15}$$

($\boldsymbol{x}(l)$ — lattice vector), from which originates the property that the Fourier transform of a response tensor such as $\sigma(\boldsymbol{q}, \boldsymbol{q}', \omega)$ is non-zero only if $\boldsymbol{q}$ and $\boldsymbol{q}'$ differ by a reciprocal lattice vector. That is, we have

$$\sigma(\boldsymbol{q}, \boldsymbol{q}', \omega) = \sigma(\boldsymbol{k} + \boldsymbol{K}, \boldsymbol{k} + \boldsymbol{K}', \omega), \tag{A2.16}$$

where $\boldsymbol{k}$ is restricted to the first Brillouin zone (cf. (1.3.33)) and $\boldsymbol{K}, \boldsymbol{K}'$ are reciprocal lattice vectors. In the following we use the convention $\boldsymbol{q} = \boldsymbol{k} + \boldsymbol{K}$, $\boldsymbol{q}' = \boldsymbol{k} + \boldsymbol{K}'$.

As already mentioned in section 1.5., in studying electrodynamics in crystals, a macroscopic quantity is obtained by averaging the corresponding microscopic quantity over the volume of a primitive unit cell, which is equivalent to neglecting all Fourier components with $\boldsymbol{K} \neq 0$.

For weak fields the induced current density and the macroscopic total field $\boldsymbol{E}$ are connected by the macroscopic conductivity tensor:

$$\boldsymbol{j}^{\text{ind}}(\boldsymbol{k}, \omega) = \sigma^{\text{M}}(\boldsymbol{k}, \omega) \, \boldsymbol{E}(\boldsymbol{k}, \omega). \tag{A2.17}$$

On averaging (A2.8) we obtain

$$\boldsymbol{j}^{\text{ind}}(\boldsymbol{k}, \omega) = \sum_{\boldsymbol{K}'} \boldsymbol{\sigma}(\boldsymbol{k}, \boldsymbol{k} + \boldsymbol{K}', \omega) \boldsymbol{E}^{\text{int}}(\boldsymbol{k} + \boldsymbol{K}', \omega). \tag{A2.18}$$

When we neglect local field corrections (i.e. replace the actual internal field by its average) and take into account that the macroscopic electric field is given by the average of $\boldsymbol{E}^{\text{int}}$, since the average of $\boldsymbol{E}^{\text{eq}}$ vanishes, we obtain from (A2.17), (A2.18)

$$\boldsymbol{\sigma}^{\text{M}}(\boldsymbol{k}, \omega) = \boldsymbol{\sigma}(\boldsymbol{k}, \boldsymbol{k}, \omega). \tag{A2.19}$$

Local field corrections can be taken into account by expressing the Fourier components $\boldsymbol{E}^{\text{int}}(\boldsymbol{k} + \boldsymbol{K}', \omega)$ in (A2.18) by $\boldsymbol{E}(\boldsymbol{k}, \omega)$. To do this we rewrite (A2.5) and (A2.6) as

$$\boldsymbol{E}^{\text{int}}(\boldsymbol{q}, \omega) = -\frac{4\pi i}{\omega} \boldsymbol{K}(\boldsymbol{q}, \omega) [\boldsymbol{j}^{\text{ext}}(\boldsymbol{q}, \omega) + \boldsymbol{j}^{\text{ind}}(\boldsymbol{q}, \omega)] \tag{A2.20}$$

and the corresponding equations for the external field as

$$\boldsymbol{E}^{\text{ext}}(\boldsymbol{q}, \omega) = -\frac{4\pi i}{\omega} \boldsymbol{K}(\boldsymbol{q}, \omega) \boldsymbol{j}^{\text{ext}}(\boldsymbol{q}, \omega). \tag{A2.21}$$

Here $\boldsymbol{K}(\boldsymbol{q}, \omega)$ is the tensor defined by (1.5.40), and we have used the same procedure as in section 1.5. to obtain (1.5.39) from (1.5.31), (1.5.32). Inserting (A2.21) and (A2.8) into (A2.20), we get

$$\boldsymbol{E}^{\text{int}}(\boldsymbol{q}, \omega) = \sum_{\boldsymbol{q}'} \boldsymbol{\Theta}^{-1}(\boldsymbol{q}, \boldsymbol{q}', \omega) \boldsymbol{E}^{\text{ext}}(\boldsymbol{q}', \omega), \tag{A2.22}$$

where $\boldsymbol{\Theta}^{-1}$ is the reciprocal of the tensor

$$\boldsymbol{\Theta}(\boldsymbol{q}, \boldsymbol{q}', \omega) = \frac{4\pi i}{\omega} \boldsymbol{K}(\boldsymbol{q}, \omega) \boldsymbol{\sigma}(\boldsymbol{q}, \boldsymbol{q}', \omega) + \delta_{\boldsymbol{q}\boldsymbol{q}'} \boldsymbol{1}, \tag{A2.23}$$

defined by

$$\sum_{\boldsymbol{q}''} \boldsymbol{\Theta}^{-1}(\boldsymbol{q}, \boldsymbol{q}'', \omega) \boldsymbol{\Theta}(\boldsymbol{q}'', \boldsymbol{q}', \omega) = \delta_{\boldsymbol{q}\boldsymbol{q}'} \boldsymbol{1}. \tag{A2.24}$$

Using two times (A2.22) and taking into account that $\boldsymbol{E}^{\text{ext}}$ is a macroscopic field, we obtain from (A2.17) and (A2.18)

$$\boldsymbol{\sigma}^{\text{M}}(\boldsymbol{k}, \omega) = \sum_{\boldsymbol{K}} \boldsymbol{\sigma}(\boldsymbol{k}, \boldsymbol{k} + \boldsymbol{K}, \omega) \boldsymbol{\Theta}^{-1}(\boldsymbol{k} + \boldsymbol{K}, \boldsymbol{k}, \omega) [\boldsymbol{\Theta}^{-1}(\boldsymbol{k}, \boldsymbol{k}, \omega)]^{-1}. \tag{A2.25}$$

This formula formally takes into account local field correction.

Now we introduce another two response tensors by

$$\boldsymbol{j}^{\text{ind}}(\boldsymbol{q}, \omega) = \sum_{\boldsymbol{q}'} \tilde{\boldsymbol{\sigma}}(\boldsymbol{q}, \boldsymbol{q}', \omega) \boldsymbol{E}^{\text{ext}}(\boldsymbol{q}', \omega), \tag{A2.26}$$

$$\boldsymbol{j}^{\text{ext}}(\boldsymbol{q}, \omega) = -\frac{\omega}{4\pi i} \sum_{\boldsymbol{q}'} \boldsymbol{T}(\boldsymbol{q}, \boldsymbol{q}', \omega) \boldsymbol{E}^{\text{int}}(\boldsymbol{q}', \omega). \tag{A2.27}$$

$\tilde{\boldsymbol{\sigma}}$ is called microscopic quasi-conductivity tensor. To relate the tensors $\boldsymbol{\sigma}$, $\tilde{\boldsymbol{\sigma}}$ and $\boldsymbol{T}$ to each other, we insert (A2.27) and (A2.8) into (A2.20), which yields

$$\boldsymbol{T}(\boldsymbol{q}, \boldsymbol{q}', \omega) = \frac{4\pi i}{\omega} \boldsymbol{\sigma}(\boldsymbol{q}, \boldsymbol{q}', \omega) + \delta_{\boldsymbol{q}\boldsymbol{q}'} \boldsymbol{K}^{-1}(\boldsymbol{q}, \omega), \tag{A2.28}$$

and with (A2.23) we obtain

$$\boldsymbol{T}(\boldsymbol{q}, \boldsymbol{q}', \omega) = \boldsymbol{K}^{-1}(\boldsymbol{q}, \omega)\, \boldsymbol{\Theta}(\boldsymbol{q}, \boldsymbol{q}', \omega). \tag{A2.29}$$

A relation between $\boldsymbol{T}$ and $\tilde{\boldsymbol{\sigma}}$ can be obtained by means of (A2.27), (A2.20), (A2.21) and (A2.26):

$$\frac{4\pi i}{\omega}\, \tilde{\boldsymbol{\sigma}}(\boldsymbol{q}, \boldsymbol{q}', \omega) = \delta_{\boldsymbol{q}\boldsymbol{q}'} \boldsymbol{K}^{-1}(\boldsymbol{q}, \omega) - \boldsymbol{K}^{-1}(\boldsymbol{q}, \omega)\, \boldsymbol{T}^{-1}(\boldsymbol{q}, \boldsymbol{q}', \omega)\, \boldsymbol{K}^{-1}(\boldsymbol{q}', \omega). \tag{A2.30}$$

Thus, $\boldsymbol{\sigma}$ and $\boldsymbol{T}$ can be traced back to $\tilde{\boldsymbol{\sigma}}$.

Now we outline the way of calculating $\tilde{\boldsymbol{\sigma}}$ by linear response theory. To this end we consider the induced current density, which is given by

$$\boldsymbol{j}^{\text{ind}}(\boldsymbol{r}, t) = \text{Tr}\, \{\hat{\varrho}(t)\, \boldsymbol{J}(\boldsymbol{r}, t)\}. \tag{A2.31}$$

Here $\hat{\varrho}(t)$ is the statistical operator of the crystal in the presence of an external electromagnetic field. It satisfies the equation of motion

$$\dot{\hat{\varrho}} = \frac{i}{\hbar}\, [\hat{\varrho}, \mathscr{H}], \tag{A2.32}$$

where $\mathscr{H}$ is the Hamiltonian of the crystal in the external field; $\boldsymbol{J}(\boldsymbol{r}, t)$ is the current density operator in the external field:

$$\boldsymbol{J}(\boldsymbol{r}, t) = \boldsymbol{j}(\boldsymbol{r}) - \frac{e}{mc}\, \varrho(\boldsymbol{r})\, \boldsymbol{A}^{\text{ext}}(\boldsymbol{r}, t), \tag{A2.33}$$

where e is the electron charge, m the electron mass and c the speed of light; $\boldsymbol{j}(\boldsymbol{r})$ is the current density operator for absent field

$$\boldsymbol{j}(\boldsymbol{r}) = \frac{e}{2m} \sum_n \{\boldsymbol{p}_n, \delta(\boldsymbol{r} - \boldsymbol{r}_n)\} \tag{A2.34}$$

($\boldsymbol{p}_n$, $\boldsymbol{r}_n$ — momentum operator and position vector of the nth electron, respectively); $\varrho(\boldsymbol{r})$ is the charge density operator

$$\varrho(\boldsymbol{r}) = e \sum_n \delta(\boldsymbol{r} - \boldsymbol{r}_n), \tag{A2.35}$$

and $\boldsymbol{A}^{\text{ext}}(\boldsymbol{r}, t)$ is the vector potential of the external electromagnetic field. Using Coulomb calibration (vanishing scalar potential), which is adopted in the following, we have

$$\boldsymbol{A}^{\text{ext}}(\boldsymbol{r}, t) = -c \int_{-\infty}^{t} \mathrm{d}t'\, \boldsymbol{E}^{\text{ext}}(\boldsymbol{r}, t') \tag{A2.36}$$

To solve equation (A2.32), we decompose $\mathscr{H}$ as

$$\mathscr{H}(t) = H + H'(t), \tag{A2.37}$$

where H is the Hamiltonian of the crystal and $H'(t)$ is a small term describing the interaction of the crystal with the external field. H' is assumed to be turned on adiabatically at $t \to -\infty$. We write

$$\hat{\varrho}(t) \simeq \hat{\varrho}_0 + \hat{\varrho}_1(t), \tag{A2.38}$$

where $\hat{\varrho}_0 \sim (H')^0$ is the equilibrium statistical operator, and $\hat{\varrho}_1 \sim H'$ is the first order correction satisfying

$$\dot{\hat{\varrho}}_1 = \frac{i}{\hbar} [\hat{\varrho}_0, H'] + \frac{i}{\hbar} [\hat{\varrho}_1, H]. \tag{A2.39}$$

This equation can be readily solved by making use of the interaction representation $\hat{\varrho}_1{}^{\mathrm{I}} = \exp\left(\frac{i}{\hbar} Ht\right) \hat{\varrho}_1 \exp\left(-\frac{i}{\hbar} Ht\right)$. Thus we find

$$\hat{\varrho}(t) = \hat{\varrho}_0 + \frac{i}{\hbar} \int\limits_{-\infty}^{t} \mathrm{d}t' \left[\hat{\varrho}_0, \mathrm{e}^{\frac{i}{\hbar} H(t'-t)} H'(t')\, \mathrm{e}^{-\frac{i}{\hbar} H(t'-t)}\right]. \tag{A2.40}$$

The interaction operator H' is given by

$$\left.\begin{aligned} H'(t) &= \frac{1}{2m} \sum_n \left[\left(\boldsymbol{p}_n - \frac{e}{c} \boldsymbol{A}^{\mathrm{ext}}(\boldsymbol{r}_n, t)\right)^2 - \boldsymbol{p}_n{}^2\right] \\ &\simeq -\frac{1}{c} \int \mathrm{d}\boldsymbol{r}\, \boldsymbol{j}(\boldsymbol{r})\, \boldsymbol{A}^{\mathrm{ext}}(\boldsymbol{r}, t), \end{aligned}\right\} \tag{A2.41}$$

where the last form has been obtained by retaining the term linear in the field only.
Inserting (A2.33), (A2.36), (A2.40) and (A2.41) into (A2.31) we get

$$\begin{aligned} \boldsymbol{j}^{\mathrm{ind}}(\boldsymbol{r}, t) &= \frac{e}{m} \langle \varrho(\boldsymbol{r}) \rangle \int\limits_{-\infty}^{t} \mathrm{d}t' \boldsymbol{E}^{\mathrm{ext}}(\boldsymbol{r}, t') \\ &+ \frac{1}{i\hbar} \int\limits_{-\infty}^{t} \mathrm{d}t' \int\limits_{-\infty}^{t'} \mathrm{d}t'' \int \mathrm{d}\boldsymbol{r}' \langle [\boldsymbol{j}(\boldsymbol{r}, t), \boldsymbol{j}(\boldsymbol{r}', t')] \rangle\, \boldsymbol{E}^{\mathrm{ext}}(\boldsymbol{r}', t''). \end{aligned} \tag{A2.42}$$

Here

$$\boldsymbol{j}(\boldsymbol{r}, t) = \mathrm{e}^{\frac{i}{\hbar} Ht} \boldsymbol{j}(\boldsymbol{r})\, \mathrm{e}^{-\frac{i}{\hbar} Ht} \tag{A2.43}$$

and

$$\langle \ldots \rangle = \mathrm{Tr}\, \{\hat{\varrho}_0 \ldots\} \tag{A2.44}$$

denotes the thermal average in equilibrium. Accounting for $\int\limits_{-\infty}^{t} \mathrm{d}t' \int\limits_{-\infty}^{t'} \mathrm{d}t'' = \int\limits_{-\infty}^{t} \mathrm{d}t'' \int\limits_{t''}^{t} \mathrm{d}t'$ and taking the Fourier transform from (A2.42), we find

$$\tilde{\sigma}(\boldsymbol{q}, \boldsymbol{q}', \omega) = \frac{i}{\omega + i\varepsilon} \left[\frac{e}{m} \frac{1}{V} \langle \varrho(\boldsymbol{q} - \boldsymbol{q}') \rangle \mathbf{1} - \boldsymbol{\chi}(\boldsymbol{q}, \boldsymbol{q}', \omega + i\varepsilon)\right], \qquad \varepsilon \to +0. \tag{A2.45}$$

Here $\boldsymbol{\chi}(\boldsymbol{q}, \boldsymbol{q}', \omega)$ is the Fourier transform of the current-density current-density response tensor

$$\chi_{\alpha\alpha'}(\boldsymbol{r}, \boldsymbol{r}', t) = \frac{i}{\hbar}\, \Theta(t)\, \langle [j_\alpha(\boldsymbol{r}, t), j_{\alpha'}(\boldsymbol{r}', 0)] \rangle, \tag{A2.46}$$

where $\Theta(t)$ is the Heaviside step function

$$\Theta(t) = \begin{cases} 1 \\ 0 \end{cases} \text{for} \quad t \gtrless 0. \tag{A2.47}$$

Furthermore,

$$\varrho(\boldsymbol{q}) = e \sum_n \mathrm{e}^{-i\boldsymbol{q}\boldsymbol{r}_n} \tag{A2.48}$$

is the Fourier transform of the charge density operator (A2.35).

The Fourier transform of the charge-density charge-density response function

$$\chi_\varrho(\boldsymbol{r}, \boldsymbol{r}', t) = \frac{i}{\hbar} \Theta(t) \langle [\varrho(\boldsymbol{r}, t), \varrho(\boldsymbol{r}', 0)] \rangle, \tag{A2.49}$$

where the t-dependence of $\varrho(\boldsymbol{r}, t)$ is to be understood as in (A2.43), is connected with χ as

$$\sum_{\alpha\alpha'} q_\alpha \chi_{\alpha\alpha'}(\boldsymbol{q}, \boldsymbol{q}', \omega)\, q'_{\alpha'} = \frac{e}{m} \frac{1}{V} \langle \varrho(\boldsymbol{q} - \boldsymbol{q}') \rangle\, \boldsymbol{q}\boldsymbol{q}' + \omega^2 \chi_\varrho(\boldsymbol{q}, \boldsymbol{q}', \omega). \tag{A2.50}$$

This can be shown by using two times the continuity equation for replacing the current density operators (appearing in χ) by charge density operators.

Now we introduce a tensor $\hat{\chi}$ by

$$\sigma(\boldsymbol{q}, \boldsymbol{q}', \omega) = \frac{i}{\omega + i\varepsilon} \left[\frac{e}{m} \frac{1}{V} \langle \varrho(\boldsymbol{q} - \boldsymbol{q}') \rangle\, \mathbf{1} - \hat{\chi}(\boldsymbol{q}, \boldsymbol{q}', \omega) \right], \quad \varepsilon \to +0, \tag{A2.51}$$

and the polarizability function $\hat{\chi}_\varrho$ by

$$\sum_{\alpha\alpha'} q_\alpha \hat{\chi}_{\alpha\alpha'}(\boldsymbol{q}, \boldsymbol{q}', \omega)\, q'_{\alpha'} = \frac{e}{m} \frac{1}{V} \langle \varrho(\boldsymbol{q} - \boldsymbol{q}') \rangle\, \boldsymbol{q}\boldsymbol{q}' + \omega^2 \hat{\chi}_\varrho(\boldsymbol{q}, \boldsymbol{q}', \omega). \tag{A2.52}$$

Below we give explicit expressions for $\hat{\chi}$ and $\hat{\chi}_\varrho$ by making use of the Hartree approximation.

From (A2.51) and (A2.52) we obtain

$$\sum_{\alpha\alpha'} q_\alpha \sigma_{\alpha\alpha'}(\boldsymbol{q}, \boldsymbol{q}', \omega)\, q'_{\alpha'} = -i\omega \hat{\chi}_\varrho(\boldsymbol{q}, \boldsymbol{q}', \omega). \tag{A2.53}$$

Defining the microscopic dielectric tensor $\boldsymbol{\varepsilon}$ by

$$\boldsymbol{\varepsilon}(\boldsymbol{q}, \boldsymbol{q}', \omega) = \delta_{\boldsymbol{q}\boldsymbol{q}'} \mathbf{1} + \frac{4\pi i}{\omega} \sigma(\boldsymbol{q}, \boldsymbol{q}', \omega) \tag{A2.54}$$

and the dielectric matrix ϵ by

$$\epsilon(\boldsymbol{q}, \boldsymbol{q}', \omega) = \sum_{\alpha\alpha'} \hat{q}_\alpha \varepsilon_{\alpha\alpha'}(\boldsymbol{q}, \boldsymbol{q}', \omega)\, \hat{q}'_{\alpha'} \tag{A2.55}$$

($\hat{\boldsymbol{q}} = \boldsymbol{q}/|\boldsymbol{q}|$), then we have

$$\epsilon(\boldsymbol{q}, \boldsymbol{q}', \omega) = \delta_{\boldsymbol{q}\boldsymbol{q}'} + \frac{4\pi}{|\boldsymbol{q}|\, |\boldsymbol{q}'|} \hat{\chi}_\varrho(\boldsymbol{q}, \boldsymbol{q}', \omega). \tag{A2.56}$$

Assuming longitudinal fields, the fields can be expressed by scalar potentials as

$$\boldsymbol{E}^{\text{int}}(\boldsymbol{q}, \omega) = -i\boldsymbol{q}\varphi^{\text{int}}(\boldsymbol{q}, \omega). \tag{A2.57}$$

With (A2.57), (A2.8), the continuity equation $\omega \varrho^{\text{ind}}(\boldsymbol{q}, \omega) = \boldsymbol{q} \boldsymbol{j}^{\text{ind}}(\boldsymbol{q}, \omega)$, and (A2.53) we obtain

$$\varrho^{\text{ind}}(\boldsymbol{q}, \omega) = -\sum_{\boldsymbol{q}'} \chi_\varrho(\boldsymbol{q}, \boldsymbol{q}', \omega) \; \varphi^{\text{int}}(\boldsymbol{q}', \omega). \tag{A2.58}$$

Using the Poisson equation

$$\varphi^{\text{ind}}(\boldsymbol{q}, \omega) = \frac{4\pi}{|\boldsymbol{q}|^2} \varrho^{\text{ind}}(\boldsymbol{q}, \omega) \tag{A2.59}$$

and

$$\varphi^{\text{int}}(\boldsymbol{q}, \omega) = \varphi^{\text{ext}}(\boldsymbol{q}, \omega) + \varphi^{\text{ind}}(\boldsymbol{q}, \omega), \tag{A2.60}$$

where φ^{ext} and φ^{ind} respectively are the scalar potentials of $\boldsymbol{E}^{\text{ext}}$ and $\boldsymbol{E}^{\text{ind}}$ we get from (A2.58) and (A2.56)

$$\varphi^{\text{int}}(\boldsymbol{k} + \boldsymbol{K}, \omega) = \sum_{\boldsymbol{K}'} \frac{|\boldsymbol{k} + \boldsymbol{K}'|}{|\boldsymbol{k} + \boldsymbol{K}|} \epsilon^{-1}(\boldsymbol{k} + \boldsymbol{K}, \boldsymbol{k} + \boldsymbol{K}', \omega) \; \varphi^{\text{ext}}(\boldsymbol{k} + \boldsymbol{K}', \omega), \tag{A2.61}$$

where ϵ^{-1} is the inverse of the matrix ϵ (A2.56).

On the other hand we obtain (A2.61), with

$$\epsilon^{-1}(\boldsymbol{q}, \boldsymbol{q}', \omega) = \delta_{\boldsymbol{q}\boldsymbol{q}'} - \frac{4\pi}{|\boldsymbol{q}| \; |\boldsymbol{q}'|} \chi_\varrho(\boldsymbol{q}, \boldsymbol{q}', \omega), \tag{A2.62}$$

when we start from

$$\boldsymbol{E}^{\text{ext}}(\boldsymbol{q}, \omega) = -i\boldsymbol{q}\varphi^{\text{ext}}(\boldsymbol{q}, \omega), \tag{A2.63}$$

and use (A2.26), the continuity equation, (A2.45), (A2.50), the Poisson equation and (A2.60).

In a homogeneous isotropic medium the macroscopic dielectric tensor

$$\boldsymbol{\varepsilon}(\boldsymbol{k}, \omega) = \mathbf{1} + \frac{4\pi i}{\omega} \boldsymbol{\sigma}(\boldsymbol{k}, \omega) \tag{A2.64}$$

has the form

$$\boldsymbol{\varepsilon}(\boldsymbol{k}, \omega) = \mathbf{1}_t(\hat{\boldsymbol{k}}) \; \epsilon_t(\boldsymbol{k}, \omega) + \mathbf{1}_l(\hat{\boldsymbol{k}}) \; \epsilon_l(\boldsymbol{k}, \omega), \tag{A2.65}$$

where ϵ_t and ϵ_l respectively are the macroscopic transverse and longitudinal dielectric functions which are scalar quantities. Note that in the limit $\boldsymbol{k} \to \mathbf{0}$ (A2.65) is true also for an arbitrary direction of propagation in a cubic material, but, in general, it will not be valid in the case of crystals of lower symmetry.

The relation between a macroscopic longitudinal electrical field $\boldsymbol{E}(\boldsymbol{k}, \omega) = -i\boldsymbol{k}\varphi(\boldsymbol{k}, \omega)$ and a macroscopic longitudinal displacement field $\boldsymbol{D}(\boldsymbol{k}, \omega) = -i\boldsymbol{k}\varphi^{\text{ext}}(\boldsymbol{k}, \omega)$ reads

$$\varphi^{\text{ext}}(\boldsymbol{k}, \omega) = \epsilon_l(\boldsymbol{k}, \omega) \; \varphi(\boldsymbol{k}, \omega), \tag{A2.66}$$

where we have used

$$\epsilon_l(\boldsymbol{k}, \omega) = \sum_{\alpha\alpha'} \hat{k}_\alpha \varepsilon_{\alpha\alpha'}(\boldsymbol{k}, \omega) \; \hat{k}_{\alpha'}. \tag{A2.67}$$

On the other hand, when we average (A2.61) and take into account that φ^{ext} is a macroscopic quantity, i.e. put $\boldsymbol{K} = \boldsymbol{K}' = \mathbf{0}$, we obtain for the averaged internal potential

$$\varphi(\boldsymbol{k}, \omega) = \epsilon^{-1}(\boldsymbol{k}, \boldsymbol{k}, \omega) \; \varphi^{\text{ext}}(\boldsymbol{k}, \omega). \tag{A2.68}$$

On comparing (A2.66) and (A2.68) we find

$$\epsilon_1(\boldsymbol{k}, \omega) = [\epsilon^{-1}(\boldsymbol{k}, \boldsymbol{k}, \omega)]^{-1}. \tag{A2.69}$$

When we neglect local field corrections, i.e. put $\chi_\varrho(\boldsymbol{k} + \boldsymbol{K}, \boldsymbol{k} + \boldsymbol{K}', \omega) \approx \delta_{\boldsymbol{K}\boldsymbol{K}'} \times \chi_\varrho(\boldsymbol{k} + \boldsymbol{K}, \boldsymbol{k} + \boldsymbol{K}, \omega)$, then we have $[\epsilon^{-1}(\boldsymbol{k}, \boldsymbol{k}, \omega)]^{-1} = \epsilon(\boldsymbol{k}, \boldsymbol{k}, \omega)$, i.e. $\epsilon_1(\boldsymbol{k}, \omega) = \epsilon(\boldsymbol{k}, \boldsymbol{k}, \omega)$.

Now we calculate $\hat{\chi}$ and $\hat{\chi}_\varrho$ on account of electron-electron interaction in self-consistent field approximation (SCF) or random phase approximation (RPA), respectively (cf. Brener and Fry, 1979, 1980, and references therein). In this approximation we can decompose the Hamiltonian $\mathscr{H}$ of the crystal in the external field into

$$H_{\text{SCF}} = \sum_n H_0(\boldsymbol{p}_n, \boldsymbol{r}_n) = \sum_n \left(\frac{\boldsymbol{p}_n^2}{2m} + V_0(\boldsymbol{r}_n)\right) \tag{A2.70}$$

and

$$H'_{\text{SCF}}(t) = \int_0^t \mathrm{d}t' \int \mathrm{d}\boldsymbol{r} \boldsymbol{j}(\boldsymbol{r}) \boldsymbol{E}^{\text{int}}(\boldsymbol{r}, t'). \tag{A2.71}$$

H_{SCF} describes noninteracting electrons in a potential V_0, which is composed of the ionic potential and the self-consistent potential caused by the electrons in absent field; H'_{SCF}, which is composed of the interaction Hamiltonian H' (A2.41) and the field induced change of the self-consistent potential, describes the interaction of the electrons with $\boldsymbol{E}^{\text{int}}$ and thus takes into account a part of the electron-electron interaction through $\boldsymbol{E}^{\text{ind}}$.

In self-consistent field approximation $\hat{\chi}$ and $\hat{\chi}_\varrho$ are given by expressions (A2.46) and (A2.49), respectively, if H is replaced there by H_{SCF}. To calculate in this approximation the trace occurring in these expressions, we use the Bloch representation defined by

$$H_0 |\boldsymbol{k}\nu\rangle = E_{\boldsymbol{k}\nu} |\boldsymbol{k}\nu\rangle \tag{A2.72}$$

(ν — band index) and the statistical operator

$$\hat{\varrho}_0 = [\mathrm{e}^{\beta(H_0 - \mu)} + 1]^{-1} \tag{A2.73}$$

with

$$\hat{\varrho}_0 |\boldsymbol{k}\nu\rangle = f_0(E_{\boldsymbol{k}\nu}) |\boldsymbol{k}\nu\rangle, \tag{A2.74}$$

where

$$f_0(E_{\boldsymbol{k}\nu}) = [\mathrm{e}^{\beta(E_{\boldsymbol{k}\nu} - \mu)} + 1]^{-1} \tag{A2.75}$$

is the Fermi distribution function. Here $\beta = 1/kT$, and μ is the chemical potential determined by $\sum_{\boldsymbol{k}\nu} f_0(E_{\boldsymbol{k}\nu}) = n$ (n — total number of electrons in the crystal). Thus we obtain

$$\hat{\chi}^{\text{SCF}}_{\alpha\alpha'}(\boldsymbol{q}, \boldsymbol{q}', \omega) = -\left(\frac{e}{2m}\right)^2 \frac{1}{V} \sum_{\substack{\boldsymbol{k}\nu \\ \boldsymbol{k}'\nu'}} \langle \boldsymbol{k}\nu| \{\mathrm{e}^{i\boldsymbol{q}'\boldsymbol{r}}, p_{\alpha'}\} |\boldsymbol{k}'\nu'\rangle \langle \boldsymbol{k}'\nu'| \{\mathrm{e}^{-i\boldsymbol{q}\boldsymbol{r}}, p_\alpha\} |\boldsymbol{k}\nu\rangle$$
$$\times \frac{f_0(E_{\boldsymbol{k}'\nu'}) - f_0(E_{\boldsymbol{k}\nu})}{E_{\boldsymbol{k}'\nu'} - E_{\boldsymbol{k}\nu} + \hbar(\omega + i\varepsilon)}, \quad \varepsilon \to +0, \tag{A2.76}$$

$$\hat{\chi}^{\text{SCF}}_{\varrho}(\boldsymbol{q}, \boldsymbol{q}', \omega) = -\frac{e^2}{V} \sum_{\substack{\boldsymbol{k}\nu \\ \boldsymbol{k}'\nu'}} \langle \boldsymbol{k}\nu| \mathrm{e}^{i\boldsymbol{q}'\boldsymbol{r}} |\boldsymbol{k}'\nu'\rangle \langle \boldsymbol{k}'\nu'| \mathrm{e}^{-i\boldsymbol{q}\boldsymbol{r}} |\boldsymbol{k}\nu\rangle$$
$$\times \frac{f_0(E_{\boldsymbol{k}'\nu'}) - f_0(E_{\boldsymbol{k}\nu})}{E_{\boldsymbol{k}'\nu'} - E_{\boldsymbol{k}\nu} + \hbar(\omega + i\varepsilon)}, \quad \varepsilon \to +0, \tag{A2.77}$$

and

$$\langle \varrho(\boldsymbol{q} - \boldsymbol{q}') \rangle^{\mathrm{SCF}} = e \sum_{\boldsymbol{k}\nu} f_0(E_{\boldsymbol{k}\nu}) \langle \boldsymbol{k}\nu | \, \mathrm{e}^{-i(\boldsymbol{q}-\boldsymbol{q}')\boldsymbol{r}} \, | \boldsymbol{k}\nu \rangle . \tag{A2.78}$$

Now we derive sum rules connecting the charge density response function, the ionic potential and the average electrical charge density. To do this we subject the crystal, in which the atoms are displaced by $\boldsymbol{u}(l\varkappa)$ from the equilibrium position, to an infinitesimal rigid body displacement $\boldsymbol{v}$. The corresponding translation operator $T_{\boldsymbol{v}}$ transforms the Hamiltonian H of the electrons in the field V_i of the ions,

$$V_i(\boldsymbol{r}) = \sum_{l\varkappa} v_\varkappa(\boldsymbol{r} - \boldsymbol{x}(l) - \boldsymbol{x}(\varkappa) - \boldsymbol{u}(l\varkappa)), \tag{A2.79}$$

which is assumed to be additively composed of contributions $v_\varkappa$ of the single ions as follows:

$$T_{\boldsymbol{v}} H T_{\boldsymbol{v}}^{-1} = H + H', \tag{A2.80}$$

where

$$\begin{aligned} H' &= \sum_n [V_i(\boldsymbol{r}_n + \boldsymbol{v}) - V_i(\boldsymbol{r}_n)] \simeq \sum_n \boldsymbol{v} \nabla_{\boldsymbol{r}_n} V_i(\boldsymbol{r}_n) \\ &= \frac{1}{eV} \sum_{l\varkappa\boldsymbol{q}} i\boldsymbol{q}\boldsymbol{v}\varrho(-\boldsymbol{q}) \, \mathrm{e}^{-i\boldsymbol{q}(\boldsymbol{x}(l)+\boldsymbol{x}(\varkappa)+\boldsymbol{u}(l\varkappa))} v_\varkappa(\boldsymbol{q}) . \end{aligned} \tag{A2.81}$$

Inserting $T_{\boldsymbol{v}}^{-1} T_{\boldsymbol{v}} = 1$ into the expression for $\langle \varrho(\boldsymbol{q}) \rangle$ (cf. (A2.44)) and making use of the invariance of the trace against a cyclic permutation of the operators in it, we obtain

$$\langle \varrho(\boldsymbol{q}) \rangle = \mathrm{Tr} \, \{ \hat{\varrho}_0 (H + H') \, \varrho(\boldsymbol{q}) \} \, \mathrm{e}^{-i\boldsymbol{q}\boldsymbol{v}}, \tag{A2.82}$$

where we have accounted for

$$T_{\boldsymbol{v}} \varrho(\boldsymbol{q}) \, T_{\boldsymbol{v}}^{-1} = \mathrm{e}^{-i\boldsymbol{q}\boldsymbol{v}} \varrho(\boldsymbol{q}) . \tag{A2.83}$$

Using for $\hat{\varrho}_0$ the Gibbs statistical operator

$$\hat{\varrho}_0 = \mathrm{e}^{-\beta(H - \mu\hat{N})} / \mathrm{Tr} \, \{ \mathrm{e}^{-\beta(H - \mu\hat{N})} \} \tag{A2.84}$$

($\hat{N}$ — particle number operator) and the formula

$$\mathrm{e}^{-(A+B)} \simeq \mathrm{e}^{-A} \left[1 - \int_0^1 \mathrm{d}\lambda \, \mathrm{e}^{\lambda A} B \, \mathrm{e}^{-\lambda A} \right], \tag{A2.85}$$

where B is a small operator as compared with A, we obtain by (A2.82), from the term linear in $\boldsymbol{v}$,

$$\begin{aligned} -e\boldsymbol{q} \langle \varrho(\boldsymbol{q}) \rangle &= \frac{1}{V} \sum_{l\varkappa\boldsymbol{q}'} \int_0^\beta \mathrm{d}\lambda \langle \varrho(-\boldsymbol{q}') \, [\mathrm{e}^{-\lambda(H - \mu\hat{N})} \varrho(\boldsymbol{q}) \, \mathrm{e}^{\lambda(H - \mu\hat{N})} - \langle \varrho(\boldsymbol{q}) \rangle] \rangle \\ &\quad \times \boldsymbol{q}' \mathrm{e}^{-i\boldsymbol{q}'(\boldsymbol{x}(l)+\boldsymbol{x}(\varkappa)+\boldsymbol{u}(l\varkappa))} v_\varkappa(\boldsymbol{q}') \\ &= \sum_{l\varkappa\boldsymbol{q}'} \chi_\varrho(\boldsymbol{q}, \boldsymbol{q}', \omega = 0) \, \boldsymbol{q}' \mathrm{e}^{-i\boldsymbol{q}'(\boldsymbol{x}(l)+\boldsymbol{x}(\varkappa)+\boldsymbol{u}(l\varkappa))} v_\varkappa(\boldsymbol{q}') . \end{aligned} \tag{A2.86}$$

To obtain the last line of equation (A2.86), one has to substitute in the first line the integration variable as $i\hbar\lambda = t$, and to take into account that in the complex t-plane the resulting integral from 0 to $i\hbar\beta$ can be replaced by an integral from 0 to ∞

minus another one from $i\hbar\beta$ to $i\hbar\beta + \infty$. Furthermore, one must account for $[\hat{N}, \varrho(\boldsymbol{q})] = 0$ (for more detail see, for example, BRAUER and STREITWOLF, 1977, appendix 6).

For an ideal crystal with $\boldsymbol{u}(l\varkappa) = 0$ (with $\boldsymbol{q}' = \boldsymbol{k} + \boldsymbol{K}'$ and $\boldsymbol{q} = \boldsymbol{k} + \boldsymbol{K}$, and by (1.3.65)) we obtain from (A2.86) the so-called Keating sum rules

$$N \sum_{\boldsymbol{K}'\varkappa} \chi_\varrho(\boldsymbol{K}, \boldsymbol{K}', 0)\, \boldsymbol{K}'\, \mathrm{e}^{-i\boldsymbol{K}'\boldsymbol{x}(\varkappa)} = -\mathrm{e}\boldsymbol{K}\langle\varrho(\boldsymbol{K})\rangle. \tag{A2.87}$$

For small $\boldsymbol{u}(l\varkappa)$ we obtain (with $\boldsymbol{q} = \boldsymbol{k} + \boldsymbol{K}$) the dominant contribution to the $\boldsymbol{q}'$-sum in (A2.86) from the terms with $\boldsymbol{q}' \approx \boldsymbol{k} + \boldsymbol{K}'$. Thus, we find from (A2.86) in the limit $\boldsymbol{k} \to 0$, $\boldsymbol{u}(l\varkappa) \to \boldsymbol{0}$ the so-called acoustic sum rules

$$\lim_{\boldsymbol{k}\to 0} \left[N \sum_{\boldsymbol{K}'\varkappa} \chi_\varrho(\boldsymbol{q}, \boldsymbol{q}', 0)\, \boldsymbol{q}'\, \mathrm{e}^{-i\boldsymbol{K}'\boldsymbol{x}(\varkappa)} \boldsymbol{v}_\varkappa(\boldsymbol{q}') + \mathrm{e}\boldsymbol{q}\langle\varrho(\boldsymbol{K})\rangle\right] = \boldsymbol{0}. \tag{A2.88}$$

For a review of the present state of the theory of the dielectric matrix of matter the reader is referred to DOLGOV et al. (1981) (see also CAR et al., 1981, and references therein).

APPENDIX 3
EXPERIMENTAL QUANTITIES

A3.1. Interaction of phonons with infrared radiation

Let us consider a system consisting of electrons and ions, which interacts with an electromagnetic plane wave. On choosing the Lorentz gauge and vanishing scalar potential

$$\nabla \boldsymbol{A} = 0, \qquad \varphi = 0 \tag{A3.1}$$

($\boldsymbol{A}$ — vector potential, φ — scalar potential) the interaction of the radiation with the electrons and ions is described by the Hamiltonian

$$H'(t) = -\frac{e}{mc} \sum_j \boldsymbol{A}_0 \boldsymbol{p}_j e^{i(\boldsymbol{q}\boldsymbol{r}_j - \omega t)} - \frac{Z}{Mc} \sum_l \boldsymbol{A}_0 \, e^{i(\boldsymbol{q}\boldsymbol{R}_l - \omega t)} \boldsymbol{P}_l, \tag{A3.2}$$

where we have accounted for the term linear in the field only. Here $\boldsymbol{r}_j, \boldsymbol{p}_j$ respectively denote the position and the momentum of the electrons and $\boldsymbol{R}_l, \boldsymbol{P}_l$ the corresponding quantities of the ions; e, m respectively are the charge and the mass of an electron, and Z, M are the corresponding quantities of an ion; $\boldsymbol{k}$ is the wave vector and ω the frequency of the electromagnetic wave with vector amplitude $\boldsymbol{A}_0$.

Assuming the electrons in the system to be strictly localized at ion sites, we replace in (A3.2) $\boldsymbol{r}_j$ by $\boldsymbol{R}_l$ and the index j on the electron variables by l. Thus, H' becomes

$$H'(t) = -\frac{1}{c} \sum_l \boldsymbol{A}_0 \, e^{i(\boldsymbol{q}\boldsymbol{R}_l - \omega t)} \left(\frac{e}{m} \boldsymbol{p}_l + \frac{Z}{M} \boldsymbol{P}_l\right). \tag{A3.3}$$

Introducing the dipole moment operator $\boldsymbol{\mu}_l$ by

$$\boldsymbol{\mu}_l = e\boldsymbol{r}_l + Z\boldsymbol{R}_l, \tag{A3.4}$$

then we can write

$$\frac{e}{m} \boldsymbol{p}_l + \frac{Z}{M} \boldsymbol{P}_l = \frac{d}{dt} \mu_l = \frac{i}{\hbar} [H, \mu_l], \tag{A3.5}$$

where H is the Hamiltonian of the electron-ion system. Time-dependent perturbation theory provides the transition rate $W_{i \to f}$ for a transition from an initial state Ψ_i to a final state Ψ_f produced by H'

$$W_{i \to f} = \frac{2\pi}{\hbar} \sum_f |M_{fi}|^2 \, \delta(E_f - E_i - \hbar\omega), \tag{A3.6}$$

where

$$M_{fi} = \langle \Psi_f | \sum_l \left(\frac{1}{c}\right) e^{i\boldsymbol{q}\boldsymbol{R}_l} \boldsymbol{A}_0 \left(\frac{e}{m} \boldsymbol{p}_l + \frac{Z}{M} \boldsymbol{P}_l\right) | \Psi_i \rangle. \tag{A3.7}$$

Here E_i, E_f are the energy eigenvalues of H to Ψ_i and Ψ_f, respectively. Making use of (A3.5), M_{fi} can be rewritten as

$$M_{fi} = \frac{i}{\hbar} (E_f - E_i) A_0 \frac{1}{c} \sum_l e^{iqR_l} \langle \Psi_f | \mu_l | \Psi_i \rangle. \tag{A3.8}$$

In (A3.8) we express Ψ_f and Ψ_i by adiabatic wave functions (cf. (1.1.10)) and we assume that these functions are exact eigenfunctions of H. Since infrared processes leave the electronic state unchanged, we have

$$\left.\begin{aligned} \Psi_i(\boldsymbol{r}, \boldsymbol{R}) &= \chi_{nv}(\boldsymbol{R})\, \psi_n(\boldsymbol{r}, \boldsymbol{R}), \\ \Psi_f(\boldsymbol{r}, \boldsymbol{R}) &= \chi_{nv'}(\boldsymbol{R})\, \psi_n(\boldsymbol{r}, \boldsymbol{R}), \end{aligned}\right\} \tag{A3.9}$$

The corresponding energies are E_{nv} and $E_{nv'}$, respectively (the notation used here agrees with that in section 1.1.). With (A3.9) we obtain from (A3.8)

$$M_{fi} = \frac{i}{\hbar} (E_{nv'} - E_{nv}) A_0 \frac{1}{c} \int d\boldsymbol{R} \chi^*_{nv'}(\boldsymbol{R})\, \boldsymbol{M}(\boldsymbol{R}, \boldsymbol{q})\, \chi_{nv}(\boldsymbol{R}), \tag{A3.10}$$

where

$$\boldsymbol{M}(\boldsymbol{R}, \boldsymbol{q}) = \sum_l e^{i\boldsymbol{q}\boldsymbol{R}_l^0} \int d\boldsymbol{r} \psi_n^*(\boldsymbol{r}, \boldsymbol{R})\, \mu_l(\boldsymbol{r}, \boldsymbol{R})\, \psi_n(\boldsymbol{r}, \boldsymbol{R}) \tag{A3.11}$$

is the "phased" dipole moment. $\int d\boldsymbol{R}$ and $\int d\boldsymbol{r}$ imply integration over all electron and all ion coordinates, respectively. $\boldsymbol{R}_l^0$ is the equilibrium position of the lth ion (assumption of small displacements of the ions about definite equilibrium positions). Notice that the dipole moment (A3.11) depends on the ion coordinates via μ_l and via the electron functions $\psi_n(\boldsymbol{r}, \boldsymbol{R})$. The latter dependence reflects the deformability of the electron clouds of the atoms.

Before we proceed further, we discuss some aspects of the derivation of (A3.10) and (A3.11). Introducing the parameter of smallness in the adiabatic theory $\varkappa = (m/M)^{1/4}$ (cf. section 1.1.) we see that in (A3.2) the first term on the right-hand side is proportional to $\varkappa^0$, but the second one to $\varkappa^3$. However, it turns out that the zeroth order contribution to the dipole moment vanishes and that the first term on the right-hand side of (A3.2) also gives a contribution proportional to $\varkappa^3$ to the dipole moment (cf. Walnut and Nafie, 1977). Furthermore, the complete following of an ion by an electron orbital assumed in (A3.3) is realized, strictly speaking, only if the orbital can be expressed completely in terms of the relative position of the ion and the electron. This is, however, not the case, when the position of the orbital depends on more than one ion. Usually one puts in (A3.11) $\boldsymbol{q} \approx \boldsymbol{0}$, i.e. one assumes any dimension of the system to be smaller than the wavelength of the radiation. For simplicity, in the following we also do so here.

On account of the equation

$$\boldsymbol{E} = -\frac{1}{c} \dot{\boldsymbol{A}}, \tag{A3.12}$$

where $\boldsymbol{E}$ is the electric field, we see that the transition probability given by (A3.10), (A3.11)

$$W_{i\to f} = \frac{2\pi}{\hbar} \sum_{v'} |\langle \chi_{nv'}(\boldsymbol{R})| \frac{\omega}{c} \boldsymbol{A}_0 \boldsymbol{M} |\chi_{nv}(\boldsymbol{R})\rangle|^2\, \delta(E_{nv'} - E_{nv} - \hbar\omega) \tag{A3.13}$$

can be obtained also by assuming the radiation interacting with lattice vibrations by an interaction Hamiltonian

$$H'(t) = -\boldsymbol{M}\boldsymbol{E}(t) = -\boldsymbol{M}\boldsymbol{E}_0\, e^{-i\omega t + \varepsilon t}, \quad \varepsilon \to +0, \tag{A3.14}$$

where $\boldsymbol{M} = \boldsymbol{M}(\boldsymbol{R}, 0)$ and where in the last form the electric field is assumed to be adiabatically turned on at $t \to -\infty$. For weak radiation fields we derive the dielectric susceptibility $\chi_{\alpha\alpha'}(\omega)$ whose imaginary part describes the infrared absorption by means of the linear response theory (cf. (A2.40)). Then we find for the field induced change of the electric dipole moment

$$\Delta\boldsymbol{M} = \langle\boldsymbol{M}\rangle - \langle\boldsymbol{M}\rangle_0 = -\frac{i}{\hbar}\int\limits_{-\infty}^{\infty} dt'\Theta(t-t')\,\langle[\boldsymbol{M}(t), H'(t')]\rangle_0, \tag{A3.15}$$

where

$$\langle\ldots\rangle_0 = \mathrm{Tr}\,\{e^{-\beta H}\ldots\}/\mathrm{Tr}\,\{e^{-\beta H}\}, \quad \beta = \frac{1}{kT}. \tag{A3.16}$$

Here H is the Hamiltonian of the vibrational system, and $\boldsymbol{M}(t)$ and $H'(t)$ respectively denote $\boldsymbol{M}$ and H' in interaction representation

$$\boldsymbol{M}(t) = e^{\frac{i}{\hbar}Ht}\boldsymbol{M}e^{-\frac{i}{\hbar}Ht}. \tag{A3.17}$$

Using (A3.14) we obtain from (A3.15) after simple manipulations

$$\Delta\boldsymbol{M} = V\boldsymbol{\chi}(\omega)\,\boldsymbol{E}_0\,e^{-i\omega t+\varepsilon t}, \tag{A3.18}$$

where

$$\chi_{\alpha\alpha'}(\omega) = -\frac{1}{V}\int\limits_{-\infty}^{\infty} dt\left(-\frac{i}{\hbar}\right)\Theta(t)\,\langle[M_\alpha(t), M_{\alpha'}(0)]\rangle\,e^{i\omega t-\varepsilon t} \tag{A3.19}$$

is the electronic susceptibility and V is the volume of the system. For a cubic system we have $\chi_{\alpha\alpha'} = \delta_{\alpha\alpha'}\chi_{\alpha\alpha}$, and the absorption coefficient $\alpha(\omega)$ is given by

$$\alpha(\omega) = \frac{4\pi\omega}{c'}\,\mathrm{Im}\,\chi_{\alpha\alpha}(\omega), \tag{A3.20}$$

where c' is the velocity of light in the medium. Note that (A3.20) can be obtained also directly from (A3.13) by representing the δ-function by an integral and using the completeness of the vibrational functions χ_{nv}.

$\boldsymbol{M}$ can be expanded in a power series with respect to the displacements

$$\boldsymbol{u}(l) = \boldsymbol{R}_l - \boldsymbol{R}_l^0 \tag{A3.21}$$

as

$$M_\alpha = M_\alpha^{(0)} + \sum_{l\beta} M^{(1)}_{\alpha,\beta}(l)\,u_\beta(l) + \frac{1}{2}\sum_{ll'\beta\beta'} M^{(2)}_{\alpha,\beta\beta'}(ll')\,u_\beta(l)\,u_{\beta'}(l') + \cdots. \tag{A3.22}$$

If we only account for the linear term of equation (A3.22), then we see from (A3.19) that in this case the absorption is described by a displacement-displacement Green function (cf. section 2.1.3.).

For more details on the theory of infrared vibrational spectra, especially for application of group theory for studying selection rules in crystals, the reader is referred to Birman (1974) (in particular, for incommensurate systems see Rasing et al., 1982).

A3.2. Raman scattering by phonons

Consider a finite crystal exposed to incident light. In the crystal the light induces charge and current densities, which on their part are sources of an electromagnetic field $\boldsymbol{E}(\boldsymbol{r}, t)$, $\boldsymbol{B}(\boldsymbol{r}, t)$ (labelled $\boldsymbol{E}^{\text{ind}}(\boldsymbol{r}, t)$, $\boldsymbol{B}^{\text{ind}}(\boldsymbol{r}, t)$ in the appendix 2). The radiation emitted by the crystal is given by the average value of the Poynting vector[1]

$$\boldsymbol{S}(\boldsymbol{r}, t) = \frac{c}{4\pi} \langle \boldsymbol{E}(\boldsymbol{r}, t) \times \boldsymbol{B}(\boldsymbol{r}, t)\rangle. \tag{A3.23}$$

On choosing the Lorentz gauge and a vanishing scalar potential (cf. (A3.1)) the vector potential satisfies the equation

$$\Delta \boldsymbol{A}(\boldsymbol{r}, t) - \frac{1}{c^2} \ddot{\boldsymbol{A}}(\boldsymbol{r}, t) = -\frac{4\pi}{c} \boldsymbol{j}(\boldsymbol{r}, t), \tag{A3.24}$$

where $\boldsymbol{j}(\boldsymbol{r}, t)$ is the current density induced by the incident light. The solution of (A3.24) can be written

$$\boldsymbol{A}(\boldsymbol{r}, t) = \frac{1}{c} \int \mathrm{d}\boldsymbol{r}' \boldsymbol{j} \left(\boldsymbol{r}', t - \frac{|\boldsymbol{r} - \boldsymbol{r}'|}{c}\right) \Big/ |\boldsymbol{r} - \boldsymbol{r}'|. \tag{A3.25}$$

Assuming a large distance between crystal and detector, i.e. $|\boldsymbol{r}| \gg |\boldsymbol{r}'|$ (origin of $\boldsymbol{r}, \boldsymbol{r}'$ inside the crystal), then (A3.25) reads approximately

$$\boldsymbol{A}(\boldsymbol{r}, t) \approx \frac{1}{c\,|\boldsymbol{r}|} \int \mathrm{d}\boldsymbol{r}' \boldsymbol{j}(\boldsymbol{r}', t - |\boldsymbol{r} - \boldsymbol{r}'|/c) \tag{A3.26}$$

or Fourier transformed with respect to the time (cf. (A2.10))

$$\boldsymbol{A}(\boldsymbol{r}, \omega_s) = \frac{1}{c} \frac{1}{|\boldsymbol{r}|} \int \mathrm{d}\boldsymbol{r}' \int\limits_{-\infty}^{\infty} \mathrm{d}t\, \mathrm{e}^{i\omega_s(t + |\boldsymbol{r} - \boldsymbol{r}'|/c)} \boldsymbol{j}(\boldsymbol{r}', t). \tag{A3.27}$$

According to (A3.27), far from the crystal $\boldsymbol{A}(r, \omega_s)$ behaves approximately as a plane wave with wave vector $\boldsymbol{q}_s = \frac{\omega_s}{c} \boldsymbol{n}_s$, where $\boldsymbol{n}_s = \boldsymbol{r}/|\boldsymbol{r}|$ (wave zone approximation). In this approximation ($|\boldsymbol{r}| \gg |\boldsymbol{r}'|$, $|\boldsymbol{q}_s|\,|\boldsymbol{r}| \ll 1$), we obtain from Maxwell's equations (cf. (A2.3) to (A2.6); note that $\boldsymbol{j}(\boldsymbol{r}, t) = \boldsymbol{0}$ outside the crystal)

$$\boldsymbol{E}(\boldsymbol{r}, \omega_s) = -i \frac{\omega_s}{c} \boldsymbol{n}_s \times [\boldsymbol{n}_s \times \boldsymbol{A}(\boldsymbol{r}, \omega_s)], \tag{A3.28}$$

$$\boldsymbol{B}(\boldsymbol{r}, \omega_s) = i \frac{\omega_s}{c} \boldsymbol{n}_s \times \boldsymbol{A}(\boldsymbol{r}, \omega_s). \tag{A3.29}$$

[1] On deriving the Raman scattering cross section here we follow the approach of Enderlein et al. (1979). Note that in the latter paper light scattering and other phenomena of secondary radiation are clearly associated with statistical fluctuations $\Delta \boldsymbol{E}(\boldsymbol{r}, t) = \boldsymbol{E}(\boldsymbol{r}, t) - \langle \boldsymbol{E}(\boldsymbol{r}, t)\rangle$, $\Delta \boldsymbol{B}(\boldsymbol{r}, t) = \boldsymbol{B}(\boldsymbol{r}, t) - \langle \boldsymbol{B}(\boldsymbol{r}, t)\rangle$. Hence, in this paper these fluctuations figure in the part of the Poynting vector which governs scattering.

Experimentally, the time average of $\boldsymbol{S}(\boldsymbol{r}, t)$ is observed:

$$\boldsymbol{S}(\boldsymbol{r}) = \frac{1}{\tau} \int\limits_{-\tau/2}^{\tau/2} \mathrm{d}t \boldsymbol{S}(\boldsymbol{r}, t) = \int\limits_{0}^{\infty} \mathrm{d}\omega_s \boldsymbol{S}(\boldsymbol{r}, \omega_s), \tag{A3.30}$$

where τ is the time of observation and

$$\boldsymbol{S}(\boldsymbol{r}, \omega_s) = \frac{c}{(2\pi)^3} \frac{1}{2} \int\limits_{0}^{\infty} \mathrm{d}\omega_s' \eta_\tau(\omega_s - \omega_s') \times \langle \boldsymbol{E}^*(\boldsymbol{r}, \omega_s) \times \boldsymbol{B}(\boldsymbol{r}, \omega_s') + \boldsymbol{E}(\boldsymbol{r}, \omega_s) \times \boldsymbol{B}^*(\boldsymbol{r}, \omega_s') \rangle \tag{A3.31}$$

with

$$\eta_\tau(\omega) = (\sin(\omega\tau/2))/(\omega\tau/2). \tag{A3.32}$$

Here in the second line in (A3.30) the term proportional to $\eta_\tau(\omega_s + \omega_s')$ has been omitted, since it vanishes exactly in the limit $\tau \to \infty$. Inserting $\boldsymbol{E}$, $\boldsymbol{B}$ from (A3.28), (A3.29) into (A3.31), we obtain in the limit $\tau \to \infty$ for that part $\boldsymbol{S}(\boldsymbol{r}, \omega_s, \boldsymbol{e}_s)$ of $\boldsymbol{S}(\boldsymbol{r}, \omega_s)$ that corresponds to a definite polarization direction $\boldsymbol{e}_s \perp \boldsymbol{n}_s$ of the electric field $\boldsymbol{E}$ and to a corresponding polarization direction $\boldsymbol{e}_s \times \boldsymbol{n}_s$ of the magnetic field $\boldsymbol{B}$,

$$\boldsymbol{S}(\boldsymbol{r}, \omega_s, \boldsymbol{e}_s) = \frac{1}{(2\pi)^3} \frac{1}{c^3} \frac{\boldsymbol{n}_s}{|\boldsymbol{r}|^2} \int\limits_{0}^{\infty} \mathrm{d}\omega_s' \omega_s \omega_s' \eta_\tau(\omega_s - \omega_s') \times \int \mathrm{d}\boldsymbol{r}'' \int \mathrm{d}\boldsymbol{r}' \int\limits_{-\infty}^{\infty} \mathrm{d}t'' \int\limits_{-\infty}^{\infty} \mathrm{d}t' \, e^{i[\omega_s(t''-t') - \boldsymbol{q}_s(\boldsymbol{r}''-\boldsymbol{r}')]} \times \langle (\boldsymbol{e}_s \boldsymbol{j}(\boldsymbol{r}', t')) (\boldsymbol{e}_s \boldsymbol{j}(\boldsymbol{r}'', t'')) \rangle. \tag{A3.33}$$

On deriving (A3.33), we have used the approximation $(\omega_s/c) |\boldsymbol{r} - \boldsymbol{r}'| \approx (\omega_s/c) |\boldsymbol{r}| - \boldsymbol{r}'\boldsymbol{q}_s$, which is valid when the correlation length of the current-current correlation function in (A3.33) is smaller than the size of the crystal.

Following Enderlein et al. (1979), in (A3.33) $\boldsymbol{j}(\boldsymbol{r}, t)$ is now regarded as a quantum mechanical operator and, correspondingly, the angular brackets as a quantum statistical average to be performed with the Gibbs statistical operator (cf. (A2.84)).

The current densities occurring in (A3.33) are induced by the incident external field $\boldsymbol{A}^{\text{ext}}$. Here we calculate the current densities in linear approximation with respect to $\boldsymbol{A}^{\text{ext}}$. Note that in calculating the ensemble average of any operator O (cf. (A2.31)) the time dependence of the statistical operator $\hat{\varrho}$ can be shifted to 0. To do so, we write

$$\hat{\varrho}(t) = U(t, -\infty) \, \hat{\varrho}_0 U^+(t, -\infty), \tag{A3.34}$$

where $\hat{\varrho}_0$ is the equilibrium statistical operator and $U(t, -\infty)$ is the time evolution operator. The external field is assumed to be adiabatically turned on at $t \to -\infty$. The ensemble average of O can be expressed as

$$\langle O \rangle = \mathrm{Tr}\,\{\hat{\varrho}(t)\, O\} = \mathrm{Tr}\,\{\hat{\varrho}_0 O(t)\}, \tag{A3.35}$$

where

$$O(t) = U^+(t, -\infty)\, O U(t, -\infty). \tag{A3.36}$$

To obtain the equation of motion for $U(t, -\infty)$, we differentiate (A3.34) with respect to time and account for (A2.32). Integrating the resulting equation and solving it by iteration, we find in linear approximation

$$U(t, -\infty) = e^{-\frac{i}{\hbar}Ht} \left\{ 1 + \frac{1}{i\hbar} \int_{-\infty}^{t} dt' \, e^{\frac{i}{\hbar}Ht'} H'(t') \, e^{-\frac{i}{\hbar}Ht'} \right\}, \tag{A3.37}$$

where H is the Hamiltonian of the crystal and $H'(t)$ describes the interaction of the crystal with the external field. By (A3.36) and (A3.37) the linear contribution to the current density operator becomes

$$\boldsymbol{j}^{(1)}(\boldsymbol{r}, t) = -\frac{e}{mc} \varrho(\boldsymbol{r}, t) \boldsymbol{A}^{\text{ext}}(\boldsymbol{r}, t) + \frac{ic}{\hbar} \int d\boldsymbol{r}' \int_{-\infty}^{t} dt' [\boldsymbol{j}(\boldsymbol{r}, t), \boldsymbol{j}(\boldsymbol{r}', t')] \boldsymbol{A}^{\text{ext}}(\boldsymbol{r}', t'), \tag{A3.38}$$

where

$$\boldsymbol{j}(\boldsymbol{r}, t) = e^{\frac{i}{\hbar}Ht} \boldsymbol{j}(\boldsymbol{r}) \, e^{-\frac{i}{\hbar}Ht}. \tag{A3.39}$$

Here we have used (A2.33) and (A2.41).

The first term of (A3.38) gives rise to scattering on density fluctuations. This kind of scattering will be ignored in the following.

Now we assume an incident monochromatic plane wave with frequency ω_{i} and wave vector $\boldsymbol{q}_{\mathrm{i}}$

$$\boldsymbol{E}^{\text{ext}}(\boldsymbol{r}, t) = \boldsymbol{e}_{\mathrm{i}} E_0 \cos(\omega_{\mathrm{i}} t - \boldsymbol{q}_{\mathrm{i}} \boldsymbol{r}). \tag{A3.40}$$

The differential scattering cross section (per unit frequency and per unit angle)

$$\frac{d^2\sigma}{d\omega_s \, d\Omega} = \frac{\omega_{\mathrm{i}}}{\omega_s} \frac{|\boldsymbol{r}|^2}{S_0} |\boldsymbol{S}(\boldsymbol{r}, \omega_s, \boldsymbol{e}_s)| \tag{A3.41}$$

($S_0 = \frac{c}{8\pi} E_0^2$ — Poynting vector averaged over a period of the incident wave) then becomes (cf. Enderlein et al., 1979)

$$\frac{d^2\sigma}{d\omega_s \, d\Omega} = \frac{\omega_s \omega_{\mathrm{i}}}{c^4} \int_0^\infty d\omega_s' \eta_\tau(\omega_s - \omega_s') \langle A^+(\omega_s') \, A(\omega_s) \rangle, \tag{A3.42}$$

where

$$A(\omega_s) = \frac{i}{\hbar\omega_{\mathrm{i}}} \int_{-\infty}^{\infty} \frac{dt}{2\pi} \int_{-\infty}^{t} dt' \, e^{i(\omega_s t - \omega_{\mathrm{i}} t')} [\boldsymbol{e}_{\mathrm{i}} \boldsymbol{j}(t', -\boldsymbol{q}_{\mathrm{i}}), \boldsymbol{e}_s \boldsymbol{j}(t, \boldsymbol{q}_s)]. \tag{A3.43}$$

Here $\boldsymbol{j}(t, \boldsymbol{q})$ denotes the Fourier transform of $\boldsymbol{j}(t, \boldsymbol{r})$ (cf. (A2.10)). We have neglected all terms resulting from the above by replacing ω_{i} by $-\omega_{\mathrm{i}}$. The latter terms are commonly ascribed to induced emission.

Raman scattering can be investigated on the basis of equation (A3.42) by using perturbation theory with respect to the interaction part of the crystal Hamiltonian H.

On the other hand, we can assume that the complete system of eigenfunctions $|m\rangle$ of H

$$H\,|m\rangle = E_m\,|m\rangle \tag{A3.44}$$

is given. Then, by

$$\boldsymbol{j}(\boldsymbol{q}, t) = \sum_{mm'} |m\rangle \langle m|\, \boldsymbol{j}(\boldsymbol{q})\, |m'\rangle \langle m'|\, \mathrm{e}^{-i\omega_{m'm}t}, \tag{A3.45}$$

$$\omega_{mm'} = (E_m - E_{m'})/\hbar, \tag{A3.46}$$

we can rewrite (A3.42) as follows:

$$\frac{\mathrm{d}^2\sigma}{\mathrm{d}\omega_s\,\mathrm{d}\Omega} = \frac{\omega_s\omega_\mathrm{l}}{c^4} \sum_{mm''} \hat{\varrho}^0_{m''m''}\, |A_{mm''}(\omega_\mathrm{l})|^2\, \delta(\omega_s - \omega_\mathrm{l} + \omega_{mm''}), \tag{A3.47}$$

where

$$A_{mm''}(\omega_\mathrm{l}) = \frac{1}{\hbar\omega_\mathrm{l}} \sum_{m'} \left\{ \frac{\langle m|\, j_s\, |m'\rangle \langle m'|\, j_\mathrm{l}\, |m''\rangle}{\omega_\mathrm{l} - \omega_{m'm''} + i\varepsilon} - \frac{\langle m|\, j_\mathrm{l}\, |m'\rangle \langle m'|\, j_s\, |m''\rangle}{\omega_\mathrm{l} + \omega_{m'm} + i\varepsilon} \right\}. \tag{A3.48}$$

Here $j_s = \boldsymbol{e}_s \boldsymbol{j}(\boldsymbol{q}_s)$, $j_\mathrm{l} = \boldsymbol{e}_\mathrm{l} \boldsymbol{j}(-\boldsymbol{q}_\mathrm{l})$; $\hat{\varrho}^0_{mm} = \langle m|\, \hat{\varrho}_0\, |m\rangle$. In accordance with the adiabatic turning on of the external field, ω_l was replaced here by $\omega_\mathrm{l} + i\varepsilon$, $\varepsilon \to +0$, in order to make the integrals defined. Equations (A3.47) and (A3.48) represent the Kramer-Heisenberg formula for Raman scattering (Kramer and Heisenberg, 1925).

To proceed we assume, as in the case of infrared absorption, a strict localization of the electrons at ion sites. Then, by (A3.45), (A2.34), (A3.5), we obtain for small $\boldsymbol{q}_s$

$$\langle m|\, j_s\, |m'\rangle \simeq i\omega_{mm'} \langle m|\, \mu_s\, |m'\rangle \tag{A3.49}$$

with

$$\mu_s = \boldsymbol{e}_s \sum_l \mathrm{e}^{-i\boldsymbol{q}_s\boldsymbol{R}_l} \mu_l, \tag{A3.50}$$

where μ_l is the dipole moment operator (A3.4) and $\boldsymbol{R}_l$ is the position vector of the lth ion. A corresponding expression results for $\langle m|\, j_\mathrm{l}\, |m'\rangle$. Furthermore, in (A3.49) we use adiabatic functions (cf. (1.1.10), (A3.9)) instead of the eigenfunctions $|m\rangle$:

$$|m\rangle \to \chi_{nv}(\boldsymbol{R})\, \psi_n(\boldsymbol{r}, \boldsymbol{R}). \tag{A3.51}$$

In (A3.47) the initial and final states $|m''\rangle$ and $|m\rangle$, respectively, belong to the same electronic state ψ_0 (ground state), but to different vibrational states $\chi_{0v''}$ and χ_{0v}, respectively. In (A3.48) the intermediate states belong to electronic states, $\psi_{n'}$, which differ from the initial one, i.e. $n' \neq 0$. In the case of an insulator the separation between states with $n' \neq n$ is much greater than a phonon energy. Therefore, phonon energies may be neglected in energy differences occurring in (A3.47) and (A3.48). Inserting now (A3.49) into (A3.48) and using the completeness of the functions $\chi_{n'v'}(\boldsymbol{R})$

$$\sum_{v'} |\chi_{n'v'}(\boldsymbol{R})\rangle \langle \chi_{n'v'}(\boldsymbol{R}')| = \sum_{v'} \chi_{n'v'}(\boldsymbol{R})\, \chi^*_{n'v'}(\boldsymbol{R}') = \delta(\boldsymbol{R} - \boldsymbol{R}') \tag{A3.52}$$

we obtain

$$\frac{\mathrm{d}^2\sigma}{\mathrm{d}\omega_s\,\mathrm{d}\Omega} = \frac{\omega_s}{\omega_\mathrm{l}} \left(\frac{e}{mc}\right)^4 \sum_{vv''} \frac{\mathrm{e}^{-\beta\mathcal{E}_{0v''}}}{\sum_{\bar{v}} \mathrm{e}^{-\beta\mathcal{E}_{0\bar{v}}}}\, \delta\left((\omega_s - \omega_\mathrm{l} + \frac{1}{\hbar}\,(\mathcal{E}_{0v} - \mathcal{E}_{0v''})\right)$$
$$\times\, |\langle \chi_{0v}|\, \boldsymbol{e}_s \boldsymbol{P}(\boldsymbol{R}, \boldsymbol{q}_\mathrm{l}, \boldsymbol{q}_s)\, \boldsymbol{e}_\mathrm{l}\, |\chi_{0v''}\rangle|^2, \tag{A3.53}$$

where $\boldsymbol{P}$ is a tensor field function of $\boldsymbol{R}$ (polarizability operator) defined by

$$\boldsymbol{P}(\boldsymbol{R}, \boldsymbol{q}_{\text{i}}, \boldsymbol{q}_s) = \sum_{n'} \frac{(\tilde{\omega}_{0n'})^2}{(\omega_{\text{i}} - \tilde{\omega}_{n'0} + i\varepsilon)} \boldsymbol{M}^{0n'}(\boldsymbol{R}, -\boldsymbol{q}_s) \circ \boldsymbol{M}^{n'0}(\boldsymbol{R}, \boldsymbol{q}_{\text{i}}) \tag{A3.54}$$

with the 'phased' off-diagonal electric dipole operator (defined in analogy to (A3.11))

$$\boldsymbol{M}^{0n}(\boldsymbol{R}, -\boldsymbol{q}_s) = \sum_l \mathrm{e}^{-i\boldsymbol{q}_s\boldsymbol{R}_l^0} \int \mathrm{d}\boldsymbol{r}\psi_0^*(\boldsymbol{r}, \boldsymbol{R})\, \mu_l(\boldsymbol{r}, \boldsymbol{R})\, \psi_n(\boldsymbol{r}, \boldsymbol{R}). \tag{A3.55}$$

Here $\mathscr{E}_{0v}$ denotes the vibrational energy corresponding to χ_{0v}; $\hbar\tilde{\omega}_{0n}$ is the difference of the electronic energies of the states and ψ_n; $\boldsymbol{R}_l^0$ is the equilibrium position vector of the lth ion; the sign $\circ$ denotes the dyadic product. On deriving (A3.53), the second term on the right-hand side of (A3.48) was neglected, since it is small compared with the first term. Equation (A3.53) is in essence the result first obtained by PLACZEK (1934).

Equation (A3.52) can be rewritten as follows:

$$\frac{\mathrm{d}^2\sigma}{\mathrm{d}\omega_s\, \mathrm{d}\Omega} = \frac{\omega_s}{\omega_{\text{i}}} \left(\frac{e}{mc}\right)^4 \sum_{\alpha\alpha'\bar{\alpha}\bar{\alpha}'} e_\alpha^{s} e_{\alpha'}^{\text{i}}\, I_{\alpha\alpha'\bar{\alpha}\bar{\alpha}'} e_{\bar{\alpha}}^{s} e_{\bar{\alpha}'}^{\text{i}}, \tag{A3.56}$$

where

$$I_{\alpha\alpha'\bar{\alpha}\bar{\alpha}'}(\omega_s - \omega_{\text{i}}, T) = \frac{1}{2\pi} \int_{-\infty}^{\infty} \mathrm{d}t\mathrm{e}^{-i(\omega_s - \omega_{\text{i}})t} \langle P_{\alpha\alpha'}(t)\, P_{\bar{\alpha}\bar{\alpha}'}(0)\rangle, \tag{A3.57}$$

$$\langle\ldots\rangle = \mathrm{Tr}\,\{\mathrm{e}^{-\beta H_{\text{ph}}}\ldots\}/\mathrm{Tr}\,\{\mathrm{e}^{-\beta H_{\text{ph}}}\}, \tag{A3.58}$$

$$P_{\alpha\alpha'}(t) = \mathrm{e}^{\frac{i}{\hbar}H_{\text{ph}}t} P_{\alpha\alpha'} \mathrm{e}^{-\frac{i}{\hbar}H_{\text{ph}}t}. \tag{A3.59}$$

Here $e_\alpha^{s,\text{i}}$ is the αth cartesian component of $\mathbf{e}_{s,\text{i}}$; $P_{\alpha\alpha'}$ is the $\alpha\alpha'$ component of the polarizability tensor (A3.54), and H_{ph} is the phonon Hamiltonian corresponding to the electronic ground state. The equivalence of (A3.56) and (A3.53) can be readily proved by using in (A3.57) the completeness relation (A3.52) and the Fourier representation of the δ function:

$$\delta(t) = \frac{1}{2\pi} \int_{-\infty}^{\infty} \mathrm{d}t\; \mathrm{e}^{i\omega t}. \tag{A3.60}$$

The components of the polarizability tensor (A3.54) can be expanded in a power series with respect to the displacements (A3.21) as

$$P_{\alpha\alpha'} = P_{\alpha\alpha'}^{(0)} + \sum_{l\beta} P_{\alpha\alpha',\beta}^{(1)}(l)\, u_\beta(l) + \frac{1}{2} \sum_{ll'\beta\beta'} P_{\alpha\alpha',\beta\beta'}^{(2)}(ll')\, u_\beta(l)\, u_{\beta'}(l') + \cdots. \tag{A3.61}$$

If only the linear term of (A3.61) is accounted for, then the Raman scattering formula (A3.57) can be written in terms of the displacement-displacement Green function (cf. section 2.1.3.). For application of group theory for studying selection rules of Raman spectra in crystals the reader is referred to BIRMAN (1974). Note that formula (A3.42) is suitable for applying modern many-body technique (see, e.g., BÖTTGER et al., 1978, 1979, QUANG et al., 1980).

A3.3. Interaction of phonons with thermal neutrons

Energy and wavelength of a room temperature neutron are comparable with phonon energies and interatomic spacing in a crystal. Thus, scattering of thermal neutrons at crystals can be expected to give valuable information on lattice vibrations.

In a non-magnetic material incident neutrons interact with the nuclei only. In general, this interaction is described by the Fermi pseudopotential

$$V(\boldsymbol{r}) = \frac{2\pi\hbar^2}{m} \sum_l a_l \delta(\boldsymbol{r} - \boldsymbol{R}(l)), \tag{A3.62}$$

where a_l is the scattering length of the lth nucleus, $\boldsymbol{r}$ is the position vector of the neutron, $\boldsymbol{R}(l)$ denotes the instantaneous position of the lth nucleus and m is the neutron mass. The scattering length a_l consists in general of a spin-dependent and a spin-independent part. The former part can give rise to a spin-flip of the neutron spin during the scattering process. For different isotopes both parts scatter differently.

By the scattering process the crystal changes from a (vibrational) state $|v\rangle$ with energy $\mathscr{E}_v$ before interaction with the neutron to a state $|v'\rangle$ with energy $\mathscr{E}_{v'}$ afterwards. Before and after the scattering the neutron is described by a plane wave with wave vector $\boldsymbol{k}$ and $\boldsymbol{k}'$, respectively. The change in energy and momentum of the neutron due to the scattering is

$$\hbar\omega = \frac{\hbar^2}{2m}(k^2 - k'^2), \qquad \boldsymbol{\kappa} = \boldsymbol{k} - \boldsymbol{k}'. \tag{A3.63}$$

In the first Born approximation the differential scattering cross section is proportional to the square modulus of the matrix element of the potential (A3.62) between the initial and final states of the neutron-crystal system. In this approximation the differential scattering cross section per unit energy of the outgoing neutron and per unit angle reads

$$\frac{d^2\sigma}{d\epsilon\, d\Omega} = \frac{1}{\hbar}\frac{k'}{k} \sum_{vv'} p_v \left|\langle v'| \sum_l a_l\, e^{i\boldsymbol{R}\boldsymbol{\kappa}(l)}\, |v\rangle\right|^2 \delta\left(\omega + \frac{1}{\hbar}(\mathscr{E}_v - \mathscr{E}_{v'})\right). \tag{A3.64}$$

Here

$$p_v = e^{-\beta\mathscr{E}_v} \sum_v e^{-\beta\mathscr{E}_v}, \qquad \beta = 1/kT, \tag{A.3.65}$$

is the statistical weight of the initial state of the crystal. Equation (A3.64) can be rewritten as

$$\frac{d^2\sigma}{d\epsilon\, d\Omega} = \frac{1}{2\pi\hbar}\frac{k'}{k} \sum_{ll'} a_l a_{l'} \int_{-\infty}^{\infty} dt\, e^{i\omega t} \langle e^{-i\boldsymbol{\kappa}\boldsymbol{R}(l,t)}\, e^{i\boldsymbol{\kappa}\boldsymbol{R}(l',0)}\rangle, \tag{A3.66}$$

where $\langle\ldots\rangle$ denotes the thermal average (cf. (A3.58)) and $\boldsymbol{R}(l, t)$ is $\boldsymbol{R}(l)$ in Heisenberg representation (cf. (A3.59)). The equivalence of (A3.66) and (A3.64) can be readily shown by using in (A3.66) (A.3.59), (A3.58), the completeness relation $\sum_v |v\rangle\langle v| = 1$ (which is to be placed between the two factors in the brackets in (A3.66)) and the Fourier representation of the δ function (A3.60).

We do not know the particular spin state of the lth nucleus, and we do not know to which isotopic species this nucleus belongs. We assume that the nuclei with different scattering lengths are distributed statistically and independently over the crystal. After averaging over all possible distributions, thus we obtain the following two contributions to the scattering cross section of a monatomic system

$$\frac{d^2 \sigma_{coh}}{d\epsilon\, d\Omega} = \frac{a_{coh}^2}{2\pi\hbar} \frac{k'}{k} \sum_{ll'} \int_{-\infty}^{\infty} dt\, e^{i\omega t} \langle e^{-i\kappa R(l,t)}\, e^{i\kappa R(l',0)} \rangle, \tag{A3.67}$$

$$\frac{d^2 \sigma_{inc}}{d\epsilon\, d\Omega} = \frac{a_{inc}^2}{2\pi\hbar} \frac{k'}{k} \sum_{l} \int_{-\infty}^{\infty} dt\, e^{i\omega t} \langle e^{-i\kappa R(l,t)}\, e^{i\kappa R(l,0)} \rangle, \tag{A3.68}$$

where

$$a_{coh}^2 \equiv \langle a \rangle_c^{\,2}, \qquad a_{inc}^2 \equiv \langle a^2 \rangle_c - \langle a \rangle_c^{\,2}. \tag{A3.69}$$

Here $\langle a \rangle_c$ and $\langle a^2 \rangle_c$ respectively denote the configuration average of the scattering length and its square over the different spin directions and isotopes. The contribution (A3.67) is called coherent scattering cross section. It includes all effects from interference of neutron waves scattered from different nuclei. The contribution (A3.68) is called incoherent scattering cross section. It arises from random fluctuations of the scattering length.

Introducing the correlation functions

$$G(\boldsymbol{r}, t) = \frac{1}{N} \int d\boldsymbol{r}' \Big\langle \sum_{ll'} \delta(\boldsymbol{r}' - \boldsymbol{R}_l(0))\, \delta(\boldsymbol{r}' + \boldsymbol{r} - \boldsymbol{R}_{l'}(t)) \Big\rangle, \tag{A3.70}$$

$$G_s(\boldsymbol{r}, t) = \frac{1}{N} \int d\boldsymbol{r}' \Big\langle \sum_{l} \delta(\boldsymbol{r}' - \boldsymbol{R}_l(0))\, \delta(\boldsymbol{r}' + \boldsymbol{r} - \boldsymbol{R}_l(t)) \Big\rangle \tag{A3.71}$$

(N — number of nuclei in the system) and their Fourier transforms

$$S(\kappa, \omega) = \frac{1}{2\pi} \int d\boldsymbol{r} \int_{-\infty}^{\infty} dt\, e^{i(\kappa \boldsymbol{r} - \omega t)}\, G(\boldsymbol{r}, t), \tag{A3.72}$$

$$S_s(\kappa, \omega) = \frac{1}{2\pi} \int d\boldsymbol{r} \int_{-\infty}^{\infty} dt\, e^{i(\kappa \boldsymbol{r} - \omega t)} G_s(\boldsymbol{r}, t), \tag{A3.73}$$

the scattering cross section ((A.3.67), (A3.68)) can be expressed as

$$\frac{d^2 \sigma_{coh}}{d\epsilon\, d\Omega} = N \frac{a_{coh}^2}{\hbar} \frac{k'}{k} S(\kappa, \omega), \tag{A3.74}$$

$$\frac{d^2 \sigma_{inc}}{d\epsilon\, d\Omega} = N \frac{a_{inc}^2}{\hbar} \frac{k'}{k} S_s(\kappa, \omega). \tag{A3.75}$$

$S(\kappa, \omega)$ is called scattering function or dynamical structure factor.

By means of the particle-density operator

$$n(\boldsymbol{r}, t) = \sum_{l} \delta(\boldsymbol{r} - \boldsymbol{R}(l, t)) \tag{A3.76}$$

(A3.70) can be rewritten as

$$G(\boldsymbol{r}, t) = \frac{1}{N} \int \mathrm{d}\boldsymbol{r}' \langle n(\boldsymbol{r}', 0)\, n(\boldsymbol{r}' + \boldsymbol{r}, t)\rangle, \tag{A.3.77}$$

which in a homogeneous system (including an ideal crystal) can be simplified to

$$G(\boldsymbol{r}, t) = \frac{V}{N} \langle n(\boldsymbol{0}, 0)\, n(\boldsymbol{r}, t)\rangle \tag{A3.78}$$

(V — volume of the system). Hence, $G(\boldsymbol{r}, t)$ is the microscopic density correlation function. Classically, it describes the correlation between the density at two different positions and times separated from each other by a distance $\boldsymbol{r}$ and a time t. $G_s(\boldsymbol{r}, t)$ describes the correlation between the positions of one particle at different times.

On deriving (A3.67), (A3.68) and (A3.74), (A3.75) no special assumptions on the system have been made. Hence, these equations hold for crystals, including such with disorder and/or anharmonicity, as well as for other systems such as fluids. Now we consider a crystal and put $\boldsymbol{R}(l) = \boldsymbol{x}(l) + \boldsymbol{u}(l)$ (cf. (A3.21)). Then (A3.67) and (A3.68) become

$$\frac{\mathrm{d}^2\sigma_{\mathrm{coh}}}{\mathrm{d}\epsilon\, \mathrm{d}\Omega} = \frac{a_{\mathrm{coh}}^2}{2\pi\hbar} \frac{k'}{k} \sum_{ll'} \mathrm{e}^{-i\kappa(\boldsymbol{x}(l)-\boldsymbol{x}(l'))} \int_{-\infty}^{\infty} \mathrm{d}t\, \mathrm{e}^{i\omega t} \langle \mathrm{e}^{-i\kappa\boldsymbol{u}(l,t)}\, \mathrm{e}^{i\kappa\boldsymbol{u}(l',0)}\rangle, \tag{A3.79}$$

$$\frac{\mathrm{d}^2\sigma_{\mathrm{inc}}}{\mathrm{d}\epsilon\, \mathrm{d}\Omega} = \frac{a_{\mathrm{inc}}^2}{2\pi\hbar} \frac{k'}{k} \sum_{l} \int_{-\infty}^{\infty} \mathrm{d}t\, \mathrm{e}^{i\omega t} \langle \mathrm{e}^{-i\kappa\boldsymbol{u}(l,t)}\, \mathrm{e}^{i\kappa\boldsymbol{u}(l,0)}\rangle. \tag{A3.80}$$

In a harmonic crystal (A3.79) and (A3.80) can be rewritten as

$$\frac{\mathrm{d}^2\sigma_{\mathrm{coh}}}{\mathrm{d}\epsilon\, \mathrm{d}\Omega} = \frac{a_{\mathrm{coh}}^2}{2\pi\hbar} \frac{k'}{k} \sum_{ll'} \mathrm{e}^{-(W(l)+W(l'))}\, \mathrm{e}^{-i\kappa(\boldsymbol{x}(l)-\boldsymbol{x}(l'))} \times \int_{-\infty}^{\infty} \mathrm{d}t\, \mathrm{e}^{i\omega t}\, \mathrm{e}^{\langle(\kappa\boldsymbol{u}(l,t))(\kappa\boldsymbol{u}(l',0))\rangle}, \tag{A3.81}$$

$$\frac{\mathrm{d}^2\sigma_{\mathrm{inc}}}{\mathrm{d}\epsilon\, \mathrm{d}\Omega} = \frac{a_{\mathrm{inc}}^2}{2\pi\hbar} \frac{k'}{k} \sum_{l} \mathrm{e}^{-2W(l)} \int_{-\infty}^{\infty} \mathrm{d}t\, \mathrm{e}^{i\omega t}\, \mathrm{e}^{\langle(\kappa\boldsymbol{u}(l,t))(\kappa\boldsymbol{u}(l,0))\rangle} \tag{A3.82}$$

where

$$W(l) = \frac{1}{2} \langle(\kappa\boldsymbol{u}(l))^2\rangle \tag{A3.83}$$

is the Debye-Waller exponent.

To prove the validity of (A3.81) and (A3.82), let us consider the correlation function

$$f(t) = \langle \mathrm{e}^{-i\kappa\boldsymbol{u}(l,t)}\, \mathrm{e}^{i\kappa\boldsymbol{u}(l',0)}\rangle. \tag{A3.84}$$

On expressing the displacement operator by creation and annihilation operators by means of (1.3.14), it can be readily shown that the commutator of $-i\kappa\boldsymbol{u}(l, t)$ and $i\kappa\boldsymbol{u}(l, 0)$ is a c-number. Hence we can apply to (A3.84) the operator identity

$$\mathrm{e}^{A}\, \mathrm{e}^{B} = \mathrm{e}^{A+B}\, \mathrm{e}^{[A,B]/2}, \tag{A3.85}$$

which is valid for operators A and B whose commutator is a c-number. This gives

$$f(t) = \langle e^{-i\kappa(u(l,t)-u(l',0))}\rangle \, e^{\frac{1}{2}\langle[-i\kappa u(l,t), i\kappa u(l',0)]\rangle} \tag{A3.86}$$

where we have used the fact that the thermal average of a c-number is the c-number itself. In the first factor on the right-hand side of equation (A3.86) we express the displacement operators by creation and annihilation operators by means of (1.3.14) as

$$\kappa(\boldsymbol{u}(l, t) - \boldsymbol{u}(l', 0)) = \sum_s (C_s b_s + C_s^* b_s^+), \tag{A3.87}$$

where

$$C_s = \left(\frac{\hbar}{2\omega_s}\right)^{1/2} \left\{\frac{\kappa B^{(s)}(l)}{M_l^{1/2}} e^{-i\omega_s t} - \frac{\kappa B^{(s)}(l')}{M_{l'}^{1/2}}\right\}. \tag{A3.88}$$

Thus, we have

$$f(t) = \exp \frac{1}{2} \langle[\kappa \boldsymbol{u}(l, t), \kappa \boldsymbol{u}(l', 0)]\rangle \prod_s \frac{1}{Z_s} \sum_{n_s=0}^{\infty} \exp\{-\beta\hbar\omega_s(n_s + 1/2)\}$$
$$\times \langle n_s| \exp\{-i(C_s^* b_s^+ + C_s b_s)\} |n_s\rangle, \tag{A3.89}$$

where (cf. (1.4.2))

$$Z_s = \mathrm{Tr}\,\{e^{-\beta\hbar\omega_s(b_s^+ b_s + 1/2)}\}. \tag{A3.90}$$

To evaluate the last factor on the right-hand side of equation (A3.89), we use again (A3.85). Then this factor becomes

$$\langle n_s| \exp(-iC_s^* b_s^+) \exp(-iC_s b_s) |n_s\rangle \exp\left(-\frac{1}{2}|C_s|^2\right). \tag{A3.91}$$

Putting the completeness relation $\sum_s |n_s\rangle\langle n_s| = 1$ between the two factors in the brackets in (A3.91) and using (1.3.27), we obtain

$$\langle n_s| \exp(-iC_s^* b_s^+) \exp(-iC_s b_s) |n_s\rangle = n_s! \sum_{n=0}^{n_s} (-1)^n \frac{|C_s|^{2n}}{(n!)^2} \frac{1}{(n_s - n)!}. \tag{A3.92}$$

By (A3.91) and (A3.92), (A3.89) becomes

$$f(t) = \exp \frac{1}{2} \langle[\kappa \boldsymbol{u}(l, t), \kappa \boldsymbol{u}(l', 0)]\rangle \exp\left[-\frac{1}{2} \sum_s |C_s|^2 \coth(\beta\hbar\omega_s/2)\right]. \tag{A3.93}$$

On account of C_s from (A3.88), after some straightforward but simple manipulations (A3.93) takes the following form:

$$f(t) = \exp\left(-\frac{1}{2}\langle(\kappa \boldsymbol{u}(l, t))^2\rangle\right) \exp\left(-\frac{1}{2}\langle(\kappa \boldsymbol{u}(l', 0))^2\rangle\right)$$
$$\times \exp\langle(\kappa \boldsymbol{u}(l, t))(\kappa \boldsymbol{u}(l', 0))\rangle. \tag{A3.94}$$

By (A3.94) and (A3.84) (A3.81) and (A3.82) can be obtained from (A3.79) and (A3.80), respectively.

The displacement-displacement correlation functions present in (A3.81) and (A3.82) can be expressed by the displacement-displacement Green function, as indicated in section 2.1.3. Expanding (A3.81) and (A3.82) with respect to $\langle(\kappa\boldsymbol{u}(l,t))\,(\kappa\boldsymbol{u}(l',0))\rangle$ and $\langle(\kappa\boldsymbol{u}(l,t))\,(\kappa\boldsymbol{u}(l,0))\rangle$, respectively, the nth term in each expansion describes scattering processes, in which n phonons take part. The zeroth order terms describe elastic scattering. The one-phonon coherent and incoherent scattering cross sections, respectively, are given by

$$\left(\frac{\mathrm{d}^2\sigma_{\mathrm{coh}}}{\mathrm{d}\epsilon\,\mathrm{d}\Omega}\right)_{\text{one phonon}} = \frac{a_{\mathrm{coh}}^2}{2\pi\hbar}\frac{k'}{k}\sum_{ll'} \mathrm{e}^{-(W(l)+W(l'))}\,\mathrm{e}^{-i\kappa(\boldsymbol{x}(l)-\boldsymbol{x}(l'))} \times \int_{-\infty}^{\infty} \mathrm{d}t\; \mathrm{e}^{i\omega t}\langle(\kappa\boldsymbol{u}(l,t))\,(\kappa\boldsymbol{u}(l',0))\rangle, \tag{A3.95}$$

$$\left(\frac{\mathrm{d}^2\sigma_{\mathrm{inc}}}{\mathrm{d}\epsilon\,\mathrm{d}\Omega}\right)_{\text{one phonon}} = \frac{a_{\mathrm{inc}}^2}{2\pi\hbar}\frac{k'}{k}\sum_{l} \mathrm{e}^{-2W(l)}\int_{-\infty}^{\infty} \mathrm{d}t\; \mathrm{e}^{i\omega t}\langle(\kappa\boldsymbol{u}(l,t))\,(\kappa\boldsymbol{u}(l,0))\rangle. \tag{A3.96}$$

If we consider an ideal crystal (with one atom per unit cell, for simplicity) and evaluate the correlation functions occurring in (A3.95) and (A3.96) by means of (2.1.22) and (2.1.39), we obtain for the one-phonon cross sections

$$\left(\frac{\mathrm{d}^2\sigma_{\mathrm{coh}}}{\mathrm{d}\epsilon\,\mathrm{d}\Omega}\right)_{\text{one phonon}} = \frac{a_{\mathrm{coh}}^2 N}{2M}\frac{k'}{k}\,\mathrm{e}^{-2W}\sum_{\boldsymbol{k}j}\Delta(\boldsymbol{k}-\kappa)\,\frac{|\kappa\boldsymbol{e}(\boldsymbol{k}j)|^2}{\omega_j(\boldsymbol{k})} \times \left[(\bar{n}_j(\boldsymbol{k})+1)\,\delta(\omega-\omega_j(\boldsymbol{k})) + \bar{n}_j(\boldsymbol{k})\,\delta(\omega+\omega_j(\boldsymbol{k}))\right], \tag{A3.97}$$

$$\left(\frac{\mathrm{d}^2\sigma_{\mathrm{inc}}}{\mathrm{d}\epsilon\,\mathrm{d}\Omega}\right)_{\text{one phonon}} = \frac{a_{\mathrm{inc}}^2}{M}\,\frac{\operatorname{sgn}\omega}{1-\mathrm{e}^{-\beta\hbar\omega}}\,\frac{k}{k'}\,\mathrm{e}^{-2W} \times \sum_{\boldsymbol{k}j}|\kappa\boldsymbol{e}(\boldsymbol{k}_j)|^2\,\delta(\omega^2-\omega_j^2(\boldsymbol{k})) \tag{A3.98}$$

with the Debye-Waller exponent

$$W = \frac{1}{2NM}\sum_{\boldsymbol{k}j}\frac{|\kappa\boldsymbol{e}(\boldsymbol{k}_j)|^2}{\omega_j^2(\boldsymbol{k})}\,\bar{\varepsilon}_j(\boldsymbol{k}), \tag{A3.99}$$

where $\Delta(\boldsymbol{k})$ is defined by (1.3.65), $\bar{n}_j(\boldsymbol{k})$ by (1.4.50) and $\bar{\varepsilon}_j(\boldsymbol{k})$ by (1.4.62).

According to (A3.97), the coherent scattering cross section and $S(\kappa,\omega)$ have δ-peaks at $\omega_j(\boldsymbol{k})$ (phonon absorption), from which the dispersion relations $\omega = \omega_j(\boldsymbol{k})$ can be determined.

For an arbitrary function $f(\omega_j(\boldsymbol{k}))$ in a monatomic cubic crystal the following relation holds:

$$\sum_{\boldsymbol{k}j} e_\alpha(\boldsymbol{k}j)\,e_{\alpha'}(\boldsymbol{k}j)\,f(\omega_j(\boldsymbol{k})) = \frac{1}{3}\,\delta_{\alpha\alpha'}\sum_{\boldsymbol{k}j} f(\omega_j(\boldsymbol{k})), \tag{A3.100}$$

since in such a crystal a second rank tensor is proportional to the unit tensor. By (A3.98) and (A3.100) we find that the incoherent scattering cross section of a monatomic cubic crystal is proportional to the phonon frequency spectrum. For more details on investigations of phonons by neutron spectroscopy see, e.g., LUDWIG (1967) and for basic facts about neutron scattering by condensed matter see, e.g., SCHERM (1972).

A3.4. Lattice thermal conductivity

The use of many-body techniques for evaluating the thermal conductivity requires that we express the thermal conductivity in terms of correlation functions of the energy-flux operator. A crux of such a programme lies in the determination of the functional dependence of the energy-flux operator on the dynamical variables of the system. Following Enz (1968), we derive the energy-flux operator as follows:

We define the energy-flux density $\mathbf{s}(\boldsymbol{r})$ by the local energy conservation

$$\dot{H}(\boldsymbol{r}) + \frac{\partial}{\partial \boldsymbol{r}}\, \mathbf{s}(\boldsymbol{r}) = 0, \tag{A3.101}$$

where $H(\boldsymbol{r})$ is the energy density, for which we write (cf. (1.2.2.), (1.3.1))

$$H(\boldsymbol{r}) = \frac{1}{2} \sum_l \{H_l, \delta(\boldsymbol{r} - \boldsymbol{R}(l))\}, \tag{A3.102}$$

$$H_l = \frac{\boldsymbol{p}^2(l)}{2M(l)} + \sum_{n=2}^{\infty} \frac{1}{n!} \sum_{\substack{\alpha_2 \dots \alpha_n \\ l_2 \dots l_n \\ \alpha}} \Phi_{\alpha\alpha_2 \dots \alpha\alpha_n}(ll_2 \dots l_n)\, u_\alpha(l)\, u_{\alpha_2}(l_2) \dots u_{\alpha_n}(l_n). \tag{A3.103}$$

Here $\boldsymbol{R}(l)$ is the instantaneous position of the lth atom and $\boldsymbol{u}(l)$ is the displacement of the lth atom from its equilibrium position $\boldsymbol{x}(l)$ (cf. (1.2.1)).

Fourier transforming (A3.101) with respect to the space variable (cf. (A2.10)) we obtain

$$\dot{H}(\boldsymbol{q}) + i\boldsymbol{q}\mathbf{s}(\boldsymbol{q}) = 0. \tag{A3.104}$$

This equation determines the longitudinal part of the energy-flux density

$$\mathbf{s}^{\mathrm{lg}}(\boldsymbol{q}) = i\, \frac{\boldsymbol{q}}{|\boldsymbol{q}|^2}\, \dot{H}(\boldsymbol{q}). \tag{A3.105}$$

To evaluate the Fourier transform

$$H(\boldsymbol{q}) = \frac{1}{2} \sum_l \{H_l, \mathrm{e}^{-i\boldsymbol{q}(\boldsymbol{x}(l)+\boldsymbol{u}(l))}\}, \tag{A3.106}$$

we perform the "multipole" expansion

$$\mathrm{e}^{-i\boldsymbol{q}(\boldsymbol{x}(l)+\boldsymbol{u}(l))} \approx \mathrm{e}^{-i\boldsymbol{q}\boldsymbol{x}(l)}[1 - i\boldsymbol{q}\boldsymbol{u}(l) + \cdots]. \tag{A3.107}$$

On treating low temperature thermal conductivity, anharmonic corrections to operator densities are generally assumed to be small. Hence, neglecting the anharmonic terms of (A3.103) and taking into account only the first term of the expansion (A3.107), we find (calculating $\dot{H}(\boldsymbol{q})$ by the Heisenberg equation)

$$\dot{H}(\boldsymbol{q}) = \sum_{l'l\alpha\alpha'} \frac{1}{2M(l')}\, \Phi_{\alpha\alpha'}(ll')\, p_{\alpha'}(l')\, u_\alpha(l)\, (\mathrm{e}^{-i\boldsymbol{q}\boldsymbol{x}(l)} - \mathrm{e}^{-i\boldsymbol{q}\boldsymbol{x}(l')}). \tag{A3.108}$$

The total energy-flux $\boldsymbol{S}$ is defined by

$$\boldsymbol{S} = \int \mathrm{d}\boldsymbol{r}\mathbf{s}(\boldsymbol{r}) \tag{A3.109}$$

and equals $\boldsymbol{s}(\boldsymbol{q}=\mathbf{0})$. From (A3.105) and (A3.108) the longitudinal component of $\boldsymbol{S}$ can be readily obtained. Dropping the longitudinal projection, the total energy-flux becomes

$$\boldsymbol{S} = i\left[\frac{\partial}{\partial \boldsymbol{q}} \dot{H}(\boldsymbol{q})\right]_{\boldsymbol{q}=0} = \sum_{ll'\alpha\alpha'} \frac{1}{2M(l')} (\boldsymbol{x}(l) - \boldsymbol{x}(l'))\, \Phi_{\alpha\alpha'}(ll')\, p_{\alpha'}(l')\, u_\alpha(l). \quad \text{(A3.110)}$$

In particular, for a crystal (for brevity we treat a Bravais lattice) by (A3.110), (1.3.60), (1.3.61), (1.3.37) to (1.3.40) and (1.3.65) we find

$$\boldsymbol{S} = \frac{\hbar}{8} \sum_{\alpha\boldsymbol{k}\boldsymbol{k}'jj'} (\omega_j(\boldsymbol{k})/\omega_{j'}(\boldsymbol{k}'))^{1/2} (\omega_j^2(\boldsymbol{k}) - \omega_{j'}^2(\boldsymbol{k}'))\, e_\alpha^*(\boldsymbol{k}_j)\, e_\alpha(\boldsymbol{k}'j')$$
$$\times \{(b_{-\boldsymbol{k}j} - b^+_{\boldsymbol{k}j}), (b_{\boldsymbol{k}'j} + b^+_{-\boldsymbol{k}'j'})\} \frac{\partial}{\partial \boldsymbol{k}} \delta_{\boldsymbol{k}-\boldsymbol{k}',-\boldsymbol{K}}, \quad \text{(A3.111)}$$

where the derivative on the δ-function (note $\delta_{\boldsymbol{k}\boldsymbol{k}'} = \frac{(2\pi)^3}{V}\,\delta(\boldsymbol{k}-\boldsymbol{k}')$) is understood to be eliminated by partial integration and the braces denote an anticommutator. Performing the partial integration and noting that only the term $\boldsymbol{K}=\mathbf{0}$ contributes to $\boldsymbol{S}$, then we find

$$\boldsymbol{S} = -\frac{\hbar}{8} \sum_{\alpha\boldsymbol{k}jj'} \{(b_{-\boldsymbol{k}j} - b^+_{\boldsymbol{k}j}), (b_{\boldsymbol{k}j} + b^+_{-\boldsymbol{k}j})\}$$
$$\times \left[e_\alpha^*(\boldsymbol{k}j)\, e_\alpha(\boldsymbol{k}j) \left(\frac{\partial}{\partial \boldsymbol{k}} (\omega_j(\boldsymbol{k})/\omega_{j'}(\boldsymbol{k}'))^{1/2} (\omega_j^2(\boldsymbol{k}) - \omega_{j'}^2(\boldsymbol{k}'))\right)_{\boldsymbol{k}'=\boldsymbol{k}} \right.$$
$$\left. + (\omega_j(\boldsymbol{k})/\omega_{j'}(\boldsymbol{k}))^{1/2} (\omega_j^2(\boldsymbol{k}) - \omega_{j'}^2(\boldsymbol{k})) \left(\frac{\partial}{\partial \boldsymbol{k}} e_\alpha^*(\boldsymbol{k}j)\, e_\alpha(\boldsymbol{k}'j)\right)_{\boldsymbol{k}'=\boldsymbol{k}} \right]. \quad \text{(A3.112)}$$

By (1.3.44) and (1.3.64) the diagonal part of $\boldsymbol{S}$ (i.e. the part of $\boldsymbol{S}$ arising from the first term in the brackets on the right-hand side of (A3.112)) takes the well known form

$$\boldsymbol{S}^{\mathrm{d}} = \sum_{\boldsymbol{k}j} \frac{\partial \omega_j(\boldsymbol{k})}{\partial \boldsymbol{k}}\, \hbar\omega_j(\boldsymbol{k})\, b^+_{\boldsymbol{k}j} b_{\boldsymbol{k}j}. \quad \text{(A3.113)}$$

This term represents the sum of the one-phonon energy-fluxes multiplied by the phonon numbers.

The thermal conductivity in terms of correlation functions of the energy-flux operator may be obtained by means of the method of the nonequilibrium statistical operator (ZUBAREV, 1971, cf. ENZ, 1968), which allows the thermal perturbations to be accounted for. This method leads to the following expression for the expectation value, $\overline{\boldsymbol{s}(\boldsymbol{r})}$, of the energy-flux density:

$$\overline{\boldsymbol{s}(\boldsymbol{r})} = \langle \boldsymbol{s}(\boldsymbol{r})\rangle_l + \int \mathrm{d}\boldsymbol{r}' \int_0^1 \mathrm{d}\lambda \int_{-\infty}^0 \mathrm{d}t\, \mathrm{e}^{\varepsilon t} \{\langle \boldsymbol{s}(\boldsymbol{r}) \circ \boldsymbol{s}(\boldsymbol{r}', t, i\lambda)\rangle_l$$
$$- \langle \boldsymbol{s}(\boldsymbol{r})\rangle_l \circ \langle \boldsymbol{s}(\boldsymbol{r}', t)\rangle_l\}\, \nabla \beta(\boldsymbol{r}'), \qquad \varepsilon \to +0, \quad \text{(A3.114)}$$

where

$$\boldsymbol{s}(\boldsymbol{r}, t, i\lambda) = \exp\left[-\lambda \int \mathrm{d}\boldsymbol{r}\beta(\boldsymbol{r})\, H(\boldsymbol{r})\right] \boldsymbol{s}(\boldsymbol{r}, t) \exp\left[\lambda \int \mathrm{d}\boldsymbol{r}\beta(\boldsymbol{r})\, H(\boldsymbol{r})\right], \quad \text{(A3.115)}$$

$$\left.\begin{aligned} s(\boldsymbol{r}, t) &= \mathrm{e}^{\frac{i}{\hbar}Ht}\, \boldsymbol{s}(\boldsymbol{r})\, \mathrm{e}^{-\frac{i}{\hbar}Ht}, \\ \langle \ldots \rangle_l &= \frac{\mathrm{Tr}\left\{\exp\left[-\int \mathrm{d}\boldsymbol{r}\beta(\boldsymbol{r})\, H(\boldsymbol{r}) \ldots\right]\right\}}{\mathrm{Tr}\left\{\exp\left[-\int \mathrm{d}\boldsymbol{r}\beta(\boldsymbol{r})\, H(\boldsymbol{r})\right]\right\}}. \end{aligned}\right\} \quad \text{(A3.116)}$$

In (A3.114) to (A3.116), $\beta(\boldsymbol{r}) = 1/kT(\boldsymbol{r})$ ($T(\boldsymbol{r})$ — space dependent temperature) and the symbol $\circ$ denotes the dyadic product. Accounting for the fact that the local equilibrium value $\langle \mathbf{s}(\boldsymbol{r})\rangle_l$ vanishes and using the approximations

$$\exp\left[-\int \mathrm{d}\boldsymbol{r}\beta(\boldsymbol{r})\, H(\boldsymbol{r})\right] \approx \mathrm{e}^{-\beta H} \tag{A3.117}$$

($\beta = 1/kT$, T — average temperature of the system),

$$\nabla\beta(\boldsymbol{r}) \approx -\frac{1}{kT^2}\,\nabla T(\boldsymbol{r}), \tag{A3.118}$$

$$\langle s(\boldsymbol{r})\rangle \to \frac{1}{V}\int \mathrm{d}\boldsymbol{r}\langle s(\boldsymbol{r})\rangle, \tag{A3.119}$$

we obtain from (A3.114), for the case of a constant temperature gradient ∇T, the following expression for the expectation value $\bar{\boldsymbol{S}}$ of the total energy-flux:

$$\bar{\boldsymbol{S}} = \frac{1}{kT^2V}\int_0^1 \mathrm{d}\lambda \int_{-\infty}^0 \mathrm{d}t\, \mathrm{e}^{\varepsilon t}\langle \boldsymbol{S}\circ\boldsymbol{S}(t + i\hbar\beta\lambda)\rangle\,(-\nabla T), \qquad \varepsilon\to +0, \tag{A3.120}$$

where $\langle\ldots\rangle = \mathrm{Tr}\,(\mathrm{e}^{-\beta H}\ldots)/\mathrm{Tr}\,(\mathrm{e}^{-\beta H})$. The double integration on the right-hand side of (A3.120) can be reduced to a single one (cf. Verboven, 1960). The conductivity tensor (factor beside $(-\nabla T)$ in (A3.120)) then becomes

$$\varkappa_{\alpha\alpha'} = \frac{1}{2kT^2V}\int_{-\infty}^{\infty} \mathrm{d}t\langle S_\alpha(t)\, S_{\alpha'}(0)\rangle. \tag{A3.121}$$

Note that for a disordered harmonic solid, by (A3.121) and (A3.110), the thermal conductivity may be expressed by a configuration average of products of two Green functions (2.1.17), which leads to the problem of solving a Bethe-Salpeter equation (Böttger, 1973, Flicker and Leath, 1973).

For a review of evaluations of the lattice thermal conductivity the reader is referred to Beck (1975) (see also Kirczenow, 1980, Armstrong, 1981).

REFERENCES

ABOU-CHACRA, R., P. W. ANDERSON and D. J. THOULESS (1973), J. Phys. C **6**, 1734.

ABRAHAMS, E., P. W. ANDERSON, D. C. LICCIARDELLO and T. V. RAMAKRISHNAN (1979), Phys. Rev. Lett. **42**, 673.

ACKERMAN, C. C., B. BERTMAN, H. A. FAIRBANK and R. A. GUYER (1966), Phys. Rev. Lett. **16**, 789.

AGRAWAL, B. K., S. TRIPATHI, A. K. MISRA and D. N. TALWAR (1979), Phys. Rev. **B 19**, 5277.

ALBEN, R., and G. BURNS (1977), Phys. Rev. **B 16**, 3746.

ALBEN, R., D. WEAIRE, J. E. SMITH and M. H. BRODSKY (1975), Phys. Rev. **B 11**, 2271.

ALLENDER, D., J. W. BRAY and J. BARDEEN (1974), Phys. Rev. **B 9**, 119.

AMINI, M., D. FINCHAM and R. W. HOCKNEY (1979), J. Phys. **C 12**, 4707.

ANDERSON, P. W. (1958), Phys. Rev. **109**, 1492.

ANDERSON, P. W. (1978), Rev. Mod. Phys. **50**, 191.

ANDERSON, P. W., B. I. HALPERIN and C. M. VARMA (1972), Phil. Mag. **25**, 1.

ANDERSON, A. B., J. C. RAICH and L. B. KANNEY (1977), Phys. Rev. **B 15**, 5804.

ANTHONY, P. J., and A. C. ANDERSON (1979), Phys. Rev. **B 20**, 763.

ANTONIOU, P. D., and E. N. ECONOMOU (1977), Phys. Rev. **B 16**, 3768.

AUBRY, S. (1975), J. Chem. Phys. **62**, 3217.

AUBRY, S. (1976), J. Chem. Phys. **64**, 3392.

AUBRY, S. (1980), Seminar on the Rieman Problem, Spectral Theory and Complete Integrability 1978/79; ed. D. G. CHUDNOVSKIJ. Lecture Notes in Mathematics, Springer-Verlag, Berlin, Heidelberg, New York.

AXE, J. D., M. IIZUMI and G. SHIRANE (1980), Phys. Rev. **B 22**, 3408.

BAK, P., and S. A. BRAZOVSKY (1978), Phys. Rev. **B 17**, 3154.

BARKER, A. S., and A. J. SIEVERS (1975), Rev. Mod. Phys. **47**, Suppl. No. 2, S 1.

BARKER, A. S. (1975), Phys. Rev. **B** 10, 4071.

BARRON, T. H. K., J. G. COLLINS and G. K. WHITE (1980), Adv. Phys. **29**, 609.

BARRON, T. H. K., and M. L. KLEIN (1974), in: Dynamical Properties of Solids, Vol. I; eds. G. K. HORTON and A. A. MARADUDIN, North-Holland, Amsterdam, p. 391.

BATTEH, J. H., and J. D. POWELL (1979), Phys. Rev. **B 20**, 1398.

BECK, H. (1975), in: Dynamical Properties of Solids, Vol. II; eds. G. K. HORTON and A. A. MARADUDIN, North-Holland, Amsterdam, p. 205.

BECK, H., P. F. MEIER and A. THELLUNG (1974), phys. stat. sol. (a) **24**, 11.

BEEMAN, D., and R. ALBEN (1977), Adv. Phys. **26**, 339.

BELL, R. J. (1972), J. Phys. **C 5**, L 315.

BELL, R. J. (1976), Meth. Comp. Phys. **15**, 215.

BELL, R. J., A. CARNEVALE, C. R. KURKJIAN and G. E. PETERSON (1980), J. Non-Cryst. Solids **35–36**, 1185.

BERLINSKY, A. J. (1979), Rep. Prog. Phys. **42**, 1243.

Bertoni, C. M., V. Bortolani, C. Calandra and E. Tosatti (1974), Phys. Rev. **B 9**, 1710.
Beyeler, H. U., L. Pietronero and S. Strässler (1980), Phys. Rev. **B 22**, 2988.
Bilz, H., B. Gliss and W. Hanke (1974), in: Dynamical Properties of Solids, Vol. I; eds. G. K. Horton and A. A. Maradudin, North-Holland, Amsterdam, p. 343.
Birman, J. L. (1974), Space Group Theory and Infra-Red and Raman Optical Processes in Crystals, Handbuch d. Physik, Vol. XXV/2b; ed. S. Flügge, Springer-Verlag, Berlin, Heidelberg, New York.
Bishop, A. R. (1978), Physica **93 A**, 82.
Bishop, A. R., and J. A. Krumhansl (1975), Phys. Rev. **B 12**, 2824.
Bishop, A. R., J. A. Krumhansl and S. E. Trullinger (1980), Physica **1 D**, 1.
Blunck, M., (1978), Z. Physik **B 31**, 1.
Borland, R. E. (1963), Proc. Roy. Soc. Lond. **A 274**, 529.
Born, M., and K. Huang (1954), Dynamical Theory of Crystal Lattices, Clarendon Press, Oxford.
Born, M., and R. Oppenheimer (1927), Ann. Phys. (Leipzig) **84**, 457.
Böttger, H. (1968), phys. stat. sol. **28**, 611.
Böttger, H. (1972), phys. stat. sol. (b) **51**, 139.
Böttger, H. (1973a), phys. stat. sol. (b) **59**, 517.
Böttger, H. (1973b), phys. stat. sol. (b) **56**, 107.
Böttger, H. (1974), phys. stat. sol. (b) **62**, 9.
Böttger, H. (1976), phys. stat. sol. (b) **77**, 561.
Böttger, H., and P. Kleinert (1981), phys. stat. sol. (b) **103**, 221.
Brauer, W., and H. W. Streitwolf (1977), Theoretische Grundlagen der Halbleiterphysik, Akademie-Verlag, Berlin.
Brener, N. E., and J. L. Fry (1979), Phys. Rev. **B 19**, 1720.
Brenig, W., and A. J. Martin (1975), Z. Physik **B 21**, 203.
Brovman, E. G., and Yu. M. Kagan (1974), in: Dynamical Properties of Solids, Vol. I; eds. G. K. Horton and A. A. Maradudin, North-Holland, Amsterdam, p. 191.
Bruce, A. D. (1978), in: Solitons in Condensed Matter Physics, Proc. sympos. on nonlinear (soliton) structure and dynamics in condensed matter, Oxford; eds. A. R. Bishop and T. Schneider, Springer-Verlag, Berlin, Heidelberg, New York, p. 116.
Bruce, A. D. (1980a), Adv. Phys. **29**, 111.
Bruce, A. D., and R. A. Cowley (1973), J. Phys. **C 6**, 2422.
Bruce, A. D., and R. A. Cowley (1980), Adv. Phys. **29**, 219.
Bruce, A. D., R. A. Cowley and A. F. Murray (1978), J. Phys. **C 11**, 3591.
Bruce, A. D., K. A. Müller and W. Berlinger (1979), Phys. Rev. Lett. **42**, 185.
Bruce, A. D., W. Taylor and A. F. Murray (1980), J. Phys. **C 13**, 483.
Bulajewskij, L. N., and D. I. Khomskij (1978) (Булаевский, Л. Н., Д. И. Хомский), ЖЭТФ **74**, 1863.
Bullough, R. K. (1977), in: Interaction of Radiation with Condensed Matter, Vol. I, Vienna, p. 381.
Burns, G., R. Alben, F. H. Dacol and M. W. Shafer (1979), Phys. Rev. **B 20**, 638.
Burstein, E., A. A. Maradudin and R. Minnick (1971), in: Phonons; ed. M. A. Nusimovici, Flammarion, Paris, p. 48.
Buzdin, A. I., and L. N. Bulajewskij (1977) (Буздин, А. I., Л. Н. Булаевский), ФТТ **19**, 3435.
Caron, L. G. (1976), Phys. Rev. **B 13**, 4545.
Chang, I. F. (1976), Phys. Rev. **B 14**, 4318.
Chang, I. F., and S. S. Mitra (1968), Phys. Rev. **172**, 924.
Chang, I. F., and S. S. Mitra (1971), Adv. Phys. **20**, 359.
Cochran, W. (1971), CRC Crit. Rev. Solid State Sci. **2**, 1.

CONNELL, G. A. N., and G. LUCOVSKY (1978), J. Non-Cryst. Solids **31**, 123.
CONNELL, G. A. N., and R. J. TEMKIN, (1974), Phys. Rev. **B 9**, 5323.
COPLEY, J. R. D., and S. W. LOVESEY (1975), Rep. Prog. Phys. **38**, 461.
COUCHMAN, P. R. (1979), Phil. Mag. **40**, 637.
COWLEY, R. A. (1963), Adv. Phys. **12**, 421.
COWLEY, R. A. (1965), Phil. Mag. **11**, 673.
COWLEY, R. A. (1980), Adv. Phys. **29**, 1.
COWLEY, R. A., and G. J. COOMBS (1973), J. Phys. **C 6**, 121, 143.
COWLEY, R. A., and A. D. WOODS (1971), Canad. J. Phys. **49**, 177.
CURRIE, J. F., J. A. KRUMHANSL, A. R. BISHOP and S. E. TRULLINGER (1980), Phys. Rev. **B 22**, 477.
DAWBER, P. G., and R. J. ELLIOTT (1963a), Proc. Roy. Soc. **273**, 222.
DAWBER, P. G., and R. J. ELLIOTT (1963b), Proc. Phys. Soc. **81**, 453.
DEAN, P. (1964), Proc. Phys. Soc. **84**, 727.
DEAN, P. (1972), Rev. Mod. Phys. **44**, 127.
DEAN, P., and M. D. BACON (1963), Proc. Phys. Soc. **81**, 642.
DEBYE, P. (1912), Ann. Phys. (Leipzig) **39**, 789.
DEDERICHS, P. H., and R. ZELLER (1976), Phys. Rev. **B 14**, 2314.
DIEHL, H. W. (1977), Z. Physik **B 27**, 189.
DIEHL, H. W., and P. L. LEATH (1979), Phys. Rev. **B 19**, 587.
DIETERICH, W. (1976), Adv. Phys. **25**, 615.
DIETSCHE, W., and H. KINDER (1979), Phys. Rev. Lett. **43**, 1413.
DJORDJEVIC, R., R. J. ELLIOTT and T. FUJIWARA (1979), J. Phys. **C 12**, 2519.
DOLLING, G. (1963), in: Inelastic Scattering of Neutrons in Solids and Liquids, Vol. II, International Atomic Energy Agency, Vienna, p. 37.
DOLLING, G. (1974), in: Dynamical Properties of Solids, Vol. I; eds. G. K. HORTON and A. A. MARADUDIN, North-Holland, Amsterdam, p. 541.
DUCASTELLE, F. (1974), J. Phys. **C 7**, 1795.
DUTHLER, C. J. (1976), Phys. Rev. **B 14**, 4606.
ECONOMOU, E. N. (1971), Sol. State Commun. **9**, 1317.
ECONOMOU, E. N., and M. H. COHEN (1972), Phys. Rev. **B 5**, 2931.
ELLIOTT, R. J., J. A. KRUMHANSL and P. L. LEATH (1974), Rev. Mod. Phys. **46**, 465.
ELLIOTT, R. J., and D. W. TAYLOR (1967), Proc. Roy. Soc. **296**, 161.
EMERY, V. J., and J. D. AXE (1978), Phys. Rev. Lett. **40**, 1507.
ENDERLEIN, R., K. PEUKER and F. BECHSTEDT (1979), phys. stat. sol. (b) **92**, 149.
ENZ, C. P. (1968), Ann. Phys. **46**, 114.
ENZ, C. P. (1974), Rev. Mod. Phys. **46**, 705.
FERMI, E., J. R. PASTA and S. M. ULAM (1955), Los Alamos Sci. Lab. Rep. LA-1940; also: Collected works of Enrico Fermi, Vol. II, Univ. Chicago Press, Chicago, p. 978.
FERTEL, J. H., and C. H. PERRY (1969), Phys. Rev. **184**, 874.
FETTER, A. L., and J. D. WALECKA (1971), Quantum Theory of Many-Particle Systems, McGraw-Hill, New York.
FLEMING, R. M., D. E. MONCTON, D. B. MCWHAN and F. J. DI SALVO (1980), Phys. Rev. Lett. **45**, 576.
FLICKER, J. K., and P. L. LEATH (1973), Phys. Rev. **B 7**, 2296.
FOGEL, M. B., S. E. TRULLINGER, A. R. BISHOP and J. A. KRUMHANSL (1977), Phys. Rev. **B 15**, 1578.
FRENKEL, J., and T. KONTOROVA (1939), J. Phys. USSR **1**, 37.
FRIEND, R. H., and D. JEROME (1979), J. Phys. **C 12**, 1441.
FRÖHLICH, H. (1954), Proc. Roy. Soc. **A 223**, 296.
FUJITA, T., and J. HORI (1972), J. Phys. **C 5**, 1059.

Galeener, F. L., and P. N. Sen (1978), Phys. Rev. **B 17**, 1928.
Gaskell, P. H. (1979), J. Phys. **C 12**, 4337.
Geick, R. (1962), Z. Physik **166**, 122.
Genzel, L., and W. Bauhofer (1976), Z. Physik **B 25**, 13.
Genzel, L., H. Happ and R. Weber (1959), Z. Physik **154**, 13.
Gillis, N. S., and T. R. Koehler (1971), Phys. Rev. **B 4**, 3971.
Gillis, N. S., and T. R. Koehler (1972), Phys. Rev. **B 5**, 1925.
Gläser, W. (1974), Festkörperprobleme XIV; ed. O. Madelung, Verlag Vieweg und Sohn GmbH, Braunschweig, p. 205.
Golding, B. (1980), J. Non-Cryst. Solids **35–36**, 1125.
Gonis, A., and J. W. Garland (1978), Phys. Rev. **B 18**, 3999.
Gurevich, V. L. (1980) (Гуревич, В. Л.), Кинетика фононных систем, Наука, Moscow.
Guyer, R. (1969), Sol. State Phys. **23**, 413.
Handrich, K., and S. Kobe (1980), Amorphe Ferro- und Ferrimagnetika, Akademie-Verlag, Berlin.
Hanke, W. (1973), Phys. Rev. **B 8**, 4585.
Hanke, W. (1978), Adv. Phys. **27**, 287.
Hansen, F. Y. (1978), Phys. Rev. **B 18**, 4015.
Halperin, B. I., and C. M. Varma (1976), Phys. Rev. **B 14**, 4030.
Hardy, J. R. (1974), in: Dynamical properties of solids, Vol. I; eds. G. K. Horton and A. A. Maradudin, North-Holland, Amsterdam, p. 157.
Hasenfratz, W., and R. Klein (1977), Physica **89 A**, 191.
Heimendahl, L. V., and M. F. Thorpe (1975), J. Phys. **F 5**, L 87.
von Heimendahl, L. (1979), J. Phys. **F 9**, 161.
Heinrichs, J. (1977), Phys. Rev. **B 16**, 4365.
Hirota, T. (1973), Prog. Theor. Phys. **50**, 1240.
Hirsch, J. E. (1978), Phys. Rev. **B 18**, 3976.
Horner, H. (1974), in: Dynamical Properties of Solids, Vol. I.; eds. G. K. Horton and A. A. Maradudin, North-Holland, Amsterdam, p. 451.
Horovitz, B. (1980), Sol. State Commun. **34**, 61.
Horovitz, B., and J. A. Krumhansl (1978), Sol. State Commun. **26**, 81.
Horovitz, B., J. L. Murray and J. A. Krumhansl (1978), Phys. Rev. **B 18**, 3549.
van Hove, L. (1953), Phys. Rev. **89**, 1189.
Hunklinger, S. (1974), in: Proc. satell. symp. on microwave acoust. of the 8th intern. congr. acoust.; eds. E. R. Dobb and J. K. Wigmore, Lancaster.
Iizumi, M., J. D. Axe, G. Shirane and K. Shimaoka (1977), Phys. Rev. **B 15**, 4392.
Ikezi, H., R. J. Taylor and R. D. Baker (1970), Phys. Rev. Lett. **25**, 11.
Ishii, K. (1973), Prog. Theor. Phys. Suppl. **53**, 77.
Jackson, H. E., C. T. Walker and T. F. McNelly (1970), Phys. Rev. Lett. **25**, 26.
Jacucci, G., M. L. Klein and R. Taylor (1978), Phys. Rev. **B 18**, 3782.
Jahne, E., and B. Ulrici (1980), phys. stat. sol. (b) **101**, 169.
Janner, A., and T. Janssen (1977), Phys. Rev. **B 15**, 643.
Janssen, T. (1979), J. Phys. **C 12**, 5381.
Jaswal, S. S. (1965), Phys. Rev. **137**, A 302.
Joannopoulos, J. D. (1979), Non-Cryst. Solids **32**, 241.
Joannopoulos, J. D., and F. Yndurain (1974), Phys. Rev. **B 10**, 5146.
Jones, D. P., N. Thomas and W. A. Phillips (1978), Phil. Mag. **38**, 271.
Kamitakahara, W. A., and D. W. Taylor (1974), Phys. Rev. **B 10**, 1190.
Kamitakahara, W. A., and J. R. D. Copley (1978), Phys. Rev. **B 18**, 3772.
Kaneyoshi, T. (1978), J. Phys. Soc. Japan **45**, 1835.

KAPLAN, T., and L. J. GRAY (1976), Phys. Rev. B 14, 3462.
KAPLAN, T., P. L. LEATH, L. J. GRAY and H. W. DIEHL (1980), Phys. Rev. B 21, 4230.
KIM, K., and M. NELKIN (1973), Phys. Rev. B 7, 2762.
KIRCZENOW, G. (1980), Ann. Phys. (N.Y.) 125, 1.
KIRZHNITS, D. A. (1967), Field Theoretical Methods in Many-Body Systems, Pergamon Press, Oxford.
KLEIN, M. L., and G. K. HORTON (1972), J. Low Temp. Phys. 9, 151.
KLEINERT, P. (1979), phys. stat. sol. (b) 95, 163.
KLEINERT, P., and R. LEIHKAUF (1980), phys. stat. sol. (b) 97, 491.
KLUG, D. D., and E. WHALLEY (1979), J. Chem. Phys. 71, 2903.
KOEHLER, T. R., A. R. BISHOP, J. A. KRUMHANSL and J. R. SCHRIEFFER (1975), Sol. State Commun. 17. 1515.
KOEHLER, T. R. (1975), in: Dynamical Properties of Solids, Vol. II: eds. G. K. HORTON and A. A. MARADUDIN, North-Holland, Amsterdam, p. 1.
KOEHLER, T. R., and P. A. LEE (1977), Phys. Rev. B 16, 5263.
KONDO, J., and K. YAMADJI (1977), J. Phys. Soc. Jap. 43, 424.
KOTANI, A. (1977), J. Phys. Soc. Jap. 42, 408, 416.
KRAMER, H. A., and W. HEISENBERG (1925), Z. Physik 31, 681.
KRUMHANSL, J. A., and J. A. D. MATTEW (1968), Phys. Rev. 166, 856.
KRUMHANSL, J. A., and J. R. SCHRIEFFER (1975), Phys. Rev. B 11, 3535.
KUROSAWA, T. (1961), J. Phys. Soc. Japan 16, 1298.
LACINA, W. B., and P. S. PERSHAN (1970), Phys. Rev. B 1, 1765.
LANDAU, L. D., and E. M. LIFSHITZ (1959), Statistical Physics, Pergamon Press, London, p. 482.
LANNIN, J. S. (1977), Phys. Rev. B 16, 1510.
LAUGHLIN, R. B., and J. D. JOANNOPOULOS (1977), Phys. Rev. B 16, 2942.
LAUGHLIN, R. B., and J. D. JOANNOPOULOS (1978), Phys. Rev. B 17, 4922.
LEADBETTER, A. J., and D. LITCHINSKY (1970), Disc. Faraday Soc. 50, 62.
LEADBETTER, A. J., and M. W. STRINGFELLOW (1972), Neutron Inelastic Scattering, Proceedings of the Grenoble Conference, IAEA, Vienna, p. 501.
LEATH, P. L., and B. GOODMAN (1969), Phys. Rev. 181, 1062.
LEE, P. A., T. M. RICE and P. W. ANDERSON (1974), Phys. Rev. Lett. 31, 703.
LEIBFRIED, G., and W. LUDWIG (1961), Sol. State Phys. 12, 275.
LEIGH, R. S., B. SZIGETI and V. K. TEWARY (1971), Proc. Roy. Soc. A 320, 505.
LEUNG, K. M., D. W. HONE, D. L. MILLS, P. S. RISEBOROUGH and S. E. TRULLINGER (1980), Phys. Rev. B 21, 4017.
LICCIARDELLO, D. C., and E. N. ECONOMOU (1975), Phys. Rev. B 11, 3697.
LINDEMANN, F. A. (1910), Z. Physik 11, 609.
LITTLEWOOD, P. B. (1979), J. Phys. C 12, 4441.
LOWNDES, R. P., and A. RASTOGI (1976), Phys. Rev. B 14, 3598.
LUCOVSKY, G., and R. M. MARTIN (1972), J. Non-Cryst. Solids 8—10, 185.
LUCOVSKY, G., R. D. BURNHAM and A. S. ALIMONDA (1976), Phys. Rev. B 14, 2503.
LUDWIG, W. (1967), Recent Developments in Lattice Theory, in: Springer Tracts in Modern Physics, Vol. 43; ed. G. HÖHLER, Springer-Verlag, Berlin.
LUKIN, V. P. (1977), (Лукин, В. П.), ФТТ 19, 379.
LYDDANE, R. H., R. G. SACHS and E. TELLER (1941), Phys. Rev. 59, 673.
MARADUDIN, A. A. (1966), Solid State Phys. 19, 1.
MARADUDIN, A. A. (1966a), Solid State Phys. 18, 273.
MARADUDIN, A. A. (1966b), Solid State Phys. 19, 1.
MARADUDIN, A. A., and A. E. FEIN (1962), Phys. Rev. 128, 2589.
MARADUDIN, A. A., E. W. MONTROLL, G. H. WEISS and I. P. IPATOVA (1971), Theory

of Lattice Dynamics in the Harmonic Approximation, 2 nd. Ed., Academic Press, New York.
Maradudin, A. A., and S. H. Vosko (1968), Rev. Mod. Phys. **40**, 1.
Meissner, G. (1968), Phys. Rev. Lett. **21**, 435.
Mc Millan, W. L. (1976), Phys. Rev. **B 14**, 1496.
Mc Millan, W. L. (1977), Phys. Rev. **B 16**, 4655.
Monga, M. R., V. K. Jindal and K. N. Pathak (1979), Phys. Rev. **B 19**, 1230.
Montgomery, C. G. (1980), J. Low Temp. Phys. **39**, 13.
Mook, H. A., N. Wakabayashi and D. Pan (1975), Phys. Rev. Lett. **34**, 1029.
Mostoller, M., and T. Kaplan (1977), Phys. Rev. **B 16**, 2350.
Mostoller, M., and T. Kaplan (1979), Phys. Rev. **B 19**, 3938.
Mott, N. F., and W. D. Twose (1961), Adv. Phys. **10**, 107.
Müller, K. A., and H. Burkard (1979), Phys. Rev. **B 19**, 3593.
Myles, C. W., and J. D. Dow (1979), Phys. Rev. **B 19**, 4939.
Nakamura, A. (1978), Progr. Theor. Phys. **59**, 1447.
Narayanamurti, V., and R. C. Dynes (1972), Phys. Rev. Lett. **28**, 1461.
Mc Nelly, T. F., S. J. Rogers, P. J. Channin, R. J. Rollefson, W. M. Goubau, G. F. Schmidt, J. A. Krumhansl and R. O. Pohl (1970), Phys. Rev. Lett. **24**, 100.
Nemanich, R. J. (1977), Phys. Rev. **B 16**, 1655.
Nilsson, G., and G. Nelin (1972), Phys. Rev. **B 6**, 3777.
Novaco, A. D. (1980), Phys. Rev. **B 22**, 1645.
Odagaki, T., and F. Yonezawa (1978), Sol. State Commun. **27**, 1203, 1207.
Odagaki, T., and F. Yonezawa (1979), J. Phys. Soc. Jap. **47**, 379, 388.
Painter, R. D., and W. M. Hartmann (1976), Phys. Rev. **B 13**, 479.
Papatriantafillou, C., E. N. Economou and T. P. Eggarter (1976), Phys. Rev. **B 13**, 910.
Peierls, R. E. (1955), Quantum theory of solids, Oxford University Press, Oxford, p. 108.
Petschek, R. G. (1980), Phys. Rev. **B 22**, 1409.
Phillips, J. C. (1956), Phys. Rev. **104**, 1263.
Phillips, W. A. (1972), J. Low-Temp. Phys. **7**, 351.
Phillips, W. A. (1978), J. Non-Cryst. Solids **31**, 267.
Pick, R. M. (1971), in: Phonons; ed. M. A. Nusimovici, Flammarion, Paris, p. 20.
Pick, R., M. H. Cohen and R. M. Martin (1970), Phys. Rev. **B 1**, 910.
Pietronero, L., and S. Strässler (1978), Sol. State Commun. **27**, 1041.
Pietronero, L., and S. Strässler (1979), Phys. Rev. Lett. **42**, 188.
Placzek, G. (1934), in: Handbuch der Radiologie, Vol. 6; ed. E. Marx, Teubner, Leipzig.
Plakida, N. M., and T. Siklos (1969), phys. stat. sol. **33**, 103.
Plakida, N. M. (1973) (Плáкида, Н. М.), in: Статистическая физика и квантовая теория поля; ed. N. N. Bogoljubov, (под. ред. Н. Н. Боголюбова), Наука, Moscow, p. 205.
Pohl, D. W., and V. Irniger (1976), Phys. Rev. Lett. **36**, 480.
Pokrovskij, V. L., and A. L. Talapov (1978) (Покровский, В. Л., А. Л.Талапов), ЖЭТФ **75**, 1151.
Pollard, W. B., and J. D. Joannopoulos (1978), Phys. Rev. **B 17**, 1778.
Pollard, W. B., and J. D. Joannopoulos (1980), Phys. Rev. **B 21**, 760.
Polk, D. E. (1971), J. Non-Cryst. Solids **5**, 365.
Rahman, A. (1964), Phys. Rev. **136, A** 405.
Rebbi, C. (1979), Scientific American **240**, 76.
Rehr, J. J., and R. Alben (1977), Phys. Rev. **B 16**, 2400.

RENKER, B., H. RIETSCHEL, L. PINTSCHOVIUS, W. GLÄSER, W. BRUESCH, D. KUSE and M. J. RICE (1973), Phys. Rev. Lett. **30**, 1144.
RICE, M. J., and S. STRÄSSLER (1973), Sol. State Commun. **13**, 125.
RISTE, T., E. J. SAMUELSEN, K. OTNES and J. FEDER (1971), Sol. State Commun. **9**, 1455.
ROSENSTOCK, H. B. (1955), Phys. Rev. **97**, 290.
ROTH, L. M. (1979), J. Phys. **C 12**, 4879.
SAHNI, P. S., and G. F. MAZENKO (1979), Phys. Rev. **B 20**, 4674.
SARBACH, S. (1977), Phys. Rev. **B 15**, 2694.
SARKAR, S. K., and S. SENGUPTA (1977), phys. stat. sol. (b) **83**, 263.
SCALAPINO, D. J., M. SEARS and R. A. FERRELL (1972), Phys. Rev. **B 6**, 3409.
SCOTT, A. C., F. Y. F. CHU and D. W. MC LAUGHLIN (1973), Proc. IEEE **61**, 1443.
SCHMIDT, V. H. (1979), Phys. Rev. **B 11**, 4397.
SCHNEIDER, T., H. BECK and E. STOLL (1976), Phys. Rev. **B 13**, 1123.
SCHNEIDER, T., and E. STOLL (1976), Phys. Rev. **B 13**, 1216.
SCHNEIDER, T., and E. STOLL (1978a), Phys. Rev. **B 18**, 6468.
SCHNEIDER, T., and E. STOLL (1978b), Phys. Rev. **B 17**, 1302.
SCHULZ, H. J. (1978), Phys. Rev. **B 18**, 5756.
SCHUSTER, H. G. (1975), Phys. Rev. **B 11**, 613.
SEN, P. N., and W. M. HARTMAN (1974), Phys. Rev. **B 9**, 367.
SEN, P. N., and M. F. THORPE (1977), Phys. Rev. **B 15**, 4030.
SHAM, L. J. (1974), in: Dynamical Properties of Solids, Vol. I; eds. G. K. HORTON and A. A. MARADUDIN, North-Holland, Amsterdam, p. 301.
SHERRINGTON, D. W. (1971), J. Phys. **C 4**, 2771.
SINHA, S. K. (1973), CRC Crit. Rev. Solid State Sci. **3**, 273.
SINHA, S. K. (1978), in Proc. into conf. on lattice dynamics, Flammarion, Paris, p. 7.
SINHA, S. K. (1980), in: Dynamical Properties of Solids, Vol. III; eds. G. K. HORTON and A. A. MARADUDIN, North-Holland, Amsterdam, p. 1.
SMITH, H. G., and W. GLÄSER (1970), Phys. Rev. Lett. **25**, 1611.
SNYMAN, A. J., and J. H. VAN DER MERWE (1974), Surf. Sci. **42**, 190.
SOKOLOFF, J. B. (1977), Phys. Rev. **B 16**, 3367.
SOUKOULIS, C. M., and E. N. ECONOMOU (1980), Phys. Rev. Lett. **45**, 1590.
SOVEN, P. (1967), Phys. Rev. **156**, 809.
STEINHARDT, P., R. ALBEN and D. WEAIRE (1974), J. Non-Cryst. Solids **15**, 199.
SUITS, B. H., S. COUTURIE and C. P. SLICHTER (1980), Phys. Rev. Lett. **45**, 194.
SUSSMANN, J. A., and A. THELLUNG (1963), Proc. Phys. Soc. **81**, 1122.
SZABO, N. (1976), J. Phys. **C 9**, 259.
TAKAYAMA, H., and K. MAKI (1979), Phys. Rev. **B 20**, 3223, 5002, 5009.
TAKAYAMA, H., and K. MAKI (1980), Phys. Rev. **B 21**, 4558.
TAKENO, S. (1970), Progr. Theor. Phys. Suppl. **45**, 137.
TAKENO, S., and M. GODA (1971), Prog. Theor. Phys. **45**, 331.
TAYLOR, D. W. (1967), Phys. Rev. **156**, 1017.
TAYLOR, D. W. (1973), Sol. State Commun. **13**, 117.
TAYLOR, D. W. (1975), in: Dynamical properties of solids, Vol. II; eds. G. K. HORTON and A. A. MARADUDIN, North-Holland, Amsterdam, p. 285.
TAYLOR, D. W. (1976), J. Phys. **C 9**, 453.
THELLUNG, A., and K. WEISS (1969), Phys. kondens. Materie **9**, 300.
THEODORAKOPOULOS, N. (1979), Z. Physik **B 33**, 385.
THEODOROU, G., and T. M. RICE (1978), Phys. Rev. **B 18**, 2840.
THORPE, M. F. (1972), Phys. Rev. **B 8**, 5352.
THORPE, M. F. (1973), Phys. Rev. **B 8**, 5352.
THORPE, M. F., and F. L. GALEENER (1980), Phys. Rev. **B 22**, 3078.

TSUKADA, M. (1972), J. Phys. Soc. Jap. **32**, 1475.
TODA, M. (1967), J. Phys. Soc. Jap. **22**, 431; **23**, 501.
TODA, M. (1970), Progr. Theor. Phys. Suppl. **45**, 174.
TODA, M. (1975), Phys. Rep. **18 C**, 1.
TODA, M. (1976), Progr. Theor. Phys. Suppl. **59**, 1.
TOOMBS, G. A. (1978), Phys. Rep. **40 C**, 182.
TSUNODA, Y., N. KUITOMI, N. WAKABAYASHI, R. M. NICKLOW and H. G. SMITH (1979), Phys. Rev. **B 19**, 2876.
TURNER, R. D., and J. C. INKSON (1978), J. Phys. **C 11**, 3961.
VALKERING, T. P., and C. DE LANGE (1980), J. Phys. **A 13**, 1607.
VARMA, C. M. (1976), Phys. Rev. **B 14**, 244.
VARMA, C. M., and W. WEBER (1979), Phys. Rev. **B 12**, 6142.
VARSHNEY, S. C., J. F. VETELINO, S. S. MITRA and I. F. CHANG (1975), Phys. Rev. **B 12**, 5912.
VELICKY, B., S. KIRKPATRICK and H. EHRENREICH (1968), Phys. Rev. **175**, 747.
VERBOVEN, E. (1960), Physica (Utrecht) **26**, 1091.
VERLET, L. (1967), Phys. Rev. **159**, 98.
VERLEUR, H. W., and A. S. BARKER (1966), Phys. Rev. **149**, 715.
WALNUT, T. H., and L. A. NAFIE (1977), J. Chem. Phys. **67**, 1491.
WEBER, W. (1977), Phys. Rev. **B 15**, 4789.
WEBER, W., R. ROEDHAMMER, L. PINTSCHOVIUS, W. REICHARDT, F. GOMPF and A. N. CHRISTENSEN (1979), Phys. Rev. Lett. **43**, 868.
WEILACHER, K. H., and H. BROSS (1977), J. Phys. **F 11**, 2253.
DE WETTE, F. W., and B. R. A. NIJBOER (1965), Phys. Letters **18**, 19.
WERTHAMER, N. R. (1970), Phys. Rev. **B 1**, 572.
WHITHAM, F. R. S. (1974), Linear and Nonlinear Waves, John Wiley and Sons, New York.
WIGNER, E. P. (1959), Group Theory and its Application to the Quantum Theory of Atomic Spectra, Academic Press, New York, London.
WOODS, A. D. B., and R. A. COWLEY (1973), Rep. Prog. Phys. **36**, 1135.
WORLTON, T. G., and J. L. WARREN (1972), Comp. Phys. Commun. **3**, 88.
WU, S. Y., S. P. BOWEN and K. S. DY (1980), CRC Crit. Rev. Solid State Sci. **10**, 43.
YAMAMOTO, R., K. HAGA, T. MIHARA and M. DOYAMA (1980), J. Phys. **F 10**, 1389.
YNDURAIN, F., and P. N. SEN (1976), Phys. Rev. **B 14**, 531.
YNDURAIN, F. (1978), Phys. Rev. B **18**, 2876.
YONEZAWA, F., and K. MORIGAKI (1973), Progr. Theor. Phys. Suppl. **53**, 1.
YOSHIDA, F., and S. TAKENO (1978), J. Phys. **C 11**, 2895.
ZABUSKY, N. J., and M. D. KRUSKAL (1965), Phys. Rev. Lett. **15**, 240.
ZACHAROV, V. E., C. V. MANAKOV, S. P. NOVIKOV and L. P. PITAEVSKIJ (1980) (Захаров, В. Е., С. В. Манаков, С. П. Новиков, Л. П. Питаевский), Теория солитонов: Метод обратной задачи, Наука, Moscow.
ZEYHER, R. (1978), in: Proc. int. conf. on lattice dynamics, Flammarion, Paris, p. 17.
ZEYHER, R., and W. KRESS (1979), Phys. Rev. B **20**, 2850.
ZELLER, R. C., and R. O. POHL (1971), Phys. Rev. **B 4**, 2029.
ZIMAN, J. M. (1972), Principles of the Theory of Solids, 2nd. Ed., Cambridge University Press, London.
ZUBAREV, D. N. (1971) (Зубарев, Д. Н.), Неравновесная статистическая термодинамика, Наука, Moscow, chapter IV; German translation: D. N. SUBAREW, Statistische Thermodynamik des Nichtgleichgewichts, Akademie-Verlag, Berlin 1976.
ZWANZIG, R. (1966), Phys. Rev. **144**, 170.

References added in proof

AKESENOV, V. L., A. YU. DIDYK, V. YU. YUSHANKHAI (1981), Sol. State Commun. **40**, 825.

ALLEN, P. B., (1980), in: Dynamical Properties of Solids, Vol. III; eds. G. K. HORTON and A. A. MARADUDIN, North-Holland, Amsterdam.

ALOISIO, M. A., V. A. SINGH and L. M. ROTH (1981), J. Phys. **F 11**, 1823.

ANDERSON, P. W., D. J. THOULESS, E. ABRAHAMS and D. S. FISHER (1980), Phys. Rev. **B 22**, 3519.

ARMSTRONG, B. H. (1981), Phys. Rev. **B 23**, 883.

AUBRY, S. (1980a), in: Lecture Notes in Mathematics; ed. D. G. CHUDNOVSKIJ, Springer Verlag, Berlin, Heidelberg, New York.

AUBRY, S. (1980b), in: Proceedings of the International Workshop of Intrinsic Stochasticity in Plasmas; eds. G. LAVAL and B. GRESILON, Edition Francaise de Physique.

BEALE, P. D., S. SARKER and J. A. KRUMHANSL (1981), Phys. Rev. **B 24**, 266.

BHATTACHARYYA, A. (1981), Contemp. Phys **22**, 117.

BILZ, H., H. BÜTTNER, A. BUSSMANN-HOLDER, W. KRESS and U. SCHRÖDER (1982), Phys. Rev. Lett. **48**, 264.

BISHOP, A. R. (1981), in: Physics in One Dimension, Springer Series in Solid State Sciences, Vol. 23; eds. J. BERNASCONI, T. SCHNEIDER, Springer Verlag, Berlin, Heidelberg, New York, p. 27.

BONNEVILLE, R. (1980), Phys. Rev. **B 21** 368.

BONNEVILLE, R. (1981), Phys. Rev. **B 24**, 1987.

BÖTTGER, H., V. V. BRYKSIN and P. KLEINERT (1978), phys. stat. sol. (b) **89**, 675.

BÖTTGER, H., V. V. BRYKSIN and P. KLEINERT (1979) (Беттгер, Х., В. В. Брыксин и П. Клейнерт), ФТТ **21**, 808.

BÖTTGER, H., and V. V. BRYKSIN (1982), phys. stat. sol. (b) **113**, 9.

BRAWER, S. A. (1981), Phys. Rev. Lett. **46**, 778.

BRENER, N. E., and J. L. FRY (1980), Phys. Rev. **B 22**, 2737.

BRUCE, D. (1980b), J. Phys. **C 13**, 4615.

BUTLER, H. (1973), Phys. Rev. **B 8** 4499.

BRÜESCH, P. (1982), Phonons: Theory and Experiments I; Lattice Dynamics and Models of Interatomic Forces, Springer Series in Solid State Sciences, Vol. 34; Springer Verlag, Berlin, Heidelberg, New York.

CAR, R., E. TOSATTI, S. BARONI and S. LEELAPRUTE (1981), Phys. Rev. **B 24**, 985.

CARLES, R., A. ZWICK, M. A. RENUCCI and J. B. RENUCCI (1982), Sol. State Commun. **41**, 557.

CHUI, S. T., and J. D. WEEKS (1980), Phys. Rev. **B 21**, 4413.

CLARK, R., J. N. GRAY, H. HOMMA and M. J. WINOKUR (1981), Phys. Rev. Lett. **47**, 1407.

COUTINHO, S., P. PITANGA and P. LEDERER (1981), Phys. Rev. **B 23**, 4567.

DIEDERICH, S. (1981), Phys. Rev. **B 24**, 3186, 3193.

DOLGOV, O. V., D. A. KIRZHNITS and E. G. MAKSIMOV (1981), Rev. Mod. Phys. **53**, 81.

EILENBERGER, G. (1981), Solitons, Mathematical Methods for Physicists, Springer Series in Solid State Sciences, Vol. 19; Springer Verlag, Berlin, Heidelberg, New York.

FAIN, S. C., M. D. CHINN and R. DIEHL (1980), Pys. Rev. **B 21**, 4170.

FALTER, C., and M., SELMKE (1980), Phys. Rev. **B 21**, 2078.

FALTER, C., and M. SELMKE (1981), Phys. Rev. **B 24** 586.

FENTON, E. W. (1981), J. Phys. **C 14**, 4217.

FERTIS, A. C., A. N. ANDRIOTIS and E. N. ECONOMOU (1981), Phys. Rev. **B 24**, 5806
GILROY, K. S., and W. A. PHILLIPS (1981), Phil. Mag. **B 43**, 735.
GODA, M. (1982), J. Phys. **C 15**, 435.
GONÇALVES DA SILVA, C. E. T., and B. KOILLER (1981), Sol. State Commun. **40**, 215.
GRAY, L. J., and T. KAPLAN (1981), Phys. Rev. **B 24**, 1872.
HAMMER, C. L., J. E. SHRAUNER and B. DE FACIO (1981), Phys. Rev. **B 23**, 5890.
HARRIS, A. B., and T. C. LUBENSKY (1981), Phys. Rev. **B 23**, 2640.
HEEGER, A. J. (1982), Comm. Sol. State Phys. **10**, 133
ITOH, M., K. NIIZEKI and M. WATABE (1981), J. Phys. **F 11**, 1605.
JOOS, B. (1982), Sol. State Commun. **42**, 709.
KAVEH, M. (1981), Phys. Rev. **B 24**, 619.
KIMBALL, J. C., C. Y. FONG and Y. R. SHEN (1981), Phys. Rev. **B 23**, 4946.
KLEINERT, P., and R. LEIHKAUF (1981), phys. stat. sol. (b) **105** K 67.
KOBAYASHI, S., and S. TAKEUCHI (1980), J. Phys. **C 13**, L 969.
KRAMER, B., A. MAC KINNON and D. WEAIRE (1981), Phys. Rev. **B 23**, 6357.
KUMAR, V., and S. K. JOSHI (1981), Phys. Rev. **B 18**, 2515.
KUNC, K., and R. M. MARTIN (1982), Phys. Rev. Lett. **48**, 406.
DE LANGE, C., and T. JANSSEN (1981), J. Phys. **C 14**, 5269.
MC LAUGHLIN, D. W. (1981), in: Physics in One Dimension, Springer Series in Solid State Sciences, Vol. 23; eds. J. BERNASCONI and T. SCHNEIDER, Springer Verlag, Berlin, Heidelberg, New York, p. 13.
LÖHNEYSEN, H. v., and H. J. SCHINK (1982), Phys. Rev. Lett. **48**, 1121.
LOPONEN, M. T., R. C. DYNES, V. NARAYANAMURTI, J. P. GARNO (1982), Phys. Rev. **B 25**, 1161.
MATSUBARA, T., and K. KAMIYA (1977), Prog. Theor. Phys. **58**, 767.
MATSUSHITA, E., and T. MATSUBARA (1978), Prog. Theor. Phys. **59**, 15.
MATTIS, D. C. (1981), The Theory of Magnetism I., Statics and Dynamics, Springer Series in Solid State Sciences, Vol. 17; Springer Verlag, Berlin, Heidelberg, New York, p. 206.
MERTENS, F. G., and H. BÜTTNER (1981), Phys. Rev. Lett. **84 A**, 335.
MOKROSS, F., and H. BÜTTNER (1981), Phys. Rev. **A 24**, 2826.
NAKANISHI, H., K. MACHIDA and T. MATSUBARA (1982) Prog. Theor. Phys. **67**, 1305.
NAKAO, M., and M. DOYAMA (1981), J. Phys. **F 11**, 1585.
NATTERMANN, T. (1980), J. Phys. **C 13**, L 265.
NISHIOKA, K., and J. K. LEE (1981), Phil. Mag. **A 44**, 779.
ODAGAKI, T., and M. LAX (1981), Phys. Rev. **B 24**, 5284.
PASKIN, A., A. M. L. DE KREINER, K. SHUKLA, D. O. WELCH and G. J. DIENES (1982), Phys. Rev. **B 25**, 1297.
PATTON, B. R., and L. J. SHAM (1979), Phys. Rev. Lett. **33**, 638.
PIETRONERO, L., W. R. SCHNEIDER and S. STRÄSSLER (1981), Phys. Rev. **B 24**, 2187.
PLAKIDA, N. M., and T. SIKLOŠ (1978), Acta phys. Hungar. **45**, 37.
PLUMELLE, P., D. N. TALWARE, M. VANDEVYVER, K. KUNC and M. ZIGONE (1979), Phys. Rev. **B 20**, 4199.
POKROVSKIJ, V. L., and A. L. TALAPOV (1979), Phys. Rev. Lett. **42**, 65.
POLLARD, W. B., and J. D. JOANNOPOULOS (1981), Phys. Rev. **B 23**, 5263.
PRESS, W., B. RENKER, H. SCHULZ and H. BÖHM (1980), Phys. Rev. **B 21**, 1250.
PURI, A., and T. ODAGAKI (1981), Phys. Rev. **B 24**, 5541.
QUANG, D. N., B. ESSER and R. KEIPER (1980), phys. stat. sol. (b) **99**, 103.
RASING, TH., P. WYDER, A. JANNER and T. JANSSEN (1982), Sol. State Commun. **41** 715.
RUFFA, A. R. (1981), Phys. Rev. **B 24**, 6915.

SAHNI, P. S. (1981), Phys. Rev. **B 23, 1325.**
SAITS, B. H., S. CONTURIE and C. P. SLICHTER (1981), Phys. Rev. **B 23**, 5142.
SARKER, S., and E. DOMANY (1981), Phys. Rev. **B 23**, 6018.
SCHERM, R. (1972), Ann. Physique (Paris) **7**, 349.
SCHMELTZER, D., and R. BESERMAN (1981), Phys. Rev. Letters **47**, 860.
SCHNEIDER, T., and E. STOLL (1980), Phys. Rev. **B 22**, 5317.
SCHNEIDER, T., and E. STOLL (1981), Phys. Rev. **B 23**, 4631.
SETHNA, J. P. (1981), Phys. Rev. **B 24**, 698.
SHUKLA, K., A. PASKIN, D. O. WELCH and G. J. DIENES (1981), Phys. Rev. **B 24**, 724.
SINGH, R. K. (1982), Phys. Reports **85**, 259.
SOKOLOFF, J. B. (1981), Phys. Rev. **B 23**, 6422.
SOUKOULIS, C. M., and E. N. ECONOMOU (1981), J. Phys. **C 14**, L 221.
SOUKOULIS, C. M., and E. N. ECONOMOU (1982), Phys. Rev. Lett. **48**, 1043.
STEPHEN, M. J. (1981), J. Stat. Phys. **25**, 663.
STEPHENS, P. W., P. HEINEY, R. J. BIRGENEAU and P. M. HORN (1979), Phys. Rev. Lett. **43**, 47.
SUCK, J. B., H. RUDIN, H. J. GÜNTHERODT and H. BECK (1980), J. Phys. **C 13**, L 1045.
SUCK, J. B., H. RUDIN, H. J. GÜNTHERODT and H. BECK (1981), J. Phys. **F 11**, 1375.
SUGIYAMA, M., and OKAMOTO, H. (1980), Prog. Theor. Phys. **64**, 1945.
TODA, M. (1981), Theory of Nonlinear Lattices, Springer Series in Solid State Sciences, Springer Verlag, Berlin, Heidelberg, New York.
TOLPYGO, K. B. (1950) (Толпыго, К. Б.), ЖЭТФ **20**, 497.
TOLPYGO, K. B. (1959) (Толпыго, К. Б.), Укр. физ. Журн. **4**, 72.
VENTURA, I. (1981), Phys. Rev. **B 24**, 2812.
VILLAIN, J. (1980), in: Ordering in Strongly Fluctuating Condensed Matter Systems; ed. T. Riste, Plenum, New York, p. 222.
VISSCHER, W. M., and J. E. GUBERNATIS (1980), in: Dynamical Properties of Solids, Vol. IV; eds. G. K. HORTON and A. A. MARADUDIN, North-Holland, Amsterdam, p. 63.
VOLLHADRT, D., and P. WÖLFLE (1982), Phys. Rev. Lett. **48**, 699.
WEAIRE, D., and P. C. TAYLOR (1980), in: Dynamical Properties of Solids, Vol. IV; eds. G. K. HORTON and A. A. MARADUDIN, North-Holland, Amsterdam, p. 1.
WEGNER, F. (1976), Z. Physik. **B 25**, 327.
WHITELAW, D. J. (1981), J. Phys. **C 14**, 2871.
WOJTCZAK, L., T. SIKLOŚ, B. MRYGON and J. MIELNICKI (1982), phys. stat. sol. (b) **109**, 483.
YOSHIDA, F., and T. SAKUMA (1982), Prog. Theor. Phys. **67**, 1379.

INDEX

Errata

page 27, equ. (1.3.58)

read: $\left[\sum_{\alpha\alpha'} \ldots\right]^{1/2}$ instead of: $\sum_{\alpha\alpha'} [\ldots]^{1/2}$

page 31, equ. (1.3.69)

read: $n_j(\boldsymbol{k}) = 0, 1, 2, \ldots$ instead of: $n_j(\boldsymbol{k}) = 1, 2, \ldots$

page 56, equ. (1.5.71)

read: $j = 5, 6$ instead of: $j = 5$

762 968 2 (6628) H. Böttger, Principles of the Theory of Lattice Dynamics

Bei Fragen zur Produktsicherheit wenden Sie sich bitte an:
If you have any questions regarding product safety,
please contact:

Walter de Gruyter GmbH
Genthiner Straße 13
10785 Berlin
productsafety@degruyterbrill.com